D0780215

Signal Processing Methods for Audio, Images and Telecommunications

Signal Processing and its Applications

Signal Processing Methods for Audio, Images and Telecommunications

Edited by

Peter M. Clarkson
Biomedical Engineering Center, The Ohio State University, Columbus, Ohio, USA

Henry Stark
ECE Department, Illinois Institute of Technology, Chicago, Illinois, USA

ACADEMIC PRESS
Harcourt Brace & Company, Publishers
London San Diego New York Boston Sydney Tokyo Toronto

ACADEMIC PRESS LIMITED
24-28 Oval Road
LONDON NW1 7DX

U.S. Edition Published by
ACADEMIC PRESS INC.
San Diego, CA 92101

This book is printed on acid-free paper

A catalogue record for this book is available from the British Library

ISBN 0-12-175790-0

Printed in Great Britain by the University Press, Cambridge

Series Preface

Signal processing applications are now widespread. Relatively cheap consumer products through to the more expensive military and industrial systems extensively exploit this technology. This spread was initiated in the 1960s by introduction of cheap digital technology to implement signal processing algorithms in real-time for some applications. Since that time semiconductor technology has developed rapidly to support the spread. In parallel, an ever increasing body of mathematical theory is being used to develop signal processing algorithms. The basic mathematical foundations, however, have been known and well understood for some time.

Signal Processing and its Applications addresses the entire breadth and depth of the subject with texts that cover the theory, technology and applications of signal processing in its widest sense. This is reflected in the composition of the Editorial Board, who have interests in:

(i) Theory – The physics of the applications and the mathematics to model the system;
(ii) Implementation – VLSI/ASIC design, computer architecture, numerical methods, systems design methodology, and CAE;
(iii) Applications – Speech, sonar, radar, seismic, medical, communications (both audio and video), guidance, navigation, remote sensing, imaging, survey, archiving, non-destructive and non-intrusive testing, and personal entertainment.

Signal Processing and its Applications will typically be of most interest to post-graduate students, academics, and practising engineers who work in the field and develop signal processing applications. Some texts may also be of interest to final year undergraduates.

Richard C. Green
The Engineering Practice,
Farnborough, UK

List Of Contributors

Jezekiel Ben-Arie, Electrical and Computer Engineering Department Illinois Institute of Technology

Eric R. Buhrke, AT&T Bell Laboratories, Services and Speech Technology Department

Peter M. Clarkson, Biomedical Engineering Center, Ohio State University

Nikolas P. Galatsanos, Electrical and Computer Engineering Department, Illinois Institute of Technology

William E. Jacklin, Electrical and Computer Engineering Department, Illinois Institute of Technology

Biquan Lin, Department of Mathematics, Illinois Institute of Technology

Joseph L. LoCicero, Electrical and Computer Engineering Department, Illinois Institute of Technology

Dibyendu Nandy, Electrical and Computer Engineering Department, Illinois Institute of Technology

Bao Nguyen, Department of Mathematics, Illinois Institute of Technology

Elwood T. Olsen, Department of Mathematics, Illinois Institute of Technology

K. Raghunath Rao, Electrical and Computer Engineering Department, Illinois Institute of Technology

Henry Stark, Electrical and Computer Engineering Department, Illinois Institute of Technology

Donald R. Ucci, Electrical and Computer Engineering Department, Illinois Institute of Technology

Geoffrey A. Williamson, Electrical and Computer Engineering Department, Illinois Institute of Technology

Yongi Yang, Electrical and Computer Engineering Department, Illinois Institute of Technology

Contents

Foreword

In recent years rapid advances in computer hardware technology, including the development of specialized digital signal processors, has facilitated the development of sophisticated algorithms whose applications would have been unthinkable only a short time ago. These algorithms typically allow for real-time or non-real-time application, are nonlinear, make use of prior knowledge, tend to adapt in response to a changing environment, and are designed to achieve near-optimum performance under a broad range of operating conditions. Moreover, the reader familiar with recent research in audio, video, and telecommunications will observe that, despite superficial differences, the algorithms in use in these areas draw upon the same mathematical and optimization strategies such as constrained extremization, vector space methods, gradient descent algorithms, projection onto subspaces, self-organizing systems and the like.

The two editors, both active researchers in their respective fields, felt that a single volume describing the application of such algorithms to audio, video and telecommunications would be an invaluable aid to students, researchers, and practicing engineers in these fields. Because their colleagues at the Illinois Institute of Technology have played a leading role in signal processing research in these fields, the editors invited them to collaborate in the creation of this text.

Broadly speaking, the book is divided into four parts, namely, *methods* and applications to *audio*, *video* and *telecommunications*. Below we furnish a brief description of each chapter.

In Chapter 1, Lin, Nyugen and Olsen review *wavelet transforms* and their applications. The chapter provides a rigorous, but highly readable tutorial introduction to the subject. Following the development of the theoretical foundation for wavelets (multiresolution analysis), the authors describe continuous and discrete transforms before discussing the *fast wavelet transform* and its relation to the FFT. Two well-known wavelets (the Haar and the Meyer wavelets) are discussed in detail. The chapter concludes with applications to multiscale edge detection, nonstationary signal analysis, and the innovative Coifman–Wickerhauser data compression algorithm.

In Chapter 2, the author, Williamson, lays out the fundamental ideas of adaptive filter design and provides a unifying framework for *finite impulse response* (FIR), *infinite impulse response* (IIR) and orthogonal polynomial filters through the concept of *vector space adaptive filters* (VSAF). VSAFs are shown to have much of the performance potential of IIR systems, but without the undesirable stability problems manifest in these filters. The performance potential for the VSAF filters is illustrated via important applications to channel equalization and to long-distance echo cancellation.

Order statistic (OS) filters are an important class of nonlinear digital filters that have been used in a wide range of applications. In Chapter 3, Clarkson and Williamson continue the theme of Chapter 2 by examining the role of adaptation in the context of OS filters. After a review of the basic properties of median and other OS filters, the authors discuss *order statistic adaptive filters* (OSAFs) (in which OS operations are embedded into the mechanism of a conventional adaptive filter), and *adaptive order statistic filters* (AOSFs) (in which the coefficients of an OS filter are adaptively updated). The chapter includes discussion of the optimality and robustness in OS filters, and concludes with an application to automatic cardiac arrhythmia detection.

Chapter 4 provides a review and tutorial introduction to multilayer feedforward neural networks. Neural nets have been applied to a great many problems in areas such as audio and image processing. Buhrke and LoCicero introduce the subject through a series of simple examples before outlining the major considerations in the design of neural networks. The latter part of the chapter deals with the concepts of *statistical training* (developed as an alternative to the conventional back-propagation algorithm) and *network fusion* intended to facilitate the design of large networks by subdividing the design into a series of smaller subproblems. The practical application of these techniques is demonstrated via the design of a speaker-independent speech recognition system.

Chapter 5 provides an interesting and innovative look at auditory localization – the perception of sound sources at specific locations relative to one's position. Traditional models of localization rely on the use of correlation to exploit time and amplitude differences between sounds recorded at the two ears. Nandy and Ben-Arie examine the use of the *head related transfer function* (HRTF) – the transfer function incorporating the effects of the external ears, head and torso on an incident sound field as a tool for localization. They demonstrate that directional information in the HRTF can provide superior localization to conventional correlation based strategies.

In Chapter 6, the reader is introduced to a vector space method called *projections onto convex sets* (POCS). Following a brief tutorial of what POCS is and how it is used, the authors Yang and Stark apply the method to three important problems in image processing, namely color matching, resolution enhancement and blind deconvolution. Numerical examples illustrating the method in each case are furnished. The method of projections is applied again in Chapter 7 which deals with the optimum recovery of compressed images under the *joint pictures expert group* (JPEG) standard. A significant problem that results in JPEG compression is the appearance of so-called "blocking artifacts." The authors, Yang and Galatsanos, use POCS in a novel way to remove such artifacts, and improve the quality of the recovered image in other ways as well.

Rao and Ben-Arie, in Chapter 8, introduce the concept of *non-orthogonal expansion* and derive a method based on it, to automatically detect and locate objects in an image field. The fundamental idea here is that shifted replicas of the object of interest can be used as a basis for expanding any image in a finite dimensional space. Experimental results are presented to show the superiority of the method over classical correlation.

In Chapter 9, Jacklin and Ucci furnish a review of classical optimum detection, direct-sequence spread spectrum, and several methods for extracting probability density functions from data. They then discuss their own efforts in deriving a nonlinear, locally optimum detection strategy in telecommunications. A number of carefully worked-out examples, related to these subjects, are furnished for the reader's edification.

The book concludes with Chapter 10, by Yang and Stark, which continues upon the theme introduced in Chapter 9 of estimating probability density functions (pdfs). As readers of this book are probably aware, virtually every optimum design in signal processing and communications requires knowledge of the pdf. Chapter 10 again makes use of projection onto convex sets to combine prior knowledge with available data to furnish pdf estimates that are superior to several algorithms widely available in existing software packages.

The editors wish to thank the authors and their students for their informative and well-written contributions. Many of the results furnished in the book have not been published previously and we are grateful to those authors who agreed to choose this book as the instrument to disseminate their original work. The editors are grateful to Bonnie Dow for help with the typing, and a special thanks is due to Professor Geoff Williamson, the resident LaTeX expert, for his assistance in the typesetting of this book.

On behalf of all the authors, the editors wish to thank the staff and officers of Academic Press for encouraging this work, and the Illinois Institute of Technology for providing the stimulating environment in which this work could be done.

P.M.C.
H.S.

About the Editors

Peter M. Clarkson

Peter Clarkson received the B.Sc.(Hons) from Portsmouth Polytechnic in Mathematics, and the Ph.D. degree from the Institute of Sound and Vibration Research at the University of Southampton where he specialized in digital signal processing and its application to problems in sound and vibration. From 1984 to 1987 he remained at the University of Southampton working first as a Postdoctoral Fellow of the Science and Engineering Research Council of Great Britain, and from 1985 to 1987 as a Research Lecturer. From 1987 to 1993 he was with the Illinois Institute of Technology serving as Associate Chairman of the ECE Department from 1991 to 1993. In 1993 he left IIT to join the Ohio State University in order to pursue his growing interest in medical signal analysis. Dr. Clarkson has interests in all aspects of signal processing. He is the author of some 60 journal and conference articles. In addition to co-editing this work, he is the author of the text *Optimal and Adaptive Signal Processing* (CRC Press, 1993).

Henry Stark

Henry Stark received the B.S. degree from the City College of New York and the M.S. and Dr. Eng. Sc. from Columbia University, all in electrical engineering. He is the Bodine Professor of Electrical and Computer Engineering at the Illinois Institute of Technology and the department chairman. His interests are in image recovery, signal processing, pattern recognition, medical imaging and optical information processing. He has written some 130 papers in these areas and received an outstanding paper award for the IEEE Engineering in Medicine and Biology Society in 1980. In addition to co-editing this text, Dr. Stark is a co-author of *Modern Electrical Communications* (Prentice-Hall, 1988) and *Probability, Random Processes, and Estimation Theory for Engineers* (Prentice-Hall, 1994), and the editor of two books on *Optical Fourier Transforms* and *Image Recovery* , both published by Academic Press. Together with his students, he has written numerous book chapters on topics ranging from neural nets to sampling theory.

Dr. Stark is a former Associate Editor of the *IEEE Transactions on Medical Imaging* and a former Topical Editor on Imaging Processing for the *Journal of the Optical Society of America.* He is a consulting editor for the Oxford University Series on Optics. He is a Fellow of the IEEE and the Optical Society of America.

Chapter 1

Orthogonal Wavelets and Signal Processing

BIQUAN LIN, BAO NGUYEN and ELWOOD T. OLSEN

Introduction

The term "wavelet" was introduced by Grossman and Morlet [1-1] to describe a square integrable function, appropriate translations and dilations of which form a basis for $L^2(R)$. Thus, the study of wavelets is the study of bases for $L^2(R)$ which have the property that all of the basis functions are self-similar – they differ only by translation and change of scale from one another.

While there are non-orthogonal wavelets, the signal processing applications make use almost exclusively of orthogonal wavelets: the translations of the wavelet by integer multiples of some number b are orthogonal, and the dilations of the wavelet at one scale are orthogonal to the dilations at all other scales. A wavelet can be visualized as a damped sine wave whose amplitude is very small, perhaps zero, outside some bounded interval, and which is distorted somewhat in shape to guarantee that the orthogonality conditions all hold. Some wavelets damp so rapidly that there is only a single sign change, and wavelets used in practice typically vanish outside a bounded interval (that is, have compact support).

While the term "wavelet" is recent, the subject of wavelet analysis has historical roots. The Haar basis for $L^2(R)$ is a basis consisting of dilations and translations of dilations of a single function, which has long been known. It is discussed in Section 1.2.

The discrete wavelet transform of a discrete signal is the signal multiplied by a unitary matrix. The unitary matrix arises from a wavelet basis for $L^2(R)$ in a manner which is addressed at the beginning of Section 1.3. Discrete wavelet transformations have a historical antecedent of their own: the fast Fourier transform (FFT). The unitary matrix of a discrete wavelet transform factors naturally into unitary matrices which give transformations from each scale to the next. As a result, the wavelet transformation algorithm has a structure which is identical to the structure of the FFT algorithm. This is addressed in Section 1.4. The fact that the wavelet transform is fast undoubtedly accounts for some of the recent interest in wavelets.

The chapter begins with a review of the theoretical foundation for wavelets, called multiresolution analysis. Multiresolution analysis investigates the conditions under which a complete orthonormal basis for $L^2(R)$ can be obtained from translations and dilations of a single function. Some grasp of multiresolution analysis is needed in order to understand where the unitary matrices of the discrete wavelet transform come from. Section 1.1 gives a rigorous introduction to multiresolution analysis, and Section 1.2 gives two examples of wavelet bases for $L^2(R)$ together with an analysis of them which relies on and also illustrates the theoretical discussion of Section 1.1.

The recent interest in wavelets can be ascribed in part to the discovery of large classes of orthogonal families of wavelets. In particular, Daubechies [1-2] addressed the equations which the unitary matrices of the discrete wavelet transform must satisfy and showed how to construct a huge family of solutions directly, without first constructing a multiresolution analysis of $L^2(R)$. We will discuss her work in Section 1.3. Daubechies' work offers the tantalizing possibility of constructing orthogonal wavelet bases, and hence unitary wavelet transform matrices, tailored to specific applications.

Sections 1.5, 1.6 and 1.7 are devoted to three signal processing applications of wavelet transforms.

For reasons which will appear later, the unitary matrix of the discrete wavelet transform at each scale partitions naturally into a high-pass filter and a low-pass filter. If the underlying wavelet is appropriately chosen, the high-pass part of the transformed signal at each scale is the derivative of a smoothed version of the original signal, and where the derivative of the smoothed signal has a large value, we expect an edge in the original signal. This interpretation has been exploited to make use of wavelet transforms in multiscale edge detection. It is discussed in Section 1.5.

Since the underlying idea of wavelet analysis is to expand a given signal as a linear combination of translations and dilations of a single wavelet function, the subject is like Fourier analysis – a signal is decomposed into components. It is unlike Fourier analysis in that the components have (typically) compact support, whereas the components of the Fourier transform – sine waves – do not, and as a result "local" properties of a time-varying signal can in principle be isolated and studied using wavelets. The wavelet transform therefore has a natural application to time-varying signal analysis. This is discussed in Section 1.6.

Because the unitary matrix of the wavelet transform factors into the product of unitary matrices giving the transformation from each scale to the next, not one but many representations of a signal are obtained from an application of the wavelet transform algorithm. Because the factor matrices partition nicely at each scale, pieces of the representation at one scale can be combined with pieces of the representation at other scales to obtain new representations. The result is that a very large number of different representations of the same signal are obtained via a single wavelet transformation. This can be exploited in lossless data compression applications. Data compression via wavelets is discussed in Section 1.7.

The material in this chapter is intended to be a self-contained and rigorous introduction to orthogonal wavelets. It emphasizes applications to signal processing and also emphasizes the derivation and structure of the discrete wavelet transform algorithm (see Section 1.4). The material in this chapter requires an understanding of some basic ideas from the mathematics of abstract vector spaces (see [1-3], [1-4]), of real and complex analysis (see [1-5], [1-6] or [1-7]), and of elementary Hilbert space theory (see [1-6], [1-7] or [1-8]).

1.1 Multiresolution Analysis

We first introduce some terminology. By a *wavelet* (or *mother wavelet*), we shall mean a function $\psi \in L^2(R)$, appropriate translations and dilations of which form an orthonormal basis for $L^2(R)$. We will normally assume that the wavelet ψ is a complex-valued function of a real argument, though in some cases we limit consideration to real-valued wavelets. The underlying idea of wavelet analysis is thus to choose ψ, and "scales" $s_n > 0$ and "translations" b_{mn} in such a way that any function $f \in L^2(R)$ has a series expansion of the form

$$f(x) = \sum_{m,n \in Z} c_{mn} \psi_{mn}(x), \tag{1.1-1}$$

where Z denotes the set of integers and

$$\psi_{mn}(x) = C_m \psi((x - b_{mn})/s_m). \tag{1.1-2}$$

C_m (which depends only on the scale s_m and not on the translation b_{mn}) is chosen so that the ψ_{mn}'s are normalized in some convenient manner; usually $C_m = s_m^{-1/2}$, so that $\|\psi_{mn}\|_2 = \|\psi\|_2$ for all m, n; $\|\cdot\|_2$ denotes the L^2 norm. The family $\{\psi_{mn}\}$ will be called the family of wavelets derived from (or generated by) the wavelet ψ; this somewhat circular terminology follows common usage and should cause no confusion. We restrict attention to families of orthonormal wavelets, that is, to wavelets satisfying

$$\langle \psi_{mn}, \psi_{k\ell} \rangle = \int_{-\infty}^{\infty} \psi_{mn}^* \psi_{k\ell}\, dx = \delta_{mk}\delta_{n\ell}, \tag{1.1-3}$$

since they are the most important in applications. When we assert that a family is an "orthonormal family of wavelets" we shall normally imply that it is a basis for $L^2(R)$. If we have a decomposition of the form (1.1-1), we will call the coefficients $\{c_{mn}\}$ the *wavelet transform* of the function f we started with.

In wavelet analysis, the translations b_{mn} are always taken to have the form $b_{mn} = ns_m b$ where $b = b_{01} > 0$, and the scales s_m are usually taken to be $s_m = 2^m$, $m \in Z$, so that ψ_{mn} has the form

$$\psi_{mn}(x) = 2^{-m/2}\psi(2^{-m}x - nb). \tag{1.1-4}$$

Thus, if we have an orthonormal family of wavelets, the wavelet transform c_{mn} of a function f is given by

$$c_{mn} = \langle \psi_{mn}, f \rangle = 2^{-m/2} \int_{-\infty}^{\infty} \psi^*(2^{-m}t - nb) f(t)\, dt. \tag{1.1-5}$$

We will sometimes use also the convenient notations

$$\begin{aligned} \psi_s(x) &= s^{-1/2}\psi(s^{-1}x) \\ [W_s f](x) &= \int_{-\infty}^{\infty} \psi_s^*(t - x) f(t)\, dt \end{aligned} \tag{1.1-6}$$

so that $c_{mn} = [W_{2^m} f](n2^m b)$. In the convolution-like integral (1.1-6) ($W_s f(x) = f * \tilde{\psi}^*(x)$, where $*$ denotes convolution and where $\tilde{\psi}_s(x) = \psi_s(-x)$). Both s and x are taken to be continuous parameters. The expression (1.1-6) is often called the "wavelet transform" of the function f. This terminology is convenient and standard. When necessary to avoid confusion, $W_s f$ of (1.1-6) is referred to as the continuous wavelet transform with continuous parameters, and the c_{mn}'s of (1.1-5), as the continuous wavelet transform with discrete parameters. This terminology anticipates the fact that there is a third wavelet transform, the discrete wavelet transform, which will be introduced in Section 1.3.

It is clear that if we want to construct an orthogonal wavelet basis, we will have to choose ψ and b carefully. The translations are usually $b_{mn} = n2^m b$ and the scales are always powers of 2, so these are not free to be chosen. It may be surprising that there are any choices for which the ψ_{mn}'s are an orthonormal basis for $L^2(R)$, but there are. There is a theory which guides the choice of ψ and b, which is called multiresolution analysis. We turn to it now.

The initial goal of multiresolution analysis is to state properties of a function $\phi \in L^2(R)$, which we shall call a generating function, from which a wavelet $\psi \in L^2(R)$ can be derived. We devote most of this section to the investigation of ϕ and then at the end of the section show how the wavelet ψ can be obtained from it.

Suppose that $\phi \in L^2(R)$, and fix $b > 0$. Define

$$\begin{aligned} S_0 &= \text{span}\,\{\phi(x - nb) \,|\, n \in Z\} \\ V_0 &= \bar{S}_0 \\ V_m &= \{f(2^{-m}x) \,|\, f \in V_0\}. \end{aligned}$$

Thus, by definition, S_0 is the set of all linear combinations of translates of ϕ by integer multiples of b, V_0 is the closure of S_0 (with respect to the L^2 norm), and V_m is the set of all scaled versions, with scale factor 2^m, of elements of V_0.

We will write $\phi_{mn}(x) = 2^{-m/2}\phi(2^{-m}x - nb)$. Observe that $\phi_{mn} \in V_m$ for all n. Suppose that $\phi(x)$ has compact support, say $\mathrm{spt}\,(\phi) = [-A, A]$. (The support $\mathrm{spt}\,(f)$ of a function f is the closure of the set of points where the function is non-zero, so this means that ϕ is zero outside $[-A, A]$.) Then $\phi(2^{-m}x)$ has support $[-2^m A, 2^m A]$, as is readily verified. That is, multiplying x by 2^{-m} "stretches" the graph of ϕ by a factor of 2^m. This is why we say that the scale factor is 2^m rather than 2^{-m}. The function $\phi(2^{-m}x - nb) = \phi(2^{-m}(x - n2^m b))$ has a graph which is the same as $\phi(2^{-m}x)$, translated $2^m nb$ (rather than nb) units to the right. Finally, observe that $\|\phi_{mn}\| = \|\phi\|$ for all integers m and n, since

$$\begin{aligned}\int_{-\infty}^{\infty} |\phi_{mn}(x)|^2\,dx &= \int_{-\infty}^{\infty} 2^{-m}|\phi(2^{-m}x - nb)|^2\,dx \\ &= \int_{-\infty}^{\infty} |\phi(\xi)|^2\,d\xi\end{aligned}$$

making the substitution $\xi = 2^{-m}x - nb$. This explains the normalization factor $2^{-m/2}$ included in the definition of ϕ_{mn}: it is put there so that the norms of all the normalized scaled and translated versions of ϕ have the same norm as ϕ itself.

It can be shown that V_m is a closed subspace of L^2 for all m and that it is invariant under translations by $2^m nb$ for all n: if $f(x) \in V_m$, so is $f(x - 2^m nb)$. It can also be shown that V_m is the closure of the set of all linear combinations of translates of $\phi_{m0}(x) = 2^{-m/2}\phi(2^{-m}x)$ by integer multiples of $2^m b$. Suppose now that V_0 is the closure of span $\{\phi(x - nb)\}_{n\in Z}$, that V_m, $m \in Z$, is obtained from V_0 by dilation in the manner indicated above and that, in addition, the following conditions hold:

(a) $V_m \subset V_{m-1}$ for all $m \in Z$,

(b) $\cap_{m\in Z} V_m = \{0\}$,

(c) $\cup_{m\in Z} V_m$ is dense in $L^2(R)$, and

(d) there are constants $A > 0$ and B such that for all square summable sequences c_n of complex numbers

$$A \sum_{n\in Z} |c_n|^2 \le \|\sum_{n\in Z} c_n \phi_{0n}\|_2^2 \le B \sum_{n\in Z} |c_n|^2.$$

If these conditions are satisfied, we will call $\{V_m\}$ a *multiresolution analysis* of L^2, and we will say ϕ *generates* the multiresolution analysis.

Basically, then, a multiresolution analysis is a sequence of nested ($V_m \subset V_{m-1}$) subspaces of L^2, each of which is invariant under translations by integer multiples of $2^m b$, and each of which is generated by translates of a single function ϕ_{m0}. We require also, however, that the spaces satisfy additional technical conditions (b) through (d) above. We wish to explore these conditions, which are basic to the subject of wavelets.

First, if $\phi \in V_{-1}$, that is, ϕ has a series expansion of the form

$$\phi(x) = \sum_{n \in Z} a_n \phi_{-1,n}(x),$$

then condition (a) holds. This is true for the following reasons: If ϕ has a series expansion as above, then for any ℓ,

$$\phi_{0\ell}(x) = \phi(x - \ell b) = \sum_{n \in Z} a_n \phi_{-1,n}(x - \ell b).$$

Since $\phi_{-1,n}(x - \ell b) = \phi_{-1,2\ell+n}(x)$, it follows that $\phi_{0\ell} \in V_{-1}$. Since V_{-1} is a subspace, it follows that linear combinations of the ϕ_{0n}'s are in V_{-1}, that is, $S_0 \subset V_{-1}$. Since V_{-1} is a closed set, the closure of S_0, which is just V_0, is contained in V_{-1}. Finally, since $f(2^m x) \in V_0$ if and only if $f(x) \in V_m$, we have, for any $f \in V_m$,

$$f(2^m x) = \sum_{n \in Z} a_n \phi_{-1,n}(x),$$

using $V_0 \subset V_{-1}$, so that

$$f(x) = \sum_{n \in Z} a_n \phi_{-1,n}(2^{-m} x) = \sum_{n \in Z} a_n \sqrt{2} \phi(2^{-m+1} x - nb) = \sum_{n \in Z} 2^{m/2} a_n \phi_{m-1,n}(x).$$

That is, $f \in V_{m-1}$, and since $f \in V_m$ was arbitrary, $V_m \subset V_{m-1}$. The upshot is that to satisfy condition (a) of a multiresolution analysis, it is necessary and sufficient that ϕ and b be chosen so that ϕ can be expanded as a series in its dilations and translations at scale 2. This is not so easy, as the reader can verify by trying it, but it can be done, as examples in Section 1.2 will show.

Next, if condition (d) holds, then for any m and for all square summable sequences c_n of complex numbers

$$A \sum_{n \in Z} |c_n|^2 \leq \| \sum_{n \in Z} c_n \phi_{mn} \|_2^2 \leq B \sum_{n \in Z} |c_n|^2.$$

This holds by a substitution of variables argument:

$$\begin{aligned}
\| \sum_{n \in Z} c_n \phi_{mn} \|^2 &= \int_{-\infty}^{\infty} | \sum_{n \in Z} c_n \phi_{mn}(x) |^2 \, dx \\
&= \int_{-\infty}^{\infty} 2^{-m} | \sum_{n \in Z} c_n \phi(2^{-m} x - nb) |^2 \, dx \\
&= \int_{-\infty}^{\infty} | \sum_{n \in Z} c_n \phi(\xi - nb) |^2 \, d\xi \\
&= \| \sum_{n \in Z} c_n \phi_{0n} \|^2
\end{aligned}$$

making the substitution $\xi = 2^{-m}x$. Thus, if condition (d) holds in V_0, an analogous condition necessarily holds in every subspace V_m.

If ϕ can be chosen in such a way that its translates ϕ_{0n} are orthogonal, then condition (d) holds:

Theorem 1.1-1 If $\phi \in L^2$ and $\|\phi\|_2 > 0$, and translates

$$\phi_{0n}(x) = \phi(x - nb)$$

are orthogonal, then condition (d) above is satisfied with $A = B = \|\phi\|_2^2$.

Proof: Orthogonality implies

$$\begin{aligned}\|\sum_{n\in Z} c_n\phi_{0n}\|^2 &= \sum_{n,k\in Z} c_n^* c_k \int_{-\infty}^{\infty} \phi_{0n}^*(x)\phi_{0k}(x)\,dx \\ &= \|\phi\|^2 \sum_{n,k\in Z} c_n^* c_k \delta_{nk} \\ &= \|\phi\|^2 \sum_{n\in Z} |c_n|^2.\end{aligned}$$

This implies (d) with $A = B = \|\phi\|_2^2$. ∎

Condition (d) may hold even when orthogonality of translates of ϕ fails, but we may interpret condition (d) as meaning that the translates are "nearly" orthogonal.

Theorem 1.1-1 says that if translates of ϕ are orthogonal, then condition (d) of a multiresolution analysis holds. In fact, if ϕ possesses orthonormal translates and if in addition ϕ integrates to 1, and if finally condition (a) ($V_m \subset V_{m-1}$) holds, then also conditions (b) and (c) of a multiresolution analysis hold, so long as the Fourier transform $\hat{\phi}(\omega)$ is continuous at $\omega = 0$.

Theorem 1.1-2 Suppose $\phi \in L^2$, and let $b > 0$ be given. Suppose also that

$$\int_{-\infty}^{\infty} \phi(x)\,dx = 1.$$

Set

$$\phi_{mn}(x) = 2^{-m/2}\phi(2^{-m}x - nb)$$

and let V_m denote the closure of the span of $\{\phi_{mn}\}_{n\in Z}$. If $V_0 \subset V_{-1}$, if the translates of ϕ are orthonormal, that is $\langle\phi_{0n}, \phi_{0\ell}\rangle = \delta_{n\ell}$, and if in addition $\hat{\phi}$ is continuous at $\omega = 0$, then conditions (b) and (c) of a multiresolution analysis hold, that is, $\cap_{m\in Z} V_m = \{0\}$ and $\cup_{m\in Z} V_m$ is dense in L^2.

Proof: For convenience, we take $b = 1$ in the proof. The proof that $\cap V_m$ is the zero vector will be based on the claim that if g is a continuous and compactly supported function defined on $(-\infty, \infty)$ then the orthogonal projection $P_m g$ of g on V_m satisfies $\|P_m g\|_2 \to 0$ as $m \to \infty$. Assume for the moment that the claim is true. Fix any function $f \in \cap_{m\in Z} V_m$. Let $\epsilon > 0$ be given. Choose a continuous and compactly supported function g such that $\|f - g\|_2 < \epsilon$. Then for every $m \in Z$ we have

$$\|f - P_m g\|_2 = \|P_m(f - g)\|_2 \leq \|f - g\|_2 < \epsilon,$$

where in the first equality we have used the assumption that $f \in \cap_{m\in Z} V_m$. Letting $m \to \infty$, we have $\|f\|_2 \leq \epsilon$ since $\|P_m g\|_2 \to 0$ and hence $\|f - P_m g\|_2 \to \|f\|_2$. Since ϵ was arbitrary, condition (b) of a multiresolution analysis follows.

Returning to the claim, we now show that $\|P_m g\|_2 \to 0$ for any given continuous function g with support in $[-A, A]$. We have

$$\begin{aligned} \|P_m g\|_2^2 &= \sum_{n\in Z} |\langle g, \phi_{mn}\rangle|^2 \\ &= 2^{-m} \sum_{n\in Z} \left|\int_{-A}^{A} g(x)\phi(2^{-m}x - n)\,dx\right|^2 \\ &\leq 2^{-m}\|g\|_2^2 \sum_{n\in Z} \int_{-A}^{A} |\phi(2^{-m}x - n)|^2\,dx \\ &\leq \|g\|_2^2 \sum_{n\in Z} \int_{-2^{-m}A}^{2^{-m}A} |\phi(u - n)|^2\,du, \end{aligned} \tag{1.1-7}$$

where the first equality uses the property that $\{\phi_{mn}\}$ is an orthonormal basis of V_m, the third step uses the Cauchy–Schwartz inequality, and the last uses the substitution $u = 2^{-m}x$. Let m be so large that $2^{-m}A < 1/2$ and let I_m be the indicator function of $\cup_{n\in Z}[n - 2^{-m}A, n + 2^{-m}A]$. Then, using (1.1-7),

$$\|P_m g\|_2^2 \leq \frac{1}{2\pi}\|g\|_2^2 \int_{-\infty}^{\infty} |\phi(u)|^2 I_m(u)\,du. \tag{1.1-8}$$

Since $I_m(u) \to 0$ pointwise almost everywhere as $m \to \infty$, by the Lebesgue dominated convergence theorem the right-hand side of (1.1-8) converges to zero. This completes the proof of the claim.

As to the last condition of a multiresolution analysis, that $\cup_{m\in Z} V_m$ be dense in $L^2(R)$, observe that this condition follows if we can show that for any function g with the property that its Fourier transform $\hat{g}$ is compactly supported (say, $\hat{g}$ vanishes for $|\omega| > A$) $\|P_m g\|_2 \to \|g\|_2$ as $m \to -\infty$, since the set of continuous functions with compact support is dense in $L^2(R)$. (A subspace is dense in L^2 in the frequency domain if and only if the inverse Fourier transforms of functions in the subspace are dense in L^2 in the time domain, by Plancherel's theorem.) We have by orthonormality of ϕ_{mn}

$$\|P_m g\|_2^2 = \sum_{n\in Z} |\langle g, \phi_{mn}\rangle|^2$$

$$
\begin{aligned}
&= \sum_{n\in Z} |\langle \hat{g}, \hat{\phi}_{mn}\rangle|^2 \\
&= 2^m \sum_{n\in Z} |\int_{-\infty}^{\infty} \hat{g}^*(\omega)\hat{\phi}(2^m\omega)e^{-j2^m n\omega}\, d\omega|^2 \\
&= 2^m \sum_{n\in Z} |\int_{0}^{2\pi 2^{-m}} U(\omega)e^{-j2^m n\omega}\, d\omega|^2. \qquad (1.1\text{-}9)
\end{aligned}
$$

Here, we have made use of the fact that

$$
\begin{aligned}
\hat{\phi}_{mn}(\omega) &= \frac{2^{-m/2}}{\sqrt{2\pi}} \int_{-\infty}^{\infty} \phi(2^{-m}x - n)e^{-j\omega x}\, dx \\
&= \frac{2^{m/2}}{\sqrt{2\pi}} e^{-jn2^m\omega} \int_{-\infty}^{\infty} \phi(u)e^{-j2^m\omega u}\, du \\
&= 2^{m/2}e^{-jn2^m\omega}\hat{\phi}(2^m\omega)
\end{aligned}
$$

via the substitution $u = 2^{-m}x - n$. The function U in (1.1-9) is the $2\pi 2^{-m}$-periodic function defined by

$$
U(\omega) = \sum_{k\in Z} \hat{g}^*(\omega + 2\pi 2^{-m}k)\hat{\phi}(2^m\omega + 2\pi k).
$$

Apply Parseval's theorem to the right-hand side of (1.1-9), which differs from the sum of squares of the Fourier coefficients in the Fourier series expansion of U on $[0, 2\pi 2^{-m}]$ by a constant multiple, to obtain

$$
\begin{aligned}
\|P_m g\|_2^2 &= 2\pi \int_0^{2\pi 2^{-m}} |U(\omega)|^2\, d\omega \\
&= 2\pi \int_0^{2\pi 2^{-m}} |\sum_{k\in Z} \hat{g}^*(\omega + 2\pi 2^{-m}k)\hat{\phi}(2^m\omega + 2\pi k)|^2\, d\omega \\
&= 2\pi \int_0^{2\pi 2^{-m}} \sum_{k,\ell\in Z} \hat{g}(\omega + 2\pi 2^{-m}k)\hat{\phi}^*(2^m\omega + 2\pi k) \\
&\qquad \hat{g}^*(\omega + 2\pi 2^{-m}\ell)\hat{\phi}(2^m\omega + 2\pi\ell)\, d\omega \\
&= 2\pi \int_{-\infty}^{\infty} \sum_{k\in Z} \hat{g}(\omega + 2\pi 2^{-m}k)\hat{\phi}^*(2^m\omega + 2\pi k)\hat{g}^*(\omega)\hat{\phi}(2^m\omega)\, d\omega. \qquad (1.1\text{-}10)
\end{aligned}
$$

Observe, that if $\hat{g}$ vanishes for $|\omega| > A$, and m is large in magnitude and negative, then the product

$$
\hat{g}^*(\omega)\hat{g}(\omega + 2\pi 2^{-m}k) = 0
$$

for all k except $k = 0$. In fact, this occurs for all m such that

$$A < \pi 2^{-m}.$$

For all such m, therefore, the last term in (1.1-10) is equal to

$$2\pi \int_{-A}^{A} |\hat{g}(\omega)\hat{\phi}(2^m\omega)|^2 \, d\omega. \tag{1.1-11}$$

Since $\hat{\phi}$ is continuous at 0, $\hat{\phi}(2^m\omega) \to \hat{\phi}(0) = 1/\sqrt{2\pi}$ uniformly on $[-A, A]$ as $m \to -\infty$. It follows that the expression in (1.1-11) converges to $\|\hat{g}\|_2^2 = \|g\|_2^2$ as $m \to -\infty$. Therefore, $\cup V_m$ is dense in L^2. ∎

Theorems 1.1-1 and 1.1-2 together suggest the following procedure for constructing a multiresolution analysis: Find a non-zero function ϕ and a number b such that translates of ϕ by integer multiples of b are orthogonal, such that ϕ integrates to 1 and such that condition (a) of a multiresolution analysis, that is, $V_0 \subset V_{-1}$, is satisfied. Then by Theorems 1.1-1 and 1.1-2 all of the other conditions of a multiresolution analysis are satisfied. This is still a non-trivial task; note in particular that the condition that ϕ have orthonormal translates and that also ϕ integrate to 1 cannot simultaneously be satisfied by normalization. Still, as we shall see, the task can be carried out. The examples of Section 1.2 below and Daubechies' construction in Section 1.3 below both use basically this approach.

We next prove a surprising fact: If the function ϕ has a Fourier transform $\hat{\phi}$ which is sufficiently well behaved, then there is a function $\tilde{\phi}$ for which $\tilde{V}_0 = \text{span}\,\{\tilde{\phi}_{0n}\}_{n\in Z}$ is identical to V_0 (so that also $\tilde{V}_m = V_m$ for all m) and in addition the $\tilde{\phi}_{0m}$'s are orthogonal. This fact, together with the remarks above, says roughly that orthogonality of translates of ϕ as such is not so difficult to achieve, and that condition (a) itself, that is, $V_0 \subset V_{-1}$, is the nub of the multiresolution analysis notion.

Theorem 1.1-3 Suppose $\phi \in L^2$, and let $b > 0$ be given. Set $\phi_{0n}(x) = \phi(x - nb)$. Suppose that ϕ has a Fourier transform $\hat{\phi}$ for which the following conditions hold:

(a) $\Phi(\omega) = \sum_{k\in Z} |\hat{\phi}(\omega + 2k\frac{\pi}{b})|^2$ exists and is non-zero for almost all $\omega \in (-\infty, \infty)$, and furthermore both Φ and $1/\Phi$ are Lebesgue integrable over $[0, b]$.

(b) $\hat{g} = \hat{\phi}/\sqrt{\Phi}$ is Lebesgue integrable.

Then there is a function $\tilde{\phi} \in L^2$ such that $\tilde{V}_0 = V_0$, where $\tilde{V}_0$ denotes the closure of $\tilde{S}_0 = \text{span}\,\{\tilde{\phi}_{0n}\}$ and V_0 denotes the closure of $S_0 = \text{span}\,\{\phi_{0n}\}$. Furthermore $\{\tilde{\phi}_{0n}\}$ is an orthogonal set.

Proof: Observe that since

$$\Phi(\omega + 2\frac{\pi}{b}) = \sum_{k\in Z} |\hat{\phi}(\omega + 2(k+1)\frac{\pi}{b})|^2 = \sum_{\ell\in Z} |\hat{\phi}(\omega + 2\ell\frac{\pi}{b})|^2 = \Phi(\omega) \tag{1.1-12}$$

(making the substitution $\ell = k+1$ in the summation), it follows that Φ is periodic of period $2\pi/b$. We define $\tilde{\phi}(x)$ to be the inverse Fourier transform of $\hat{g} = \hat{\phi}/\sqrt{\Phi}$ and note that the existence of $\tilde{\phi}$ is guaranteed by the fact that $\hat{g}$ is integrable, condition (b).

We show first that the $\tilde{\phi}_{0m}$'s are square integrable and orthogonal. To that end, observe that

$$\begin{aligned}
\int_{-\infty}^{\infty} \tilde{\phi}_{0\ell}^*(x)\tilde{\phi}_{0n}(x)\,dx &= \int_{-\infty}^{\infty} \widehat{\tilde{\phi}_{0\ell}}^*(\omega)\widehat{\tilde{\phi}_{0n}}(\omega)\,d\omega \\
&= \int_{-\infty}^{\infty} \hat{g}^*(\omega)e^{j\omega\ell b}\hat{g}(\omega)e^{-j\omega nb}\,d\omega \\
&= \int_{-\infty}^{\infty} \frac{|\hat{\phi}|^2}{\Phi}e^{-j\omega(n-\ell)b}\,d\omega
\end{aligned} \tag{1.1-13}$$

using Plancherel's theorem (which says that $\langle f, g\rangle = \langle \hat{f}, \hat{g}\rangle$ for any L^2 functions f and g) and the fact that $\widehat{\tilde{\phi}_{0\ell}}(\omega) = e^{-j\ell b\omega}\hat{\tilde{\phi}}(\omega)$, which can be verified by a simple calculation. We now break up the integral in the last expression of (1.1-13) above into a sum of integrals over intervals of length $2\pi/b$ to obtain:

$$\begin{aligned}
\int_{-\infty}^{\infty} \tilde{\phi}_{0\ell}^*(x)\tilde{\phi}_{0n}(x)\,dx &= \sum_{k\in Z}\int_{2k\pi/b}^{2(k+1)\pi/b} |\hat{\phi}(\omega)|^2\frac{1}{\Phi(\omega)}e^{-j\omega(n-\ell)b}\,d\omega \\
&= \sum_{k\in Z}\int_{0}^{2\pi/b} |\hat{\phi}(u+2k\pi/b)|^2\frac{1}{\Phi(u)}e^{-ju(n-\ell)b}\,du \\
&= \int_{0}^{2\pi/b} \sum_{k\in Z}|\hat{\phi}(u+2k\pi/b)|^2\frac{1}{\Phi(u)}e^{-ju(n-\ell)b}\,du \\
&= \int_{0}^{2\pi/b} e^{-ju(n-\ell)b}\,du \\
&= \begin{cases} 2\pi/b\,; & n=\ell \\ 0\,; & \text{otherwise.}\end{cases}
\end{aligned}$$

The second step uses the substitution $u = \omega - 2k\pi/b$ and the periodicity of Φ (from Eq. (1.1-12)) and of $e^{-ju(n-\ell)b}$. Inverting the order of summation of the series and integration is justified by the monotone convergence theorem and the observation that the partial sums

$$S_N(u) = \sum_{-N\le k\le N} |\hat{\phi}(u+2k\pi/b)|^2\frac{1}{\Phi(u)}$$

are monotone increasing and bounded above by 1 almost everywhere. Thus, both the square integrability of the $\tilde{\phi}_{0n}$'s and their orthogonality is established.

It remains to show that $V_0 = \tilde{V}_0$. Observe that since $\Phi^{-1/2}$ is periodic of period $2\pi/b$ and square integrable on $[0, b]$, it has a Fourier series expansion

$$\Phi^{-1/2}(\omega) = \sum_{k \in Z} c_k e^{-jkb\omega}$$

which converges to $\Phi^{-1/2}$ pointwise almost everywhere. It follows that

$$\begin{aligned}
\tilde{\phi}(x) &= \frac{1}{\sqrt{2\pi}} \int_{-\infty}^{\infty} \hat{\phi}(\omega) \Phi^{-1/2}(\omega) e^{j\omega x}\, d\omega \\
&= \sum_{k \in Z} c_k \frac{1}{\sqrt{2\pi}} \int_{-\infty}^{\infty} \hat{\phi}(\omega) e^{j\omega(x-kb)}\, d\omega \\
&= \sum_{k \in Z} c_k \phi(x - kb) \\
&= \sum_{k \in Z} c_k \phi_{0k}(x).
\end{aligned}$$

It follows that $\tilde{S}_0 \subset V_0$ and thus that $\tilde{V}_0 \subset V_0$. A similar argument (expanding $\Phi^{1/2}$ in a Fourier series) shows that $V_0 \subset \tilde{V}_0$. The desired result follows. ∎

We can ask what happens if we apply the construction of Theorem 1.1-3 to a function ϕ whose translates are already orthonormal. The answer is interesting and useful: $\tilde{\phi} = \phi$, that is, the original orthogonal wavelet is recovered. This is because the function Φ of Theorem 1.1-3 is necessarily almost certainly 1 when the translates of $\phi \in L^2$ are orthonormal:

$$\Phi(\omega) = \sum_{k \in Z} |\hat{\phi}(\omega + 2k\frac{\pi}{b})|^2 = 1 \quad \text{a.e.} \tag{1.1-14}$$

In fact, translates of ϕ are orthogonal if and only if (1.1-14) holds. To see this, we take $b = 1$ for convenience and observe that (since Φ is 2π-periodic, Eq. (1.1-12)) we can expand Φ in a Fourier series:

$$\Phi(\omega) = \sum_{\ell \in Z} c_\ell e^{-j\omega\ell},$$

where

$$\begin{aligned}
c_\ell &= \frac{1}{2\pi} \int_0^{2\pi} \Phi(\omega) e^{j\omega\ell}\, d\omega \\
&= \frac{1}{2\pi} \sum_{k \in Z} \int_0^{2\pi} |\hat{\phi}(\omega + 2k\pi)|^2 e^{j(\omega+2k\pi)\ell}\, d\omega
\end{aligned}$$

$$= \int_{-\infty}^{\infty} |\hat{\phi}(\omega)|^2 e^{j\omega\ell}\, d\omega$$

$$= \int_{-\infty}^{\infty} \phi^*(x-\ell)\phi(x)\, dx.$$

If the integer translates of ϕ are orthogonal, the last expression above vanishes for $\ell \neq 0$, so Φ is constant; on the other hand, if Φ is constant, $c_\ell = 0$ for $\ell \neq 0$, so that the last expression above vanishes for $\ell \neq 0$, that is, translates of ϕ are orthogonal. This uses that fact that if $g(x) = \phi(x - \ell)$, then the Fourier transform $\hat{g}(\omega) = e^{-j\omega\ell}\hat{\phi}(\omega)$, as is readily verified by direct calculation.

Theorem 1.1-3 says that, given a sufficiently well-behaved function ϕ, there is an equivalent function $\tilde{\phi}$ (equivalent in the sense that $\tilde{V}_m = V_m$ for all m) for which the $\tilde{\phi}_{mn}$'s are orthogonal for each m. Observe that it still will not follow that $\tilde{\phi}_{mn}$ and $\tilde{\phi}_{kj}$ will be orthogonal when $k \neq m$. In fact, orthogonality of $\tilde{\phi}_{mn}$ with $\tilde{\phi}_{m-1,j}$ for all n and j would obviously imply orthogonality of every element of V_m with every element of V_{m-1}, and since $V_m \subset V_{m-1}$, orthogonality of every element of V_m with every element of V_m. This would imply that $V_m = \{0\}$. Since m is arbitrary, this would imply that $\cup V_m = \{0\}$, contradicting the fact that, since the V_k's are a multiresolution analysis, their union is dense. *Accordingly, we cannot have orthogonality among the basis elements at different scales in a multiresolution analysis.*

It is precisely the multiresolution analysis framework, however, which gives us a general method of constructing a wavelet ψ whose dilations and their translates are a basis for L^2 with the desired orthogonality properties. This is the crux of multiresolution analysis.

We introduce some terminology which will be useful in the sequel. We will say that a Hilbert space V is the *orthogonal direct sum* of closed subspaces S_1 and S_2, written $V = S_1 \oplus S_2$, if for every $v \in V$, there is an $s_1 \in S_1$ and an $s_2 \in S_2$ such that $v = s_1 + s_2$, and if in addition the subspaces S_1 and S_2 are mutually orthogonal, that is, $\langle s_1, s_2\rangle = 0$ for all $s_1 \in S_1$ and $s_2 \in S_2$. We define also for any subspace $S \subset V$ the orthogonal complement of S in V to be the set $S^\perp$ of all vectors $u \in V$ orthogonal to every vector $s \in S$:

$$S^\perp = \{u \in V \mid \langle u, s\rangle = 0 \text{ for all } s \in S\}.$$

Then, as is easy to verify, necessarily $S^\perp$ is a subspace of V (i.e., it is closed under addition and scalar multiplication), $S^\perp$ is closed, and $V = S \oplus S^\perp$. The notion of the orthogonal direct sum extends easily to finite and countable collections of mutually orthogonal subspaces: We say that V is the orthogonal direct sum of closed subspaces S_m, $m \in Z$, written

$$V = \oplus_{m\in Z} S_m,$$

if for every $v \in V$, there is an $s_m \in S_m$, such that

$$v = \sum_{m \in Z} s_m$$

(in the usual sense that the partial sums of the series converge in the norm to v) and if in addition the subspaces S_m and S_ℓ are mutually orthogonal for every $m \neq \ell$.

Using the notation of the previous paragraph, we can write $V_{m-1} = V_m \oplus W_m$, where W_m is the orthogonal complement of V_m in V_{m-1}. Similarly, $V_m = V_{m+1} \oplus W_{m+1}$, where W_{m+1} is the orthogonal complement of V_{m+1} in V_m, and so forth. Observe that since $W_{m+1} \subset V_m$ and W_m is orthogonal to V_m, necessarily W_{m+1} and W_m are orthogonal. This is the key observation; to restate it: if we start with a multiresolution analysis, the orthogonal complements W_m of the spaces V_m in V_{m-1} form a family of mutually orthogonal subspaces of L^2. It is this fact which is exploited to construct a wavelet ψ whose dilations and their translates form an orthonormal basis for L^2. The remainder of this section is devoted to exploring the properties of the W_m's and the construction of the wavelet ψ.

The mutual orthogonality of subspaces W_m of L^2 implies that the intersection of any two W_m's is the zero function and hence that

$$\cap_{m \in Z} W_m = \{0\}.$$

This much is obvious. Also, since $V_m = V_{m+1} \oplus W_{m+1} = V_{m+2} \oplus W_{m+2} \oplus W_{m+1}$ and so forth, we have for any m and any $k \geq 0$

$$V_m = V_{m+k} \oplus W_{m+k-1} \oplus W_{m+k-2} \cdots \oplus W_{m+1}. \tag{1.1-15}$$

Theorem 1.1-4 If the V_m's are a multiresolution analysis, so that $\cup V_m$ is dense in L^2, then

$$L^2 = \oplus_{m \in Z} W_m.$$

Proof: This follows from (1.1-15). Since the subspaces W_m are mutually orthogonal, it suffices to show that for any m

$$V_m = \oplus_{k=1}^{\infty} W_{m+k} \tag{1.1-16}$$

for this implies that $\oplus_{m \in Z} W_m$ is dense in L^2, and hence that L^2 is the orthogonal direct sum of the W_m's.

Since V_m is closed, the right-hand side of (1.1-16), which is the closure of a union of subspaces $\oplus_{1 \leq k \leq N} W_{m+k} \subset V_m$, is necessarily contained in the left-hand side. On the other hand, if $v \in V_m$ and Π denotes the orthogonal projection operator on $\oplus_{k \geq 1} W_{m+k}$, then $v - \Pi v$ is orthogonal to the right-hand side of (1.1-16), by the Hilbert projection theorem, so that in particular $v - \Pi v$ is orthogonal to W_{m+k} for all $k \geq 0$. But then (1.1-15) implies that $v - \Pi v \in V_{m+k}$ for all $k \geq 1$, whence

$$v - \Pi v \in \cap_{k=1}^{\infty} V_{m+k} = \cap_{\ell \in Z} V_\ell = \{0\}$$

using the fact that V_k is a multiresolution analysis. It follows that $\Pi v = v$, that is, $V_m \subset \oplus_{k\geq 1} W_{m+k}$. This establishes Eq. (1.1-16), and the fact that $\oplus_{k\in Z} W_k$ is dense in L^2 follows. ▮

We turn now to the construction of the wavelet ψ. This construction presumes that the ϕ_{0n}'s are orthonormal and real-valued, and we shall henceforth assume that this is the case. The function $\psi \in W_0$ which we shall construct has the property that the ψ_{0n}'s (defined by $\psi_{0n}(x) = \psi(x - nb)$) are orthogonal real-valued functions and furthermore that W_0 is the closure of $\text{span}\{\psi_{0n}\}_{n\in Z}$. As noted above, this is the heart of multiresolution analysis; it gives a general algorithm for the construction of wavelet bases for L^2. We outline why the procedure works and then state the result as a formal theorem.

Since $V_0 \subset V_{-1}$, we can expand ϕ with respect to the $\phi_{-1,n}$'s:

$$\phi = \sum_{n\in Z} h_n \phi_{-1,n}.$$

Note that since both ϕ and $\phi_{-1,n}$ are real functions, the coefficients h_n are real. Set

$$\psi = \sum_{n\in Z} g_n \phi_{-1,n}, \tag{1.1-17}$$

where $g_n \triangleq (-1)^n h_{-n+1}$. Then we claim that $\psi \in W_0$, that the ψ_{0n}'s are orthogonal and that W_0 is the closure of $\text{span}\{\psi_{0n}\}_{n\in Z}$.

Since ϕ is square integrable, the sequence h_n is square summable, and using the orthonormality of the $\phi_{-1,n}$'s,

$$\sum_{n\in Z} h_n^2 = \int_{-\infty}^{\infty} \phi^2 \, dx = 1. \tag{1.1-18}$$

We observe first that ψ is orthogonal to ϕ, since

$$\begin{aligned}
\int_{-\infty}^{\infty} \psi\phi \, dx &= \sum_{n,\ell} (-1)^n h_{-n+1} h_\ell \int_{-\infty}^{\infty} \phi_{-1,n}\phi_{-1,\ell} \, dx \\
&= \sum_{n,\ell} (-1)^n h_{-n+1} h_\ell \delta_{n\ell} \\
&= \sum_{n\in Z} (-1)^n h_{-n+1} h_n \\
&= 0.
\end{aligned}$$

The conclusion follows from the fact that, making the substitution $\ell = -n + 1$ in the sum in the second from last expression, we obtain

$$\sum_{n\in Z} (-1)^n h_{-n+1} h_n = \sum_{\ell\in Z} (-1)^{-\ell+1} h_\ell h_{-\ell+1} = \sum_{\ell\in Z} (-1)^\ell h_\ell h_{-\ell+1}.$$

But $a = -a$ if and only if $a = 0$; it follows that the expression must vanish.

Next, since by hypothesis the ϕ_{0n}'s are orthonormal, we must have, for $k \neq 0$,

$$\begin{aligned}
0 &= \int_{-\infty}^{\infty} \phi_{00}\phi_{0k}\,dx \\
&= \sum_{n,\ell} h_n h_\ell \int_{-\infty}^{\infty} \phi_{-1,n}(x)\phi_{-1,\ell}(x-kb)\,dx \\
&= \sum_{n,\ell} h_n h_\ell \int_{-\infty}^{\infty} \phi_{-1,n}(x)\phi_{-1,\ell+2k}(x)\,dx \\
&= \sum_{n,\ell} h_n h_\ell \delta_{n,\ell+2k} \\
&= \sum_{n\in Z} h_n h_{n-2k}. \qquad (1.1\text{-}19)
\end{aligned}$$

Using this, we calculate that

$$\begin{aligned}
\int_{-\infty}^{\infty} \psi_{0q}\psi_{0k}\,dx &= \sum_{n,\ell}(-1)^{n+\ell} h_{-n+1}h_{-\ell+1}\int_{-\infty}^{\infty}\phi_{-1,n}(x-qb)\phi_{-1,\ell}(x-kb)\,dx \\
&= \sum_{n,\ell}(-1)^{n+\ell} h_{-n+1}h_{-\ell+1}\int_{-\infty}^{\infty}\phi_{-1,n+2q}(x)\phi_{-1,\ell+2k}(x)\,dx \\
&= \sum_{n,\ell}(-1)^{n+\ell} h_{-n+1}h_{-\ell+1}\delta_{n+2q,\ell+2k} \\
&= \sum_{n\in Z}(-1)^{n+n+2(q-k)} h_n h_{n-2(k-q)} \\
&= \begin{cases} 1, & k = q \\ 0, & k \neq q \end{cases}
\end{aligned}$$

using (1.1-18) and (1.1-19). Thus, the ψ_{0n}'s are necessarily orthonormal.

Let U_0 denote the closure of $\text{span}\{\psi_{0n}\}_{n\in Z}$. Since each ψ_{0n} is in V_{-1} and is orthogonal to all of the ϕ_{0n}'s, and thus to V_0, we have $U_0 \subset W_0$. We now show that $U_0 = W_0$. To that end, we construct the best approximation to $\phi_{-1,\ell}$ from $V_0 \oplus U_0$; it has the form

$$\Pi\phi_{-1,\ell} = \sum_{k\in Z} c_{k\ell}\phi_{0k} + \sum_{k\in Z} d_{k\ell}\psi_{0k}, \qquad (1.1\text{-}20)$$

where

$$\begin{aligned}
c_{k\ell} &= \langle \phi_{-1,\ell}, \phi_{0k}\rangle \\
&= \sum_{n\in Z} h_n \int_{-\infty}^{\infty} \phi_{-1,\ell}(x)\phi_{-1,n}(x-kb)\,dx
\end{aligned}$$

$$
\begin{aligned}
&= \sum_{n\in Z} h_n \int_{-\infty}^{\infty} \phi_{-1,\ell}(x)\phi_{-1,n+2k}(x)\,dx \\
&= \sum_{n\in Z} h_n \delta_{\ell,n+2k} \\
&= h_{\ell-2k}
\end{aligned}
$$

and $d_{k\ell}$ is given by

$$
\begin{aligned}
d_{k\ell} &= \langle \phi_{-1,\ell}, \psi_{0k} \rangle \\
&= \sum_{n\in Z} (-1)^n h_{-n+1} \int_{-\infty}^{\infty} \phi_{-1,\ell}(x)\phi_{-1,n}(x-kb)\,dx \\
&= \sum_{n\in Z} (-1)^n h_{-n+1} \int_{-\infty}^{\infty} \phi_{-1,\ell}(x)\phi_{-1,n+2k}(x)\,dx \\
&= \sum_{n\in Z} (-1)^n h_{-n+1}\delta_{\ell,n+2k} \\
&= (-1)^\ell h_{2k+1-\ell}.
\end{aligned}
$$

Inserting these results in Eq. (1.1-20), we obtain the formula

$$
\Pi\phi_{-1,\ell} = \sum_{k\in Z} h_{\ell-2k}\phi_{0k} + (-1)^\ell \sum_{k\in Z} h_{2k+1-\ell}\psi_{0k}. \tag{1.1-21}
$$

It follows from (1.1-21) (observing that for each ℓ, one of the sums on the right contains h_n for all even n and the other, h_n for all odd n) and the orthonormality of the set $\{\phi_{0k}\}_{k\in Z} \cup \{\psi_{0\ell}\}_{\ell\in Z}$ that

$$
||\Pi\phi_{-1,\ell}||^2 = \sum_{k\in Z} h_k^2 = 1
$$

and thus (by a simple calculation we omit) that

$$
||\phi_{-1,\ell} - \Pi\phi_{-1,\ell}||^2 = 0.
$$

We have shown that $\phi_{-1,\ell} \in V_0 \oplus U_0$, for all $\ell \in Z$, so that $V_{-1} \subset V_0 \oplus U_0$. The converse inclusion is trivial, and we have established that V_{-1} is the orthogonal direct sum of V_0 and U_0, so that necessarily $U_0 = W_0$, as claimed.

We have proved the following important theorem:

Theorem 1.1-5 Suppose that V_m is a multiresolution analysis and that the function ϕ which generates it is real-valued and has the property that its translates ϕ_{0n} are orthogonal. Let h_n, $n \in Z$, denote the coefficients in the expansion of $\phi = \phi_{00}$ with respect to the $\phi_{-1,n}$'s:

$$
\phi_{00} = \sum_{n\in Z} h_n \phi_{-1,n}.
$$

Then the function ψ defined by

$$\psi = \sum_{n \in Z} (-1)^n h_{-n+1} \phi_{-1,n}$$

has the following properties:

1. $\psi \in W_0$, that is, ψ is orthogonal to V_0.
2. The translates $\psi_{0n}(x) = \psi(x - nb)$ of ψ are orthogonal.
3. W_0 is the closure of the set of linear combinations of the ψ_{0n}'s.
4. $\{\psi_{mn}\}_{m,n \in Z}$ is an orthonormal basis for L^2. ∎

It is important to observe that multiresolution analysis gives us not one but two orthonormal bases for each space V_m. In the case of V_{-1}, one is the set $\{\phi_{-1,n}\}_{n \in Z}$ and the other is the set $\{\phi_{0n}\}_{n \in Z} \cup \{\psi_{0n}\}_{n \in Z}$. We can expand the elements of any orthonormal basis in terms of the other. In the present case, the expansions follow from the expressions we have just been using. If we write $g_n = (-1)^n h_{1-n}$, we have

$$\phi_{0\ell}(x) = \sum_{n \in Z} h_n \phi_{-1,n}(x - \ell b) = \sum_{n \in Z} h_n \phi_{-1,n+2\ell}(x) = \sum_{n \in Z} h_{n-2\ell} \phi_{-1,n}(x) \quad (1.1\text{-}22)$$

$$\psi_{0\ell}(x) = \sum_{n \in Z} g_n \phi_{-1,n}(x - \ell b) = \sum_{n \in Z} g_{n-2\ell} \phi_{-1,n}(x) \quad (1.1\text{-}23)$$

$$\phi_{-1,\ell}(x) = \sum_{k \in Z} h_{\ell-2k} \phi_{0k} + \sum_{k \in Z} g_{\ell-2k} \psi_{0k}. \quad (1.1\text{-}24)$$

It follows that the unitary matrix which represents the orthogonal change of basis is known, once we know the sequences $\boldsymbol{h}$ and $\boldsymbol{g}$ of coefficients in (1.1-22) and (1.1-23). It is this unitary matrix which yields the discrete wavelet transform, and we shall investigate it in Section 1.3 and Section 1.4 below.

This completes the discussion of multiresolution analysis. It is the theoretical framework for generating wavelet bases. In the next section, we give two examples of wavelet bases.

1.2 Two Orthogonal Wavelet Bases

A. The Haar wavelet

The Haar basis is an example of an orthonormal wavelet basis for $L^2(R)$. We analyze it in the framework of multiresolution analysis.

Set the translation parameter $b = 1$ and set

$$\phi(x) = \chi_{[0,1)}(x) = \begin{cases} 1\,; & x \in [0, 1) \\ 0\,; & \text{otherwise.} \end{cases}$$

Then

$$\begin{aligned}\phi_{mn}(x) &= 2^{-m/2}\chi_{[0,1)}(2^{-m}x-n)\\ &= 2^{-m/2}\chi_{[n,n+1)}(2^{-m}x)\\ &= 2^{-m/2}\chi_{[n2^m,(n+1)2^m)}(x). \qquad (1.2\text{-}1)\end{aligned}$$

It follows that, for example,

$$\phi_{10}(x) = \frac{1}{\sqrt{2}}\chi_{[0,2)}(x) = \begin{cases} 1/\sqrt{2}\,; & x \in [0,2) \\ 0\,; & \text{otherwise}\end{cases}$$

$$\phi_{12}(x) = \frac{1}{\sqrt{2}}\chi_{[4,6)}(x) = \begin{cases} 1/\sqrt{2}\,; & x \in [4,6) \\ 0\,; & \text{otherwise}\end{cases}$$

$$\phi_{-1,0}(x) = \sqrt{2}\chi_{[0,1/2)}(x) = \begin{cases} \sqrt{2}\,; & x \in [0,1/2) \\ 0\,; & \text{otherwise}\end{cases}$$

$$\phi_{-1,2}(x) = \sqrt{2}\chi_{[1,3/2)}(x) = \begin{cases} \sqrt{2}\,; & x \in [1,3/2) \\ 0\,; & \text{otherwise.}\end{cases}$$

Observe in particular that $\phi(2^{-m}x)$ is non-zero on an interval of length 2^m – multiplying x by 2^{-m} "stretches" the graph of ϕ by a factor of 2^m, as noted above. Observe also that $\phi(2^{-m}x-n) = \phi(2^{-m}(x-n2^m))$ has a graph which is the same as $\phi(2^{-m}x)$, translated $2^m n$ (rather than n) units to the right.

With this choice of ϕ, V_0 is the set of all functions whose values are constant on each interval of the form $[m, m+1)$, $m \in Z$, and which are square integrable. We can represent any such function in the form

$$f(x) = \sum_{n\in Z} c_n\phi_{0n}(x) = \sum_{n\in Z} c_n\chi_{[n,n+1)}(x),$$

where the sequence c_n is square summable. Note that in this case the translates ϕ_{0n} are orthonormal:

$$\int_{-\infty}^{\infty}\phi_{0n}(x)\phi_{0\ell}(x)\,dx = \int_{-\infty}^{\infty}\chi_{[n,n+1)}(x)\chi_{[\ell,\ell+1)}(x)\,dx = \begin{cases} 1\,; & \ell = n \\ 0\,; & \ell \neq n\end{cases}$$

(Orthogonality would not hold in this case, if we had chosen $b < 1$, say $b = 1/2$, and in general cannot be expected to hold unless ϕ and b are chosen carefully.) Thus, in this case we need not resort to the procedure of Theorem 1.1-3 in order to obtain orthogonal $\tilde{\phi}_{0n}$'s.

To verify that the conditions for a multiresolution analysis hold for this example, we need only verify condition (a) that $V_m \subset V_{m-1}$ for all $m \in Z$, since the other conditions hold via Theorems 1.1-1 and 1.1-2, using the fact that the integer translates of ϕ are orthogonal. It is easy to verify that $\hat{\phi}$ is continuous, indeed analytic, at $\omega = 0$, so that the hypotheses of Theorem 1.1-2 are satisfied here. As to condition (a), $V_m \subset V_{m-1}$ holds when $m = 0$ because

$$\begin{aligned}\phi(x) &= \chi_{[0,1)}(x) \\ &= \chi_{[0,1/2)}(x) + \chi_{[1/2,1)}(x) \\ &= \frac{1}{\sqrt{2}}\phi_{-1,0}(x) + \frac{1}{\sqrt{2}}\phi_{-1,1}(x) \qquad (1.2\text{-}2)\end{aligned}$$

using Eq. (1.2-1). Since the right-hand side is in V_{-1}, it follows that $\phi \in V_{-1}$ and, since V_{-1} is invariant under translations by integer multiples of 1/2, and thus by integer multiples of 1, $\phi_{0n} \in V_{-1}$ for all n. Since V_{-1} is closed under linear combinations, all linear combinations of the ϕ_{0n}'s are contained in V_{-1}, and thus $S_0 \subset V_{-1}$. Finally, since V_{-1} is a closed set, the closure of any subset of V_{-1} is contained in V_{-1}, hence $V_0 \subset V_{-1}$. This completes the proof for $m = 0$. The proof for arbitrary m is the same. It follows that the subspaces V_m generated by $\phi = \chi_{[0,1)}$ with $b = 1$ are a multiresolution analysis of L^2.

We conclude that the family of multiresolution analyses is not empty; we shall see that in fact there are many multiresolution analyses, and that in fact there is enough flexibility so that multiresolution analyses can be constructed which possess desirable properties in addition to the satisfaction of the four basic conditions (a) through (d) above.

We now obtain the spaces W_m and the wavelet ψ for this example. We already have an expansion

$$\phi = \sum_{n \in Z} h_n \phi_{-1,n}.$$

It is given by (1.2-2): $h_0 = h_1 = 1/\sqrt{2}$ and $h_n = 0$ for all other values of n. Then

$$\begin{aligned}\psi &= \sum_{n \in Z} (-1)^n h_{1-n} \phi_{-1,n} \\ &= (-1)^0 h_1 \phi_{-1,0} + (-1)^1 h_0 \phi_{-1,1} \\ &= \frac{1}{\sqrt{2}}(\phi_{-1,0} - \phi_{-1,1}) \\ &= \chi_{[0,1/2)} - \chi_{[1/2,1)}.\end{aligned}$$

We set W_m equal to the closure of span$\{\psi_{mn}\}_{n \in Z}$, where $\psi_{mn}(x) = 2^{-m/2}\psi(2^{-m} - n)$. Then, by the general results in Section 1.1, W_m is orthogonal to V_m, $V_{m-1} = V_m \oplus W_m$ for all m, the ψ_{mn}'s are orthonormal, and $L^2 = \oplus_{m \in Z} W_m$.

We verify some of these facts, in order to illustrate the general arguments made above. It is obvious that ϕ_{0n} is orthogonal to ψ, since this holds by construction when $n = 0$ and, when $n \neq 0$, by the fact that the support $[n, n+1)$ of ϕ_{0n} and the support $[0,1)$ of ψ are disjoint sets. But then for all m, n and j we have

$$\begin{aligned}\int_{-\infty}^{\infty} \phi_{mn}(x)\psi_{mj}(x)\,dx &= \int_{-\infty}^{\infty} 2^{-m}\phi(2^{-m}x-n)\psi(2^{-m}x-j)\,dx \\ &= \int_{-\infty}^{\infty} \phi(\xi+j-n)\psi(\xi)\,d\xi \\ &= 0\end{aligned}$$

making the substitution $\xi = 2^{-m}x - j$ and using the observation above.

Verifying that $V_{m-1} = V_m \oplus W_m$ is also trivial in this case. For observe that

$$\phi(x) + \psi(x) = 2\chi_{[0,1/2)}(x) = \sqrt{2}\phi_{-1,0}(x).$$

It follows that every translate $\phi_{-1,n}$ is a scalar multiple of $\phi_{0n} + \psi_{0n}$, and, since the direct sum of closed subspaces is necessarily closed, that

$$V_{-1} \subset V_0 \oplus W_0.$$

But $V_0 \subset V_{-1}$ and also $W_0 \subset V_{-1}$, so that necessarily $V_0 \oplus W_0 \subset V_{-1}$. This gives the desired result for $m = 0$. The like result for arbitrary m follows similarly.

The orthogonality of the translates ψ_{0n} (another general property guaranteed by the construction) can easily be verified by direct calculation in this example. We omit the calculation. The orthogonality of the ψ_{mn}'s for all m and n then follows from the orthogonality of the ψ_{mn}'s for fixed m and the fact that for $m \neq k$, the subspaces W_m and W_k are orthogonal.

From the general result in Section 1.1 that

$$L^2 = \oplus_{m \in Z} W_m$$

(Theorem 1.1-4) we have that the ψ_{mn} are an orthonormal basis for L^2. Thus, for any $v \in L^2$,

$$v = \sum_{m,n \in Z} c_{mn}\psi_{mn} \quad \text{a.e.},$$

where $c_{mn} = \langle v, \psi_{mn}\rangle$, using the fact that the ψ_{mn}'s have unit norm. The set $\{\psi_{mn}\}$ is called the Haar basis, and the function $\psi = \chi_{[0,1/2)} - \chi_{[1/2,1)}$ is sometimes called the Haar wavelet. It is one of the most important wavelets in applications.

Observe that since $V_0 = \oplus_{m \geq 1} W_m$, in particular ϕ_{00}, which integrates to 1, can be written as a limit of (finite) linear combinations of the ψ_{mn}'s, $m \geq 1$, and each such linear combination integrates to 0. If this seems strange, it may be instructive to expand ϕ_{00} explicitly in terms of the ψ_{mn}'s to investigate how it can happen. We leave this as an exercise for the reader.

B. The Meyer Wavelet

We turn to a wavelet basis which was introduced by Meyer in 1985 [1-9], discussed in [1-10]. This basis derives from a multiresolution analysis generated by a function $\phi \in L^2(R)$ whose Fourier transform $\hat{\phi}$ is a smooth function supported in $(-4\pi/3, 4\pi/3)$ with the following properties:

(i) $\hat{\phi}(\omega) = 1$ if $\omega \in [0, 2\pi/3]$;

(ii) $|\hat{\phi}(\omega)|^2 + |\hat{\phi}(2\pi - \omega)|^2 = 1$ if $\omega \in (2\pi/3, 4\pi/3)$; and

(iii) $\hat{\phi}(\omega) = \hat{\phi}(-\omega)$ for every $\omega \in (-\infty, \infty)$.

Observe that since $\hat{\phi}$ is even, bounded, and compactly supported, necessarily its inverse Fourier transform ϕ is real and C^∞ differentiable. It is also easy to verify that

$$\|\hat{\phi}\|_2^2 = \int_{-\infty}^{\infty} |\hat{\phi}(\omega)|^2 \, d\omega = 2\pi.$$

To construct $\hat{\phi}$, proceed as follows: Observe that only the values of $\hat{\phi}(\omega)$ for $\omega \in (2\pi/3, 4\pi/3)$ need to be specified. To that end, define

$$v(\omega) = \begin{cases} \exp(-\dfrac{1}{(3\omega - \pi)^2(3\omega + \pi)^2}) \; ; & |\omega| < \pi/3 \\ 0 \; ; & |\omega| \geq \pi/3. \end{cases}$$

Then define

$$\hat{\phi}(\omega) = \begin{cases} 1 \; ; & |\omega| \in [0, 2\pi/3] \\ \cos \lambda(|\omega|) \; ; & |\omega| \in (2\pi/3, \, 4\pi/3) \\ 0 \; ; & |\omega| \geq 4\pi/3, \end{cases}$$

where

$$\lambda(\omega) = \frac{\pi}{4} + \kappa_0 \int_0^{\omega - \pi} v(t) \, dt$$

and κ_0 is such that

$$\kappa_0 \int_0^{\pi/3} v(t) \, dt = \frac{\pi}{4}.$$

It can be shown that the function $\hat{\phi}$ so defined is a C^∞ function.

Now we want to check that (ii) holds. For $\omega \in (2\pi/3, \, 4\pi/3)$ we have

$$\begin{aligned} \lambda(\omega) + \lambda(2\pi - \omega) &= \frac{\pi}{2} + \kappa_0 \int_0^{\omega - \pi} v(t) \, dt + \kappa_0 \int_0^{\pi - \omega} v(t) \, dt \\ &= \frac{\pi}{2} \end{aligned}$$

since the sum of the two integrals is equal to zero (because v is an even function). Therefore, we have

$$|\hat{\phi}(\omega)|^2 + |\hat{\phi}(2\pi - \omega)|^2 = \cos^2 \lambda(\omega) + \cos^2 \lambda(2\pi - \omega)$$

$$
\begin{aligned}
&= \cos^2 \lambda(\omega) + \cos^2(\frac{\pi}{2} - \lambda(\omega)) \\
&= \cos^2 \lambda(\omega) + \sin^2 \lambda(\omega) \\
&= 1
\end{aligned}
$$

for every $\omega \in (2\pi/3, 4\pi/3)$. This establishes (ii). Clearly $\hat{\phi}$ is an even function, so that (iii) holds. We conclude that at least one function $\hat{\phi}$ with properties (i), (ii), and (iii) can be constructed.

The purpose of having (i), (ii), and (iii) is to guarantee that

$$
\Phi(\omega) \equiv \sum_{k \in Z} |\hat{\phi}(\omega + 2k\pi)|^2 = 1 \tag{1.2-3}
$$

for every $\omega \in (-\infty, \infty)$. Since the function Φ is 2π-periodic, as is readily verified, we only need to check (1.2-3) on an interval of length 2π, say $[-2\pi/3, 4\pi/3]$. On the subinterval $[-2\pi/3, 2\pi/3]$, only the $k = 0$ term in the summation in (1.2-3) is non-zero, so $\Phi(\omega) = |\hat{\phi}(\omega)|^2 = 1$ by condition (i). On the subinterval $[2\pi/3, 4\pi/3]$ only the $k = 0$ and $k = -1$ terms in the summation in (1.2-3) are non-zero, and in this subinterval

$$
\begin{aligned}
\Phi(\omega) &= |\hat{\phi}(\omega)|^2 + |\hat{\phi}(\omega - 2\pi)|^2 \\
&= |\hat{\phi}(\omega)|^2 + |\hat{\phi}(2\pi - \omega)|^2 \\
&= 1.
\end{aligned}
$$

This uses conditions (ii) and (iii). It follows that $\Phi(\omega) = 1$ for every $\omega \in (-\infty, \infty)$.

An immediate consequence of this result and Theorem 1.1-3 (and following remarks) is that the translates ϕ_{0n} of ϕ are orthogonal. It is also true that $\phi/\sqrt{2\pi}$ integrates to 1 (since $\hat{\phi}(0) = 1$), and the reader can verify that $\{\phi_{0n}/\sqrt{2\pi}\}_{n \in Z}$ is an orthonormal set. Thus, assuming that condition (a) of a multiresolution analysis holds, that is $V_m \subset V_{m-1}$, we have by Theorems 1.1-1 and 1.1-2 all of the other conditions of a multiresolution analysis, exactly as in the case of the Haar wavelet.

Set

$$
V_0 = \text{the closure span of } \{\phi(x - nb) | n \in Z\} \text{ with } b = 1
$$

and define V_m in the same manner as discussed in Section 1.1. We wish to verify that $V_m \subset V_{m-1}$ for all m. To that end, observe first that any function $f \in V_m$ can be represented by

$$
f(x) = \sum_{n \in Z} c_n \phi_{mn}(x) = 2^{-m/2} \sum_{n \in Z} c_n \phi(2^{-m}x - n) \tag{1.2-4}
$$

for some square summable complex sequence $\{c_n\}$. Taking the Fourier transform of both sides of (1.2-4) we have

$$
\hat{f}(\omega) = \frac{1}{\sqrt{2\pi}} \int_{-\infty}^{\infty} f(x) e^{-j\omega x}\, dx
$$

$$\begin{aligned}
&= \frac{1}{\sqrt{2\pi}} \sum_{n\in Z} c_n 2^{-m/2} \int_{-\infty}^{\infty} \phi(2^{-m}x - n) e^{-j\omega x}\, dx \\
&= \frac{1}{\sqrt{2\pi}} \sum_{n\in Z} c_n e^{-j2^m n\omega} 2^{m/2} \int_{-\infty}^{\infty} \phi(u) e^{-j2^m \omega u}\, du \\
&= 2^{m/2} \hat{c}(2^m\omega)\hat{\phi}(2^m\omega) \qquad (1.2\text{-}5)
\end{aligned}$$

using the change of variable $u = 2^{-m}x - n$ in the integral. Here $\hat{c}$ is the discrete Fourier transform of $\{c_n\}$ given by

$$\hat{c}(\omega) = \sum_{n\in Z} c_n e^{-j\omega n}.$$

Note that $\hat{c}$ is 2π-periodic. By reversing the steps leading to Eq. (1.2-5), it is easy to see that if the Fourier transform $\hat{f}$ of a function satisfies $\hat{f}(\omega) = M(2^n\omega)\hat{\phi}(2^n\omega)$ for some 2π-periodic function M, then $f \in V_m$.

We want to show that $V_m \subset V_{m-1}$ for all $m \in Z$. We shall show $V_0 \subset V_{-1}$; the proof for any other m is similar. The desired result will follow if $\phi \in V_{-1}$, which in turn, by the observation above, will follow if we can find some 2π-periodic function $M \in L^2(0, 2\pi)$ such that

$$\begin{aligned}
\hat{\phi}(\omega) &= M(\frac{\omega}{2})\hat{\phi}(\frac{\omega}{2}) \quad \text{or equivalently} \\
\hat{\phi}(2\omega) &= M(\omega)\hat{\phi}(\omega). \qquad (1.2\text{-}6)
\end{aligned}$$

We claim that

$$M(\omega) = \sum_{n\in Z} \hat{\phi}(2\omega + 4\pi n) \qquad (1.2\text{-}7)$$

works. It is easy to check that M is a 2π-periodic function. Since the support of $\hat{\phi}(\omega)$ is contained in $(-4\pi/3, 4\pi/3)$, $M(\omega) = \hat{\phi}(2\omega) + \hat{\phi}(2\omega - 4\pi)$ for every $\omega \in [0, 2\pi]$; all of the other terms in the summation vanish. It follows that M is a finite sum of integrable functions and thus $M \in L^2([0, 2\pi])$. Finally, note that

$$M(\omega)\hat{\phi}(\omega) = \sum_{n\in Z} \hat{\phi}(\omega)\hat{\phi}(2\omega + 4\pi n) = \hat{\phi}(\omega)\hat{\phi}(2\omega) \qquad (1.2\text{-}8)$$

since the support of each $\hat{\phi}(2\omega + 4\pi n)$ is contained in $(-2\pi n - 2\pi/3, -2\pi n + 2\pi/3)$, which does not intersect with the support of $\hat{\phi}(\omega)$ unless $n = 0$. The expression on the right in (1.2-8) is always equal to $\hat{\phi}(2\omega)$: if $|\omega| \leq 2\pi/3$, $\hat{\phi}(\omega) = 1$ by conditions (i) and (iii), so $\hat{\phi}(\omega)\hat{\phi}(2\omega) = \hat{\phi}(2\omega)$ in this case, and if $|\omega| > 2\pi/3$ then $|2\omega| > 4\pi/3$, so $\hat{\phi}(2\omega) = 0$ and likewise $\hat{\phi}(\omega)\hat{\phi}(2\omega) = 0$. Thus, we have (1.2-6) for the function M of (1.2-7), and necessarily $V_0 \subset V_{-1}$.

Since, as noted above, conditions (b), (c) and (d) of a multiresolution analysis follow from condition (a) and the orthogonality of translates of ϕ, Theorems 1.1-1 and 1.1-2, we have that the inverse Fourier transform ϕ of a function $\hat{\phi}$ satisfying conditions (i) through (iii) generates a multiresolution analysis. In principle, we obtain the corresponding wavelet ψ in the manner of Section 1.1: expand ϕ in terms of the functions $\phi_{-1,n}$ to obtain

$$\phi(x) = \sum_{n \in Z} h_n \phi_{-1,n}(x)$$

and set

$$\psi(x) = \sum_{n \in Z} (-1)^n h_{1-n} \phi_{-1,n}(x).$$

We cannot write down ψ explicitly in the case of the Meyer wavelet, but it is not difficult to show that its Fourier transform $\hat{\psi}$ is given by

$$\hat{\psi}(\omega) = -e^{-j\omega/2} M(\frac{\omega}{2} + \pi) \hat{\phi}(\frac{\omega}{2}),$$

where M is defined as in (1.2-7). It follows via a short calculation that

$$\hat{\psi}(\omega) = \begin{cases} -e^{-j\omega/2} \sin \lambda(|\omega|) \; ; & 2\pi/3 \leq |\omega| \leq 4\pi/3 \\ -e^{-j\omega/2} \cos \lambda(|\omega|/2) \; ; & 4\pi/3 \leq |\omega| \leq 8\pi/3 \\ 0 \; ; & \text{otherwise.} \end{cases}$$

Observe that the Meyer wavelet has compact support in the frequency domain (i.e. is band limited) whereas the Haar wavelet has compact support in the time domain. In the next section, we describe a technique due to Daubechies which can be used to generate many wavelets with compact support in the time domain. The Daubechies wavelets are thus a generalization of the Haar-type wavelet; as we shall see, the techniques used in their construction and analysis are in many respects similar to those used above for the Meyer wavelet – most of the analysis is done in the frequency domain.

1.3 Discrete Wavelets and the Daubechies Construction

So far, we have looked at wavelets in $L^2(R)$, but in applications we are typically interested in discrete signals, elements of ℓ^2. In this section, we look at wavelets in discrete space.

Also, in Section 1.2 we gave examples of two orthogonal wavelet bases. In this section, we outline a technique, due to Daubechies, for constructing many orthogonal wavelet bases. Daubechies' construction proceeds not in $L^2(R)$ but in discrete space ℓ^2, so it is convenient to treat these two topics together.

A. Two Provisional Definitions of Discrete Wavelets

Discrete wavelets are not obtained from wavelets $\psi \in L^2$ by a process of sampling or approximation. Rather, they are sequences which arise in a change of basis in a multiresolution analysis.

We resume the discussion of multiresolution analysis where we left it at the end of Section 1.1. Suppose ϕ has orthonormal translates and generates a multiresolution analysis $\{V_m\}$ of L^2. Suppose that ψ has orthonormal translates which are a basis for the orthocomplement W_0 of V_0 in V_{-1}. Then we have two orthonormal bases for V_{-1}, namely $\{\phi_{-1,n}\}_{n\in Z}$ and $\{\phi_{0n}\}_{n\in Z} \cup \{\psi_{0n}\}_{n\in Z}$. We will usually assume that both ϕ and ψ are real.

We expand ϕ and ψ in the basis $\{\phi_{-1,n}\}_{n\in Z}$ for V_{-1} to obtain

$$\phi(x) = \sum_{n\in Z} h_n \phi_{-1,n}(x) \tag{1.3-1}$$

$$\psi(x) = \sum_{n\in Z} g_n \phi_{-1,n}(x). \tag{1.3-2}$$

Note that $h_n = \langle \phi_{-1,n}, \phi \rangle$ and $g_n = \langle \phi_{-1,n}, \psi \rangle$ because of the orthonormality properties of translates of ϕ and ψ. It follows from (1.3-1) and (1.3-2) that

$$\phi_{0\ell}(x) = \sum_{n\in Z} h_n \phi_{-1,n}(x - \ell b) = \sum_{n\in Z} h_n \phi_{-1,n+2\ell}(x) = \sum_{n\in Z} h_{n-2\ell} \phi_{-1,n}(x) \tag{1.3-3}$$

$$\psi_{0\ell}(x) = \sum_{n\in Z} g_n \phi_{-1,n}(x - \ell b) = \sum_{n\in Z} g_{n-2\ell} \phi_{-1,n}(x) \tag{1.3-4}$$

$$\phi_{-1,\ell}(x) = \sum_{k\in Z} h_{\ell-2k} \phi_{0k} + \sum_{k\in Z} g_{\ell-2k} \psi_{0k}. \tag{1.3-5}$$

These are the change of basis equations (1.1-22) through (1.1-24) derived at the end of Section 1.1.

Define a matrix $\boldsymbol{H}$ by $H_{n\ell} = h_{\ell-2n}$ and note that $\boldsymbol{H}$ has a circulant structure: the nth row is the 0th row shifted to the right by $2n$ units, if $n > 0$, or to the left by $2|n|$ units, if $n < 0$. Similarly, define a matrix $\boldsymbol{G}$ by $G_{n\ell} = g_{\ell-2n}$ and note that $\boldsymbol{G}$ has the same circulant structure as $\boldsymbol{H}$. Since the matrix $\begin{pmatrix} \boldsymbol{H} \\ \boldsymbol{G} \end{pmatrix}$ is the matrix of an orthonormal basis change in V_{-1}, from the basis $\{\phi_{-1,n}\}_{n\in Z}$ to the basis $\{\phi_{0n}\}_{n\in Z} \cup \{\psi_{0n}\}_{n\in Z}$, it is necessarily unitary and real. Thus,

$$\boldsymbol{I} = (\boldsymbol{H}^T, \boldsymbol{G}^T) \begin{pmatrix} \boldsymbol{H} \\ \boldsymbol{G} \end{pmatrix} = \boldsymbol{H}^T \boldsymbol{H} + \boldsymbol{G}^T \boldsymbol{G}$$

and

$$\boldsymbol{I} = \begin{pmatrix} \boldsymbol{H} \\ \boldsymbol{G} \end{pmatrix} (\boldsymbol{H}^T, \boldsymbol{G}^T) = \begin{pmatrix} \boldsymbol{H}\boldsymbol{H}^T & \boldsymbol{H}\boldsymbol{G}^T \\ \boldsymbol{G}\boldsymbol{H}^T & \boldsymbol{G}\boldsymbol{G}^T \end{pmatrix}.$$

We accordingly obtain the following relationships:

$$\boldsymbol{H}\boldsymbol{H}^T = \boldsymbol{I} \tag{1.3-6}$$

$$\boldsymbol{G}\boldsymbol{G}^T = \boldsymbol{I} \tag{1.3-7}$$

$$\boldsymbol{G}\boldsymbol{H}^T = \boldsymbol{H}\boldsymbol{G}^T = 0 \tag{1.3-8}$$

$$\boldsymbol{H}^T\boldsymbol{H} + \boldsymbol{G}^T\boldsymbol{G} = \boldsymbol{I}. \tag{1.3-9}$$

We now give two provisional definitions of discrete wavelets. First, we will call the coefficient sequences $\boldsymbol{h} = (\ldots, h_{-1}, h_0, h_1, \ldots)$ and $\boldsymbol{g} = (\ldots, g_{-1}, g_0, g_1, \ldots)$ of (1.3-1) and (1.3-2) a *discrete wavelet pair*; that is, a discrete wavelet pair is the two coefficient sequences obtained when we expand the functions ϕ and ψ of a multiresolution analysis in terms of the basis $\{\phi_{-1,n}\}_{n\in Z}$ for V_{-1}. This provisional definition ties discrete wavelets to the multiresolution analysis context; the properties of the discrete wavelet pair $\boldsymbol{h}$, $\boldsymbol{g}$ are whatever properties they may inherit from the multiresolution analysis context. Second, we will call the pair $\boldsymbol{h}$ and $\boldsymbol{g}$ a *discrete wavelet pair* if the matrices $\boldsymbol{H}$ and $\boldsymbol{G}$ defined above satisfy Eqs. (1.3-6) through (1.3-9). This provisional definition makes no explicit reference to the multiresolution analysis context; the properties of $\boldsymbol{h}$ and $\boldsymbol{g}$ are thus the properties they must have in order that the matrix $\begin{pmatrix} \boldsymbol{H} \\ \boldsymbol{G} \end{pmatrix}$ be unitary. Daubechies showed that the two definitions are essentially equivalent, and we outline the argument in part **D** of this section below.

The practical use of multiresolution analysis is precisely to generate discrete wavelet pairs $\boldsymbol{h}$ and $\boldsymbol{g}$ with desirable properties. Since the two provisional definitions of a discrete wavelet pair are essentially equivalent, it follows that we can obtain discrete wavelet pairs with desirable properties without going through the process of constructing a multiresolution analysis. Instead, we can solve Eqs. (1.3-6) through (1.3-9) directly. Daubechies showed how to obtain many solutions of these equations, and hence how to obtain many orthogonal wavelet bases. This is of interest in signal processing, because it offers the possibility, not yet by any means fully explored, of constructing orthogonal wavelet bases tailored to particular applications.

B. The Discrete Wavelet Equations

If we write Eqs. (1.3-6) through (1.3-9) in component form, we obtain the following four equations for $\boldsymbol{h}$ and $\boldsymbol{g}$:

$$\sum_{k\in Z} h_{k-2n} h_{k-2\ell} = \delta_{\ell n} \tag{1.3-10}$$

$$\sum_{k\in Z} g_{k-2n} g_{k-2\ell} = \delta_{\ell n}. \tag{1.3-11}$$

$$\sum_{k\in Z} h_{k-2n} g_{k-2\ell} = 0 \tag{1.3-12}$$

$$\sum_{k\in Z} h_{\ell-2k} h_{q-2k} + \sum_{k\in Z} g_{\ell-2k} g_{q-2k} = \delta_{\ell q}. \tag{1.3-13}$$

It is shifts of $\boldsymbol{h}$ and $\boldsymbol{g}$ by even integers (rather than all integers) that are orthogonal to one another, and this introduces some difficulties of notation. They can be handled by introducing auxiliary sequences

$$a_n = h_{2n} \tag{1.3-14}$$

$$b_n = h_{2n+1} \tag{1.3-15}$$

$$c_n = g_{2n} \tag{1.3-16}$$

$$d_n = g_{2n+1}. \tag{1.3-17}$$

The three equations (1.3-10) through (1.3-12) are equivalent to the equations

$$\sum_{k\in Z} a_{k-n} a_{k-\ell} + \sum_{k\in Z} b_{k-n} b_{k-\ell} = \delta_{\ell n} \tag{1.3-18}$$

$$\sum_{k\in Z} c_{k-n} c_{k-\ell} + \sum_{k\in Z} d_{k-n} d_{k-\ell} = \delta_{\ell n} \tag{1.3-19}$$

$$\sum_{k\in Z} a_{k-n} c_{k-\ell} + \sum_{k\in Z} b_{k-n} d_{k-\ell} = 0. \tag{1.3-20}$$

Equation (1.3-13) gives rise to three equations relating the auxiliary sequences $\boldsymbol{a}$, $\boldsymbol{b}$, $\boldsymbol{c}$ and $\boldsymbol{d}$. A short calculation shows that (1.3-13) holds if and only if

$$\sum_{k\in Z} a_{n-k} b_{\ell-k} + \sum_{k\in Z} c_{n-k} d_{\ell-k} = 0 \tag{1.3-21}$$

$$\sum_{k\in Z} a_{n-k} a_{\ell-k} + \sum_{k\in Z} c_{n-k} c_{\ell-k} = \delta_{\ell n} \tag{1.3-22}$$

$$\sum_{k\in Z} b_{n-k} b_{\ell-k} + \sum_{k\in Z} d_{n-k} d_{\ell-k} = \delta_{\ell n}. \tag{1.3-23}$$

Thus, we have a set of six equations in the four unknown sequences $\boldsymbol{a}$, $\boldsymbol{b}$, $\boldsymbol{c}$ and $\boldsymbol{d}$. The equations are not all independent, as will become clear below.

Taking the Fourier transform of the sequence $\boldsymbol{h}$, we obtain

$$\begin{aligned}\hat{h}(\omega) &= \sum_{n\,\text{even}} h_n e^{-j\omega n} + \sum_{n\,\text{odd}} h_n e^{-j\omega n} \\ &= \sum_{n\in Z} a_n e^{-j\omega 2n} + \sum_{n\in Z} b_n e^{-j\omega(2n+1)} \\ &= \hat{a}(2\omega) + e^{-j\omega}\hat{b}(2\omega).\end{aligned} \tag{1.3-24}$$

Similarly,

$$\hat{g}(\omega) = \hat{c}(2\omega) + e^{-j\omega}\hat{d}(2\omega). \tag{1.3-25}$$

Taking the Fourier transforms of the six equations (1.3-18) through (1.3-23) yields the following six equations:

$$|\hat{a}(\omega)|^2 + |\hat{b}(\omega)|^2 = 1 \tag{1.3-26}$$

$$|\hat{c}(\omega)|^2 + |\hat{d}(\omega)|^2 = 1 \tag{1.3-27}$$

$$\hat{a}^*(\omega)c(\omega) + \hat{b}^*(\omega)\hat{d}(\omega) = 0 \tag{1.3-28}$$

$$\hat{a}^*(\omega)b(\omega) + \hat{c}^*(\omega)\hat{d}(\omega) = 0 \tag{1.3-29}$$

$$|\hat{a}(\omega)|^2 + |\hat{c}(\omega)|^2 = 1 \tag{1.3-30}$$

$$|\hat{b}(\omega)|^2 + |\hat{d}(\omega)|^2 = 1. \tag{1.3-31}$$

It is now easy to see, by changing to polar coordinates, that Eqs. (1.3-29), (1.3-30) and (1.3-31) are implied by the preceding three equations. (This says that Eq. (1.3-13) contains no information not already present in Eqs. (1.3-10) through (1.3-12).) A simple calculation shows that Eqs. (1.3-26), (1.3-27) and (1.3-28) are satisfied by sequences $\boldsymbol{a}$, $\boldsymbol{b}$, $\boldsymbol{c}$ and $\boldsymbol{d}$ if and only if Eq. (1.3-26) is satisfied and in addition there is a real function $\theta(\omega)$ such that

$$\hat{c}(\omega) = e^{j\theta(\omega)}\hat{b}^*(\omega) \tag{1.3-32}$$

$$\hat{d}(\omega) = -e^{j\theta(\omega)}\hat{a}^*(\omega). \tag{1.3-33}$$

Putting the discussion above together, we obtain the following important result:

Theorem 1.3-1 Sequences $\boldsymbol{h}$ and $\boldsymbol{g}$ satisfy Eqs. (1.3-10) through (1.3-13) if and only if there are 2π-periodic complex-valued functions $\hat{a}(\omega)$ and $\hat{b}(\omega)$ in $L^2([0, 2\pi])$ and a real-valued function $\theta(\omega)$ satisfying $\theta(\omega + 2\pi) = \theta(\omega) + 2k\pi$ for some integer k (where possibly k varies depending on ω), such that

$$\hat{h}(\omega) = \hat{a}(2\omega) + e^{-j\omega}\hat{b}(2\omega)$$

$$\hat{g}(\omega) = e^{j\theta(\omega)}(\hat{b}^*(2\omega) - e^{-j\omega}\hat{a}^*(2\omega)) \quad \text{and}$$

$$|\hat{a}(\omega)|^2 + |\hat{b}(\omega)|^2 = 1.$$

∎

Following Daubechies, we henceforth assume that $\theta(\omega) = 0$, so that $\hat{c} = \hat{b}^*$ and $\hat{d} = -\hat{a}^*$. The effect of this assumption is that

$$g_n = \frac{1}{2\pi}\int_0^{2\pi} \hat{g}(\omega)e^{j\omega n}\, d\omega$$

$$= \frac{1}{2\pi}\int_0^{2\pi}(\sum_{\ell\in Z} b_\ell e^{j\omega(2\ell+n)} - \sum_{\ell\in Z} a_\ell e^{j\omega(2\ell+n-1)})\, d\omega$$

$$= \begin{cases} b_\ell, & n+2\ell=0 \\ -a_\ell, & n+2\ell-1=0 \end{cases}$$

$$= (-1)^n h_{1-n}.$$

Thus, the effect of the assumption is that the sequence $\boldsymbol{g}$ corresponding to $\boldsymbol{h}$ is precisely the sequence which we pulled out of the air in Section 1.1. These remarks yield the following corollary of Theorem 1.3-1:

Theorem 1.3-2 Suppose that $\boldsymbol{h}$ satisfies Eq. (1.3-10). Define $\boldsymbol{g}$ via $g_n = (-1)^n h_{1-n}$. Then the pair $\boldsymbol{h}$, $\boldsymbol{g}$ satisfies Eqs. (1.3-10) through (1.3-13). ∎

In effect, therefore, we need only consider Eq. (1.3-10).

What happens if we do not take $\theta(\omega) = 0$ in Theorem 1.3-1? It is readily verified that taking $\theta(\omega) = k\omega$, where k is an integer, results only in shifting $\boldsymbol{g}$. The interpretation of the arbitrary phase $\theta(\omega)$ of Theorem 1.3-1 is not so easy, however; this is an issue which, to our knowledge, is not explored in the literature. We will take $\theta(\omega) = 0$ in the following discussion.

C. Daubechies Wavelets

We shall refer to Daubechies' solutions of the discrete wavelet equations (1.3-10) through (1.3-13) as Daubechies wavelets. According to Theorem 1.3-1, assuming the phase function $\theta(\omega) = 0$, sequences $\boldsymbol{h}$ and $\boldsymbol{g}$ satisfy Eqs. (1.3-10) through (1.3-13) if and only if there are complex-valued functions $\hat{a}(\omega)$ and $\hat{b}(\omega)$ in $L^2([0, 2\pi])$ such that

$$\hat{h}(\omega) = \hat{a}(2\omega) + e^{-j\omega}\hat{b}(2\omega)$$

$$\hat{g}(\omega) = \hat{b}^*(2\omega) - e^{-j\omega}\hat{a}^*(2\omega) \quad \text{and}$$

$$|\hat{a}(\omega)|^2 + |\hat{b}(\omega)|^2 = 1.$$

This says that if we can find 2π-periodic functions $\hat{a}$ and $\hat{b}$ satisfying the third condition, we can construct $\hat{h}$ and $\hat{g}$ via the two foregoing equations, and then we can recover the sequences $\boldsymbol{h}$ and $\boldsymbol{g}$ satisfying Eqs. (1.3-10) through (1.3-13) as the coefficients in the Fourier transforms $\hat{h}$ and $\hat{g}$. This is Daubechies' construction.

In fact, she reduces the condition $|\hat{a}|^2 + |\hat{b}|^2 = 1$ to a condition on $\hat{h}$ by observing that, since

$$\hat{h}(\omega) = \hat{a}(2\omega) + e^{-j\omega}\hat{b}(2\omega) \quad \text{and}$$

$$\hat{h}(\omega+\pi) = \hat{a}(2\omega) - e^{-j\omega}\hat{b}(2\omega)$$

(the latter equation uses the fact that $\hat{a}$ and $\hat{b}$ are 2π-periodic), $|\hat{a}|^2 + |\hat{b}|^2 = 1$ if and only if

$$|\hat{h}(\omega)|^2 + |\hat{h}(\omega + \pi)|^2 = 2 \tag{1.3-34}$$

as is easy to verify. If we work with $\tilde{h} = \hat{h}/\sqrt{2}$, the foregoing equation assumes the form

$$|\tilde{h}(\omega)|^2 + |\tilde{h}(\omega + \pi)|^2 = 1. \tag{1.3-35}$$

Daubechies wavelets are obtained by solving (1.3-34) or (1.3-35).

Daubechies shows how to obtain many solutions of these equations [1-2]. We shall not describe her general results, for which the interested reader is referred to her paper. Rather, we shall look at two particular solutions of Eq. (1.3-35) and show by example how the sequences $\boldsymbol{h}$ and $\boldsymbol{g}$ are obtained.

Examples

1. Set

$$\tilde{h}(\omega) = \frac{1 + e^{-j\omega}}{2}.$$

Then

$$|\tilde{h}(\omega)|^2 = \cos^2 \frac{\omega}{2}$$

and

$$|\tilde{h}(\omega)|^2 + |\tilde{h}(\omega + \pi)|^2 = \cos^2 \frac{\omega}{2} + \sin^2 \frac{\omega}{2} = 1$$

so that Eq. (1.3-35) holds, and necessarily the coefficients h_n in the Fourier series

$$\hat{h}(\omega) = \sqrt{2}\tilde{h}(\omega) = \frac{1}{\sqrt{2}} + \frac{1}{\sqrt{2}} e^{-j\omega}$$

satisfy Eq. (1.3-10). By inspection, we have

$$\begin{aligned} h_0 &= \frac{1}{\sqrt{2}} \\ h_1 &= \frac{1}{\sqrt{2}} \\ h_n &= 0 \quad n > 1 \text{ or } n < 0. \end{aligned}$$

Then using $g_n = (-1)^n h_{1-n}$ we obtain

$$\begin{aligned} g_1 &= -h_0 = -\frac{1}{\sqrt{2}} \\ g_0 &= h_1 = \frac{1}{\sqrt{2}} \\ g_n &= 0 \quad n > 1 \text{ or } n < 0. \end{aligned}$$

This is the discrete Haar wavelet. It is the simplest Daubechies wavelet.

2. Set

$$\tilde{h}(\omega) = \frac{1}{2}(\frac{1+e^{-j\omega}}{2})^2((1+\sqrt{3})+(1-\sqrt{3})e^{-j\omega}). \tag{1.3-36}$$

Then

$$|\tilde{h}(\omega)|^2 = \left[\cos^4\frac{\omega}{2}\right](2-\cos\omega)$$

$$|\tilde{h}(\omega+\pi)|^2 = \left[\sin^4\frac{\omega}{2}\right](2+\cos\omega)$$

as is readily verified; summing, we conclude that Eq. (1.3-35) holds. Necessarily, therefore, the coefficients h_n in the Fourier series $\hat{h} = \sqrt{2}\tilde{h}$ satisfy Eq. (1.3-10). From Eq. (1.3-36) (multiplying by $\sqrt{2}$ and expanding), we obtain

$$\hat{h}(\omega) = \frac{1}{4\sqrt{2}}((1+\sqrt{3})+(3+\sqrt{3})e^{-j\omega}+(3-\sqrt{3})e^{-j2\omega}+(1-\sqrt{3})e^{-j3\omega}).$$

Reading off the coefficients, we obtain

$$\begin{aligned}
h_0 &= \frac{1+\sqrt{3}}{4\sqrt{2}}\\
h_1 &= \frac{3+\sqrt{3}}{4\sqrt{2}}\\
h_2 &= \frac{3-\sqrt{3}}{4\sqrt{2}}\\
h_3 &= \frac{1-\sqrt{3}}{4\sqrt{2}}\\
h_n &= 0\ ;\quad n>3 \text{ or } n<0.
\end{aligned}$$

Then using $g_n = (-1)^n h_{1-n}$ we obtain

$$\begin{aligned}
g_1 &= -h_0 = -\frac{1+\sqrt{3}}{4\sqrt{2}}\\
g_0 &= h_1 = \frac{3+\sqrt{3}}{4\sqrt{2}}\\
g_{-1} &= -h_2 = -\frac{3-\sqrt{3}}{4\sqrt{2}}\\
g_{-2} &= h_3 = \frac{1-\sqrt{3}}{4\sqrt{2}}\\
g_n &= 0\ ;\quad n>1 \text{ or } n<-2.
\end{aligned}$$

By Theorem 1.3-1 and remarks above, these sequences $\boldsymbol{h}$ and $\boldsymbol{g}$ must satisfy Eqs. (1.3-10) through (1.3-13), and it can be verified by direct calculation that they do.

If, as in these examples, only a finite number of terms in the sequence $\boldsymbol{h}$ are non-zero, then $\hat{h}(\omega)$ is a trigonometric polynomial. Daubechies shows how to construct *all* trigonometric polynomial functions $\hat{h}$ satisfying (1.3-34). There is, accordingly, a huge family of Daubechies wavelets to choose from. Daubechies actually constructs in [1-2] wavelets satisfying certain smoothness constraints as side conditions.

D. Recovering Multiresolution Analyses from Discrete Wavelets

We have two provisional definitions of discrete wavelets: $\boldsymbol{h}$, $\boldsymbol{g}$ is a discrete wavelet pair if, on the one hand, they are the coefficient sequences in the expansion of functions ϕ and ψ of a multiresolution analysis with respect to the orthonormal basis $\{\phi_{-1,n}\}_{n\in Z}$ of V_{-1} and if, on the other, they are solutions of Eqs. (1.3-10) through (1.3-13). We address next the question of whether these provisional definitions are equivalent. Clearly a discrete wavelet pair $\boldsymbol{h}$, $\boldsymbol{g}$ from a multiresolution analysis satisfies Eqs. (1.3-10) through (1.3-13), since we derived these equations (at the end of Section 1.1) by examining the properties of the functions ϕ and ψ of a multiresolution analysis. The remaining question, therefore, is whether we can go backwards; does a solution of Eqs. (1.3-10) through (1.3-13) imply the existence of a multiresolution analysis? Daubechies shows that if Eq. (1.3-10) is satisfied by an ℓ^1 sequence $\boldsymbol{h}$ satisfying certain additional assumptions, then a multiresolution analysis of $L^2(R)$ follows. This shows that Eq. (1.3-10) contains in effect all of the information contained in a multiresolution analysis. We outline Daubechies' argument in the remainder of this section.

In the following discussion, we will assume that $\boldsymbol{h}$ is real. Note that the structure of Eqs. (1.3-10) through (1.3-13) is symmetric in $\boldsymbol{h}$ and $\boldsymbol{g}$. We now (following Daubechies) break this symmetry by assuming that

$$\hat{h}(0) = \hat{a}(0) + \hat{b}(0) > 0 \tag{1.3-37}$$

$$\hat{g}(0) = \hat{b}(0) - \hat{a}(0) = 0. \tag{1.3-38}$$

Equation (1.3-37) is motivated by the fact that $\boldsymbol{h}$ should correspond to the function ϕ in a multiresolution analysis; ϕ in the examples given in Section 1.2 integrates to a positive number and corresponds to the kernel of a smoothing operator or low-pass filter. Equation (1.3-38) on the other hand is motivated by the fact that $\boldsymbol{g}$ should correspond to the function ψ in a multiresolution analysis; ψ in the examples given in Section 1.2 integrates to zero and corresponds to the kernel of a high-pass filter. Together Eqs. (1.3-26), (1.3-37) and (1.3-38) imply

$$\hat{a}(0) = \hat{b}(0) = 1/\sqrt{2}. \tag{1.3-39}$$

It follows that

$$\hat{h}(0) = \sum_{n \in Z} h_n = \sqrt{2}. \qquad (1.3\text{-}40)$$

It is worth remembering that there may be solutions of Eqs. (1.3-10) through (1.3-13) which do not satisfy Eq. (1.3-39) and that the motivation for assuming (1.3-39) is to restrict attention to solutions which "should" imply a multiresolution analysis.

Daubechies' basic idea for reconstructing ϕ (the function which generates a multiresolution analysis) from a solution $\boldsymbol{h}$ of Eq. (1.3-10) is as follows: apply the operator T_H defined by

$$[T_H f](x) = 2^{1/2} \sum_{n \in Z} h_n f(2x - n) \qquad (1.3\text{-}41)$$

to the characteristic function

$$f_0 = \chi_{[-1/2,1/2)}$$

to obtain iterates

$$\eta_k = T_H \eta_{k-1} = T_H^k f_0. \qquad (1.3\text{-}42)$$

We give the substance of Daubechies' result by way of two theorems which skirt technical details; for the technical details, we refer the reader to [1-2].

Theorem 1.3-3 Suppose that $\boldsymbol{h} \in \ell^1$ is a solution of Eq. (1.3-10), and suppose that f_0 is a non-zero L^2 function whose translates $f_{0,n}(x) = f_0(x-n)$ are orthogonal. Suppose also that the iterates $\eta_k = T_H^k f_0$ are norm Cauchy, and let $\phi = \eta_\infty$ denote their limit. Then: (1) $\|\phi\|_2 = \|f_0\|_2$, (2) translates $\phi_{0,n}(x) = \phi(x-n)$ of ϕ are orthogonal, and (3) ϕ is a fixed point of the operator T_H, that is, $\phi = T_H \phi$.

Proof: We show first that if translates of η_k are orthogonal, then so are translates of η_{k+1}. We have (assuming that the h_n's are real but that η_k is not necessarily real):

$$\begin{aligned}
\langle \eta_{k+1,\ell}, \eta_{k+1,n} \rangle &= 2 \sum_{p,q \in Z} h_p h_q \int_{-\infty}^{\infty} \eta_k^*(2(x-\ell)-p)\eta_k(2(x-n)-q)\, dx \\
&= \sum_{p,q \in Z} h_p h_q \int_{-\infty}^{\infty} \eta_k^*(u-(2\ell+p))\eta_k(u-(2n+q))\, du \\
&= \sum_{p,q \in Z} h_p h_q \delta_{2\ell+p,2n+q} \|\eta_k\|_2^2 \\
&= \sum_{p \in Z} h_p h_{p+2(\ell-n)} \|\eta_k\|_2^2 \\
&= \delta_{\ell n} \|\eta_k\|_2^2.
\end{aligned}$$

The second line uses the substitution $u = 2x$, the third uses the assumed orthogonality of translates of η_k, and the last uses the fact that $\boldsymbol{h}$ satisfies Eq. (1.3-10).

Thus, translates of η_{k+1} are orthogonal if translates of η_k are, and since orthogonality of translates of $\eta_0 = f_0$ is assumed, translates of every iterate η_k are orthogonal, by induction.

Observe next that the calculation above shows that $\|\eta_{k+1}\|_2 = \|\eta_k\|_2$ for all k, so long as translates of η_k are orthogonal. Thus by induction $\|\eta_k\|_2 = \|f_0\|_2$ for all k. Conclusion (1) follows from this and from the fact that $|\|\eta_k\|_2 - \|\phi\|_2| \leq \|\eta_k - \phi\|_2$.

As to conclusion (3), observe that T_H is a linear operator. A calculation similar to that above, but not assuming orthogonality of translates, shows that

$$\|T_H f\|_2 \leq (\sum_{n \in Z} |h_n|)\|f\|_2$$

for all $f \in L^2$. The hypothesis that $\boldsymbol{h} \in \ell^1$ thus implies that T_H is a bounded linear operator on L^2. It follows that

$$\begin{aligned}
\|\phi - T_H\phi\|_2 &= \|(\phi - \eta_{k+1}) - T_H(\phi - \eta_k)\|_2 \\
&\leq \|\phi - \eta_{k+1}\|_2 + \|T_H(\phi - \eta_k)\|_2 \\
&\leq \|\phi - \eta_{k+1}\|_2 + \|\boldsymbol{h}\|_1\|\phi - \eta_k\|_2.
\end{aligned}$$

Since the last term can be made as small as we please by choosing k large enough, and the first term is independent of k, necessarily the first term is zero; that is, ϕ is indeed a fixed point of T_H.

Conclusion (2) follows from a similar argument. Observe, on the one hand, that

$$\begin{aligned}
\langle \phi_{0,\ell} - \eta_{k,\ell}, \phi_{0,n} - \eta_{k,n} \rangle &= \langle \phi_{0,\ell}, \phi_{0,n} \rangle - \langle \eta_{k,\ell}, \phi_{0,n} \rangle - \langle \phi_{0,\ell}, \eta_{k,n} \rangle + \delta_{\ell n}\|\eta_k\|_2^2 \\
&\rightarrow -\langle \phi_{0,\ell}, \phi_{0,n} \rangle + \delta_{\ell n}\|\phi\|_2^2
\end{aligned}$$

as $k \rightarrow \infty$. Observe, on the other hand, that by Schwartz's inequality

$$\begin{aligned}
|\langle \phi_{0,\ell} - \eta_{k,\ell}, \phi_{0,n} - \eta_{k,n} \rangle| &\leq \|\phi_{0,\ell} - \eta_{k,\ell}\| \, \|\phi_{0,n} - \eta_{k,n}\| \\
&= 2\|\phi - \eta_k\| \\
&\rightarrow 0
\end{aligned}$$

as $k \rightarrow \infty$. Putting these results together, we obtain

$$\langle \phi_{0,\ell}, \phi_{0,n} \rangle = \delta_{\ell n}\|\phi\|_2^2,$$

that is, the translates of ϕ are orthogonal, as claimed. ∎

Theorem 1.3-4 Suppose $\boldsymbol{h}$ is real and satisfies Eq. (1.3-10). Suppose that

$$\sum_{n \in Z} |h_n| \, |n| < \infty$$

(so that $\hat{h}$ and $\hat{h}'$ are both continuous). Write

$$\hat{h}(\omega) = \hat{a}(2\omega) + e^{-j\omega}\hat{b}(2\omega),$$

where $\hat{a}$ and $\hat{b}$ are the Fourier transforms of the sequences a_n and b_n defined in Eqs. (1.3-14) and (1.3-15), and assume that

$$\hat{a}(0) = \hat{b}(0) = 1/\sqrt{2}$$

so that $\hat{h}(0) = \sqrt{2}$. Then the Fourier transforms $\hat{\eta}_k$ of the iterates η_k of Eq. (1.3-42) converge pointwise to

$$\hat{\eta}_\infty(\omega) = \hat{\phi}(\omega) = \frac{1}{\sqrt{2\pi}}\Pi_{k=1}^{\infty}\tilde{h}(2^{-k}\omega),$$

where $\tilde{h} = \hat{h}/\sqrt{2}$. Furthermore, convergence is uniform on compact sets.

Proof: See [1-2]. ∎

If we start, therefore, with a sequence $\boldsymbol{h}$ which satisfies Eq. (1.3-10), and if we use the iteration scheme (1.3-42), and if the iterates converge, Theorem 1.3-3 says that the limit ϕ of the iterates is a non-zero L^2 function for which $\phi = T_H\phi$ and that the integer translates of the limit function ϕ are orthogonal. Since $\phi = T_H\phi$ says precisely that $\phi \in V_{-1}$, whence $V_0 \subset V_{-1}$, we conclude that condition (a) of a multiresolution analysis holds. Theorem 1.3-4 guarantees that the Fourier transform of the limit is continuous at $\omega = 0$ (since the uniform limit of continuous functions is continuous) and that $\hat{\phi}(0) = 1/\sqrt{2\pi}$ (since $\tilde{h}(0) = 1$), so that the limit function ϕ necessarily integrates to 1. The hypotheses of Theorem 1.1-2 are accordingly satisfied, and conditions (b), (c) and (d) of a multiresolution analysis follow via Theorems 1.1-1 and 1.1-2. If, therefore, we start with a sequence $\boldsymbol{h}$ which satisfies (1.3-10) and the iterates in the iteration scheme (1.3-42) converge, Theorems 1.3-3 and 1.3-4, together with prior results, guarantee that the limit ϕ generates a multiresolution analysis, as desired.

Theorem 1.3-3 does not say when convergence of the iterates of the fixed point scheme (1.3-42) can be expected to occur. The hypotheses of Theorem 1.3-4, together with an additional regularity condition on the discrete wavelet sequence $\boldsymbol{h}$ guarantee convergence of the iterates. For a precise statement of the extra regularity condition and a proof of convergence, we refer the reader to [1-2].

We close the discussion of Daubechies wavelets with a remark about terminology. Daubechies wavelets are often referred to as wavelets with compact support. This terminology reflects the fact that the function ϕ constructed via the iterative scheme (1.3-42) from a solution $\boldsymbol{h}$ of Eq. (1.3-10) has compact support if only a finite number of terms in the sequence $\boldsymbol{h}$ are non-zero. To see this, suppose that

$h_n = 0$ for $|n| > N$ and observe that if $\eta_k = T_H^k f_0$ vanishes for $|x| > a_k$, then we can write $\eta_k = \eta_k \chi_{[-a_k, a_k]}$, so that

$$\begin{aligned} \eta_{k+1}(x) &= \sqrt{2} \sum_{n=-N}^{N} h_n \eta_k(2x-n)\chi_{[-a_k,a_k]}(2x-n) \\ &= \sqrt{2} \sum_{n=-N}^{N} h_n \eta_k(2x-n)\chi_{[(n-a_k)/2,(n+a_k)/2]}(x). \end{aligned}$$

Since $-N \leq n \leq N$, it follows that η_{k+1} vanishes for $|x| > a_{k+1} = (N + a_k)/2$. Since $\eta_0 = f_0 = \chi_{[-1/2,1/2)}$, we can take $a_0 = 1/2$, and the recursion above yields $a_k = (1 - 2^{-k})N + 2^{-(k+1)}$. When the iterative scheme converges, therefore, the function ϕ to which the η_k's converge necessarily has support contained in the interval $[-N, N]$. The term "compactly supported wavelets" is accordingly an appropriate one for Daubechies wavelets; it reflects directly the fact that only a finite number of terms in the solutions $\boldsymbol{h}$ and $\boldsymbol{g}$ of Eqs. (1.3-10) through (1.3-13) are non-zero. We will, however, continue to use the term "Daubechies wavelets" in the remainder of this exposition.

1.4 The Discrete Wavelet Transform Algorithm

Let $\boldsymbol{h}$ and $\boldsymbol{g}$ denote a (real) discrete wavelet pair, and let $\boldsymbol{H}$ and $\boldsymbol{G}$ denote the matrices constructed from $\boldsymbol{h}$ and $\boldsymbol{g}$, which satisfy Eqs. (1.3-6) through (1.3-9). Let $\boldsymbol{f}$ be a discrete signal, and let K be a positive integer. We will call the collection of signals

$$W_K \boldsymbol{f} = (\boldsymbol{H}^K \boldsymbol{f}, \boldsymbol{G}\boldsymbol{H}^{K-1}\boldsymbol{f}, \ldots, \boldsymbol{G}\boldsymbol{H}\boldsymbol{f}, \boldsymbol{G}\boldsymbol{f}) \tag{1.4-1}$$

the "discrete wavelet transform" of the signal $\boldsymbol{f}$ at scale K.

The interpretation of the discrete wavelet transform is as follows: When we use discrete wavelets to analyze a discrete signal, we in effect treat the discrete signal $\boldsymbol{f}$ as the vector of coefficients in the expansion of the projection $P_{-1}f$ of some L^2 signal f on V_{-1}, with respect to the orthogonal basis $\{\phi_{-1,n}\}$ for V_{-1}. We can write $P_{-1}f = P_0 f + Q_0 f$, where $P_0 f$ denotes the projection of f on V_0 and $Q_0 f$, the projection of f on W_0. The coefficients of $P_0 f$ and $Q_0 f$ in the bases $\{\phi_{0n}\}$ for V_0 and $\{\psi_{0n}\}$ for W_0 are given by $\boldsymbol{H}\boldsymbol{f}$ and $\boldsymbol{G}\boldsymbol{f}$, as is readily verified. Now consider the equation $P_0 f = P_1 f + Q_1 f$. The coefficients in the expansion of $P_0 f$ with respect to the basis $\{\phi_{0n}\}$ for V_0 are given by the sequence $\boldsymbol{H}\boldsymbol{f}$, by the argument above, and the coefficients of $P_1 f$ and $Q_1 f$ with respect to the bases $\{\phi_{1n}\}$ for V_1 and $\{\psi_{1n}\}$ for W_1 are given by $\boldsymbol{H}\boldsymbol{H}\boldsymbol{f}$ and $\boldsymbol{G}\boldsymbol{H}\boldsymbol{f}$, as is also readily verified. We can repeat this same analysis at other scales. In general, we obtain

$$\boldsymbol{f} \sim (\boldsymbol{H}^K \boldsymbol{f}, \boldsymbol{G}\boldsymbol{H}^{K-1}\boldsymbol{f}, \ldots, \boldsymbol{G}\boldsymbol{H}\boldsymbol{f}, \boldsymbol{G}\boldsymbol{f}) = W_K \boldsymbol{f},$$

where $\sim$ means that the sequence on the left and the collection of sequences on the right give the coefficients in the expansion of the same signal, $P_{-1}f$, with respect to different orthonormal bases for V_{-1}.

Note that the scale invariance property of basis elements in a multiresolution analysis translates in the discrete setting into the property that the basis transformation matrices $\boldsymbol{H}$ and $\boldsymbol{G}$ are the same at each stage in the analysis of a discrete signal.

In this section we discuss the discrete wavelet transform algorithm and compare it with the fast Fourier transform (FFT) algorithm. The two transforms are closely related. Both are fast ways of multiplying a signal $\boldsymbol{f}$ by a unitary matrix. The exploited structure of the matrix is essentially the same in both cases – the matrix can be factored as a product of sparse unitary matrices of very simple structure. Furthermore, the origin of the simplicity of structure of the factor matrices is the same in both cases; it is a natural "decimation" which occurs when the transforms are applied to a periodic signal of period $N = 2^M$ (that is, to a signal of length N extended periodically in both directions), as is always the case in applications.

We shall start with a schematic analysis of the FFT algorithm and write down explicitly a generalization of the algorithm, using notation which facilitates its comparison with the wavelet transform algorithm. We discuss the FFT at some length, since we shall need the resulting generalized FFT algorithm in one of the applications considered below (see Section 1.7). Then we shall describe the wavelet transform, using similar notation.

A. Generalization of the Fast Fourier Transform

The FFT algorithm is well known and widely used. We review it here, in order to highlight similarities and differences with the wavelet decomposition algorithm.

Suppose $\boldsymbol{f}$ is a signal of length N. The discrete Fourier transform (DFT) of the signal is the Fourier transform $\hat{f}(\omega)$ evaluated at the points $\omega = 2\pi\ell/N$, that is,

$$\hat{f}_\ell = \hat{f}(2\pi\ell/N) = \sum_{k=0}^{N-1} f_k e^{-j2\pi k\ell/N}. \tag{1.4-2}$$

Observe that if we define the $N \times N$ matrix $\boldsymbol{Q}$ via $Q_{\ell k} = \exp(-j2\pi k\ell/N)$, we can rewrite Eq. (1.4-2) in the form

$$\hat{\boldsymbol{f}} = \boldsymbol{Q}\boldsymbol{f}. \tag{1.4-3}$$

An elementary calculation shows that the matrix $\boldsymbol{Q}$ of Eq. (1.4-3) is, except for a scale factor, a unitary matrix. In fact,

$$\boldsymbol{Q}^\dagger \boldsymbol{Q} = N\boldsymbol{I},$$

where $\boldsymbol{Q}^\dagger$ is the complex conjugate transpose of $\boldsymbol{Q}$, and $\boldsymbol{I}$ is the $N \times N$ identity matrix. It follows that we can recover $\boldsymbol{f}$ from its discrete Fourier transform $\hat{\boldsymbol{f}}$ via

$$\boldsymbol{f} = \frac{1}{N} Q^\dagger \hat{\boldsymbol{f}}.$$

Thus, we can go from discrete values of a signal $\boldsymbol{f}$ to values of its discrete Fourier transform $\hat{\boldsymbol{f}}$ and back again at the cost of a matrix multiplication.

The FFT is a special algorithm for doing the necessary matrix multiplication very efficiently. It relies on a result sometimes called the Danielson–Lanczos lemma:

Lemma 1.4-1 (Danielson–Lanczos) Let N be an even integer. For (possibly complex) data f_ℓ, $\ell = 0, \ldots, N-1$, form the trigonometric polynomial

$$F(\omega) = \sum_{\ell=0}^{N-1} f_\ell e^{-j2\pi\ell\omega/N}.$$

Then

$$F_k = F_k^e + \exp(-j2k\pi/N)F_k^o\ ; \quad k = 0, \ldots, N-1,$$

where F^e and F^o are the trigonometric polynomials given by

$$F^e(\omega) = \sum_{\ell=0}^{N/2-1} f_{2\ell} e^{-j4\pi\ell\omega/N} \tag{1.4-4}$$

$$F^o(\omega) = \sum_{\ell=0}^{N/2-1} f_{2\ell+1} e^{-j4\pi\ell\omega/N} \tag{1.4-5}$$

and $F_k = F(k)$, $F_k^e = F^e(k)$ and $F_k^o = F^o(k)$.

Proof: The proof of the Danielson–Lanczos lemma is a straightforward calculation, which we leave to the reader. ∎

The significance of the Danielson–Lanczos lemma derives from the fact that F is periodic of period N, whereas both F^e and F^o are periodic of period $N/2$. This follows directly from Eqs. (1.4-4) and (1.4-5); for example,

$$F^e(\omega + N/2) = \sum_{\ell=0}^{N/2-1} f_{2\ell} e^{-j4\pi\ell(\omega+N/2)/N} = \sum_{\ell=0}^{N/2-1} f_{2\ell} e^{-j4\pi\ell\omega/N} = F^e(\omega).$$

It follows that we get all of the values of F^e and F^o at the points $k = 0, \ldots, N-1$ from their values at $k = 0, \ldots, N/2-1$. In fact, if we have F_k^e and F_k^o for $k = 0, \ldots, N/2-1$, and we wish to obtain F_k for $k = 0, \ldots, N-1$, we can do so via

$$F_k = F_k^e + \exp(-2k\pi j/N)F_k^o\ ; \quad k = 0, \ldots, \frac{N}{2} - 1 \tag{1.4-6}$$

$$\begin{aligned} F_{k+N/2} &= F_k^e + \exp(-2(k+N/2)\pi j/N)F_k^o \\ &= F_k^e - \exp(-2k\pi j/N)F_k^o\ ; \quad k = 0, \ldots, \frac{N}{2} - 1. \end{aligned} \tag{1.4-7}$$

Eqs. (1.4-6) and (1.4-7) are the heart of the FFT algorithm.

If N is a power of 2, that is, if $N = 2^M$ for some positive integer M, this process can be repeated. We can write, for example, using the Danielson–Lanczos lemma

$$\begin{aligned}
F_k^e &= F_k^{ee} + \exp(-2k\pi j/N_1)F_k^{eo}\ ; \quad k = 0, \dots, N_1 - 1 \\
F_k^o &= F_k^{oe} + \exp(-2k\pi j/N_1)F_k^{oo}\ ; \quad k = 0, \dots, N_1 - 1 \\
F_k^{ee} &= F_k^{eee} + \exp(-2k\pi j/N_2)F_k^{eeo}\ ; \quad k = 0, \dots, N_2 - 1, \\
&\dots
\end{aligned}$$

where $N_k = N/2^k$ and the values F_k^{ee}, F_k^{eo}, F_k^{oe} and F_k^{oo} are values of trigonometric polynomials of period $N_1/2 = N/4$, and F_k^{eee}, F_k^{eeo}, etc. are values of trigonometric polynomials of period $N_2/2 = N/8$, so that Eqs. (1.4-6) and (1.4-7) can be used at each step. After repeating this process M times, we have, at the last step, F_k^η, where η is a string of e's and o's of length M, whose values are the distinct values of $F^\eta(\omega)$, a trigonometric polynomial of period $N_{M-1}/2 = 1$. There is, accordingly, a single distinct value of this trigonometric polynomial. Using the formula of the lemma with $N = 2$, we find that $F_k^\eta = f_\ell$ for some ℓ and all $k = 0, \dots, N-1$. The relation of the sequence η of e's and o's to ℓ is not important to our purpose; it is described in [1-11]. The part of the FFT algorithm that relates F_k^η to f_ℓ is called "bit reversing" $\boldsymbol{f}$; it just reorders $\boldsymbol{f}$ and so is equivalent to multiplying $\boldsymbol{f}$ by a permutation matrix $\boldsymbol{P}$.

Quasi-code for one version (the so-called Cooley–Tukey version) of the FFT is as follows:

FFT ALGORITHM

1. Bit reverse $\boldsymbol{f}$: $\boldsymbol{f} \to \boldsymbol{f}^{(0)}$
2. For $k = 1$ to M
3. $\theta_k = 2^{1-k}\pi$
4. For $\ell = 1$ to 2^{M-k}
5. For $m = 1$ to 2^{k-1}
6. $n = m + (\ell - 1)2^k$
7. $np = n + 2^{k-1}$
8. $f_n^{(k)} = f_n^{(k-1)} + e^{-jm\theta_k}\hat{f}_{np}^{(k-1)}$
9. $f_{np}^{(k)} = f_n^{(k-1)} - e^{-jm\theta_k}\hat{f}_{np}^{(k-1)}$
10. Next m
11. Next ℓ
12. Next k

The output $\boldsymbol{f}^{(M)}$ of the algorithm is the discrete Fourier transform of $\boldsymbol{f}$. Steps 8 and 9 incorporate the key Eqs. (1.4-6) and (1.4-7).

If we write $\boldsymbol{f}_\ell^{(k)}$ for the 2^k entries $(f_{1+(\ell-1)2^k}^{(k)}, f_{2+(\ell-1)2^k}^{(k)}, \ldots, f_{\ell 2^k}^{(k)})$ of $\boldsymbol{f}^{(k)}$, then the inner loop from steps 5 through 10 of the FFT algorithm can be written in the matrix form

$$\boldsymbol{f}_\ell^{(k)} = \begin{pmatrix} \boldsymbol{I} & \boldsymbol{D}^{(k)} \\ \boldsymbol{I} & -\boldsymbol{D}^{(k)} \end{pmatrix} \begin{pmatrix} \boldsymbol{f}_{2\ell-1}^{(k-1)} \\ \boldsymbol{f}_{2\ell}^{(k-1)} \end{pmatrix},$$

where $\boldsymbol{I}$ denotes the $2^{k-1} \times 2^{k-1}$ identity matrix, and $\boldsymbol{D}^{(k)}$ is the $2^{k-1} \times 2^{k-1}$ matrix whose mth diagonal entry is $\exp(-jm\theta_k)$. It follows that $\boldsymbol{f}^{(k)} = \boldsymbol{Q}^{(k)} \boldsymbol{f}^{(k-1)}$, where $\boldsymbol{Q}^{(k)}$ has the block diagonal form

$$\boldsymbol{Q}^{(k)} = \begin{pmatrix} \boldsymbol{J}^{(k)} & & \\ & \ddots & \\ & & \boldsymbol{J}^{(k)} \end{pmatrix}. \tag{1.4-8}$$

Here $\boldsymbol{J}^{(k)}$ is the $2^k \times 2^k$ matrix

$$\boldsymbol{J}^{(k)} = \begin{pmatrix} \boldsymbol{I} & \boldsymbol{D}^{(k)} \\ \boldsymbol{I} & -\boldsymbol{D}^{(k)} \end{pmatrix}. \tag{1.4-9}$$

There are thus 2^{M-k} identical diagonal blocks in the matrix $Q^{(k)}$ and since

$$J^{(k)\dagger} J^{(k)} = \begin{pmatrix} \boldsymbol{I} & \boldsymbol{I} \\ \boldsymbol{D}^{(k)*} & -\boldsymbol{D}^{(k)*} \end{pmatrix} \begin{pmatrix} \boldsymbol{I} & \boldsymbol{D}^{(k)} \\ \boldsymbol{I} & -\boldsymbol{D}^{(k)} \end{pmatrix} = 2 \begin{pmatrix} \boldsymbol{I} & 0 \\ 0 & \boldsymbol{I} \end{pmatrix}$$

$\boldsymbol{Q}^{(k)}$ is a scalar multiple of a unitary matrix for each k. (This implies that each step of the FFT obtains the representation of the original signal $\boldsymbol{f}$ with respect to a different orthogonal basis; this is true also of wavelet transforms, and it is a fact which can be exploited, see Section 1.7.) We have accordingly

$$\boldsymbol{Q} = \boldsymbol{Q}^{(M)} \boldsymbol{Q}^{(M-1)} \ldots \boldsymbol{Q}^{(2)} \boldsymbol{Q}^{(1)} \boldsymbol{P} \tag{1.4-10}$$

$$\boldsymbol{Q}^\dagger = \boldsymbol{P} \boldsymbol{Q}^{(1)\dagger} \boldsymbol{Q}^{(2)\dagger} \ldots \boldsymbol{Q}^{(M-1)\dagger} \boldsymbol{Q}^{(M)\dagger}, \tag{1.4-11}$$

where $\boldsymbol{P}$ is a permutation matrix which denotes the bit reversal process; observe that two executions of the bit reversal process restore the original vector, so $\boldsymbol{P}^T = \boldsymbol{P}$. Equations (1.4-8) through (1.4-11) state the recursive structure of $\boldsymbol{Q}$ which the FFT algorithm exploits.

We are now in a position to write down a generalization of the FFT algorithm. It will be convenient to work with Eq. (1.4-11) rather than Eq. (1.4-10), since the structure of the wavelet transform algorithm is more similar to (1.4-11). We write Eq. (1.4-11) in a slightly different way: Set

$$\boldsymbol{J}^{(M-k+1)\dagger} = \begin{pmatrix} \boldsymbol{H}_k \\ \boldsymbol{G}_k \end{pmatrix}$$

where $\boldsymbol{H}_k$ and $\boldsymbol{G}_k$ are $2^{M-k} \times 2^{M-k+1}$ matrices given by

$$\boldsymbol{H}_k = (\boldsymbol{I} \quad \boldsymbol{I}) \quad \text{and} \quad \boldsymbol{G}_k = (\boldsymbol{D}^{(M-k+1)*} \quad -\boldsymbol{D}^{(M-k+1)*}). \tag{1.4-12}$$

Then if $\boldsymbol{f}$ is a signal and we wish to form $\boldsymbol{Q}^\dagger \boldsymbol{f}$ using Eq. (1.4-11) (as we would wish to do if we were computing the inverse discrete Fourier transform of $\boldsymbol{f}$), we can do so via

$$\begin{aligned}
\boldsymbol{f}^{(1)} &= \boldsymbol{Q}^{(M)\dagger}\boldsymbol{f} = \begin{pmatrix} \boldsymbol{H}_1\boldsymbol{f} \\ \boldsymbol{G}_1\boldsymbol{f} \end{pmatrix} \\
\boldsymbol{f}^{(2)} &= \boldsymbol{Q}^{(M-1)\dagger}\boldsymbol{f}^{(1)} = \begin{pmatrix} \begin{pmatrix} \boldsymbol{H}_2 \\ \boldsymbol{G}_2 \end{pmatrix} & \boldsymbol{H}_1\boldsymbol{f} \\ \begin{pmatrix} \boldsymbol{H}_2 \\ \boldsymbol{G}_2 \end{pmatrix} & \boldsymbol{G}_1\boldsymbol{f} \end{pmatrix} = \begin{pmatrix} \boldsymbol{H}_2\boldsymbol{H}_1\boldsymbol{f} \\ \boldsymbol{G}_2\boldsymbol{H}_1\boldsymbol{f} \\ \boldsymbol{H}_2\boldsymbol{G}_1\boldsymbol{f} \\ \boldsymbol{G}_2\boldsymbol{G}_1\boldsymbol{f} \end{pmatrix} \\
\boldsymbol{f}^{(3)} &= \boldsymbol{Q}^{(M-2)\dagger}\boldsymbol{f}^{(2)} = \begin{pmatrix} \begin{pmatrix} \boldsymbol{H}_3 \\ \boldsymbol{G}_3 \end{pmatrix} & \boldsymbol{H}_2\boldsymbol{H}_1\boldsymbol{f} \\ \begin{pmatrix} \boldsymbol{H}_3 \\ \boldsymbol{G}_3 \end{pmatrix} & \boldsymbol{G}_2\boldsymbol{H}_1\boldsymbol{f} \\ \begin{pmatrix} \boldsymbol{H}_3 \\ \boldsymbol{G}_3 \end{pmatrix} & \boldsymbol{H}_2\boldsymbol{G}_1\boldsymbol{f} \\ \begin{pmatrix} \boldsymbol{H}_3 \\ \boldsymbol{G}_3 \end{pmatrix} & \boldsymbol{G}_2\boldsymbol{G}_1\boldsymbol{f} \end{pmatrix} = \begin{pmatrix} \boldsymbol{H}_3\boldsymbol{H}_2\boldsymbol{H}_1\boldsymbol{f} \\ \boldsymbol{G}_3\boldsymbol{H}_2\boldsymbol{H}_1\boldsymbol{f} \\ \boldsymbol{H}_3\boldsymbol{G}_2\boldsymbol{H}_1\boldsymbol{f} \\ \boldsymbol{G}_3\boldsymbol{G}_2\boldsymbol{H}_1\boldsymbol{f} \\ \boldsymbol{H}_3\boldsymbol{H}_2\boldsymbol{G}_1\boldsymbol{f} \\ \boldsymbol{G}_3\boldsymbol{H}_2\boldsymbol{G}_1\boldsymbol{f} \\ \boldsymbol{H}_3\boldsymbol{G}_2\boldsymbol{G}_1\boldsymbol{f} \\ \boldsymbol{G}_3\boldsymbol{G}_2\boldsymbol{G}_1\boldsymbol{f} \end{pmatrix} \\
\boldsymbol{f}^{(4)} &= \ldots.
\end{aligned} \tag{1.4-13}$$

Using Eq. (1.4-12), we obtain the following set of equations satisfied by $\boldsymbol{H}_k$ and $\boldsymbol{G}_k$:

$$\boldsymbol{H}_k\boldsymbol{H}_k^\dagger = (\boldsymbol{I} \quad \boldsymbol{I})\begin{pmatrix} \boldsymbol{I} \\ \boldsymbol{I} \end{pmatrix} = 2\boldsymbol{I} \tag{1.4-14}$$

$$\boldsymbol{G}_k\boldsymbol{G}_k^\dagger = (\boldsymbol{D}^{(M-k+1)*} \quad -\boldsymbol{D}^{(M-k+1)*})\begin{pmatrix} \boldsymbol{D}^{(M-k+1)} \\ -\boldsymbol{D}^{(M-k+1)} \end{pmatrix} = 2\boldsymbol{I} \tag{1.4-15}$$

$$\begin{aligned}
\boldsymbol{H}_k^\dagger\boldsymbol{H}_k + \boldsymbol{G}_k^\dagger\boldsymbol{G}_k &= \begin{pmatrix} \boldsymbol{I} \\ \boldsymbol{I} \end{pmatrix}(\boldsymbol{I} \quad \boldsymbol{I}) + \begin{pmatrix} \boldsymbol{D}^{(M-k+1)} \\ -\boldsymbol{D}^{(M-k+1)} \end{pmatrix}(\boldsymbol{D}^{(M-k+1)*} \quad -\boldsymbol{D}^{(M-k+1)*}) \\
&= \begin{pmatrix} \boldsymbol{I} & \boldsymbol{I} \\ \boldsymbol{I} & \boldsymbol{I} \end{pmatrix} + \begin{pmatrix} \boldsymbol{I} & -\boldsymbol{I} \\ -\boldsymbol{I} & \boldsymbol{I} \end{pmatrix} = 2\begin{pmatrix} I & 0 \\ 0 & I \end{pmatrix}
\end{aligned} \tag{1.4-16}$$

$$\boldsymbol{G}_k\boldsymbol{H}_k^\dagger = \boldsymbol{H}_k\boldsymbol{G}_k^\dagger = 0. \tag{1.4-17}$$

Compare these relations to the wavelet equations (1.3-6) through (1.3-9). The equations are identical, except for a scalar normalization factor and except for the requirement, understood rather than expressed in Eqs. (1.3-6) through (1.3-9), that $\boldsymbol{H}$ and $\boldsymbol{G}$ should possess a certain circulant structure: each row of $\boldsymbol{H}$ should be the row above shifted by two places, and similarly for $\boldsymbol{G}$. This requirement does not hold for the matrices of the FFT algorithm.

Note next that if we take *any* $2^{M-k} \times 2^{M-k+1}$ matrices $\boldsymbol{G}_k$ and $\boldsymbol{H}_k$ satisfying Eqs. (1.4-14) through (1.4-17), $k = 1, \ldots, K \leq M$, the matrix $\boldsymbol{Q}^{(k)}$ defined by

$$\boldsymbol{Q}^{(k)} = \begin{pmatrix} \boldsymbol{J}^{(k)} & & \\ & \ddots & \\ & & \boldsymbol{J}^{(k)} \end{pmatrix},$$

where

$$\boldsymbol{J}^{(k)\dagger} = \begin{pmatrix} \boldsymbol{H}_{M-k+1} \\ \boldsymbol{G}_{M-k+1} \end{pmatrix},$$

is a scalar multiple of a unitary matrix (note that the symbol $\boldsymbol{Q}^{(k)}$ is redefined here – it is no longer associated with the FFT algorithm), and if we want to obtain the matrix product

$$\boldsymbol{Q}\boldsymbol{f} = \boldsymbol{Q}^{(M-k+1)\dagger} \ldots \boldsymbol{Q}^{(M)\dagger}\boldsymbol{f}$$

we may do so via the algorithm given below:

GENERALIZED FFT ALGORITHM

1. $\boldsymbol{f}_1^{(0)} = \boldsymbol{f}$
2. For $k = 1$ to K
3. For $\ell = 1$ to 2^{k-1}
4. $\boldsymbol{f}_{2\ell-1}^{(k)} = \boldsymbol{H}_k \boldsymbol{f}_\ell^{(k-1)}$
5. $\boldsymbol{f}_{2\ell}^{(k)} = \boldsymbol{G}_k \boldsymbol{f}_\ell^{(k-1)}$
6. Next ℓ
7. Next k

The output $\boldsymbol{u} = (\boldsymbol{f}_1^{(K)}, \boldsymbol{f}_2^{(K)}, \ldots, \boldsymbol{f}_{2^K}^{(K)})$ is the desired matrix product.

We can recover the original signal $\boldsymbol{f}$ from the output $\boldsymbol{u} = (\boldsymbol{f}_1^{(K)}, \boldsymbol{f}_2^{(K)}, \ldots, \boldsymbol{f}_{2^K}^{(K)})$ of the wavelet transform algorithm by an algorithm which makes use of the fact that $\boldsymbol{Q}^{(k)}$ is unitary:

INVERSE GENERALIZED FFT ALGORITHM

1. $\boldsymbol{u} = (\boldsymbol{u}_1^{(0)}, \boldsymbol{u}_2^{(0)}, \ldots, \boldsymbol{u}_{2^K}^{(0)}) (= (\boldsymbol{f}_1^{(K)}, \boldsymbol{f}_2^{(K)}, \ldots, \boldsymbol{f}_{2^K}^{(K)}))$
2. For $k = 1$ to K

3. For $\ell = 1$ to 2^{K-k+1}
4. $\boldsymbol{u}_\ell^{(k)} = (\boldsymbol{H}^\dagger_{K-k+1}\boldsymbol{u}_{2\ell-1}^{(k-1)} + \boldsymbol{G}^\dagger_{K-k+1}\boldsymbol{u}_{2\ell}^{(k-1)})/2$
5. Next ℓ
6. Next k

The output $\boldsymbol{u}_1^{(K)}$ is the original signal $\boldsymbol{f}$.

What is the cost of the generalized FFT algorithm? Observe that if the matrices $\boldsymbol{H}_k$ and $\boldsymbol{G}_k$ are all full matrices, the cost of steps 4 and 5 of the generalized FFT algorithm is $2^{2(M-k)+1}$ multiplication operations in each case, so that the total cost of the loops starting at steps 2 and 3 is

$$\text{COST} = \sum_{k=1}^{K}\sum_{j=1}^{2^{k-1}} 2^{2(M-k+1)} = 2\cdot 2^{2M} + \text{lower order terms} = 2N^2 + \text{lower order terms}$$

The algorithm is twice as expensive as doing a matrix multiplication with an arbitrary $N \times N$ matrix; there is no saving. On the other hand, if the matrices $\boldsymbol{H}_k$ and $\boldsymbol{G}_k$ are sparse matrices, the cost of steps 4 and 5 of the algorithm can be much reduced. If, for example, every row of $\boldsymbol{G}_k$ and $\boldsymbol{H}_k$ for each value of k has at most p non-zero entries (where $p \leq 2^{M-K+1}$), and we write an efficient program which avoids handling zero entries, the cost of steps 4 and 5 is at most $p2^{M-k}$ multiplication operations in each case, so that the total cost of the loops starting at steps 2 and 3 is

$$\text{COST} = p\sum_{k=1}^{K}\sum_{j=1}^{2^{k-1}} 2^{M-k+1} = pK2^M.$$

Observe that, with $p = 2$ (which is the case for the matrices $\boldsymbol{H}_k$ and $\boldsymbol{G}_k$ of the FFT) and with $K = M$, the cost is $2N \log N$ multiplications.

We conclude that the generalized FFT algorithm written down above is probably not of interest unless it happens that the matrices $\boldsymbol{H}_k$ and $\boldsymbol{G}_k$ are very sparse, but that, when this condition is met, the generalized FFT algorithm can be very fast.

B. The Discrete Wavelet Transform

Consider again Eqs. (1.3-6) through (1.3-9). We may ask what happens when we apply the transformations $\boldsymbol{f} \to \boldsymbol{Hf}$ and $\boldsymbol{f} \to \boldsymbol{Gf}$ to a signal $\boldsymbol{f}$ which is periodic of period N (which we obtain from a signal of length N by extending periodically in both directions).

Lemma 1.4-2 Let N be an even integer and suppose that $\boldsymbol{f}$ is periodic of period N. Let $\boldsymbol{h} \in \ell^1$ so that $\langle \boldsymbol{h}, \boldsymbol{f}\rangle$ is finite. Then the sequence $\boldsymbol{b}$ defined by

$$b_k = \sum_{n\in Z} h_{n-2k} f_n$$

is periodic of period $N/2$.

Proof: $b_{k+N/2} = \sum_{n \in Z} h_{n-2(k+N/2)} f_n = \sum_{\ell \in Z} h_{\ell-2k} f_{\ell+N} = \sum_{\ell \in Z} h_{\ell-2k} f_\ell = b_k$. This uses the substitution $\ell = n - N$ and the fact that $\boldsymbol{f}$ is periodic of period N. The proof is complete. ∎

This lemma plays a role in the wavelet transform algorithm identical to that played by the Danielson–Lanczos algorithm in the FFT: it says that if we have a periodic signal $\boldsymbol{f}$ of period $N = 2^M$ for some M, then $\boldsymbol{Hf}$ is periodic of period 2^{M-1} (so we have to compute only $N/2$ distinct values), and $\boldsymbol{H}^2\boldsymbol{f}$ is periodic of period 2^{M-2} (so we have to compute only $N/4$ distinct values), and so forth. The lemma states a natural "decimation" of a finite length signal which occurs in the wavelet transform process and which can be exploited in a manner similar to the FFT algorithm's exploitation of the Danielson–Lanczos property.

To be somewhat more precise, Lemma 1.4-2 holds for any ℓ^1 signal $\boldsymbol{h}$; if all but a finite number of terms of $\boldsymbol{h}$ are zero, say, $h_n = 0$ for $n < 0$ and $n > p-1$ for some positive integer p, and $p < 2^{M-K+1}$ for some positive integer $K \leq M$, then we can obtain all the distinct values of $\boldsymbol{Hf}$ by multiplying $\boldsymbol{f}$ by a matrix of dimension $2^{M-1} \times 2^M$, call it $\boldsymbol{H}_1$, constructed as follows: the first row is $(h_0, h_1, \ldots, h_{2^M-1})$ and the second row is the circular shift by two units of the first, the third, the circular shift by four units of the first, and so forth. We can obtain all of the distinct values of $\boldsymbol{H}^2\boldsymbol{f}$ by multiplying $\boldsymbol{H}_1\boldsymbol{f}$ by a matrix of dimension $2^{M-2} \times 2^{M-1}$, call it $\boldsymbol{H}_2$, constructed as follows: the first row is $(h_0, h_1, \ldots, h_{2^{M-1}-1})$ and the second row is the circular shift by two units of the first, the third, the circular shift by four units of the first, and so forth. Repeat the process while the number of columns in $\boldsymbol{H}_k$ (which is a $2^{M-k} \times 2^{M-k+1}$ matrix) exceeds p. Something different could perhaps be done to achieve the same effect when $2^{M-k+1} < p$, but in practice one stops at this point; the effective length p of the sequence $\boldsymbol{h}$ and the length $N = 2^M$ of the data sequence accordingly determine the maximum number of scales used in the wavelet transform. The analysis for the products $\boldsymbol{G}_1\boldsymbol{f}$, $\boldsymbol{G}_2\boldsymbol{H}_1\boldsymbol{f}$, and so forth, is the same. For further discussion of the implementation of the wavelet transform algorithm, see [1-12]; for an example of the matrices $\boldsymbol{H}_k$ and $\boldsymbol{G}_k$ for the discrete Haar wavelet and $M = 3$, see Section 1.7 below.

If $\boldsymbol{f}$ is a signal of period $N = 2^M$, therefore, and if we wish to form the discrete wavelet transform $W_K\boldsymbol{f}$, we can do so via

$$W_1\boldsymbol{f} = \begin{pmatrix} \boldsymbol{H}_1\boldsymbol{f} \\ \boldsymbol{G}_1\boldsymbol{f} \end{pmatrix}$$

$$W_2\boldsymbol{f} = \begin{pmatrix} \begin{pmatrix} \boldsymbol{H}_2 \\ \boldsymbol{G}_2 \end{pmatrix} \boldsymbol{H}_1\boldsymbol{f} \\ \boldsymbol{G}_1\boldsymbol{f} \end{pmatrix} = \begin{pmatrix} \boldsymbol{H}_2\boldsymbol{H}_1\boldsymbol{f} \\ \boldsymbol{G}_2\boldsymbol{H}_1\boldsymbol{f} \\ \boldsymbol{G}_1\boldsymbol{f} \end{pmatrix}$$

$$
\begin{aligned}
W_3 \boldsymbol{f} &= \begin{pmatrix} \begin{pmatrix} \boldsymbol{H}_3 \\ G_3 \end{pmatrix} \boldsymbol{H}_2 \boldsymbol{H}_1 \boldsymbol{f} \\ G_2 \boldsymbol{H}_1 \boldsymbol{f} \\ G_1 \boldsymbol{f} \end{pmatrix} = \begin{pmatrix} \boldsymbol{H}_3 \boldsymbol{H}_2 \boldsymbol{H}_1 \boldsymbol{f} \\ G_3 \boldsymbol{H}_2 \boldsymbol{H}_1 \boldsymbol{f} \\ G_2 \boldsymbol{H}_1 \boldsymbol{f} \\ G_1 \boldsymbol{f} \end{pmatrix} \\
W_4 \boldsymbol{f} &= \ldots.
\end{aligned} \tag{1.4-18}
$$

This should be compared with the symbolic representation of the generalized FFT algorithm, Eq. (1.4-13). Note that, in contrast to Eq. (1.4-13), here it is only $\boldsymbol{H}_k \boldsymbol{H}_{k-1} \ldots \boldsymbol{H}_1 \boldsymbol{f}$ which is operated on in the $(k+1)$st step of the algorithm. Equation (1.4-18) leads to the following algorithm:

WAVELET TRANSFORM ALGORITHM

1. $\boldsymbol{f}_1^{(0)} = \boldsymbol{f}$
2. For $k = 1$ to K
3. $\boldsymbol{f}^{(k)} = \boldsymbol{H}_k \boldsymbol{f}^{(k-1)}$
4. $\boldsymbol{g}^{(k)} = G_k \boldsymbol{f}^{(k-1)}$
5. Next k

The output $(\boldsymbol{f}^{(K)}, \boldsymbol{g}^{(K)}, \boldsymbol{g}^{(K-1)}, \ldots, \boldsymbol{g}^{(1)})$ is the discrete wavelet transform $W_K \boldsymbol{f}$ of the original signal $\boldsymbol{f}$ at scale K.

What is the cost of the wavelet transform algorithm? Observe that if the matrices $\boldsymbol{H}_k$ and $\boldsymbol{G}_k$ were full matrices, the cost of steps 3 and 4 of the algorithm would be $2^{2(M-k)+1}$ multiplication operations in each case, so that the total cost of the loop starting at step 2 would be

$$
\text{COST} = \sum_{k=1}^{K} 2^{2(M-k+1)} \leq \frac{4}{3} 2^{2M} = \frac{4}{3} N^2.
$$

The algorithm would be 4/3 times as expensive as doing a matrix multiplication with an arbitrary $N \times N$ matrix. But the matrices $\boldsymbol{H}_k$ and $\boldsymbol{G}_k$ are sparse matrices; every row of $\boldsymbol{G}_k$ and $\boldsymbol{H}_k$ for each value of k has at most p non-zero entries, so that if we write an efficient program which avoids handling zero entries, the cost of steps 3 and 4 is at most $p2^{M-k}$ multiplication operations in each case. It follows that the total cost of the loop starting at step 2 is

$$
\text{COST} = p \sum_{k=1}^{K} 2^{M-k+1} \leq \frac{2}{3} p 2^M.
$$

Observe that, with $p = 2$, which is the case for the Haar basis, the wavelet transformation algorithm costs $4N/3$ operations; *it is even faster than the FFT algorithm.* Comparison of Eq. (1.4-13) with Eq. (1.4-18) shows immediately why: only one of the intermediate products from a previous step is processed in a step of the wavelet transform algorithm, whereas all of the intermediate products from a previous step are processed in a step of the generalized FFT algorithm. It is sometimes stated that the wavelet transform algorithm is an $N \log N$ algorithm, possibly reflecting the fact that as M grows, the number of non-zero entries p per row can grow.

The inverse wavelet transform algorithm follows the inverse generalized FFT algorithm set out above. We leave its precise statement as an exercise for the reader.

1.5 Wavelets and Multiscale Edge Detection

An edge in a signal or an image is a boundary where a significant change in intensity occurs. Detecting edges in a signal with noise is a basic problem in digital signal processing. In some applications, reconstructing the signal from information about the edges is also important.

Wavelets appear to be a natural tool for edge detection and also for reconstruction of signals from information about edges. It is natural to look at different scales when attempting to detect edges. One would expect that edges of large shadow regions in an image would be detected well using large-scale information – basically the output of a low-pass filter applied to the original signal – whereas the edges of fine detail regions of an image would require fine-scale information. This suggests that it is possible to use information at different scales for different edges. In the presence of noise, it may also be very desirable to use information at different scales for different edges. If an image is noisy, one would expect that fine-scale edge information would be difficult to separate from noise, whereas large-scale edge information would not be. Thus, if what we are interested in is edges of a large shadow region, and we work at too fine a scale, we may find that the edge is undetectable due to high-frequency noise.

To analyze edges of a signal "in depth", Marr and Hildreth [1-13] used a scheme which filtered the original signal via a series of low-pass filters with different cut-off frequencies, hence different scales, and detected edges for each filtered signal. Wavelet transforms are a convenient and fast method of implementing such a scheme, and their use in edge detection applications is beginning to be addressed in the literature. In particular, Mallat *el al.* [1-14], [1-15] developed the theory of multiscale edge detection in the setting of wavelet transforms. The idea is that the high-pass part of the wavelet transform at any scale operates like a difference operator on the signal at the next lower scale (in the simplest case of the Haar basis, the filter coefficients for the high-pass part of the transform are $g_1 = -1$, $g_0 = 1$ and $g_k = 0$ for all other k, so the high-pass part is precisely a difference operator); thus, the points at which the coefficients in the wavelet transform are largest – that is, at which the modulus of the coefficient is a local maximum – are candidates for edges in the image.

In this section, we describe multiscale edge detection and address the issue of reconstruction of a signal from edge information. We will focus on one-dimensional signals.

Let $f \in L^2(R)$ be a signal with a continuous second derivative. If $|f'(x)|$ is very large then f is changing rapidly. If $|f'(x)|$ is greater than a certain threshold, the point x is a candidate edge point. The task of detecting all edges in a signal becomes one of finding all values of x such that $|f'(x)|$ is a local maximum larger than the threshold. This approach of detecting edges in a signal is referred to as the *magnitude method.* Wavelet multiscale edge detection schemes are based on this approach.

On the other hand, the local extreme values of f' are among the zero crossings of f''. Edges may accordingly be detected by looking for zero crossings of f''. This approach to edge detection is referred to as the *Laplacian method.*

For smooth functions, the two methods are not essentially different, except that the Laplacian method will detect zero crossings of f'' which do not correspond to local extreme values of f'. The details of implementation for the two approaches for discrete signals are, however, somewhat different. Both methods are easily implemented. For numerical implementation of the magnitude method, f' may be replaced by

$$f(i) - f(i-1) \tag{1.5-1}$$

or

$$\frac{f(i+1) - f(i-1)}{2}. \tag{1.5-2}$$

We calculate the magnitude of (1.5-1) or (1.5-2) for each point, sort through the values to obtain local maxima, and compare each local maximum with a threshold to determine if the point is an edge point. Similarly, f'' can be discretized by

$$f(i+1) - 2f(i) + f(i-1). \tag{1.5-3}$$

To implement the Laplacian method, one calculates (1.5-3) for each point and determines if it is an edge point by checking if it is a zero-crossing point.

In the usual implementation of both magnitude and Laplacian methods, the location of edge points is kept, but the actual value of f' at the edge points is normally not kept or used; the goal of these algorithms is to locate edges, not to reconstruct images from them. Note that it is not so clear how the image could be reconstructed from the edge information, even if the value of f' (or rather its discrete analog) were retained; one of the nice features of the wavelet technique, discussed below, is that it comes equipped, so to say, with a technique for reconstructing the signal from the edge information.

The Marr and Hildreth multiscale edge detection method is essentially as follows: For a Gaussian function h given by

$$h(x) = \frac{1}{\sigma\sqrt{2\pi}} e^{-x^2/2\sigma^2},$$

where σ is a scaling parameter, the Fourier transform (or frequency response) of h is

$$\hat{h}(\omega) = e^{-\sigma^2\omega^2/2}.$$

Since $h > 0$ and

$$\hat{h}(\omega) \to 1 \text{ as } \sigma \to 0,$$

convolving a signal with h is a low-pass filter operation. The parameter σ determines the effective cutoff frequency; the larger σ, the lower the cutoff frequency. For a signal f, Marr and Hildreth take the convolution of f and h, and obtain the smoothed signal

$$f_\sigma(x) = f * h(x) = \int_{-\infty}^{\infty} f(y)h(x-y)dy.$$

For the smoothed signal f_σ at several different cutoff frequencies σ, Marr and Hildreth use a Laplacian method to find all edge points, i.e., they take a point x to be an edge point if the second derivative of f_σ has zero-crossing at x. Thus, Marr and Hildreth's multiscale edge detection method first smooths at several scales, then uses the Laplacian method at each scale. Marr and Hildreth also do not save information about f_σ' at the locations of zero-crossings of f_σ'' or reconstruct an approximation to the original signal from the information. Their goal is identification of the edges, not reconstruction of the signal from edge information.

We outline a wavelet multiscale edge detection technique due to Mallat *et al.* [1-14], [1-15], [1-16]. Their algorithm finds modulus maxima of wavelet transform coefficients at various scales and so is in essence a magnitude method which has much in common with the Marr and Hildreth technique.

Let $\psi \in L^1 \cap L^2$ be a smooth wavelet, and assume that $\int_{-\infty}^{\infty} \psi\, dt = 0$. Define

$$\theta(x) = \int_{-\infty}^{x} \psi(t)\, dt$$

and assume that also $\theta \in L^1 \cap L^2$. This will be the case, for example, if ψ has compact support or in general if $\psi(t)$ decays rapidly enough as $|t| \to \infty$, and thus in particular is true of wavelets of interest. In general

$$\int_{-\infty}^{\infty} \theta(t)\, dt \neq 0.$$

We will assume the integral is positive (if not, replace ψ by $-\psi$ and θ by $-\theta$). Then θ is smoother than ψ and has fewer sign changes than ψ; if ψ has a single sign change, then θ has no sign changes, and the convolution $f*\theta$ gives some average value of the signal f; convolution with θ smooths and operates like a low-pass filter of some sort. The interpretation may not be so easy if ψ has many sign changes, but we still interpret the convolution $f * \theta$ as some kind of smoothed version of f. We set $\theta_s(x) = s^{-1/2}\theta(x/s)$ and $\psi_s(x) = s^{-1/2}\psi(x/s)$, the versions of θ and ψ at scale s. We have

$$[W_s f](x) = f * \psi_s(x) = f * (s\frac{d\theta_s}{dx})(x) = s\frac{d}{dx}(f * \theta_s)(x).$$

Similarly,

$$\frac{dW_s f(x)}{dx} = s^2 \frac{d^2}{dx^2}(f * \theta_s)(x).$$

Observe that if θ is smooth, then so is $f * \theta_s$, regardless of how smooth f is.

The argument of the preceding paragraph says roughly that for any well-behaved wavelet ψ, convolutions of a given signal f with scaled versions of ψ are derivatives of *some* smoothing kernel θ convolved with f. If we were to use the θ implied by the wavelet ψ in place of a Gaussian kernel, but otherwise to use Marr and Hildreth's approach, and to take for the values of s powers of 2, we would obtain edges at various scales of $f * \theta_s$. This is precisely the algorithm of Mallat *et al.*, except that they also use the magnitude approach rather than the Laplacian approach to the edge detection problem, and so maximize $|(f * \theta_s)'|$ rather than search for zero crossings of $(f * \theta_s)''$.

If the wavelet ψ comes from a multiresolution analysis, it is associated with a function ϕ which is normally a smoothing kernel. Note, however, that in general $\psi \neq \phi'$, so the smoothing kernel θ implied by the wavelet edge detection technique cannot be identified with ϕ.

Marr and Hildreth conjectured that a signal can be perfectly reconstructed from multiscale edge information. Mallat *et al.* refined the conjecture; they conjecture that a signal can be reconstructed from the locations of the modulus maxima of the wavelet transform at different scales together with the values of the wavelet transform coefficients at the modulus maxima locations (we will refer to the information as the "multiscale edge information" or the "modulus maxima locations and values," with the understanding that it is not the value of the modulus of the coefficient but of the coefficient itself which is meant). This fascinating conjecture has been disproved by Meyer [1-17]. In spite of Meyer's counterexamples, experiments indicate that the reconstruction of a signal from modulus maxima locations and values gives an excellent approximation to the original signal in many cases. The reconstruction algorithm basically fills in wavelet transform coefficients at locations other than modulus maxima locations by an interpolation technique, then recovers the approximation to the original signal by taking the inverse wavelet transform.

The technique of reconstructing a signal from multiscale edge information could be used in data compression applications, as an alternative to the technique discussed in Section 1.7 below; this was suggested by Mallat and Zhong [1-14].

We close this section with a discussion of the differences between Marr and Hildreth's approach and that of Mallat *et al.*

First, if we were actually working with L^2 functions rather then discrete signals, and if we were computing the convolution $f * h$ exactly, the Gaussian function would be a good choice for the smoothing operator. A smooth function is unlikely to introduce edges that are not present in the original signals. However, Lindeberg [1-18] pointed out that the use of Gaussian functions in a discrete setting generates undesirable spurious edges because of error in discretizing h, especially at fine scales. It is not a smooth h as such which we want in a discrete setting, but rather a discrete convolution filter which does not introduce spurious edges, and the smoothness of h does not necessarily guarantee this property. It is in our mind an open question whether the wavelet transform technique improves upon Marr and Hildreth's technique in this respect, but Daubechies' smooth wavelets with compact support, or some appropriate modification of them, appear to offer hope that this problem can be solved.

Second, while a Gaussian function is relatively localized in the sense that it achieves the minimum product of effective duration and effective bandwidth, a Gaussian function does not have compact support in either the time domain or the frequency domain. This limitation makes it undesirable in some applications. The limitation can be avoided in the wavelet multiscale edge detector by proper choice of the wavelet ψ (and hence the implied smoothing kernel θ); in particular, Daubechies wavelets have compact support in the time domain.

Third, Marr and Hildreth do not make any use of orthogonality or of the completeness of the representation of the original signal by convolutions of it at various scales; this, on the other hand, is built into wavelet transform methods. The sampled Gaussian functions which Marr and Hildreth use are not an orthogonal basis for the space of discrete signals, and perhaps not a basis at all. Orthogonality can be desirable in itself; orthogonal representations of a signal avoid redundancy and avoid introducing correlations of noise in transformed signals – if a signal is corrupted by independent identically distributed noise before transformation, the noise in components of the transformed signal after an orthogonal transformation is at least uncorrelated, whereas for non-orthogonal transformations, the noise in components of the transformed signal are in general correlated. In some applications, this feature is important. As to completeness, we mean by this that the transformation of the signal gives the representation of the signal with respect to a different basis for the vector space in which the signal lives, so that all of the information in the original signal is retained in the transformed signal. Without completeness, the reconstruction of an approximation to the original signal from edge information alone could not hope to be realized. The wavelet transform is an orthogonal basis transformation, so that completeness is automatic. This is a principal difference between the wavelet based multiscale edge detection method and that of Marr and Hildreth.

This third point, we believe, captures the real strength of the wavelet approach to multiscale edge detection. Even though Marr and Hildreth conjectured that a signal can be perfectly reconstructed by its zero-crossings of all scales, they did not provide an explicit reconstruction scheme. On the other hand, Mallat and Zhong gave a explicit algorithm which approximates a signal from the modulus maxima of its wavelet transforms [1-14]. For the details of the algorithm, we refer the reader to Mallat and Zhong's article.

1.6 Wavelets and Nonstationary Signal Analysis

A "stationary signal" $f(t)$ is one whose statistical properties do not vary with time; this means, at the very least ("weak stationarity") that the expectation is time-invariant, and that $\mathrm{Cov}\,(f(s), f(t))$ is a function only of the difference $|s-t|$, so that the covariance of the signal at a fixed time difference $|s-t|$ is time-invariant. The Fourier transform (FT) of a signal $f(t)$,

$$\hat{f}(\omega) = \frac{1}{\sqrt{2\pi}} \int_{-\infty}^{\infty} f(t) e^{-j\omega t}\, dt,$$

is clearly a time-invariant quantity and hence can be expected to be (and is) a useful tool for analyzing stationary signals. It is not so clear that the FT of a nonstationary signal gives useful information; since it is a time-independent quantity, at best it gives time averaged information about the frequency components of a nonstationary signal, when, often, it is the manner in which the frequency components vary with time which is of interest. Some time-varying transform is desirable. The question we address here is whether the wavelet transform is a good candidate for the time-varying transform to use in the analysis of nonstationary signals.

For nonstationary signals, a usual way to introduce a time-varying Fourier transform is to localize the signal in time by adding a window to each time and taking the FT of the resulting windowed signal. This approach is referred to as the *windowed Fourier transform* (WFT) or *short-time Fourier transform.* Let $f(t) \in L^2$ denote a signal and $w(t) \in L^2$ denote a window function.

$$\hat{f}(\omega, s) = \int_{-\infty}^{\infty} f(t) w(t-s) e^{-j\omega t}\, dt$$

is the WFT of $f(t)$ at time s. The Gaussian window function

$$w_g(t) = \frac{1}{\sqrt{2\pi}} e^{-t^2/2}$$

and the rectangular window function

$$w_r(t) = \begin{cases} 1/T, & t \in [-T/2, T/2] \\ 0, & \text{otherwise} \end{cases}$$

are two typical window functions. A wavelet transform (WT) also localizes the signal. This may be more or less true depending upon the wavelet we choose; it is clearly the case if we choose a wavelet with compact support in the time domain.

In this section, we will look at the following natural questions: What are the differences between a wavelet transform and a windowed Fourier transform? Does a wavelet transform have any advantages over a windowed Fourier transform? As will be seen, there are at least two differences between the WT and WFT. First, the resolution of the wavelet transform varies with frequency whereas the resolution of the windowed Fourier transform does not. Second, there does not exist an orthonormal basis with what is called "good localization" for the windowed Fourier transform but there do exist such bases for the wavelet transform. These two facts may be considered as the main advantages of the wavelet transform over the windowed Fourier transform.

A. Effective Duration and Effective Bandwidth of a Signal

We first introduce some terminology.

Definition 1.6-1 A signal $g(t)$ has "duration" T if $g(t) = 0$ for all $|t| > T/2$; $g(t)$ has "bandwidth" W if $\hat{g}(\omega) = 0$ for $|\omega| > W/2$, where $\hat{g}(\omega)$ is the Fourier transform of $g(t)$. ∎

It is well known that $f = 0$ is the only signal which has finite duration and finite bandwidth (see [1-20] for a rigorous argument). This motivates the following possibly more useful definition.

Definition 1.6-2 The "effective duration" of a signal $g(t)$ is given by

$$\Delta t = \sqrt{\int_{-\infty}^{\infty} (t-\bar{t})^2 |g(t)|^2 \, dt} \Big/ \sqrt{\int_{-\infty}^{\infty} |g(t)|^2 \, dt}$$

where

$$\bar{t} = \int_{-\infty}^{\infty} t|g(t)|^2 \, dt \Big/ \int_{-\infty}^{\infty} |g(t)|^2 \, dt.$$

∎

Definition 1.6-3 The "effective bandwidth" of a signal $g(t)$ is given by

$$\Delta\omega = \sqrt{\int_{-\infty}^{\infty} (\omega-\overline{\omega})^2 |\hat{g}(\omega)|^2 \, d\omega} \Big/ \sqrt{\int_{-\infty}^{\infty} |\hat{g}(\omega)|^2 \, d\omega}.$$

where

$$\overline{\omega} = \int_{-\infty}^{\infty} \omega|\hat{g}(\omega)|^2 \, d\omega \Big/ \int_{-\infty}^{\infty} |\hat{g}(\omega)|^2 \, d\omega.$$

∎

Many signals have both finite effective bandwidth and finite effective duration. A Gaussian function is an example. Sometimes in the literature, the effective duration Δt is referred to as the time resolution of the signal and the effective bandwidth $\Delta\omega$ is referred to as the frequency resolution of the signal.

Definition 1.6-4 An orthonormal basis $\{f_k(t)\}$, $k \in Z$, is said to have "good localization" if, for every k, $f_k(t)$ has both finite effective duration and finite effective bandwidth. ∎

There is no hope of finding a basis for L^2, every element of which possesses finite duration and finite bandwidth, since the zero function is the only such function, but there is hope of finding a basis, every element of which possesses finite effective duration and finite effective bandwidth. Indeed, as we shall see below, at least some wavelet bases have the "good localization" property.

B. Comparison of the Resolution of the Windowed Fourier and Wavelet Transforms

For the purpose of comparison, we discretize the windowed Fourier transform and wavelet transform. Let $t_0 > 0$ and $\omega_0 > 0$ be two positive constants, and let $w(t)$ denote a window function. Set

$$g_{mn}(t) = w(t - nt_0)e^{-jm\omega_0 t}$$

so that

$$\hat{f}(m\omega_0, nt_0) = \int_{-\infty}^{\infty} f(t)g_{mn}(t)\,dt,$$

where $\hat{f}(\omega, s)$ denotes the WFT of f. Similarly, let a and b be two positive constants and let $h(t)$ denote a wavelet function. As usual, set

$$h_{mn}(t) = a^{-m/2}h(a^{-m}t - nb)$$

so that

$$[W_{a^m}f](na^m b) = \int_{-\infty}^{\infty} f(t)h_{mn}(t)\,dt.$$

We wish to compare the resolution of the discretized window function $g_{mn}(t)$ with that of the discretized wavelet function $h_{mn}(t)$.

Theorem 1.6-1 The discretized window functions $g_{mn}(t)$ have the same effective duration and the same effective bandwidth for all (m, n).

Proof: For convenience, we normalize g so that

$$\int_{-\infty}^{\infty} |g(t)|^2 dt = 1.$$

We have then

$$\bar{t}_{mn} = \int_{-\infty}^{\infty} t|g_{mn}(t)|^2\,dt$$

$$
\begin{aligned}
&= \int_{-\infty}^{\infty} t|w(t - nt_0)|^2\, dt \\
&= \int_{-\infty}^{\infty} (u + nt_0)|w(u)|^2\, du \\
&= \int_{-\infty}^{\infty} u|g(u)|^2\, du + nt_0 \\
&= \bar{t}_{00} + nt_0
\end{aligned}
$$

via the substitution $t = u + nt_0$. It follows that

$$
\begin{aligned}
(\Delta t)^2_{mn} &= \int_{-\infty}^{\infty} (t - \bar{t}_{mn})^2 |g_{mn}(t)|^2\, dt \\
&= \int_{-\infty}^{\infty} (u + nt_0 - (\bar{t}_{00} + nt_0))^2 |w(u)|^2\, du \\
&= \int_{-\infty}^{\infty} (u - \bar{t}_{00})^2 |g(u)|^2\, du \\
&= (\Delta t)^2_{00}.
\end{aligned}
$$

That is, the time resolution of g_{mn} is the same as the time resolution of g_{00}. The argument that g_{mn} and g_{00} have the same frequency resolution is similar, and we omit it. ∎

Theorem 1.6-2 The effective duration Δt_{mn} and the effective bandwidth $\Delta\omega_{mn}$ of the discrete wavelet functions $h_{mn}(t)$ satisfy

$$
\Delta t_{mn} = \Delta t_{00} a^m \quad \text{and} \quad \Delta\omega_{mn} = a^{-m}\Delta\omega_{00}.
$$

Proof: For convenience, we normalize h so that

$$
\int_{-\infty}^{\infty} |h(t)|^2\, dt = 1.
$$

We have then

$$
\begin{aligned}
\bar{t}_{mn} &= \int_{-\infty}^{\infty} t|h_{mn}(t)|^2\, dt \\
&= \int_{-\infty}^{\infty} a^{-m} t|h(a^{-m}t - nb)|^2\, dt \\
&= a^m \int_{-\infty}^{\infty} (u + nb)|h(u)|^2\, du \\
&= a^m(\bar{t}_{00} + nb)
\end{aligned}
$$

via the substitution $t = a^m(u + nt_0)$. It follows that

$$\begin{aligned}(\Delta t)^2_{mn} &= \int_{-\infty}^{\infty} (t - \bar{t}_{mn})^2 |h_{mn}(t)|^2 \, dt \\ &= \int_{-\infty}^{\infty} (t - a^m(\bar{t}_{00} + nb))^2 a^{-m} |h(a^{-m}t - nb)|^2 \, dt \\ &= \int_{-\infty}^{\infty} a^{2m}(u + nb - (\bar{t}_{00} + nb))^2 |h(u)|^2 \, du \\ &= a^{2m} (\Delta t)^2_{00}\end{aligned}$$

via the same substitution of variables. The argument for the frequency resolution of the wavelet h_{mn} is similar, and we omit it. ∎

It is instructive to analyze exactly what gives rise to the differences in time and frequency resolution behavior noted in Theorems 1.6-1 and 1.6-2. We do so by way of an example.

Example

We have

$$g_{mn}(t) = w(t - nt_0)e^{-jm\omega_0 t}$$

for the windowed Fourier transform. We take $h(t) = w(t)e^{-j\omega_0 t}$ for the wavelet of Theorem 1.6-2 (translates of h need not be orthogonal, but the arguments of Theorem 1.6-2 depend in no way on orthogonality properties). Then the "effective frequency" of the wavelet is ω_0, and we take $a = 2$ and $b = t_0$ to obtain

$$h_{mn}(t) = 2^{-m/2} w(2^{-m}t - nt_0)e^{-j\omega_0(2^{-m}t - nt_0)}.$$

The differences between the expressions for g_{mn} and h_{mn} deserve study. The "effective frequency" of h_{mn} is $\omega_m = 2^{-m}\omega_0$, and the effective window length for the wavelet function changes with the effective frequency, since 2^{-m} multiplies t in the argument of w. The effective window length is independent of effective frequency in the case of the window function g_{mn}. In fact, for the wavelet function in this example, we have by Theorem 1.6-2 and the definition above of effective frequency

$$\Delta t_{mn}\omega_m = \Delta t_{00}\omega_0 \tag{1.6-1}$$

$$\frac{\Delta\omega_{m,0}}{\omega_m} = \frac{\Delta\omega_{00}}{\omega_0}. \tag{1.6-2}$$

Since $a > 1$, these relationships imply that the wavelet transform method for analyzing nonstationary signals achieves good time resolution (small Δt) but poor frequency resolution (large $\Delta\omega$) at high frequencies and achieves good frequency resolution (small $\Delta\omega$) but poor time resolution (large Δt) at low frequencies.

In the example above, it was easy to associate an "effective frequency" with a wavelet function; it is not so easy to do so in general. Nonetheless, something like Eqs. (1.6-1) and (1.6-2) will hold for wavelet transform methods in non-stationary signal analysis, whatever definition we may adopt for "effective frequency." We could achieve the same result with windowed Fourier transform methods, if we let the effective window duration vary with frequency: we could compute the windowed Fourier transform of a signal using window functions $w_m(t) = w(2^{-m}t)$ for several values of m, and we could keep only the low-frequency information for large positive m (large effective window length) and only the high-frequency information for m small or negative (small effective window length). The approach of using a different effective window duration for different frequencies is traditionally known as a "constant-Q" method. The wavelet approach is basically an efficient way to implement a "constant-Q" method.

There is a question as to whether the "constant-Q" approach is a reasonable one. It presupposes that a larger effective window length is needed to resolve low-frequency components of a signal than is needed to resolve high-frequency components. This seems to be true; a complete cycle of a low-frequency component requires a longer time, by definition, than a complete cycle of a high-frequency component of a signal.

C. Orthonormal Window and Wavelet Bases

There is another interesting comparison between the windowed Fourier transform and the wavelet transform techniques for non-stationary signals.

A "window" basis is a basis generated by translates of a window function $w(t)$. We call the functions

$$g_{mn}(t) = e^{-j\omega_0 mt} w(t - nt_0)$$

a "window" basis, if the closure of their span is dense in L^2. We will call them an orthogonal window basis if in addition they are an orthogonal basis. It is easy to see that if $w = \chi_{[0,1]}$, the g_{mn}'s defined above constitute an orthonormal window basis so long as $t_0 = 1$ and $\omega_0 = 2\pi$.

An orthogonal "wavelet" basis is a basis obtained from translates of dilations of a wavelet ψ of a multiresolution analysis, when the translates of the function ϕ which generates the multiresolution analysis are orthogonal, in the manner described in Section 1.1. Thus, the basis functions h_{mn} have the form

$$h_{mn}(t) = a^{-m/2} \psi(a^{-m}t - nt_0).$$

In this subsection, we point out an interesting difference between certain orthonormal wavelet bases and orthonormal window bases. Recall (Definition 1.6-3) that an orthonormal basis $\{f_k(x)\}_{k \in Z}$, is said to have good localization if, for any k, $f_k(t)$ has both finite effective duration and finite effective bandwidth. It is a basic fact that there exist orthonormal wavelet bases with good localization, but there does not exist an orthonormal window basis with good localization.

As to orthogonal wavelet bases, we can rely on the following result which we state without proof:

Lemma 1.6-1 If $f(t)$ is a C^1 function with compact support, then f has finite effective duration and finite effective bandwidth. Similarly, if $\hat{f}(\omega)$ is a C^1 function with compact support, then f has finite effective duration and finite effective bandwidth. ∎

The Fourier transform $\hat{\psi}$ of the Meyer wavelet of Section 1.2 is a C^∞ function with compact support. Since

$$\widehat{\psi_{mn}}(\omega) = 2^{m/2} e^{-jn2^m\omega} \hat{\psi}(2^m\omega),$$

as is readily verified, $\widehat{\psi_{mn}}$ is C^∞ with compact support for every m and n. Thus, by Lemma 1.6-1, ψ_{mn} has finite effective duration and finite effective bandwidth for every m and n. It follows that the orthonormal wavelet basis obtained from the Meyer wavelet has good localization. The Haar wavelet is not C^0 and the other example of a Daubechies wavelet given in Section 1.3 is not C^1, so both fail the hypotheses of Lemma 1.6-1. However, Daubechies' technique gives rise to many C^1 wavelets ψ with compact support in the time domain, see [1-10]. Since every translated and scaled version of any such wavelet is also a C^1 function with compact support, the family of wavelets obtained from any such wavelet is an orthogonal basis with good localization, via Lemma 1.6-1. Thus, there are many wavelet bases with good localization.

On the other hand, there does not exist an orthonormal window basis with good localization. This result is known as the Balian–Low theorem in the literature. According to [1-19], the theorem was first stated by Balian [1-21] in 1981 and by Low [1-22] in 1985, but their proofs had gaps. The first rigorous proof was given by Coifman and Semmes. Battle [1-23] gave a nice proof of the theorem in 1988, using ideas from quantum mechanics. The theorem is given below:

Theorem 1.6-3 (Balian–Low theorem) Suppose that

$$\sqrt{|t|}|w(t)| \to 0 \text{ as } |t| \to \infty. \tag{1.6-3}$$

If the window functions

$$g_{mn}(t) = w(t - nt_0)\, e^{-jm\omega_0 t} \tag{1.6-4}$$

constitute an orthonormal basis, then either

$$\int_{-\infty}^{\infty} t^2 |w(t)|^2 dt = \infty \quad \text{or} \quad \int_{-\infty}^{\infty} \omega^2 |\hat{w}(\omega)|^2 d\omega = \infty.$$

∎

The Balian–Low theorem says that we cannot obtain an orthogonal window basis with good localization. If, therefore, good localization is something we cannot live without in the context of nonstationary signal analysis, the "constant-Q" techniques which are built into the wavelet approach to signal analysis are preferable: the effective window duration should change with frequency. Whether good localization is in fact a property we cannot live without is an open issue.

1.7 Signal Compression Using Wavelets

In signal and image compression, it is desirable to find fast algorithms to code a given signal or image in such a way that the number of binary digits used is as small as possible. A recent widely publicized example concerns Federal Bureau of Investigation (FBI) fingerprint records. The FBI has more than 200 million fingerprint cards that have occupied an acre of office space. In order to save space in storing and time in transmitting fingerprint images, the FBI wanted to digitize the images and then to store the files electronically in computer discs. However, coding an entire fingerprint card at a resolution of 500 pixels per inch with 256 gray-levels in a straightforward manner would take up approximately 600 kilobytes, and furthermore, transmitting this amount of information at a standard modem rate of 9600 bits per second with 20% overhead would tie up a phone line for nearly three hours (see [1-24]). This problem lead the FBI to support efforts to develop effective algorithms to minimize cost in storing and transmitting fingerprint images. Recently, a wavelet compression approach has been approved by the FBI for coding the fingerprint images [1-24]. It is called the Coifman–Wickerhauser algorithm, and we describe it here.

The basic idea in lossless signal compression is that if $\boldsymbol{A}$ is an $N \times N$ invertible matrix then the discrete signal $\boldsymbol{f}$, which may be thought of as an element of R^N, can be recovered from its representation $\boldsymbol{b} = \boldsymbol{A}\boldsymbol{f}$. Therefore, if coding $\boldsymbol{b}$ requires less cost (fewer binary digits) than coding the original signal $\boldsymbol{f}$, the representation $\boldsymbol{b}$ can be coded and stored, and the signal $\boldsymbol{f}$ can be recovered by solving the equation $\boldsymbol{A}\boldsymbol{f} = \boldsymbol{b}$. This leads to the problem of finding an optimal matrix $\boldsymbol{A}$ for coding a set of similar signals with the minimum cost. Finding such a matrix is a complicated problem in general, since minimizing the cost of encoding the signals in the set with respect to a large number of variables (in this case there are N^2 variables – the entries of $\boldsymbol{A}$ – and N can be very large in practice) is a formidable task.

The approach used in wavelet tree library signal compression is related but different. In the first place, the cost function is not the (almost necessarily modeled) cost of coding all signals in the set but the actual cost of coding a particular signal; in the second place, the entire cost surface is not searched but rather the actual cost is determined at a large number of fixed points (more than $2^{N/2}$ of them) on the surface, and a point at which the minimum coding cost occurs is selected. The process is repeated for each signal in the set.

It may seem that the wavelet tree library approach, while avoiding a difficult optimization problem, could not possibly be done efficiently. But with wavelets it can. In the first place, if the $2^{N/2}$ candidate matrices $\boldsymbol{A}$ were arbitrary matrices, the cost of forming the matrix product $\boldsymbol{b} = \boldsymbol{Af}$ for each $\boldsymbol{A}$ would be N^2 operations, and the cost of solving $\boldsymbol{Af} = \boldsymbol{b}$ to recover $\boldsymbol{f}$ is (ignoring factorization of the matrices $\boldsymbol{A}$, which in principle could be done once and for all) is also N^2 operations; when the matrix $\boldsymbol{A}$ is a wavelet matrix, however, the cost of forming $\boldsymbol{b} = \boldsymbol{Af}$ and the cost of recovering $\boldsymbol{f}$ from $\boldsymbol{b}$ is $N \log N$ operations, as shown in Section 1.4 above. This is, however, not the principal cost saving. In fact, when the wavelet transformation $W_T\boldsymbol{f}$ is formed, each of the intermediate steps in the transformation algorithm gives rise to a distinct representation of the signal $\boldsymbol{f}$ equivalent to the original one. The Coifman–Wickerhauser algorithm exploits this fact: each intermediate representation, as well as the final one, is in the wavelet tree library. In fact, the wavelet transformation algorithm is modified slightly from that given in Section 1.4 (the generalized FFT algorithm described in Section 1.4 is used) so as to expand the library even further. One computes in the end that the library consists of more than $2^{N/2}$ orthogonal matrices $\boldsymbol{A}$, and that all of the products $\boldsymbol{A}s$ for each of the more than $2^{N/2}$ matrices A can be formed for about $pN \log N$ operations, see Section 1.4, where p is the number of non-zero entries in the discrete wavelet $\boldsymbol{h}$ or $\boldsymbol{g}$ (for the Haar wavelet, $p = 2$). Wavelet transform methods accordingly have a natural tree structure that reduces dramatically the number of operations needed to compute and compare the costs of coding a signal with respect to the many orthogonal bases and make possible the wavelet tree library approach to signal compression.

Coifman and Wickerhauser's wavelet signal compression algorithm [1-25] is basically as follows: First, use a modified wavelet transformation to obtain the representation of a signal $\boldsymbol{f}$ with respect to a large number of orthogonal bases. Second, compute the cost of coding the signal for each such orthogonal basis representation. Third, select the orthogonal basis with lowest cost, code the signal in that basis, and code also information as to what basis has been chosen.

The remainder of this section is devoted to describing in greater detail the Coifman–Wickerhauser algorithm and giving examples.

Coifman and Wickerhauser's algorithm searches for a minimum cost representation in a class of linear subspaces with tree structure, which they called a "tree library".

Definition 1.7-1 A collection $\{\mathcal{H}(\omega), \omega \in \Omega\}$, of subspaces of a Hilbert space $\mathcal{H}$ is called a tree library if

(i) each index $\omega \in \Omega$ is a finite string consisting of binary numbers 0 and 1 of length at most M;

(ii) $\mathcal{H} = \mathcal{H}(0) \oplus \mathcal{H}(1)$; and

(iii) $\mathcal{H}(\omega) = \mathcal{H}(\omega 0) \oplus \mathcal{H}(\omega 1)$ for every string ω of length at most $M - 1$.

We always take $\mathcal{H}$ itself to be an element of the tree library ($\mathcal{H} = \mathcal{H}(\omega)$ where ω is the string of length 0). ∎

Note that a tree library of subspaces is different in principle from a multiresolution analysis. If $\{V_m\}$ is a multiresolution analysis of $L^2(R)$, and we take $\mathcal{H} = V_m$ for some m, then since $V_m = V_{m+1} \oplus W_{m+1}$, we can take $\mathcal{H}(0) = V_{m+1}$ and $\mathcal{H}(1) = W_{m+1}$. We can take V_{m+2} for $\mathcal{H}(00)$ and W_{m+2} for $\mathcal{H}(01)$, but we do not appear to have natural choices for $\mathcal{H}(10)$ and $\mathcal{H}(11)$, the subspaces of which in the tree library setting $\mathcal{H}(1) = W_{m+1}$ should be a direct sum. Nonetheless, a tree library can be naturally constructed in the discrete wavelet transform context. Let $\boldsymbol{h}$ and $\boldsymbol{g}$ be a discrete wavelet pair and let $\boldsymbol{H}$ and $\boldsymbol{G}$ be the matrices obtained from them which satisfy Eqs. (1.3-6) through (1.3-9). If the sequence $\boldsymbol{f}$ gives the coefficients of the projection $P_{-1}f$ of a continuous signal f on V_{-1} with respect to the orthonormal basis $\{\phi_{-1,n}\}_{n\in Z}$ (see Section 1.4 above), then the pair of sequences $\boldsymbol{Hf}$ and $\boldsymbol{Gf}$ give the coefficients of $P_{-1}f$ with respect to the orthonormal basis $\{\phi_{0n}\}_{n\in Z} \cup \{\psi_{0n}\}_{n\in Z}$ for V_{-1}. As we have seen, $\boldsymbol{H}^2\boldsymbol{f}$ and $\boldsymbol{GHf}$ give the coefficients of $P_0 f$ (the projection of the same signal on V_0) with respect to the basis $\{\phi_{1n}\}_{n\in Z} \cup \{\psi_{1n}\}_{n\in Z}$ for V_0. We now observe that $\boldsymbol{HGf}$ and $\boldsymbol{G}^2\boldsymbol{f}$ give the coefficients of $Q_0 f$, the projection of the signal on W_0, with respect to *some* orthonormal basis for W_0, since the matrix $\begin{pmatrix} \boldsymbol{H} \\ \boldsymbol{G} \end{pmatrix}$ is unitary. The spaces $\mathcal{H}(10)$ and $\mathcal{H}(11)$ are then defined so that $\begin{pmatrix} H \\ G \end{pmatrix}$ is the matrix of the basis change, that is, the new basis W_0 is $\{\xi_n\}_{n\in Z} \cup \{\eta_n\}_{n\in Z}$ where

$$\begin{aligned} \xi_n &= \sum_{k\in Z} h_{k-2n}\psi_{0k} \\ \eta_n &= \sum_{k\in Z} g_{k-2n}\psi_{0k}. \end{aligned}$$

It is readily verified that this gives a new orthonormal basis for W_0. We set $\mathcal{H}(10) = \operatorname{span}\{\xi_n\}_{n\in Z}$ and $\mathcal{H}(11) = \operatorname{span}\{\eta_n\}_{n\in Z}$, and we say that $\mathcal{H}(10)$ and $\mathcal{H}(11)$ are the images of $\mathcal{H}$ under the projection operators implied by $\boldsymbol{HG}$ and $\boldsymbol{G}^2$, respectively. This process is repeated iteratively to obtain a tree library.

A tree library actually used in the Coifman–Wickerhauser algorithm is a tree library constructed in a finite dimensional Hilbert space. In this case, the operators $\boldsymbol{H}$ and $\boldsymbol{G}$ are replaced by operators which satisfy the finite dimensional analogs of Eqs. (1.3-6) through (1.3-9): For each $J \in \{1, ..., M\}$ let $\boldsymbol{H}_J$ and $\boldsymbol{G}_J$ be $2^{M-J} \times 2^{M-J+1}$ dimensional matrices such that

$$\boldsymbol{H}_J\boldsymbol{H}_J^\dagger = \kappa\,\boldsymbol{I}_J \tag{1.7-1}$$

$$\boldsymbol{G}_J\boldsymbol{G}_J^\dagger = \kappa\,\boldsymbol{I}_J \tag{1.7-2}$$

$$\boldsymbol{H}_J^\dagger\boldsymbol{H}_J + \boldsymbol{G}_J^\dagger\boldsymbol{G}_J = \kappa\,\boldsymbol{I}_{J+1} \tag{1.7-3}$$

$$\boldsymbol{H}_J\boldsymbol{G}_J^{\dagger} = \boldsymbol{G}_J\boldsymbol{H}_J^{\dagger} = 0 \tag{1.7-4}$$

where $\boldsymbol{I}_J$ is the $2^{M-J} \times 2^{M-J}$ unit matrix and $\kappa > 0$ is a scalar normalization factor. Now for any string of binary numbers of length $J \leq M$, say $\omega = (\omega_1, ..., \omega_J)$, let $\boldsymbol{F}_\omega = \boldsymbol{F}_J^{\omega_J} ... \boldsymbol{F}_1^{\omega_1}$, where $\boldsymbol{F}_k^0 = \boldsymbol{H}_k$ and $\boldsymbol{F}_k^1 = \boldsymbol{G}_k$. Let $\mathcal{H}(\omega)$ be the image of $\mathcal{H} = R^{2^M}$ under the projection operator implied by F_ω, in the manner of the preceding paragraph. Then the collection of such $\mathcal{H}(\omega)$ is a tree library for R^{2^M}. We give an example, using the discrete Haar wavelet and $N = 2^M = 8$, below.

Observe that if we have a tree library $\{\mathcal{H}(\omega)\}$ for a Hilbert space $\mathcal{H}$, we have many direct sum decompositions of $\mathcal{H}$ of the form

$$\mathcal{H} = \mathcal{H}(\omega_1) \oplus ... \oplus \mathcal{H}(\omega_m). \tag{1.7-5}$$

We address next the question of just how many there are. Let M be the maximum length of a binary sequence ω, and let n_M denote the number of direct sum decompositions of $\mathcal{H}$ of the form (1.7-5). Then $n_M = 1 + (n_{M-1})^2$. To see this, observe that since $\mathcal{H} = \mathcal{H}(0) \oplus \mathcal{H}(1)$, and since the number of direct sum decompositions of $\mathcal{H}(0)$ and of $\mathcal{H}(1)$ from the tree library is n_{M-1}, and since any direct sum decomposition of $\mathcal{H}(0)$ combined with any direct sum decomposition of $\mathcal{H}(1)$ gives a direct sum decomposition of $\mathcal{H}$, the number of direct sum decompositions of $\mathcal{H}$ from the tree library is at least as large as $1+(n_{M-1})^2$ (where we have counted as a direct sum decomposition the trivial decomposition $\mathcal{H} = \mathcal{H}$). The number of direct sum decompositions of $\mathcal{H}$ from the tree library cannot exceed $1 + (n_{M-1})^2$ because every subspace in the tree library which is not either $\mathcal{H}$ or $\mathcal{H}(0)$ or $\mathcal{H}(1)$ is necessarily itself a subspace either of $\mathcal{H}(0)$ or $\mathcal{H}(1)$, so that if we take any direct sum decomposition of the form (1.7-5), other than $\mathcal{H} = \mathcal{H}$, and collect the subspaces $\mathcal{H}(\omega_j) \subset \mathcal{H}(0)$, we must obtain a direct sum decomposition of $\mathcal{H}(0)$, and similarly the subspaces $\mathcal{H}(\omega_j) \subset \mathcal{H}(1)$ must yield a direct sum decomposition of $\mathcal{H}(1)$. Thus, every direct sum decomposition of $\mathcal{H}$ using subspaces in the tree library, except the trivial one $\mathcal{H} = \mathcal{H}$, comes from a direct sum decomposition of $\mathcal{H}(0)$ and a direct sum decomposition of $\mathcal{H}(1)$. Observe next that if $M = 2$, so that the tree library is $\{\mathcal{H}, \mathcal{H}(0), \mathcal{H}(1), \mathcal{H}(00), \mathcal{H}(01), \mathcal{H}(10), \mathcal{H}(11)\}$, the direct sum decompositions of $\mathcal{H}$ from the tree library are:

$$\begin{aligned}
\mathcal{H} &= \mathcal{H} \\
\mathcal{H} &= \mathcal{H}(0) \oplus \mathcal{H}(1) \\
\mathcal{H} &= \mathcal{H}(00) \oplus \mathcal{H}(01) \oplus \mathcal{H}(1) \\
\mathcal{H} &= \mathcal{H}(0) \oplus \mathcal{H}(10) \oplus \mathcal{H}(11) \\
\mathcal{H} &= \mathcal{H}(00) \oplus \mathcal{H}(01) \oplus \mathcal{H}(10) \oplus \mathcal{H}(11).
\end{aligned}$$

There are no others. It follows that $n_2 = 5$, and, from the recursion formula developed above, that $n_3 = 26$, $n_4 = 677$, $n_5 = 338,330$ and so forth. A convenient estimate for how many distinct direct sum decompositions one obtains from a tree library is $n_M \geq 2^{2^{M-1}}$, or setting $N = 2^M$ (the dimension of the smallest finite dimensional Hilbert space containing a tree library $\{\mathcal{H}(\omega)\}$ using binary strings ω of length M or less), $n_M \geq 2^{N/2}$. This estimate follows immediately from the recursion formula developed above and the observation that for $M = 2$, $n_M = 5 \geq 4 = 2^{2^{M-1}}$.

Assume that we have a direct sum decomposition

$$\mathcal{H} = \mathcal{H}(\omega_1) \oplus \ldots \oplus \mathcal{H}(\omega_m)$$

of the form (1.7-5) from the tree library, and assume that the cost of coding the projection $P(\omega_j)\boldsymbol{f}$ of a signal $\boldsymbol{f}$ on $\mathcal{H}(\omega_j)$ is $C(\omega_j)$. We will take the cost to be the number of binary digits required to code the projection $P(\omega_j)\boldsymbol{f}$, though one could imagine using some other cost function. Then the total cost $C(\omega_1, \ldots \omega_m)$ of coding the representation of the signal $\boldsymbol{f}$ via its projections on each of the $\mathcal{H}(\omega_j)$'s in the direct sum decomposition is

$$C(\omega_1, \ldots \omega_m) = \sum_{j=1}^{m} C(\omega_j).$$

For any signal $\boldsymbol{f}$, the Coifman–Wickerhauser algorithm minimizes $C(\omega_1, \ldots \omega_m)$ over all the n_M direct sum decompositions of $\mathcal{H}$ allowed by the tree library. If $\omega_j = (\omega_{j1}, \ldots, \omega_{j\ell})$, where each ω_{jk} is a 0 or a 1, the projection $P(\omega_j)\boldsymbol{f}$ is just $F_{\omega_j}\boldsymbol{f} = F_\ell^{\omega_{j\ell}} \ldots F_1^{\omega_{j1}}\boldsymbol{f}$, where $F_k^0 = H_k$ and $F_k^1 = G_k$.

The algorithm is as follows:

1. Execute the generalized FFT algorithm of Section 1.4, saving all intermediate products.

2. Sort the intermediate products to obtain the collection of intermediate products giving a representation of $\boldsymbol{f}$ which can be coded at least cost; save the lowest cost representation and information allowing the original signal $\boldsymbol{f}$ to be reconstructed from it.

For implementational details, we refer the reader to [1-25]; here we give instead an illustrative example.

Example

Suppose that we want to use Coifman and Wickerhauser's algorithm to code the following signal:

$$f = 20, 6, 4, 2, 10, 6, 8, 4$$

using the tree library obtained from the discrete Haar wavelet and $M = 3$:

$$\boldsymbol{H}_1 = \frac{1}{2}\begin{pmatrix} 1 & 1 & 0 & 0 & 0 & 0 & 0 & 0 \\ 0 & 0 & 1 & 1 & 0 & 0 & 0 & 0 \\ 0 & 0 & 0 & 0 & 1 & 1 & 0 & 0 \\ 0 & 0 & 0 & 0 & 0 & 0 & 1 & 1 \end{pmatrix}$$

$$G_1 = \frac{1}{2}\begin{pmatrix} 1 & -1 & 0 & 0 & 0 & 0 & 0 & 0 \\ 0 & 0 & 1 & -1 & 0 & 0 & 0 & 0 \\ 0 & 0 & 0 & 0 & 1 & -1 & 0 & 0 \\ 0 & 0 & 0 & 0 & 0 & 0 & 1 & -1 \end{pmatrix}$$

$$H_2 = \frac{1}{2}\begin{pmatrix} 1 & 1 & 0 & 0 \\ 0 & 0 & 1 & 1 \end{pmatrix}$$

$$G_2 = \frac{1}{2}\begin{pmatrix} 1 & -1 & 0 & 0 \\ 0 & 0 & 1 & -1 \end{pmatrix}$$

$$H_3 = \frac{1}{2}\begin{pmatrix} 1 & 1 \end{pmatrix}$$

$$G_3 = \frac{1}{2}\begin{pmatrix} 1 & -1 \end{pmatrix}.$$

(Note that these matrices are normalized so that, for example, $\boldsymbol{H}_k\boldsymbol{H}_k^\dagger = \frac{1}{2}\boldsymbol{I}$ rather than $\boldsymbol{I}$, so as to avoid entries which are not powers of 2.) In $\boldsymbol{H}_1$, coefficients h_1 and h_2 are obtained by applying the decomposition (1.3-1) to the Haar wavelet defined in Eq. (1.2-1). The coefficients g_1 and g_2 in $\boldsymbol{G}_1$ follow from the definition in (1.1-17). We decompose the signal $\boldsymbol{f}$ into $\boldsymbol{H}_1\boldsymbol{f}, \boldsymbol{G}_1\boldsymbol{f}$ and then into $\boldsymbol{H}_2\boldsymbol{H}_1\boldsymbol{f}, \boldsymbol{G}_2\boldsymbol{H}_1\boldsymbol{f}, \boldsymbol{H}_2\boldsymbol{G}_1\boldsymbol{f}, \boldsymbol{G}_2\boldsymbol{G}_1\boldsymbol{f}$, and so on, using the generalized FFT algorithm of Section 1.4. Replacing the signal f by the vector $\boldsymbol{f}$, this sequence of decompositions can be described by the following tree:

<table>
<tr><td colspan="8" align="center">f</td></tr>
<tr><td colspan="4" align="center">H_1f</td><td colspan="4" align="center">G_1f</td></tr>
<tr><td colspan="2" align="center">H_2H_1f</td><td colspan="2" align="center">G_2H_1f</td><td colspan="2" align="center">H_2G_1f</td><td colspan="2" align="center">G_2G_1f</td></tr>
<tr><td>$H_3H_2H_1f$</td><td>$G_3H_2H_1f$</td><td>$H_3G_2H_1f$</td><td>$G_3G_2H_1f$</td><td>$H_3H_2G_1f$</td><td>$G_3H_2G_1f$</td><td>$H_3G_2G_1f$</td><td>$G_3G_2G_1f$</td></tr>
</table>

By direct calculation, the elements in the tree above take the following values:

<table>
<tr><td colspan="8">20 6 4 2 10 6 8 4</td></tr>
<tr><td colspan="4">13 3 8 6</td><td colspan="4">7 1 2 2</td></tr>
<tr><td colspan="2">8 7</td><td colspan="2">5 1</td><td colspan="2">4 2</td><td colspan="2">3 0</td></tr>
<tr><td>15/2</td><td>1/2</td><td>3</td><td>2</td><td>3</td><td>1</td><td>3/2</td><td>3/2</td></tr>
</table>

This completes step 1 of the algorithm. In step 2, we analyze coding costs starting with external nodes of the tree (the last row in the array above). We label intermediate steps $k = 0$ through $k = 3$, going from bottom to top in the array above.

At step $k = 0$ we have the following sequence:

$$(\frac{15}{2}), (\frac{1}{2}), (3), (2), (3), (1), (\frac{3}{2}), (\frac{3}{2}).$$

This sequence can be coded with binary digits, dots, and commas as

$$111.1,,.1,,11,,10,,11,,1,,1.1,,1.1$$

We have used the more convenient double comma notation ",," to replace punctuations of the form "),(" in the original data sequence. The string $,,c_1, c_2, ..., c_\ell,,$ should be thought of as a block representing the expansion coefficients of $P(\omega_j)f$ with respect to the basis for $\mathcal{H}(\omega_j)$ for some binary sequence ω_j. Since each direct sum decomposition of $\mathcal{H}$ has its own block decomposition, there is a unique punctuation pattern that determines the decomposition, and, in the end, the record of what decomposition was used for the signal $\boldsymbol{f}$ is retained in the pattern of commas separating the symbols. Clearly, we only need at most $2N, N = 2^M$, commas. For the sake of illustrating the algorithm we do not count commas in the cost of our codes; also for simplicity, we ignore the cost of using dots (in separating the integer and the fractional parts of a binary number). In practice, for large N the cost of commas used in giving the record of the decomposition is relatively small in comparison to the cost of binary digits used in coding the coefficients.

At step $k = 1$, we compare

$$(8, 7) = (1000, 111)_2 \quad \text{with} \quad (\frac{15}{2}), (\frac{1}{2}) = (111.1,,.1)_2$$

$$(5, 1) = (101, 1)_2 \quad \text{with} \quad (3), (2) = (11,,10)_2$$

$$(4, 2) = (100, 10)_2 \quad \text{with} \quad (3), (1) = (11,,1)_2$$

$$(3, 0) = (11, 0)_2 \quad \text{with} \quad (\frac{3}{2}), (\frac{3}{2}) = (1.1,,1.1)_2$$

taking always the alternative which uses the fewest binary digits. We choose

$$(\frac{15}{2}), (\frac{1}{2}), (5,1), (3), (1), (3,0)$$

Notice that coding $(5,1)$ by $(101,1)$ requires the same number of binary digits as coding $(3),(2)$ by $(11,,10)$; but the former requires less punctuations so it is preferable. Thus at the end of step 1 we select the tree with external nodes that correspond to the following coefficients (in binary)

$$111.1,,.1,,101,1,,11,,1,,11,0$$

At step $k = 2$, by comparing the cost of coding

$$(13,3,8,6) = (1101,11,1000,110)_2 \quad \text{with} \quad ((\frac{15}{2}),(\frac{1}{2}),(5,1)) = (111.1,,.1,,101,1)_2$$

and

$$(7,1,2,2) = (111,1,10,10)_2 \quad \text{with} \quad ((3),(1),(3,0)) = (11,,1,,11,0)_2$$

we choose the latter ones. Thus at the end of step $k = 2$, we select the same tree as in step $k = 1$.

At the last step $k = 3$, we compare the cost of our code selected in step $k = 2$ with the cost of coding the original signal in binary by

$$10100, 110, 100, 10, 1010, 110, 1000, 100$$

Since our previous choice requires only 15 binary digits, compared to 27 binary digits for direct coding of the original signal $\boldsymbol{f}$, we choose the former. Thus, ignoring the costs associated with commas and dots, we have saved 44%.

We observe that the Coifman–Wickerhauser algorithm easily extends to two-dimensional images; in this application, we can, for example, treat the image as a collection of one-dimensional signals, one for each row of pixels in the two-dimensional image.

We close the discussion of Coifman and Wickerhauser's algorithm with a summary of the characteristics of the algorithm:

1. The modified wavelet transformation has a nice tree structure in which more than $2^{N/2}$ orthogonal bases can be built; and therefore, there is a large number of orthogonal bases in which the signal $\boldsymbol{f}$ can be expanded.

2. The computation is fast: it requires only $O(N \log N)$ operations to find a candidate for the minimum cost representative.

3. There is great freedom in selecting the wavelet to be used; depending on applications, one could substitute for the Haar wavelet any one of Daubechies' compactly supported wavelets, for example, or some other wavelet designed especially for the set of signals to be coded. There is a premium on using matrices $\boldsymbol{H}_k$ and $\boldsymbol{G}_k$ whose entries are even powers of 2, however.

Summary

The properties of wavelet transforms exploited in the applications to multiscale edge detection, to nonstationary signal analysis and to data compression are all different. In multiscale edge detection, the crucial property is that the wavelet transform at different scales could be interpreted as a derivative of the convolution of the signal with scaled versions of a smoothing kernel. In nonstationary signal analysis, the crucial property is that the effective window duration in a wavelet transform changes in what appears to be just the right way with scale. In data compression (at least, in the data compression algorithm we looked at), the crucial property was that the wavelet decomposition gives not one but many orthogonal decompositions of the same signal, allowing a wide variety of choices for how to code the signal.

The differences in the crucial properties for the three applications give a good list of the properties which make wavelets attractive: at each step in the wavelet transform process, we obtain a smoothed version of the smoothed version of a signal at the previous step, together with a vector which represents the difference between the new smoothed version and the smoothed version from the previous step; we view the wavelet transform process accordingly as applying orthogonal low-pass (giving a new smoothed version) and high-pass (giving the difference between the new smoothed version and the old) filters at each step. This point of view leads to applications like multiscale edge detection. Since the effective window duration of the high-pass filter becomes larger at each step in the wavelet decomposition process, we obtain automatically a "constant-Q" signal decomposition which leads to applications in nonstationary signal analysis. And since each step of the wavelet transform process gives a new orthogonal decomposition of the original signal, we obtain not one but many orthogonal decompositions in a single wavelet transform, leading to applications in data compression in which signal entropy is minimized over several alternative decompositions of the same signal.

The applications have one thing in common: the ideas involved in multiscale edge detection, in nonstationary signal analysis, and in minimum entropy coding precede wavelets. In each case, wavelets are a natural way to do efficiently something that was previously done in a less efficient manner. It is the fact that wavelet transforms are fast, more than any other single attribute, which makes them attractive in various applications.

Observe that the different properties of wavelets exploited in the three applications considered leads to the conclusion that different wavelets would be optimal in the applications. A wavelet with one sign change would be optimal for the edge detection application, whereas a wavelet which itself possessed a reasonably well-defined frequency – several sign changes at approximately equal intervals – might be optimal for the nonstationary signal analysis application, and a wavelet chosen so as to give inherently low entropy for the data being compressed – if possible, a wavelet that already looked something like a fingerprint (so long as its non-zero components were integer powers of 2, to the extent possible), in the case of the compression of the FBI fingerprint files – might be optimal for the data compression application. Because of the great flexibility offered by Daubechies' orthogonal wavelet construction technique, many wavelets are available for various applications.

References

[1-1] A. Grossman and J. Morlet, "Decomposition of Hardy functions into square integrable wavelets of constant shape," *SIAM J. Math. Anal.*, vol. 15, pp. 723-736, 1984.

[1-2] I. Daubechies, "Orthonomal bases of compactly supported wavelets," *Comm. Pure and Applied Math.*, vol. 41, pp. 909-996, 1988.

[1-3] B. Noble and J.W. Daniel, *Applied Linear Algebra*, 2nd ed., Prentice–Hall, Englewood Cliffs, 1977.

[1-4] P.R. Halmos, *Finite–Dimensional Vector Spaces*, Springer, New York, 1987.

[1-5] T.M. Apostol, *Mathematical Analysis*, 2nd ed., Addison–Wesley, Reading, 1974.

[1-6] W. Rudin, *Real and Complex Analysis*, 3nd ed., McGraw–Hill, New York, 1987.

[1-7] J. Dieudonné, *Foundations of Modern Analysis*, Academic Press, New York, 1960.

[1-8] P.R. Halmos, *Introduction to Hilbert Space and the Theory of Spectral Multiplicity*, Chelsea, New York, 1951.

[1-9] Y. Meyer, "Principe d'incertitude, bases hilbertiennes et algèbres d'opérateurs," *Seminaire Bourbaki*, no. 662, 1985.

[1-10] I. Daubechies, *Ten Lectures on Wavelets*, SIAM, Philadelphia, 1992.

[1-11] W.H. Press, B.P. Flannery, S.A. Teukolsky and W.T. Vetterling, *Numerical Recipes: The Art of Scientific Computing*, Cambridge University Press, London, 1986.

[1-12] J. Gubner and W.-B. Chang, "Wavelet transforms for discrete-time periodic signals," *IEEE Trans. Signal Proc.*, to appear, 1993.

[1-13] D. Marr and E. Hildreth, "Theory of edge detection," *Proc. Roy. Soc. Lon.*, vol. 207, pp. 187-217, 1980.

[1-14] S. Mallat and S. Zhong, "Complete signal representation with multiscale edges," Technical Report, New York University, 1991.

[1-15] S. Mallat and S. Zhong, "Characterization of signals from multiscale edges," *IEEE Trans. Patt. Anal. Machine Intell.*, vol. PAMI-14, No. 7, pp. 710-732, July 1992.

[1-16] S. Mallat and W.L. Hwang, "Singularity detection and processing with wavelets," *IEEE Trans. Inform. Theory*, vol. IT-38, No. 2, pp. 617-645, Mar. 1992.

[1-17] Y. Meyer (translated by R. Ryan), *Wavelets Algorithms and Applications*, SIAM, Philadelphia, 1993.

[1-18] T. Lindeberg, "Scale-space for discrete signals," *IEEE Trans. Patt. Anal. Machine Intell.*, vol. PAMI-12, No. 3, pp. 234-254, Mar. 1990.

[1-19] I. Daubechies, "The wavelet transform, time-frequency localization and signal analysis," *IEEE Trans. Inform. Theory*, vol. IT-36, pp. 961-1005, Sept. 1990.

[1-20] D. Slepian, "Some comments on Fourier analysis, uncertainty and modeling," *SIAM Review*, vol. 25, No. 3, pp. 379-393, July 1983.

[1-21] R. Balian, "Un principe d'incertitude fort en théorie du signal ou en mécanique quantique," *C.R. Acad. Sc. Paris*, vol. 292, série 2, 1981.

[1-22] F. Low, "Complete sets of wave-packets," in *A Passion for Physics – Essays in Honor of Geoffrey Chew*, pp. 17-22, World Scientific, Singapore, 1985.

[1-23] G. Battle, "Heisenberg proof of Balian–Low theorem," *Letter Math. Phys.*, vol. 15, pp. 175-177, 1988.

[1-24] B. Cipra, "Wavelet applications come to the fore," *SIAM News*, Nov. 1993.

[1-25] R. Coifman and M.V. Wickerhauser, "Entropy-based algorithms for best basis selection," *IEEE Trans. Inform. Theory*, vol. IT-38 No. 2, pp. 713-718, 1992.

Chapter 2

Adaptive Filtering Using Vector Spaces of Systems

GEOFFREY A. WILLIAMSON

Introduction

In a variety of practical situations demanding the electronic processing of signals, one has insufficient information in order to specify *a priori* an optimal filter. Additionally, there are instances in which the operating environment for the filter will change, and thus to maintain optimality the filter specification must track these changes. Adaptive filters are time-varying filters which autonomously assess the operating environment and then adjust themselves in response to this assessment in order to attain and retain optimality. Hence, they are candidate solutions for application situations of the type noted above. Applications in telecommunications were among the first uses of adaptive filters. Lucky introduced adaptive processing to the equalization of telephone transmission paths [2-1], and now adaptive processing in channel equalization evidences a great deal of success [2-2]. The use of adaptive filters in telecommunications occurs as well for echo cancellation [2-3], [2-4] and adaptive differential pulse code modulation [2-5]. These examples represent several among many applications within audio, image, and telecommunications processing that are amenable to adaptive filtering solutions.

Traditionally, adaptive filters have been placed in two categories, those with finite impulse responses (FIR filters), and those with infinite impulse responses (IIR filters). Adaptive FIR filters possess a key feature which makes them the adaptive filter of choice in practice: their adaptation mechanism brings the filter close or exactly to the optimal setting in a wide variety of circumstances. Adaptive IIR filters are not so well behaved, and their lack of guaranteed convergence to the optimal filter is to a large extent responsible for their limited use in applications. Nonetheless, there remains a great deal of interest in adaptive IIR filters because, with adequate adaptation properties, they would achieve much improved performance, relative to their FIR counterparts, for a given level of computational complexity.

In this chapter, we reconcile the convergence properties of adaptive FIR filters with the performance capabilities of adaptive IIR filters by approaching adaptive filtering from a vector space viewpoint. From one perspective, adaptive FIR filters work well because their adaptivity is based on a "nice" minimization problem. These filters usually minimize either a least squares or a least mean-square cost criterion, conducting the minimization over the set of filters having a finite impulse response. These cost criteria are quadratic functions of the filters' impulse response coefficients. Furthermore, the filter set has a vector space structure: summing two FIR filters of length N or less yields a similar filter, as does a scalar multiple of such a filter. With the impulse response coefficients as a parameterization of this filter vector space, the adaptation problem is now seen as minimization of a quadratic cost function over a vector space, a problem which is readily solved. The iterative techniques that are common in adaptive FIR filtering work well because they effectively descend the quadratic cost function surface to the optimal value at the bottom [2-8].

These same techniques would work just as well for minimizing a quadratic cost over any vector space. Hence, the same convergence properties accrue to adaptive filters that minimize least squares or least mean-square cost function, but that do the minimization over an arbitrary vector space of filters, including filters having an infinite impulse response. This class of adaptive filters we term *vector space adaptive filters* (VSAFs), that is filters whose adaptation varies the filter description within a vector space of systems. The advantage of switching the vector space from that of N-point, FIR filters is the potential to achieve a smaller cost for the same number of adapted parameters. Basing one's choice of the filter vector space, and its parameterization or basis, on available *a priori* knowledge of the expected optimal filter will enable significant performance advantages over adaptive FIR filters to be obtained. Thus, the VSAF formalism creates two new design parameters in the selection of an adaptive filter structure: choice of the vector space, and choice of its basis.

Note that for a given adaptive filter complexity, the modeling capabilities of VSAFs remain less than those of adaptive IIR filters, and thus their achievable performance will not be as great as for IIR filters. However, the performance advantage of VSAFs are obtained while maintaining guaranteed convergence in the adaptation, in contrast to adaptive IIR filters.

Below, we review the fundamentals of gradient-descent based adaptive filtering, using a standard algorithm for the filter adaptation. Our use of the adaptive FIR filter as the archetype for the development of this review then allows the generalization to VSAFs to come naturally. We also describe an important special form of the VSAF: the *fixed pole adaptive filter* (FPAF). FPAFs include adaptive FIR filters as special cases, and some newer filter structures based on orthogonal polynomials. We show that the FPAF framework allows for user choice in the adaptive filter specification without increasing the computational burden relative to that of the adaptive FIR filter. The possibility for more general VSAF formulations, including VSAFs based on wavelets (see Chapter 1 for a general introduction to wavelets), is also noted.

In what follows, we develop a basic paradigm for vector space selection to improve the adaptive filter performance, based on *a priori* information about the filter's operating circumstances. This paradigm converts the task of vector space selection to an optimization problem. The choice of basis for the filter vector space can affect the ability of the adaptive filter to track time variations in the optimal filter parameters; one may optimize the choice of basis based on this measure of filter performance. The choice of basis relates as well to effects of the filter excitation and the algorithm "step size" parameters, which are characteristics known to affect filter performance.

The chapter is concluded with application examples to channel equalization and to long-distance echo cancellation, which evidence the ability of the VSAF formulation to improve adaptive filter performance.

2.1 Fundamentals of Adaptive Filtering

The basic structure for an adaptive filter is depicted in Fig. 2.1-1. The filter is indicated by $G(z, \boldsymbol{b})$, an element of a set of possible filtering operations that is parameterized by a vector $\boldsymbol{b}$. For instance, if the filtering operation is an FIR filter of length N, then Eq. (2.1-1) below indicates the particular form for $G(z, \boldsymbol{b})$ in that case. The parameter vector $\boldsymbol{b}$ is adapted in time with the goal of minimizing the mean-square error between the filter output $y(n)$ and a desired filter output $d(n)$, given that the input to the filter is $u(n)$. The fundamental elements of the adaptive filtering configuration include the structure of the class of possible filters, including the parameterization of this class, and the algorithm for the updating of the parameter vector.

It is often useful to view the adaptive filtering problem as one of system identification, as shown in Fig. 2.1-2. The desired output $d(n)$ has a component depending on $u(n)$ through a linear system H, plus an additional component $v(n)$ considered to be independent of $u(l)$ for all n and l. The adaptive filter then attempts to model the "true" system H as best it can. Typically, the adaptive filter is unable to completely model H, since the complexity of the adaptive filter is usually insufficient to describe all the dynamics of H. That is, for no $\boldsymbol{b}$ in the parameter set will $G(z, \boldsymbol{b}) = H(z)$ in Fig. 2.1-2. This is the case, for instance, when the adaptive filter is a finite dimensional system, while H is infinite dimensional.

The most widely used adaptive filter structure is the adaptive FIR, or transver-

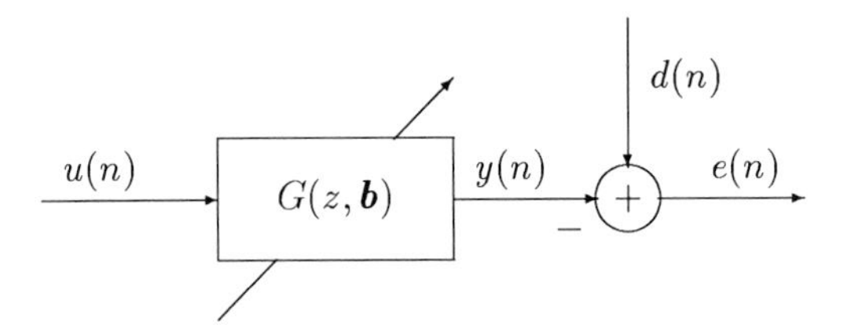

Figure 2.1-1: Basic adaptive filtering diagram.

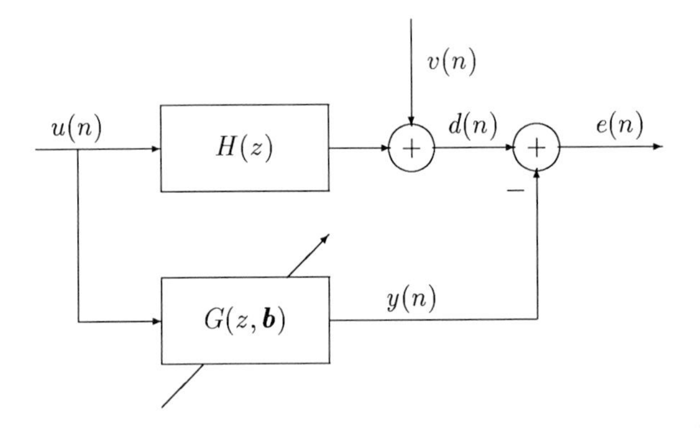

Figure 2.1-2: System identification configuration of adaptive filtering.

sal, filter. In this instance, the filter input–output relation is

$$\begin{aligned} y(n) &= b_0(n)u(n) + b_1(n)u(n-1) + \cdots + b_N(n)u(n-N) \\ &= \boldsymbol{x}^T(n)\boldsymbol{b}(n), \end{aligned}$$

where $\boldsymbol{x}^T(n) = [u(n) \ \cdots \ u(n-N)]$ and $\boldsymbol{b}(n) = [b_0(n) \ \cdots \ b_N(n)]^T$. Here, $G(z, \boldsymbol{b})$ in Fig. 2.1-2 is

$$G(z, \boldsymbol{b}) = b_0 + b_1 z^{-1} + \cdots + b_N z^{-N}. \tag{2.1-1}$$

The objective is to minimize the mean-square error (MSE)

$$E\{e^2(n)\} = E\left\{\left[d(n) - \boldsymbol{x}^T(n)\boldsymbol{b}\right]^2\right\}$$

by choice of parameters $\boldsymbol{b}$, and the optimal choice of $\boldsymbol{b}$ is given by

$$\boldsymbol{b}^* = \boldsymbol{R}^{-1}\boldsymbol{g}, \tag{2.1-2}$$

where $\boldsymbol{R}$ is the autocorrelation of $\boldsymbol{x}(n)$, and $\boldsymbol{g}$ is the crosscorrelation of $\boldsymbol{x}(n)$ and $d(n)$ [2-6].

An algorithm favored by many adaptive filtering practitioners for iteratively finding this optimal parameter vector is the *least mean squares* (LMS) algorithm [2-7], given by

$$\boldsymbol{b}(n+1) = \boldsymbol{b}(n) + \mu \boldsymbol{x}(n) e(n), \tag{2.1-3}$$

where μ is an algorithm "step size" or adaptation rate. The LMS algorithm effectively adjusts the parameters at each iteration by moving them down the gradient of the adaptive filter's "error surface" [2-8]. The error surface is the cost function $E\{e^2(n)\}$, viewed as a function of the filter parameters $\boldsymbol{b}$. This gradient descent approach works very well for the adaptive FIR filter because the error surface is quadratic in the parameters. As a consequence, there exists a unique minimum of the cost function (given by $\boldsymbol{b}^*$ in Eq. (2.1-2)), and under modest conditions (usually termed "persistent excitation" or "sufficient excitation"), one has global convergence of $\boldsymbol{b}(n)$ (in an average sense) to this optimal parameter vector [2-9], [2-10].

One measures the adaptive filter performance in terms of the MSE $E\{e^2\}$. This MSE has two components: the first is the *minimum MSE* achieved when $\boldsymbol{b}(n) \equiv \boldsymbol{b}^*$, and the second is the *misadjustment* error caused by residual motion of the parameters at convergence [2-8], [2-6]. Misadjustment arises because the adaptive filter will never be able to reconstruct $d(n)$ exactly, so that the non-zero error $e(n) = d(n) - y(n)$ causes continual adaptation of $\boldsymbol{b}(n)$ via Eq. (2.1-3). Though $\boldsymbol{b}(n)$ remains on average at $\boldsymbol{b}^*$, the residual motion increases the MSE beyond the minimum value. (Although we have presented misadjustment in the context of the LMS algorithm, any algorithm which retains adaptability for the filter will suffer from misadjustment.)

The minimum MSE can be reduced simply by increasing the number of parameters in the FIR filter. This allows longer and longer impulse responses for the filter, enabling better modeling of $H(z)$ using $G(z, \boldsymbol{b})$ in Eq. (2.1-1) for large N. However, there are two disadvantages to increasing the number of parameters. First of all, the misadjustment is proportional to the number of adapted parameters [2-11], so that a trade-off exists in this regard. Secondly, the computational complexity of the adaptive filter implementation also increases with the number of parameters. These facts motivate the desire for an adaptive filter structure which enables achievement of a reasonably small minimum MSE, while using relatively few adapted parameters.

A candidate adaptive filter structure meeting these requirements is the adaptive IIR filter. The filter input–output relationship is now

$$\begin{aligned} y(n) &= b_0(n)u(n) + b_1(n)u(n-1) + \cdots + b_N(n)u(n-N) \\ &\quad + a_1(n)y(n-1) + \cdots + a_N(n)y(n-N) \\ &= \boldsymbol{x}^T(n)\boldsymbol{b}(n), \end{aligned} \tag{2.1-4}$$

with $\boldsymbol{x}(n)$ and $\boldsymbol{b}(n)$ now redefined to include the past output values $y(n-i)$ and the parameters $a_i(n)$, respectively. Using an adaptive filter with poles and zeros increases the flexibility in modeling the relationship between $\{u(n)\}$ and $\{d(n)\}$, and hence should yield a smaller minimum MSE given an equivalent number of parameters.

The adaptation of the IIR filter parameters poses some difficulty, however. Notice that the expression for $y(n)$ in Eq. (2.1-4) is *nonlinear* in the parameters $\boldsymbol{b}(n)$ due to the dependence of $\boldsymbol{x}(n)$ on the parameter values. This fact results in an error surface that is no longer quadratic in the parameters, so that the potential for multiple minima in this surface arises. Many such minima can be local and not global, and gradient descent algorithms may converge to these local minima, resulting in suboptimal performance of the adaptive filter [2-12]. Some work has gone into the study of error surface geometry for adaptive IIR filters [2-13], [2-14], [2-15], yet a good understanding of the geometry is still lacking. In situations for which the adaptive IIR filter can *exactly* model the relationship between $u(n)$ and $d(n)$, it has been shown that the error surface is unimodal (possesses a unique local minimum), given an input sequence $u(n)$ derived as a linearly filtered white signal [2-15] (in addition, the number of zeros of the adaptive filter must be at least as great as $N+1$, where N is the number of poles in the "true" system H). However, such conditions for unimodal error surfaces rarely hold in practice, so that few guarantees for optimal performance of adaptive IIR filters have been established.

Another difficulty with gradient descent based adaptive IIR filters is the requirement to check their stability at each iteration of the adaptation [2-16]. The necessity of constraining the (instantaneous) adaptive filter specification to be stable can cause "lock up" on the stability boundary. That is, if the filter parameters consistently want to update to unstable values (with the adaptation algorithm consistently preventing this), the parameters can be frozen on the stability boundary. This effect is manifested in convergence analyses that prove only convergence either to a local minimum of the error surface or to a point on the stability boundary [2-17].

We can then see that, in spite of the potential benefits of reduced filter complexity that adaptive IIR filters provide for a given computation burden, the gradient descent approach to their adaptation has a number of shortcomings. What is desired is an adaptive filter structure with the convergence properties of adaptive FIR filters, but which possesses a broader modeling capability than FIR filters. The vector space approach to adaptive filtering, discussed here, provides such a structure.

2.2 The Vector Space Adaptive Filter

The vector space approach to adaptive filters is captured by the diagram in Fig. 2.2-1, which depicts what we term a *vector space adaptive filter* (VSAF) [2-18]. The filter output $y(n)$ is the adaptively weighted sum of N intermediate outputs $\{x_i(n)\}$. Each output $x_i(n)$ is produced by passing the input $u(n)$ through a time-invariant system denoted by $G_i(z)$.

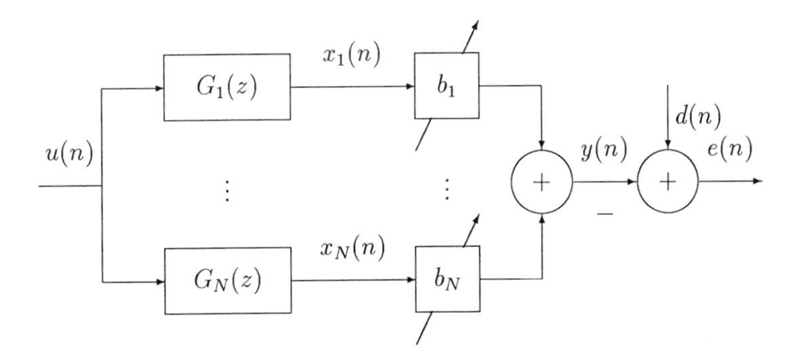

Figure 2.2-1: Vector space adaptive filter.

As usual, the weights $b_i(n)$ are adjusted in order to minimize the mean-square error (MSE) between the filter output $y(n)$ and some desired output signal $d(n)$. Since $y(n)$ is a linear function of the adapted parameters $\{b_i(n)\}$, we have preserved the linear-in-the-parameters formulation that guarantees good behavior for adaptive FIR filters. The behavior of adaptive algorithms for VSAFs will be discussed below in Section 2.3.

The structure of the VSAF shown in Fig. 2.2-1 allows a wide range of possibilities for choice of the fixed filters $\{G_i(z)\}$. Specifying these systems fixes the vector space of systems

$$\mathcal{G} = \left\{ \sum_{i=1}^{N} b_i G_i(z) : b_i \in \Re \right\} \tag{2.2-1}$$

over which the adaptation occurs. The set $\{G_i(z)\}_{i=1}^{N}$ forms a basis for the vector space $\mathcal{G}$ given in Eq. (2.2-1); that is, any member of $\mathcal{G}$ can be uniquely written as a linear combination of the basis elements. Hence, we refer to each $G_i(z)$ as a *basis system* for the vector space. It is important to stress that almost *any* choice of basis systems is admissible in this context. We will focus our attention here on choices which are linear systems; this is reflected in our notation, which implies that each basis system has a transfer function $G_i(z)$. Nonetheless, it is possible to let the G_i's be nonlinear systems. For instance, adaptive filters based on Volterra [2-19] and polynomial [2-20] models fit this paradigm. Also, some neural network structures [2-21], such as single-layer neural networks, can be placed within the VSAF framework.

The two key structural features of a VSAF are first, the choice of vector space $\mathcal{G}$ itself, and second, a choice of basis for the vector space. Selecting a vector space is equivalent to choosing the model set for the identification, and selecting the basis is equivalent to specifying the parameterization for the model set.

A. Fixed Pole Adaptive Filters

A particularly important subset of VSAFs is the *fixed pole adaptive filter* (FPAF), which we have introduced in [2-22] and [2-23]. This class of adaptive filter includes several adaptive filtering structures which have been considered to a modest extent in the literature, as discussed below. The FPAF may be implemented in the modular structure depicted in Fig. 2.2-2. Each block in Fig. 2.2-2 is either a single, real pole section, or a complex-conjugate pole pair section, as shown in Figs. 2.2-3 and 2.2-4, respectively. The adjustable parameters in the FPAF are the b_i coefficients; the pole locations, given by a_i, are fixed and remain unadapted. In [2-24], it is shown that any transfer function of the form

$$H(z) = \frac{c_0 z^N + c_1 z^{N-1} + \cdots + c_N}{(z-a_1)(z-a_2)\cdots(z-a_N)} \tag{2.2-2}$$

can be realized by the FPAF. Therefore, one sees that this filter structure can adapt to any Nth order transfer function that has poles constrained to locations selected by the adaptive filter designer.

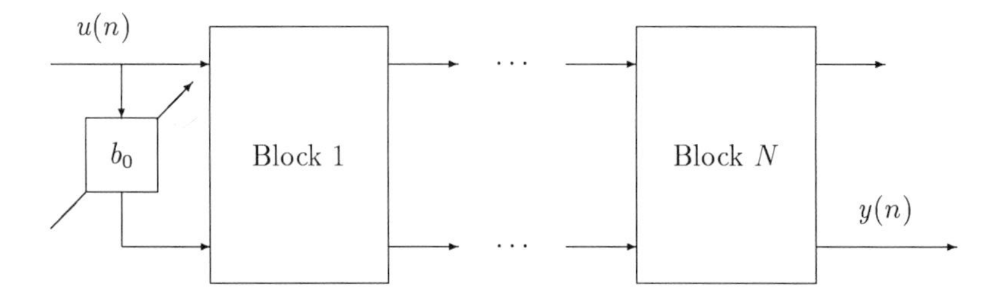

Figure 2.2-2: Block structure of the FPAF.

The motivation of the FPAF structure is as follows. If the adaptive filter's poles are chosen reasonably close to the dominant poles of the optimal filter $H(z)$, then a small mean-square prediction error can be achieved for a relatively small value of N.

A number of special cases of FPAFs have been previously studied. Each case is characterized by a particular choice of pole locations, or families of pole locations, for the filter. We enumerate these special cases below.

1. **FIR filters:** By far the most widespread and most often studied FPAF is the adaptive FIR filter. Typically one thinks of FIR filters as "all-zero filters", but in actuality an N-point FIR filter will have N poles at $z = 0$. Setting each $a_i = 0$ in Eq. (2.2-2) yields the class of FIR filter transfer functions. One can

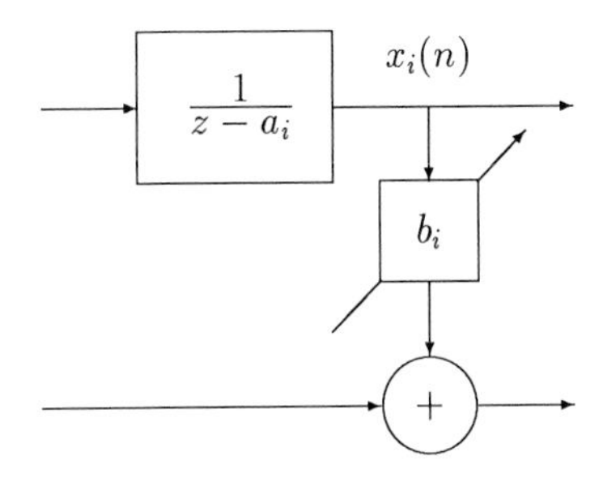

Figure 2.2-3: FPAF block with real-valued pole.

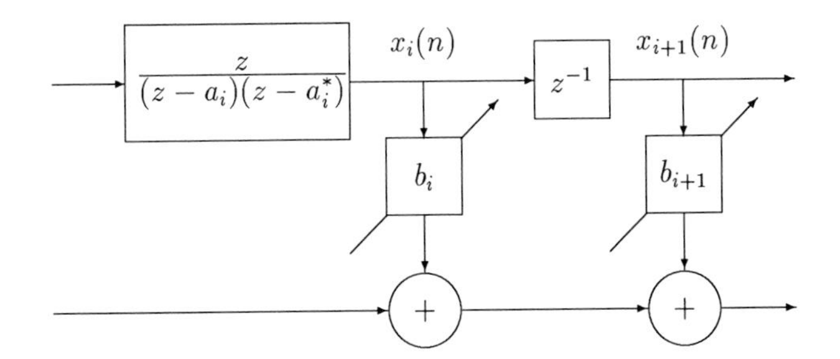

Figure 2.2-4: FPAF block with complex-conjugate pole pair.

also see in Fig. 2.2-3 that when each $a_i = 0$, we recover the tapped-delay-line structure commonly associated with adaptive FIR filters.

2. **Laguerre filters:** By setting each $a_i = a$ for some $-1 < a < 1$, we attain the class of adaptive Laguerre filters. Laguerre representations for the approximation of stable linear systems were considered in [2-25, 2-26], and also applied to echo cancellation in [2-27]. The basis for the Laguerre vector space of systems as in [2-25] is given by

$$\left\{ \frac{1}{z-a}, \frac{1-az}{(z-a)^2}, \ldots, \frac{(1-az)^{N-1}}{(z-a)^N} \right\} \tag{2.2-3}$$

(though different basis functions result from the structure in Fig. 2.2-1). The transfer functions of Eq. (2.2-3) are the z-transforms of discrete Laguerre polynomials (though we have rescaled the resulting transforms slightly). It can be shown that by taking the system order N large enough, one can arbitrarily well approximate any stable linear system as a linear combination of Laguerre filters [2-25]. Knowledge of dominant system time constants is useful in the selection of a. If a reflects the dominant time constants, then this expected system characteristic is built into the model, and a good approximation can be achieved with a relatively small value for N. Note as well that by setting $a = 0$, we recover within the Laguerre class the FIR filter formulation.

3. **Legendre filters:** The Legendre filters are similar to the Laguerre filters in the sense that they too come from transforms of complete, orthogonal polynomials, in this case the discrete Legendre polynomials. The fixed pole locations for this set of filters are given by

$$a_i = a^i, \quad i = 1, \ldots, N, \qquad 0 < a < 1.$$

Adaptive filters using the Legendre representation were studied in the context of echo cancellation in [2-27] and [2-28]. As with the Laguerre representation, the selection of a will depend on knowledge of the dominant time constant of the desired filter specification, though the other system poles are spread along the real axis of the z-plane.

4. **Kautz filters:** In the Kautz filter description, the fixed poles are all located at one complex-conjugate pair of pole positions. In this sense, Kautz filters are something like a complex-conjugate version of the Laguerre filter. Kautz representations have been studied for system identification in [2-29]. It is also mentioned in [2-25] that the possibility exists of combining different Laguerre and Kautz stages into one identification structure. One should select fixed pole locations for the complex-conjugate Kautz positions and real Laguerre positions in accordance with expected pole locations in the unknown system. This approach comes close to the general FPAF structure proposed herein.

5. **Gamma filters:** The adaptive filter structure proposed in [2-30], termed a "generalized feedforward filter", lies within the VSAF class, and a special case discussed in [2-30] reduces to an FPAF. The basis functions for the filter vector space in a generalized feedforward filter are

$$\left\{G(z), G^2(z), \ldots, G^N(z)\right\}$$

for some fixed, stable, transfer function $G(z)$. In the special case of the "gamma filter" discussed in [2-30], one has

$$G(z) = \frac{\mu}{z - (1 - \mu)},$$

which one may observe to be equivalent to a Laguerre filter with $a = 1 - \mu$, but with a coordinate transformation on the filter vector space.

B. Other VSAF Structures with Linear Models

It is readily possible to construct a range of VSAF structures, simply by specifying the basis elements $G_i(z)$ in Eq. (2.2-1). One approach to selecting the basis systems $G_i(z)$ is to optimize over all, or over a class of, linear system descriptions. An optimization procedure of this type appears below in Section 2.5. It is essential, however, to capture some or all of the features of the desired filter specification within the vector space $\mathcal{G}$.

One broad class of models for the VSAF which merits exploration is that derived from wavelet transforms. A review of wavelets in signal processing can be found in Chapter 1. For identification and its adaptive filtering analogues, one views wavelets as a way of decomposing a *system* description, rather than simply a signal, in time and frequency. A study of the use of wavelet transforms in the approximation and identification of continuous time systems appears in [2-31], and discrete-time analogues may be similarly developed. Such decompositions could give insight into appropriate choices for a vector space model set, so as to get good, low-order approximations for the class of expected optimal filters.

2.3 Algorithm Convergence and Asymptotic Performance

Of critical importance to the performance of an adaptive filter is the good behavior of the algorithms adjusting the filter parameters. One desires that the filter parameters are brought close to their optimal values by the adaptation process. Once near the optimal settings, one can assess performance via the asymptotic behavior of the parameters. These issues we consider in this section for several commonly employed adaptive algorithms. As we have noted earlier, under modest excitation conditions, one obtains algorithm convergence for VSAFs using standard algorithms; below we make this notion precise. We then consider the asymptotic behavior, and show its relationship to the VSAF vector space and its basis. These two entities are user defined, so that their selection in terms of asymptotic performance is a design option built into the VSAF structure.

First, we set up the algorithms. With respect to Fig. 2.2-1, define the parameter vector

$$\boldsymbol{b}(n) = \begin{bmatrix} b_1(n) & \cdots & b_N(n) \end{bmatrix}^T \tag{2.3-1}$$

and the regressor vector

$$\boldsymbol{x}(n) = \begin{bmatrix} x_1(n) & \cdots & x_N(n) \end{bmatrix}^T. \tag{2.3-2}$$

Further define the prediction error $e(n) = d(n) - y(n)$, where $y(n)$ is the output and $d(n)$ is the desired signal to be obtained by filtering $u(n)$.

Then the *least mean squares* (LMS) algorithm is

$$\boldsymbol{b}(n+1) = \boldsymbol{b}(n) + \mu \boldsymbol{x}(n) e(n), \tag{2.3-3}$$

the *normalized least mean squares* (NLMS) algorithm is

$$\boldsymbol{b}(n+1) = \boldsymbol{b}(n) + \frac{\mu}{\epsilon + \boldsymbol{x}^T(n)\boldsymbol{x}(n)} \boldsymbol{x}(n) e(n), \tag{2.3-4}$$

and the *recursive least squares* (RLS) algorithm is

$$\boldsymbol{b}(n+1) = \boldsymbol{b}(n) + \frac{1}{\lambda + \boldsymbol{x}^T(n)\boldsymbol{P}(n)\boldsymbol{x}(n)} \boldsymbol{P}(n)\boldsymbol{x}(n)e(n) \tag{2.3-5}$$

$$\boldsymbol{P}(n+1) = \frac{1}{\lambda}\left[\boldsymbol{P}(n) + \frac{\boldsymbol{P}(n)\boldsymbol{x}(n)\boldsymbol{x}^T(n)\boldsymbol{P}(n)}{\lambda + \boldsymbol{x}^T(n)\boldsymbol{P}(n)\boldsymbol{x}(n)}\right]. \tag{2.3-6}$$

For LMS and NLMS, μ is the adaptation step size and ϵ is a small, positive constant. In RLS, λ is the forgetting factor for RLS, and $\boldsymbol{P}(0)$ is initialized to some positive definite value.

A. Algorithm Convergence

We now show that for LMS, NLMS, and RLS, we achieve asymptotic convergence to an optimal coefficient vector given satisfaction of an excitation condition on the regressor vector $\boldsymbol{x}(n)$. The following discussion is based in part on the work in [2-24]. Define $\boldsymbol{R} = E\{\boldsymbol{x}(n)\boldsymbol{x}^T(n)\}$ and

$$\boldsymbol{R}_L(l) = \sum_{n=l+1}^{l+L} \boldsymbol{x}(n)\boldsymbol{x}^T(n). \tag{2.3-7}$$

Positive definiteness of $\boldsymbol{R}$ or $\boldsymbol{R}_L(l)$ (uniformly in l) leads to adaptive filter convergence. For instance, the following convergence results are standard.

Stochastic Convergence Result: (See, for example, [2-6].) Given a stationary random environment, let

$$\boldsymbol{b}^* = \arg\ \inf E\left\{\left(d(n) - \boldsymbol{x}^T(n)\boldsymbol{b}\right)^2\right\},$$

and assume that $\beta \boldsymbol{I} \geq \boldsymbol{R} \geq \alpha \boldsymbol{I} > 0$. Then under LMS with $0 < \mu < 2/\mathrm{tr}(\boldsymbol{R})$, under NLMS with $0 < \mu < 2$ and $\epsilon > 0$, and under RLS with $0 < \lambda \leq 1$, $E\{\boldsymbol{b}(n)\}$ converges to $\boldsymbol{b}^*$, and $E\{e^2(n)\}$ converges. ∎

Deterministic Convergence Result: (See, for example, [2-32].) Let $d(n) = \boldsymbol{x}^T(n)\boldsymbol{b}^* + w(n)$, with $\{w(n)\}$ a bounded sequence, and assume that for some L, $\beta \boldsymbol{I} \geq \boldsymbol{R}_L(l) \geq \alpha \boldsymbol{I} > 0$ holds for all l. Then under LMS with $0 < \mu < 2/\beta$, under

NLMS with $0 < \mu < 2$ and $\epsilon > 0$, and under RLS with $0 < \lambda \leq 1$, $\boldsymbol{b}(n)$ converges to a ball about $\boldsymbol{b}^*$ with radius proportional to μ (for LMS or NLMS) or $1 - \lambda$ (for RLS). ∎

It remains to establish conditions on the input signal $u(n)$ that guarantee satisfaction of the excitation condition on the regressor $\boldsymbol{x}(n)$ in the VSAF. This is accomplished using the results of [2-33].

Definition 2.3-1 [2-33] The signal $\{\boldsymbol{x}(n)\}$ is said to be *persistently exciting* if

$$\lim_{L\to\infty} \inf \frac{1}{L}\left(\sum_{n=1}^{L} \boldsymbol{x}(n)\boldsymbol{x}^T(n)\right) > 0.$$

The signal $\{\boldsymbol{x}(n)\}$ is said to be *exciting over an interval* $[l+1, l+L]$ if for some constant $K > 0$,

$$\sum_{n=l+1}^{l+L} \boldsymbol{x}(n)\boldsymbol{x}^T(n) \geq KI.$$

∎

Define the input regressor vector

$$\boldsymbol{u}(n) = [\ u(n) \quad u(n-1) \quad \cdots \quad u(n-N+1)\ ]^T. \tag{2.3-8}$$

Now our task is to relate excitation conditions on $\boldsymbol{u}(n)$ to those of $\boldsymbol{x}(n)$.

Theorem 2.3-1 [2-24] Let the $N \times 1$ transfer function

$$\boldsymbol{G}(z) = \left[\ G_1(z) \quad \cdots \quad G_N(z)\ \right]^T$$

have full row rank, and let its McMillan degree be m.[1] Under this condition, $\boldsymbol{x}(n)$ is exciting over the interval $[l+1-m, l+L]$ if $\boldsymbol{u}(n)$ is exciting over the interval $[l+1, l+L]$. Furthermore, $\boldsymbol{x}(n)$ is persistently exciting if $\boldsymbol{u}(n)$ is. ∎

The result of Theorem 2.3-1 is well known for the case when $G_i(z) = z^{-i}$, corresponding to the tapped delay line of the adaptive FIR filter. In such a situation, we have $\boldsymbol{x}(n) = \boldsymbol{u}(n)$. For a more general VSAF with linear basis elements, the translation of excitation in the input to excitation in the regressor as given in Theorem 2.3-1 essentially requires the basis elements $\{G_1(z), \ldots, G_N(z)\}$ to be linearly independent. Clearly, a lack of linear independence among $\{G_1(z), \ldots, G_N(z)\}$ implies a redundancy in the VSAF description, which would be undesirable in any case. For the FPAF case, the structure of the filter in Figs. 2.2-2–2.2-4 prevents such redundancy from occurring. We note this specialization of Theorem 2.3-1 below.

[1]The McMillan degree of a transfer function is the number of states in a state space realization of the smallest possible size (a minimal realization).

Theorem 2.3-2 [2-24] The regressor $\boldsymbol{x}(n)$ in an N-pole FPAF is exciting over the interval $[l+1-N, l+L]$, or is persistently exciting, if the same condition holds for the input regressor $\boldsymbol{u}(n)$. ∎

Remark 2.3-1 A frequency domain condition for $\boldsymbol{u}(n)$ to be exciting over all length L intervals for some L is that $\{u(n)\}$ has $n+1$ spectral lines [2-34]. That is, a deterministic $u(n)$ must have at least $(n+1)/2$ different sinusoidal components, and a stochastic $u(n)$ must have a spectrum whose support contains at least $n+1$ distinct points. Such a $\boldsymbol{u}(n)$ is also persistently exciting. ∎

Remark 2.3-2 For stationary, ergodic signals, if $\boldsymbol{x}(n)$ is persistently exciting as defined above, then $R = E\{\boldsymbol{x}(n)\boldsymbol{x}^T(n)\} > 0$. ∎

In sum, the above theory shows that if the filter input $u(n)$ is exciting, then the VSAF will converge to the optimal coefficients. The conditions under which this occurs are precisely the same as those for convergence of adaptive FIR filters, given satisfaction of the hypotheses of Theorem 2.3-1. For FPAFs, by Theorem 2.3-2, algorithm convergence conditions are identical to those for adaptive FIR filters.

B. Asymptotic Frequency Domain Performance

Properties regarding the asymptotic performance of these identification schemes are easily obtained by adapting standard results that are found, for instance, in [2-35]. Assume $\{u(n)\}$ and $\{v(n)\}$ independent, and let $E\{v(n)\} = 0$. Let $d(n)$ and $u(n)$ be related by

$$d(n) = H(q)\,[u(n)] + v(n),$$

as depicted in Fig. 2.1-2, with $H(q)$ the system operator corresponding to transfer function $H(z)$ (q^{-1} is the delay operator). Use $G(q, \boldsymbol{b}(n))$ to denote the system operator for the VSAF at time n, with parameter vector $\boldsymbol{b}(n)$. Given that $\boldsymbol{b}(n)$ converges to $\boldsymbol{b}^*$ under the adaptive algorithm, we have

$$G^*(z) = b_1^* G_1(z) + \cdots + b_N^* G_N(z)$$

as the asymptotic VSAF transfer function. Note that $\boldsymbol{b}^* = [b_1^* \ \cdots \ b_N^*]^T$ is the asymptotic parameter vector with respect to the choice of basis $\{G_1, \ldots, G_N\}$.

Definition 2.3-2 [2-36], [2-35] The asymptotic *bias error* over frequency achieved by the identification is

$$E_B(\omega) = H(e^{j\omega}) - G^*(e^{j\omega}).$$

∎

Definition 2.3-3 [2-35] The asymptotic *variance error* over frequency achieved by the identification is

$$E_V^2(\omega) = \lim_{n\to\infty} E\left\{\left|H(e^{j\omega}) - G(e^{j\omega}, \boldsymbol{b}(n))\right|^2\right\}.$$

∎

An item of interest is the manner in which user choices affect these two types of identification error.

Theorem 2.3-3 [2-18] Under the conditions given above, $E_B(\omega)$ is a property only of H, the spectrum of u, and the model set $\mathcal{G}$. ∎

Theorem 2.3-3 stresses the role of $\mathcal{G}$ in determining properties of the optimal identified model G^* relative to H. The bias error is the distribution of the minimum MSE over frequency. Hence, the choice of $\mathcal{G}$ will not only influence the value of the minimum MSE, but also its frequency dependence. In some ways it is possible to choose each G_i (which in turn define $\mathcal{G}$) so as to match the general shape of its frequency response to what is expected in the frequency response of H. This is described in [2-25], for the case when $\mathcal{G}$ arises from the class of Laguerre based model sets.

Theorem 2.3-4 [2-18] For LMS adaptation with $0 < \mu \ll 2/\lambda_{\max}(\boldsymbol{R})$, $E_V^2(\omega)$ is given approximately as

$$E_V^2(\omega) = \frac{\mu\sigma^2}{2}\sum_{i=1}^{N}|G_i(e^{j\omega})|^2, \tag{2.3-9}$$

where σ^2 is the minimum MSE for the identification. ∎

The implication of Theorem 2.3-4 is that selection of a *basis* for $\mathcal{G}$ will affect the variance error. In fact, once $\mathcal{G}$ is selected, we can see from Eq. (2.3-9) that the choice of basis is the sole determining factor for the variance error when using LMS. Using $\{G_1, \ldots, G_N\}$ as the set of fixed, linear systems in the VSAF implicitly chooses a basis for $\mathcal{G}$; one can shape the variance error by a simple coordinate change. Thus, not only the model set for the identifier configuration, but also the basis for that model set, will affect the asymptotic performance characteristics. More will be said regarding the choice of basis in Section 2.5.

2.4 Choosing the Vector Space

We have mentioned that to select the VSAF vector space advantageously, one requires *a priori* information about the filter's operating environment. The nature of such information will depend on the particular application in question. "First principles" information regarding the optimal filter may suggest a particular structure for the filter vector space. More commonly, however, information about the operating environment will arise via experimentally obtained data. In the remainder of this section, we discuss several techniques for selecting the VSAF vector space that depend on the acquisition of "example systems." An example system is an optimal filter configuration for one circumstance of the operating environment. An example system description may come from a prior, off-line, identification study of the adaptive filtering environment. Assembling many of these example systems for a variety of operating conditions provides a picture of the set of possible optimal filter specifications. From this picture, an appropriate, parsimoniously parameterized vector space may be derived.

As an example, consider the problem of echo cancellation in long-distance telecommunications. There, the adaptive filter essentially models the echo path in the telephone line. Prior to actual use of the adaptive filter, one may collect input–output data for echo paths by studying a sample long-distance connection, and applying standard system identification techniques to acquire a model for its echo path. Doing this for a number of different long-distance connections assembles a collection of example echo path models, which forms the set of example systems for this application. One then uses the example systems to specify an appropriate VSAF.

Throughout this section, we will assume that we have available M impulse responses $\{h_1(n)\}, \ldots, \{h_M(n)\}$ corresponding to M example systems. Clearly, the impulse response descriptions may be obtained from transfer function or state space models, and to a certain extent, the reverse is true, so that example systems in these forms may be used. Also, vector space selection methods that act directly on input–output data, or system descriptions other than impulse response form, may be developed as well.

A. Optimization of the Vector Space

The set of M example system impulse responses immediately defines an M-dimensional vector space, obtained as the span of the example systems. One approach to VSAF specification is to reduce this vector space from M to N dimensions in an optimal manner. Below we describe a method for doing so, as given in [2-38]. Throughout the development, we assume $h_i \in \ell^2$, with $\|h\| = \left(\sum_n |h(n)|^2\right)^{1/2}$ and $\langle g, h\rangle = \sum_n g(n)h(n)$ the usual norm and inner product for ℓ^2.

Define an orthonormal basis $V = \{v_1, \ldots, v_M\}$ for $\mathcal{H} = \text{span}\{h_1, \ldots, h_M\}$, and let

$$h_i = \sum_{j=1}^{M} c_{ij} v_j. \tag{2.4-1}$$

Denote by $\boldsymbol{c}_i = [c_{i1} \ \cdots \ c_{iM}]^T$ the coordinate vector of h_i with respect to this basis.

Theorem 2.4-1 [2-38] Let $N \leq M$. Given $h_1, \ldots, h_M$ whose coordinates with respect to V are defined by Eq. (2.4-1), define N unit norm systems $g_1, \ldots, g_N$ by the coordinate vectors (with respect to V) which are the unit eigenvectors of

$$\boldsymbol{C} = \sum_{i=1}^{M} \boldsymbol{c}_i \boldsymbol{c}_i^T$$

corresponding to the N largest eigenvalues. Then this choice of $g_1, \ldots, g_N$ minimizes

$$\sum_{i=1}^{M} \min_{\alpha_1, \ldots, \alpha_N} \left\| h_i - \sum_{j=1}^{N} \alpha_j g_j \right\|^2 \tag{2.4-2}$$

over choice of $\{g_j\}$. ∎

Theorem 2.4-1 motivates a choice of basis for a vector space adaptive filter as $\mathcal{G} = \text{span}\{G_1(z), \ldots, G_N(z)\}$, where $G_i(z)$ is the system transfer function associated with the impulse response $g_i(n)$. We can interpret this choice of $\mathcal{G}$ in terms of the system identification setup of the VSAF depicted in Figure 2.1-2. Given that the input sequence is white, the mean-square estimation error $E\{e^2(n)\}$ is given by

$$\sum_{j=1}^{N} \|h_i - \langle g_j, h_i \rangle g_j\|^2$$

when the true system is h_i and the model set is $\mathcal{G}$. Therefore, Eq. (2.4-2) represents the sum of the minimum mean-square identification errors over $\{h_1, \ldots, h_M\}$ when using the model set $\mathcal{G}$. The choice of $\mathcal{G}$ specified by Theorem 2.4-1 minimizes this total, minimum MSE.

If in the identification scenario the input signal is colored, then the optimization of $\mathcal{G}$ as in Theorem 2.4-1 is no longer appropriate as the interpretation above does not hold. Let us consider the case in which $u(n)$ arises as a linearly filtered white sequence as depicted in the identification set-up of Fig. 2.4-1, and assume that the system $F(z)$ by which the white sequence is filtered possesses a stable inverse. Let $V = \{v_1, \ldots, v_N\}$ now be the basis for $\overline{\mathcal{H}} = \text{span}\{f * h_1, \ldots, f * h_n\}$, with f the impulse response of $F(z)$, and let $\overline{\boldsymbol{c}}_i$ be the coordinate vector of $f * h_i$ with respect to V. Then we have the following:

Theorem 2.4-2 [2-38] The total minimum MSE resulting from identification of $h_1, \ldots, h_M$ is minimized by setting

$$\mathcal{G} = \left\{ f^{-1} * \overline{g}_1, \ldots, f^{-1} * \overline{g}_N \right\},$$

where $\overline{g}_1, \ldots, \overline{g}_N$ are the systems in $\overline{\mathcal{H}}$ whose coordinates with respect to V are unit eigenvectors of

$$\overline{\boldsymbol{C}} = \sum_{i=1}^{M} \overline{\boldsymbol{c}}_i \overline{\boldsymbol{c}}_i^T$$

corresponding to the N largest eigenvalues. ∎

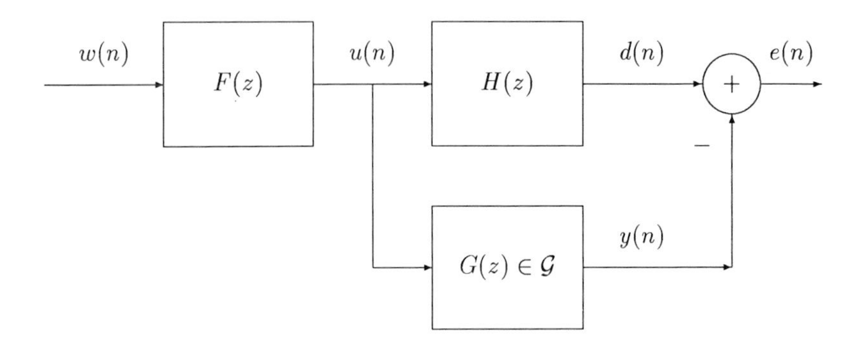

Figure 2.4-1: Parallel identification with colored input.

One possible drawback in the above scheme arises when the VSAF is to be implemented. We have motivated the VSAF approach as one which can provide potential savings in computational complexity when compared with adaptive FIR filters. The above scheme will do so if complexity is adequately measured only in terms of the number of adapted parameters. If, however, the complexity of implementation of the basis systems is a factor, then the realizations of the impulse responses $g_1, \ldots, g_N$ may be unwieldy. To avoid this problem, one can restrict consideration to basis systems that are implementable in simple structures. The FPAF considered in the following subsection meets this criterion.

B. Selecting Poles for an FPAF

The FPAF structure described in Section 2.2 (see Figs. 2.2-2–2.2-4) has the advantage of a simple implementation complexity, on the same order as an adaptive FIR filter [2-24], while permitting flexibility in terms of user choice of its pole locations. When choosing a vector space for an FPAF, one need only specify the fixed pole locations that go into that structure. In this section, we demonstrate how the example system format for *a priori* knowledge about the adaptive filtering environment enables one to intelligently select these N pole locations. A simple analytic approach to this problem via Prony analysis may be found in [2-23]; other more sophisticated ideas appear in [2-24]. Here, we summarize the contributions of [2-24] and the references therein.

A paradigm we adopt for the selection of the N poles is represented in Fig. 2.4-2. One starts with a collection of M example systems, here assumed to be in impulse response form. The objective is to extract information from these example systems in terms of the N pole locations, represented in Fig. 2.4-2 as the bottom block. Some of the techniques we discuss require an intermediate stage. One such intermediate stage in Fig. 2.4-2 represents the reduction of the M example systems into a smaller number M_1 of system descriptions. These could be obtained by the procedure of the last subsection, for instance. This intermediate stage can be seen as a second starting point for the pole selection procedure, in that all the options which were originally available are still possible at this stage.

A second intermediate stage in Fig. 2.4-2 represents one single-input, M-output, Nth order system. This intermediate system description figures prominently in the procedures to be discussed below, and we will focus on this path through the diagram. It represents the most important step in the process, as once the example systems have been reduced to this form, the N fixed poles of the FPAF are readily selected as the poles of this system. In some cases, one may move directly between the example system descriptions and the pole locations, though in general one at least implicitly defines the system in this intermediate stage of Fig. 2.4-2.

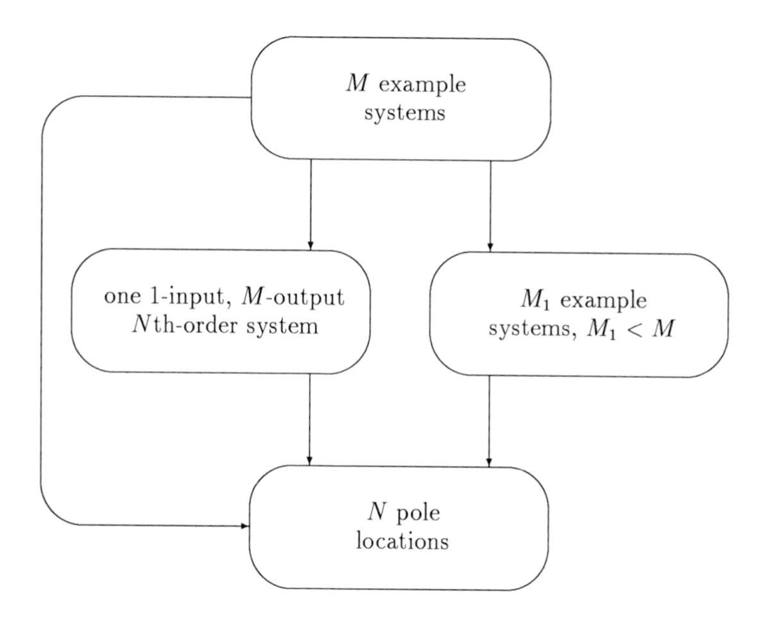

Figure 2.4-2: Paradigm for pole selection.

To make the step to the intermediate stage with the single-input, M-output, Nth order system, we are doing the following. We wish to approximate each of the M example systems with one of the input-to-output relationships of the Nth order system. All the systems share a common input; each of the outputs of the intermediate system description approximates one of the outputs of the example system. We can view this problem as one of system identification, where we are modeling the collection of example systems via one system whose dynamics we identify. Alternatively, we may think of this as a problem of model reduction. Here, the example systems are collected into one single-input, M-output system whose states are the union of the states of all the example systems. We then reduce this system description, using standard model reduction techniques, to one which has only N states. Finally, one may view this problem as one of filter design, in which we want to find a low-order single-input, M-output filter whose impulse response is close to those of the given example systems.

One may borrow a variety of techniques from the system identification, model reduction, and filter design disciplines in order to achieve pole selection via the above paradigm. Below, we present two such techniques, one based on a system identification technique as developed in [2-39], the other based on model order reduction, as developed in [2-24]. In Section 2.6, we compare the performance of these two approaches in some example situations. At the end of this section, we also indicate some alternative methods of pole selection.

1. System Identification Based Pole Selection

Reference [2-39] describes an iterative algorithm to estimate pole locations, when the set of example systems is in impulse response form. If it is applied in the single-system case ($M = N = 1$), the method corresponds to the Steiglitz–McBride approach to system identification [2-40]. We hereafter refer to this approach as the SM-method.

Let $h_i(n)$ be the impulse response of the ith example system, and let $\hat{g}_i^{a_1,\ldots,a_N}(n)$ be the impulse response of the N-pole system having poles at $a_1, \ldots, a_N$ that best approximates $h_i(n)$ in an ℓ^2 sense. The method of [2-39] is designed to minimize, by choice of poles $a_1, \ldots, a_N$, the cost function

$$J = \sum_{i=1}^{M} \rho_i^2 \left\| h_i - \hat{g}_i^{a_1,\ldots,a_N} \right\|^2 ,$$

where ρ_i^2 is a weighting assigned to the fit for the ith example system. In the computations described below, the impulse responses are truncated to length L, with L taken large enough to capture most of the energy in each h_i.

The minimization procedure has the following steps:

1. Assemble the $L \times (N+1)$ matrix

$$\boldsymbol{H}_j = \begin{bmatrix} h_i(N) & h_i(N-1) & \cdots & h_i(0) \\ h_i(N+1) & h_i(N) & \cdots & h_i(1) \\ \vdots & \vdots & \ddots & \vdots \\ h_i(L-1) & h_i(L-2) & \cdots & h_i(L-N-1) \end{bmatrix}$$

and the $L \times (L+N)$ matrix

$$\boldsymbol{Q}^T(\boldsymbol{\alpha}) = \begin{bmatrix} \alpha_N & \alpha_{N-1} & \cdots & \alpha_1 & 1 & & & 0 \\ & \alpha_N & \alpha_{N-1} & \cdots & \alpha_1 & 1 & & \\ & & \ddots & \ddots & & \ddots & \ddots & \\ 0 & & & \alpha_N & \alpha_{N-1} & \cdots & \alpha_1 & 1 \end{bmatrix}$$

and define the vectors $\boldsymbol{e}_1 = [1\ 0\ \cdots\ 0]^T$ and $\boldsymbol{\alpha} = [1\ \alpha_1\ \cdots\ \alpha_N]^T$. Initialize $\boldsymbol{\alpha}(0) = \boldsymbol{e}_1$.

2. At iteration k, find

$$\begin{aligned} \boldsymbol{F}_i(k) &= \boldsymbol{H}_i^T \left[\boldsymbol{Q}^T(\boldsymbol{\alpha}(k))\,\boldsymbol{Q}(\boldsymbol{\alpha}(k))\right]^{-1} \boldsymbol{H}_i \\ \boldsymbol{F}(k) &= \sum_{i=1}^{M} \rho_i^2 \boldsymbol{F}_i(k). \end{aligned}$$

   Now minimize the cost

$$J(k) = \boldsymbol{\alpha}^T \boldsymbol{F}(k) \boldsymbol{\alpha}$$

   by choice of $\boldsymbol{\alpha}$ (ignoring the dependence of $\boldsymbol{F}(k)$ on $\boldsymbol{\alpha}(k)$) by setting

$$\boldsymbol{\alpha}(k+1) = -\frac{\boldsymbol{F}^{-1}(k)\boldsymbol{e}_1}{\boldsymbol{e}_1^T \boldsymbol{F}^{-1}(k)\boldsymbol{e}_1}.$$

3. Reform $\boldsymbol{Q}$ in term of the new $\boldsymbol{\alpha}$, and repeat step 2. Continue until the cost $J(k)$ no longer decreases, and then find the fixed poles $\{a_i\}$ as the roots of the polynomial

$$1 - \alpha_1 z^{-1} - \cdots - \alpha_N z^{-N}.$$

The procedure described above computes the N pole locations directly from the impulse responses of the example systems. Implicit, however, is the determination of an approximating single-input, M-output, Nth-order system. The denominator coefficients of each of the M transfer functions of the approximating system are given in the vector $\boldsymbol{\alpha}$; the numerator coefficients may be determined by an additional calculation.

The above procedure works from impulse response data, and can thus be viewed as a single-input, M-output filter design problem. It is also possible to modify the procedure to take M input-output responses as input. Such a modified procedure would be the equivalent of using Steiglitz–McBride to simultaneously identify M systems having the same poles, which fits the paradigm of single-input, M-output system identification, though the input data corresponding to each output may differ.

2. Model Reduction Based Pole Selection

This method is based on the Hankel-norm approximation of linear systems [2-41], in which a good Nth order approximation is found to a high-order system. The high order system here is the amalgamation of the M example systems; the poles of the Nth order approximation will form the poles for the FPAF. We hereafter refer to this approach as the HN-method.

To proceed, an augmented single-input, M-output system is first assembled as

$$\begin{bmatrix} \boldsymbol{x}_1(n+1) \\ \boldsymbol{x}_2(n+1) \\ \vdots \\ \boldsymbol{x}_M(n+1) \end{bmatrix} = \begin{bmatrix} \boldsymbol{A}_1 & 0 & \cdots & 0 \\ 0 & \boldsymbol{A}_2 & \cdots & 0 \\ \vdots & \vdots & \ddots & \vdots \\ 0 & 0 & \cdots & \boldsymbol{A}_M \end{bmatrix} \begin{bmatrix} \boldsymbol{x}_1(n) \\ \boldsymbol{x}_2(n) \\ \vdots \\ \boldsymbol{x}_M(n) \end{bmatrix} + \begin{bmatrix} \boldsymbol{b}_1 \\ \boldsymbol{b}_2 \\ \vdots \\ \boldsymbol{b}_M \end{bmatrix} u(n)$$

$$\begin{bmatrix} y_1(n) \\ y_2(n) \\ \vdots \\ y_M(n) \end{bmatrix} = \begin{bmatrix} \boldsymbol{c}_1 & 0 & \cdots & 0 \\ 0 & \boldsymbol{c}_2 & \cdots & 0 \\ \vdots & \vdots & \ddots & \vdots \\ 0 & 0 & \cdots & \boldsymbol{c}_M \end{bmatrix} \begin{bmatrix} \boldsymbol{x}_1(n) \\ \boldsymbol{x}_2(n) \\ \vdots \\ \boldsymbol{x}_M(n) \end{bmatrix} + \begin{bmatrix} d_1 \\ d_2 \\ \vdots \\ d_M \end{bmatrix} u(n), \quad (2.4\text{-}3)$$

where $\boldsymbol{A}_j$, $\boldsymbol{b}_j$, $\boldsymbol{c}_j$ and d_j, $j = 1, \ldots, M$, are one-input, one-output state space descriptions of the example systems, and the vectors $\boldsymbol{x}_i(n)$ their states.

The augmented system is formed using the example systems in state space form. Since we originally have impulse response descriptions, we must first convert these to state space form. A direct way to arrive at a state space description is simply to truncate the impulse response and use delayed inputs as system states. However, this results in a system description of an excessively high order. Filter design techniques may be used to achieve a lower-order approximation to each example system. One useful algorithm for achieving this is presented in [2-42], and it is this approach that we use in the examples in Section 2.6. Other methods for achieving this goal are of course possible.

Now, we reduce Eq. (2.4-3) using a *Hankel-norm* (HN) approximation to obtain an N-dimensional system

$$\boldsymbol{x}(n+1) = \boldsymbol{A}\boldsymbol{x}(n) + \boldsymbol{b}u(n).$$

Procedures for computing the Hankel-norm approximation appear in a number of commercially available packages. The eigenvalues of $\boldsymbol{A}$ are then used as the fixed pole locations. This procedure easily allows an intelligent choice of the number poles for the FPAF, since Hankel singular values generated in the model reduction give an indication of the model order at which additional states (poles) no longer give significant improvement in the modeling. This represents an advantage of the HN-method over the SM-method.

If the original example system descriptions yield high-order state space models, model reduction may be applied to each individual system prior to its positioning in the augmented system. This reduces the computational requirements for the model reduction of the collection of example systems.

3. Other Approaches for Fixed Pole Selection

The above two methods represent only a sampling of the variety of different approaches to selecting pole locations given a collection of example systems. To reflect some of this variety, we briefly note below some possible alternatives.

Prony analysis [2-43] fits N pole locations $a_1, \ldots, a_N$ to one impulse response function by estimating the exponential trends in that function. This procedure can be modified to fit exponentials to a set of M example impulse responses [2-23], and the exponential rates can then be used as the FPAF's pole locations. This approach is "quick and dirty", in that it may be seen as simply the first step in the SM-method.

A second alternative is to pose the problem of finding the single-input, M-output, Nth order system of Fig. 2.4-2 from a collection of M sample input–output responses $\{y_i(n), u_i(n)\}$ as a least squares fit problem. Form the cost function

$$J = \sum_{i=1}^{M} \sum_{n=0}^{L-1} e_i^2(n) = \sum_{i=1}^{M} \sum_{n=0}^{L-1} \left(y_i(n) - \hat{y}_i(n)\right)^2,$$

where

$$\hat{y}_i(n) = \sum_{k=1}^{N} \alpha_k y_i(n-k) + \sum_{k=0}^{N} \beta_{i,k} u_i(n-k).$$

This cost is minimized by choice of the vector

$$\begin{bmatrix} \alpha_1 & \cdots & \alpha_N & \beta_{1,0} & \cdots & \beta_{1,N} & \cdots\cdots & \beta_{M,0} & \cdots & \beta_{M,N} \end{bmatrix}^T.$$

The polynomial $z^N - \alpha_1 z^{N-1} - \cdots - \alpha_N$ is then the denominator in each of the components of the single-input, M-output, Nth order system, and the polynomials $\beta_{i,0} z^N + \cdots + \beta_{i,N}$ form the numerators. The roots of the shared denominator would be the fixed poles for the FPAF.

2.5 Choosing the Basis

The previous section considered approaches for selecting the vector space over which a VSAF adapts. Once this vector space is chosen, there remains the choice of basis for the vector space. That is, given that the vector space is

$$\mathcal{G} = \text{span}\left\{G_{11}(z), \ldots, G_{1N}(z)\right\},$$

then parameterizing an element $G \in \mathcal{G}$ via

$$G(z) = \boldsymbol{G}_1^T(z)\boldsymbol{b}_1 = \begin{bmatrix} G_{11}(z) \\ \vdots \\ G_{1N}(z) \end{bmatrix}^T \begin{bmatrix} b_1 \\ \vdots \\ b_N \end{bmatrix}$$

implicitly selects $\boldsymbol{G}_1(z)$ as a vector of basis elements for $\mathcal{G}$. One may switch to a different set of coordinates for $\mathcal{G}$ by changing basis to $\boldsymbol{G}_2(z) = \boldsymbol{A}\boldsymbol{G}_1(z)$, for some invertible transformation matrix $\boldsymbol{A}$. With $\boldsymbol{b}_2$ denoting the parameter vector with respect to this basis, we have

$$G(z) = \boldsymbol{G}_2^T \boldsymbol{b}_2 = \boldsymbol{G}_1^T \boldsymbol{A}^T \boldsymbol{b}_2,$$

so that $\boldsymbol{b}_2 = \boldsymbol{A}^{-T}\boldsymbol{b}_1$. The question now arises: what criterion is used to select the basis?

Some possibilities for determining the choice of basis are as follows. In Theorem 2.3-4, we have shown that the distribution over frequency of the variance error when using LMS is determined by the frequency responses of the basis systems. Thus, one may shape this frequency distribution by choice of basis, since changing bases alters those frequency responses. Another criterion for choice of basis is the convergence speed of the adapted parameters under LMS adaptation. We discuss this criterion under A. below. Lastly, the mean-square error performance for tracking time-varying optimal parameters is a basis-dependent phenomenon. This criterion for basis selection is discussed under B. below.

A. Basis Choice for Convergence Speed

With $\boldsymbol{b}^*$ denoting the optimal parameters in a stationary environment, it is well known that for adaptive FIR filters,

$$E\left\{\boldsymbol{b}^* - \boldsymbol{b}(n+1)\right\} = [\boldsymbol{I} - \mu\boldsymbol{R}]E\left\{\boldsymbol{b}^* - \boldsymbol{b}(n)\right\}$$

governs the expected parameter convergence under the LMS algorithm [2-6], with $\boldsymbol{R} = E\left\{\boldsymbol{x}(n)\boldsymbol{x}^T(n)\right\}$ the correlation matrix of the regressor defined by Eq. (2.3-2). This result is readily seen to hold for the VSAF case as well, by virtue of its preservation of the linear-in-the-parameters property. The modes of convergence of $\boldsymbol{b}(n)$ of the VSAF thus are exponential, at rates given by $(1-\mu\lambda)^n$, where λ is an eigenvalue of $\boldsymbol{R}$. The fast mode converges at rate $(1-\mu\lambda_{\max}(\boldsymbol{R}))^n$, and the slow mode at rate $(1-\mu\lambda_{\min}(\boldsymbol{R}))^n$. Maximizing the slowest rate of convergence, while satisfying the bound on the step size μ to preserve convergence in the mean, suggests using the condition number $\gamma = \lambda_{\max}(\boldsymbol{R})/\lambda_{\min}(\boldsymbol{R})$ as a measure of convergence speed [2-44]. Minimizing γ optimizes the convergence rate.

Consider that parameterizing $G(z)$ via $\boldsymbol{b}_1$ results in a regressor correlation matrix $\boldsymbol{R}_1 = E\{\boldsymbol{x}_1\boldsymbol{x}_1^T\}$. Then, since $\boldsymbol{G}_2 = \boldsymbol{A}\boldsymbol{G}_1$, we have $\boldsymbol{R}_2 = \boldsymbol{A}\boldsymbol{R}_1\boldsymbol{A}^T$. Picking the basis transformation $\boldsymbol{A}$ to satisfy

$$\boldsymbol{A} = \boldsymbol{R}_1^{-\frac{1}{2}}$$

then achieves $\boldsymbol{R}_2 = \boldsymbol{I}$, optimizing convergence [2-45]. A limiting factor in the implementation of this optimal choice of basis, however, is that the statistics of the input signal may not be available when the adaptive filter structure is specified, or they may be expected to change.

As an alternative, a variety of orthogonal transformations, in conjunction with power normalization in the LMS algorithm, have been considered for convergence speed improvement [2-46]. These techniques may be seen to be equivalent to the change of basis noted here for VSAF specification.

Another possibility is the adaptive processing of signals in the frequency domain [2-47], which has been suggested to improve overall convergence speed. Preprocessing the input signal via a *discrete Fourier transform* (DFT), as is done in many implementations of frequency domain adaptive FIR filters, amounts to a change of basis for the vector space of systems with finite, N-point, impulse responses. Viewing frequency domain adaptive filters in this way interprets them as one option for an adaptive FIR filter within the VSAF framework.[2]

B. Tracking Time-varying Parameters

By adopting an idealized representation for the optimal filter characteristics and their time variations, one can analyze the tracking capabilities of the VSAF framework. Using this analysis, one is then able to optimize the selection of a basis for minimizing the total MSE for the adaptive filter's operation.

Let $d(n)$ in Fig. 2.1-1 be

$$d(n) = \sum_{i=1}^{N} b_i^*(n) G_i(q)\,[u(n)] + v(n); \tag{2.5-1}$$

that is, we assume that the relationship between the input $u(n)$ and the desired output $d(n)$ can be described exactly within the vector space of systems $\mathcal{G}$, plus an additive noise term assumed to be independent of $\{u(n)\}$. With respect to Fig. 2.1-2, the assumption of Eq. (2.5-1) may be interpreted as having $H \in \mathcal{G}$ at each iteration. Note that we do allow for time variations in the optimal, true parameter values $\{b_i^*(n)\}$.

[2]Note also that the VSAF framework permits a generalization of frequency domain adaptive filters, in that the basis systems $G_1(z), \ldots, G_N(z)$ can be used to preprocess the input using some frequency domain criterion in the same way that the DFT preprocesses the input for the standard frequency domain adaptive filters.

Using this framework, we have presented in [2-48] and [2-49] the tracking capabilities of VSAFs when the optimal parameters $\boldsymbol{b}^*(n)$ are assumed to follow a random walk

$$\boldsymbol{b}^*(n+1) = \boldsymbol{b}^*(n) + \boldsymbol{w}(n).$$

The forcing term $\{\boldsymbol{w}(n)\}$ is a zero mean, stationary process, independent of $\{u(n)\}$. This study is related to but extends the analysis of adaptive FIR models in [2-50]. See also [2-37] for further results on FIR models.

Define $\sigma_v^2 = E\{v^2(n)\}$, and let

$$\begin{aligned} \boldsymbol{R} &= E\left\{\boldsymbol{x}(n)\boldsymbol{x}^T(n)\right\} \\ \boldsymbol{Q} &= E\left\{\boldsymbol{w}(n)\boldsymbol{w}^T(n)\right\} \end{aligned}$$

denote covariance matrices for (stationary) $\boldsymbol{x}(n)$ and $\boldsymbol{w}(n)$. These covariance matrices are taken with respect to the particular choice of basis for the system vector space implied by Eq. (2.5-1). Let the vector of basis systems be denoted by

$$\boldsymbol{G} = \begin{bmatrix} G_1 & \cdots & G_n \end{bmatrix}^T.$$

Given this notation, the following results may be proved.

Theorem 2.5-1 [2-49] Consider LMS adaptation within $\mathcal{G}$. Then the minimum, steady state MSE achievable by LMS is

$$J^*_{\text{LMS}} = \sigma_v^2 + \sqrt{\sigma_v^2}\text{tr}\left[(\boldsymbol{Q}^{\frac{1}{2}}\boldsymbol{R}\boldsymbol{Q}^{\frac{1}{2}})^{\frac{1}{2}}\right], \tag{2.5-2}$$

and it is achieved by adapting with respect to basis

$$\boldsymbol{G}_{\text{opt}} = \boldsymbol{A}^{-1}\boldsymbol{G},$$

with $\boldsymbol{A}$ such that

$$\boldsymbol{A}\boldsymbol{A}^T = \boldsymbol{Q}^{-\frac{1}{2}}(\boldsymbol{Q}^{\frac{1}{2}}\boldsymbol{R}\boldsymbol{Q}^{\frac{1}{2}})^{\frac{1}{2}}\boldsymbol{Q}^{-\frac{1}{2}}.$$

∎

Remark 2.5-1 The optimal transformation $\boldsymbol{A}$ in Theorem 2.5-1 is not unique. However, we may desire to select the unique, symmetric $\boldsymbol{A}$ satisfying

$$\boldsymbol{A}^2 = \boldsymbol{Q}^{-\frac{1}{2}}(\boldsymbol{Q}^{\frac{1}{2}}\boldsymbol{R}\boldsymbol{Q}^{\frac{1}{2}})^{\frac{1}{2}}\boldsymbol{Q}^{-\frac{1}{2}}.$$

∎

Theorem 2.5-1 shows how to optimize the choice of basis for $\mathcal{G}$ in order to minimize the MSE when using LMS. The transformation $\boldsymbol{A}$ forms a new basis for the vector space via linear combinations of the original elements of the basis vector $\boldsymbol{G}$. For RLS adaptation, we have the following.

Theorem 2.5-2 [2-49] Consider RLS adaptation within $\mathcal{G}$. The minimum, steady state MSE achievable by RLS is

$$J^*_{\mathrm{RLS}} = \sigma_v^2 + \sqrt{N\sigma_v^2 \mathrm{tr}(\boldsymbol{RQ})}. \tag{2.5-3}$$

J^*_{RLS} is achieved using adaptation with respect to *any* basis. ∎

Note that one does not expect the choice of basis in RLS adaptation to affect the achievable MSE, since RLS optimizes an *input–output* determined cost function, which does not depend on the internal structure of the adaptive filter (though the MSE does depend on the chosen vector space for the VSAF). Finally, the following result establishes the relationship between the minimum achievable MSE for LMS and that for RLS.

Theorem 2.5-3 [2-49] The minimum achievable MSE using LMS is no greater than that for RLS: $J^*_{\mathrm{RLS}} \geq J^*_{\mathrm{LMS}}$. ∎

This last theorem shows that knowledge of the input statistics (via $\boldsymbol{R}$) and the statistics of the parameter variations (via $\boldsymbol{Q}$) enables one to choose a basis for the filter vector space so that LMS outperforms RLS in terms of MSE.

2.6 Examples

Below we present three examples which evidence the performance potential for VSAFs. The intent of the examples is to demonstrate the ability of VSAFs and FPAFs to provide improved minimum MSE performance with respect to other adaptive filter structures having the same number of adapted parameters. The test conditions under which the performance is assessed involve the use of example systems as described in Section 2.4. The first two examples involve simulation based data; the third uses measured echo path data in telecommunications channels. In all three situations, the VSAFs and FPAFs show improved performance in relation to FIR and other adaptive filter structures.

A. VSAF Performance Using Vector Space Optimization

A simulation example from [2-38] serves to illustrate the optimization method appearing under A. of Section 2.4. This example considers the problem of channel equalization, as depicted in Fig. 2.6-1. The channel input $w(n)$ is sent over the channel $C(z)$, and the distorted output $u(n)$ is received as the input to the equalizer $E(z)$. The objective in the design of $E(z)$ is so that its output $y(n)$ is close to a delayed version of the original channel input; that is, we desire $y(n) \approx w(n-\Delta)$ for some delay time Δ.

If $C(z)$ is at first unknown, $E(z)$ may be adaptively adjusted within some class of equalizers $\mathcal{G}$ so as to achieve the desire result. Typically, the structure of E is chosen as an adaptive FIR filter. Here, we will parameterize $\mathcal{G}$ using the methods of A. of Section 2.4 to obtain a better performing equalizer, relative to an adaptive FIR filter, for the same number of adapted parameters.

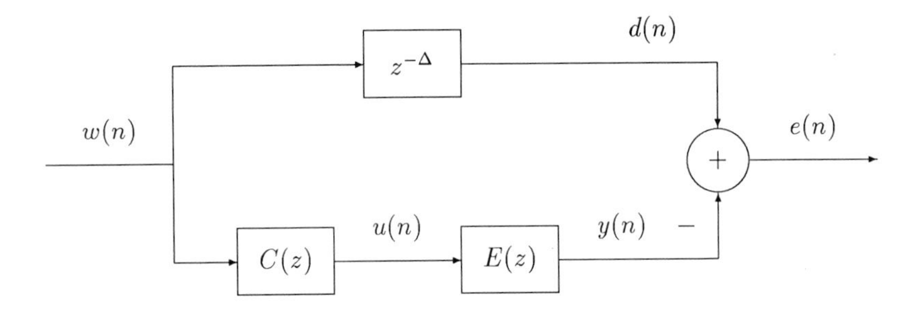

Figure 2.6-1: Channel equalization.

The channel $C(z)$ in our example is given by

$$C(z) = \frac{c_0 + c_1 z^{-1} + c_2 z^{-2} + c_3 z^{-3} + c_4 z^{-4}}{\sum_{i=0}^{4} c_i^2}$$

with each parameter taking on a random value that is uniformly distributed over a small interval. We have $c_0 \sim U(0, 0.2)$, $c_1 \sim U(0.4\sqrt{2}, 0.2{+}0.4\sqrt{2})$, $c_2 \sim U(0.8, 1.0)$, and $c_3 = c_1$ and $c_4 = c_0$. This defines a class of 5-point, FIR channels which spread a pulse input into an approximately sinusoidal shaped response. The channels are normalized so that when the channel input $w(n)$ is a white sequence with unit power, the equalizer input will have unit power as well.

Corresponding to each channel is an optimal equalizer which minimizes the MSE between the equalizer output and the delayed channel input. In the computations which are described below, we truncate the impulse responses of these optimal equalizers to 75 points in length, strictly for computational convenience. The simulation experiment proceeds as follows:

1. First we generate 50 example channels as above, and determine the optimal equalizer for each channel when $\Delta = 42$. The value $\Delta = 42$ is chosen so that the peak of the equalizer impulse response occurs at the central tap weight. This corresponds to reconstructing the channel input after a delay of half the total length of the equalizer. The input $w(n)$ is assumed white, with unit power. The 50 equalizers so determined form the set of example systems.

2. Determine as described in A. of Section 2.4 the optimal vector space $\mathcal{G}_N$, where N is the dimension of the vector space, for several values of N.

3. For each of 200 new, randomly generated channels, find the optimal equalizer within $\mathcal{G}_N$, and compute the achieved MSE $E\{(y(n) - w(n-42))^2\}$.

4. In tandem, determine the optimal FIR length equalizer of length N (and hence dimensionality N for the vector space of N-point, FIR systems), with optimality in the sense of minimizing the MSE

$$E\left\{[y(n) - w(n - (N/2) - 4)]^2\right\}.$$

   Minimizing this MSE will reconstruct the channel input after a delay of half the total length of the equalizer, as was done with the class $\mathcal{G}_N$.

5. Plot the average MSE obtained with the 200 channels as N varies.

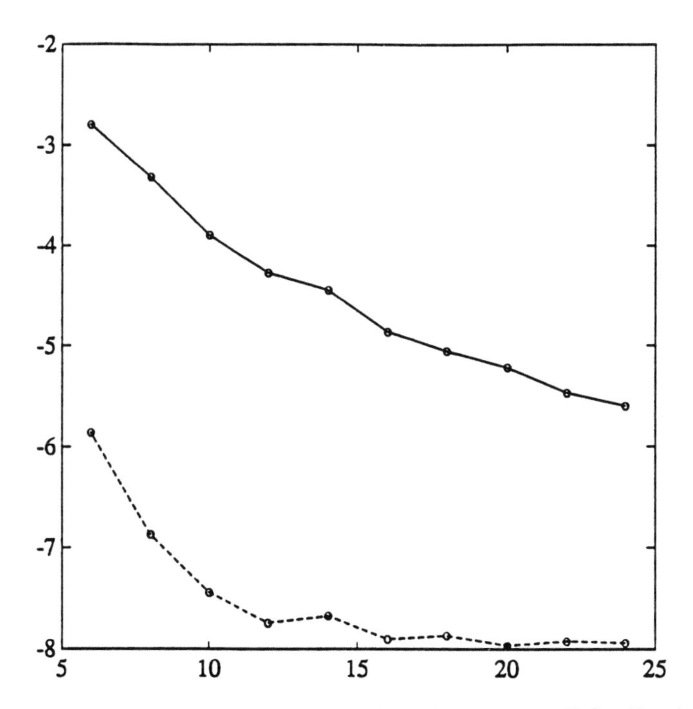

Figure 2.6-2: Average MSE in dB versus number of parameters N for N-point FIR $\mathcal{G}$ (solid line) and for optimal N-dimensional $\mathcal{G}$ (dashed line).

The results of this experiment are presented in Fig. 2.6-2. One can see from the figure that the optimization of choice of N-dimensional model set reduces the achieved

MSE, and hence improves performance. The performance improvement with the number of parameters tails off for the optimal VSAF due to the truncation of the optimal equalizer filter response.

B. FPAF Performance in Simulation

We now compare FPAF performance with that of adaptive FIR and Kautz filters, using a simulation based example from [2-24] and [2-23]. We assume the following modeling situation, based on the system identification configuration for adaptive filtering depicted in Fig. 2.1-2. The optimal filter for the adaptive filter environment is assumed to be a band-pass butterworth filter whose cutoff frequencies are variable, but lie within a restricted range. This class for the optimal filter possesses some structure, which may be incorporated into the design of the adaptive filter, but with enough variability so as to necessitate adaptivity. This scenario reflects an environment where the physical makeup and constraints limit the possibilities for the optimal filter specification, but in which a number of degrees of freedom remain.

To proceed, we randomly select $M_1 = 16$ example systems from a set of band-pass Butterworth transfer functions, whose defining parameters are as follows. The number of poles is a random integer between 6 and 12, inclusive; the normalized low-pass frequency is uniformly distributed in the interval $[0.3\pi, 0.45\pi]$; and normalized high-pass frequency uniformly distributed in the interval $[0.5\pi, 0.6\pi]$ The passband gain of each filter is normalized to a unit value. These example systems will be used to define the poles for the FPAF, and also to select a complex-conjugate pole pair for the adaptive Kautz filter. To test the performance of each adaptive filter structure, a different set of $M_2 = 50$ systems is randomly generated in the same way, and the optimal performance for each of these 50 test scenarios is measured.

The performance is computed as follows. Let $h_i(n)$ denote the impulse response of the ith test system, and let $g_i^*(n)$ denote the adaptive filter impulse response when its coefficients are adjusted to give the best mean-square fit to $h_i(n)$. For the purpose of computation, these impulse responses are truncated to length $L = 100$. Then, the sum squared error in the impulse responses, normalized by the total output power of the collection of test systems, is given by

$$J = \frac{\sum_{i=1}^{M_2} \sum_{n=0}^{L-1} \left(h_i(n) - g_i^*(n)\right)^2}{\sum_{i=1}^{M_2} \sum_{n=0}^{L-1} \left(h_i(n)\right)^2}. \tag{2.6-1}$$

The numerator of J approximates the mean-square error performance of the adaptive filter, when unit variance white noise is used as the filter input. As a performance index in the experiments below, we employ

$$I = -10\log(J). \tag{2.6-2}$$

A larger value of I denotes better performance.

Note that Eq. (2.6-1) reflects the minimum mean-square error in adaptive filter performance. On top of this, excess mean-square error due to residual parameter motion will increase the total cost of adaptive filter performance. One should note that excess mean square error, or "misadjustment", is proportional to the number of filter parameters [2-8]. Since, in relation to other adaptive filter structures, an FPAF in general requires fewer parameters to achieve a given level of minimum MSE (see the examples below), one would expect improved performance as well in excess mean-square error for the FPAF. However, here we do not pursue a detailed treatment of this effect.

As input to the pole selection procedures, we first use the impulse responses of the example filters, and second we used the impulse responses with 0.01 variance, white gaussian noise added. In both cases the impulse responses were truncated to a length $L = 100$. For each set of data, we choose N fixed pole locations using the SM-method of Section 2.4, and then compute the performance via Eq. (2.6-2) using the $M_2 = 50$ test systems.

To select poles for the FPAF, we first convert the impulse responses of the M_1 example systems to state space form using the method of [2-42], and then apply the HN-method to choose N fixed pole locations for several values of N. The achieved performance level, computed via Eq. (2.6-2) and Eq. (2.6-1), appears in Fig. 2.6-3. The similarly computed performance (using the same M_2 example systems as above) attained by adaptive FIR filters and adaptive Kautz filters also appears in Fig. 2.6-3. For the adaptive Kautz filter, we chose for the fixed complex-conjugate pole pair values which resulted from applying the SM-method to the pure impulse responses of the example systems, but estimating just two complex-conjugate poles. Using the SM-method for pole selection for the FPAF results in performance quite similar to that for the HN-method in this example [2-24].

Equating the number of adjustable coefficients in each filter allows a reasonable comparison between the FPAF, Kautz, and FIR filters. The implementation of an N-pole FPAF with arbitrary pole locations requires only $N-1$ more multiplications than a length N adaptive FIR filter, and only $N-1$ more memory locations [2-24]. Since implementing the adaptive algorithm requires at least an additional $2N$ multiplications (LMS), and up to on the order of N^2 multiplications (RLS), equating the number of adapted parameters for the filter approximately equates the complexity of implementation. One may then see from Fig. 2.6-3 that the performance of the FPAF is superior when compared with an FIR or Kautz adaptive filter of similar computational complexity. We should note that a more tuned procedure to estimate the complex-conjugate poles for the Kautz filter may result in better performance for this type of filter. However, the same may be possible for the FPAF as well.

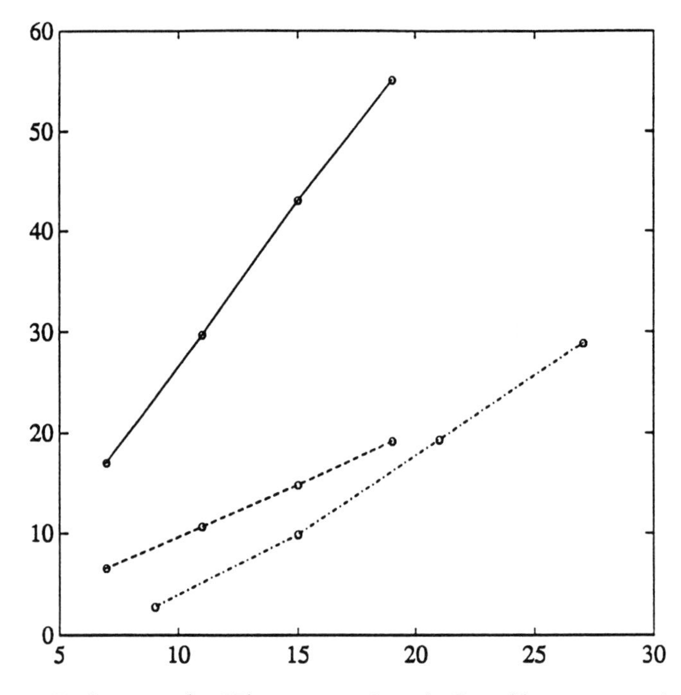

Figure 2.6-3: Performance (in dB) versus number of adjustable parameters in the Butterworth filter example. FPAF: solid; Kautz: dashed; FIR: dot-dash.

C. FPAF Performance in Long-distance Echo Cancellation

The next experiment, based on data presented in [2-24], considers the performance of FPAFs using actual, measured impulse response data for long-distance echo cancellation in telephone lines. For this experiment, we have available five impulse responses of echo paths of long-distance telephone calls.[3] They were gathered using speech as the excitation signal of a 256 tap FIR adaptive filter. An LMS update system was used to adjust the coefficients of the filter to obtain a model, and the adaptive algorithm was enabled during half-duplex intervals of user voice traffic. Therefore, if the speech signal has no support in a range of frequencies, the estimate of the impulse response is not "well matched" to the true impulse response in those same frequencies.

The typical response of an echo path such as these shows an initial delay, followed by a burst of rapid variation in the response, and finally an oscillatory but slowly decaying tail. Reference [2-27] suggests a two-stage approach for modeling

[3]The impulse response data were generously provided by AT&T Bell Laboratories.

this type of response: the first stage with a conventional FIR filter, to approximate the delay and rapid time variation, and a second stage with a Laguerre or Legendre filter, to approximate the tail. We will follow the same approach, but using FPAF, Kautz or FIR filters in the second stage. From an examination of the available impulse responses, we estimate that the FIR filter of the first stage needs 50 taps. In the performance measure of Eq. (2.6-2), we will assume perfect modeling in the first stage. We then focus on the tail of these impulse responses (samples 50 to 255).

Using the tails of the five available impulse responses as the example systems, we use the SM–method of Section 2.4 to select poles for an FPAF for the second stage. Due to the limited available data, the example systems will also serve as the test systems in the performance measure. With Eq. (2.6-2) measuring performance, we plot in Fig. 2.6-4 the effective attenuation of the echo as a function of the number of poles. For comparison, performance computations were also conducted when adaptive FIR and adaptive Kautz filters are used. We use the SM-method in the same fashion as in the preceding example to estimate the complex conjugate pair of poles for the Kautz filter. Figure 2.6-4 summarizes the results: the FPAF outperforms both the adaptive FIR and adaptive Kautz filters, for an equal number of adapted parameters.

Summary

This chapter has considered the design and use of adaptive filters from the perspective of adjusting the filter over a vector space of systems. This approach is motivated by the fact that the VSAF preserves the desirable adaptation properties of the adaptive FIR filter. Adaptive FIR filters work well because their adaptation algorithms minimize a quadratic cost function over a vector space of systems. VSAFs simply change the system vector space.

We have formalized this generalization to the VSAF, noting that the main design features that are introduced are the choice of vector space and its basis. A particularly attractive class of system vector spaces that we have discussed is that of Nth order IIR systems having variable zero locations, but fixed pole locations. These are the *fixed pole adaptive filters* (FPAFs). Their complexity of implementation is similar to that for adaptive FIR filters, but the flexibility of choice of their pole locations allows them to be tailored to specific applications environments, with a corresponding performance benefit, as seen in the examples presented above.

Critical to the success of a VSAF, or FPAF, is the appropriate selection of the system vector space. Our presentation included several techniques for choosing the system vector space based on example systems that represent the operating environment for the filter. Choices for the basis can similarly affect performance; we have presented an analysis which shows how to optimize the basis choice to reflect time variations in the optimal filter.

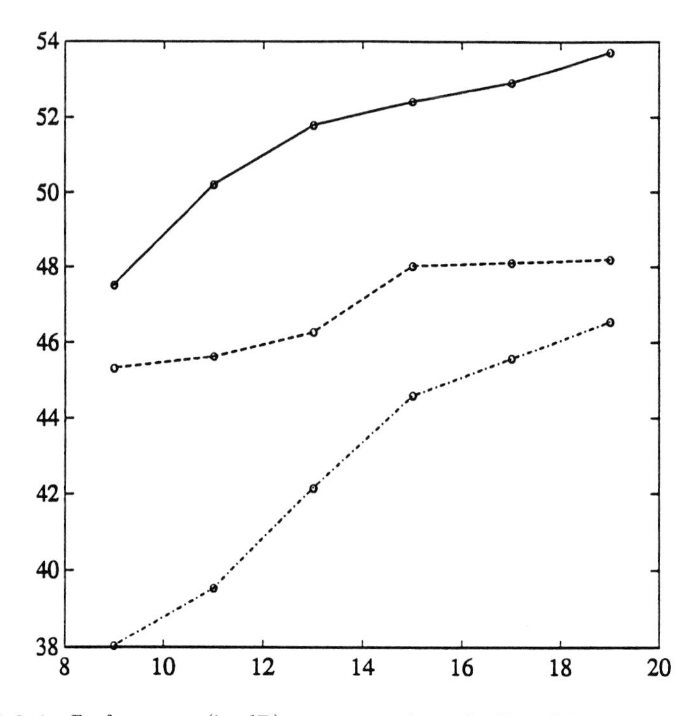

Figure 2.6-4: Performance (in dB) versus number of adjustable parameters in the telecommunication example. FPAF: solid; Kautz: dashed; FIR: dot-dash.

Most importantly, the algorithm behavior of a VSAF is essentially the same as that for adaptive FIR filters. Therefore, wherever one may desire to employ an adaptive filter, it is almost certain that an appropriately designed VSAF, based on *a priori* information regarding the application environment, will outperform an adaptive FIR filter.

References

[2-1] R.W. Lucky, "Automatic equalization for digital communication," *Bell System Technical Journal*, vol. 44, pp. 547-588, 1965.

[2-2] S.U.H. Qureshi, "Adaptive equalization," *Proceedings of the IEEE*, vol. 73, pp. 1349-1387, September 1985.

[2-3] M.M. Sondhi and D.A. Berkley, "Silencing echoes on the telephone network," *Proceedings of the IEEE*, vol. 68, pp. 948-963, August 1980.

[2-4] G. Long, D. Shwed, and D.D. Falconer, "Study of a pole-zero adaptive echo canceller," *IEEE Transactions on Circuits and Systems*, vol. CAS-34, pp. 765-769, July 1987.

[2-5] N.S. Jayant and P. Noll, *Digital coding of waveforms: principles and applications to speech and video.* Englewood Cliffs, NJ: Prentice-Hall, 1984.

[2-6] S. Haykin, *Adaptive Filter Theory.* Englewood Cliffs, NJ: Prentice-Hall, 1991.

[2-7] B. Widrow and M.E. Hoff, "Adaptive switching circuits," in *Records of the IRE WESCON Convention*, pt. 4, pp. 96-140, 1960.

[2-8] B. Widrow and S. Stearns, *Adaptive Signal Processing.* Englewood Cliffs, NJ: Prentice-Hall, 1985.

[2-9] B. Widrow, J.M. McCool, M.G. Larimore, and C.R. Johnson, Jr., "Stationary and non-stationary learning characteristics of the LMS adaptive filter," *Proceedings of the IEEE*, vol. 64, pp. 1151-1162, August 1976.

[2-10] A. Weiss and D. Mitra, "Digital adaptive filters: conditions for convergence, rates of convergence, effects of noise and errors arising from implementation," *IEEE Transactions on Information Theory*, vol. IT-25, pp. 637-652, November 1979.

[2-11] W.A. Gardner, "Learning characteristics of stochastic gradient descent algorithms: a general study, analysis, and critique," *Signal Processing*, vol. 6, pp. 113-133, 1984.

[2-12] S. Stearns, "Error surfaces of recursive adaptive filters," *IEEE Transactions on Acoustics, Speech, and Signal Processing*, vol. ASSP-29, pp. 763-766, June 1981.

[2-13] H. Fan and M. Nayeri, "On error surfaces of sufficient order adaptive IIR filters: proofs and counterexamples to a unimodality conjecture," *IEEE Transactions on Acoustics, Speech, and Signal Processing*, vol. ASSP-37, pp. 1436-1442, September 1989.

[2-14] N. Nayeri, H. Fan and W.K. Jenkins, "Some characteristics of error surfaces for insufficient order adaptive IIR filters," *IEEE Transactions on Acoustics, Speech, and Signal Processing*, vol. ASSP-38, pp. 1222-1227, July 1990.

[2-15] T. Söderström and P.G. Stoica, "Some properties of the output error method," *Automatica*, vol. 18, pp. 93-99, January 1982.

[2-16] C.R. Johnson, Jr., "Adaptive IIR filtering: current results and open issues," *IEEE Transactions on Information Theory*, vol. IT-30, pp. 237-250, March 1984.

[2-17] L. Ljung and T. Söderström, *Theory and Practice of Recursive Identification.* Boston: MIT Press, 1983.

[2-18] G.A. Williamson, "Adaptive filtering using vector spaces of systems," *1992 IEEE Digital Signal Processing Workshop*, Starved Rock State Park, IL, September 1992.

[2-19] M.V. Dokic and P.M. Clarkson, "On the performance of a second-order adaptive Volterra filter," *IEEE Transactions on Signal Processing*, vol. 41, pp. 1944-1947, May 1993.

[2-20] V.J. Mathews, "Adaptive polynomial filters," *IEEE ASSP Magazine*, vol. 8, no. 3, pp. 10-26, July 1991.

[2-21] K.S. Narendra and K. Parthasarathy, "Identification and control of dynamical systems using neural networks," *IEEE Transactions on Neural Networks*, vol. 1, pp. 4-27, March 1990.

[2-22] G.A. Williamson, "Globally convergent adaptive filters with infinite impulse responses," *Proceedings of the 1993 International Conference on Acoustics, Speech and Signal Processing*, Minneapolis MN, April 1993, pp. III 543-546.

[2-23] S. Zimmermann and G.A. Williamson, "Performance properties of fixed pole adaptive filters," *Proceedings of the 1993 International Symposium on Circuits and Systems*, Chicago IL, May 1993, pp. 56-59.

[2-24] G.A. Williamson and S. Zimmermann, "Globally convergent adaptive IIR filters based on fixed pole locations," submitted for publication in *IEEE Transactions on Signal Processing.*

[2-25] B. Wahlberg, "System identification using Laguerre models," *IEEE Transactions on Automatic Control*, vol. AC-36, pp. 551-562, May 1991.

[2-26] P.M. Mäkilä, "Approximation of stable systems by Laguerre filters," *Automatica*, vol. 26, pp. 333-345, 1990.

[2-27] G.W. Davidson and D.D. Falconer, "Reduced complexity echo cancellation using orthonormal functions," *IEEE Transactions on Circuits and Systems*, vol. 38, pp. 20-28, January 1991.

[2-28] H. Perez and S. Tsujii, "A system identification algorithm using orthogonal functions," *IEEE Transactions on Signal Processing*, vol. SP-39, pp. 752-755, March 1991.

[2-29] B. Wahlberg, "Identification of resonant systems using Kautz filters," *Proceedings of the 30th Conference on Decision and Control*, Brighton, England, December 1991, pp. 2005-2010.

[2-30] J.C. Principe, B. de Vries and P.G. de Oliveira, "The gamma filter – a new class of adaptive IIR filters with restricted feedback," *IEEE Transactions on Signal Processing*, vol. 41, pp. 649-656, February 1993.

[2-31] Y.C. Pati and P.S. Krishnaprasad, "Approximation of stable linear systems via rational wavelets," *Proceedings of the 31st Conference on Decision and Control*, Tucson AZ, December 1992, pp. 1502-1507.

[2-32] C.R. Johnson, Jr., *Lectures on Adaptive Parameter Estimation.* Englewood Cliffs, NJ: Prentice-Hall, 1988.

[2-33] M. Green and J.B. Moore, "Persistence of excitation in linear systems," *Systems and Control Letters*, vol. 7, pp. 351-360, September 1986.

[2-34] S. Boyd and S. Sastry, "On parameter convergence in adaptive control," *System and Control Letters*, vol. 3, pp. 311-319, December 1983.

[2-35] L. Ljung, *System Identification: Theory for the User.* Englewood Cliffs, NJ: Prentice-Hall, 1987.

[2-36] B. Wahlberg and L. Ljung, "Design variables for bias distribution in transfer function estimation," *IEEE Transactions on Automatic Control*, vol. AC-31, pp. 134-144, February 1986.

[2-37] L. Ljung and S. Gunnarsson, "Adaptation and tracking in system identification – a survey," *Automatica*, vol. 26, pp. 7-21, 1990.

[2-38] G.A. Williamson, "Linear in the parameters identification for classes of systems," *Proceedings of the 32nd Conference on Decision and Control*, San Antonio TX, December 1993, pp. 2607-2612.

[2-39] A. Kaelin, A.G. Lindgren and G.S. Moschytz, "Linear echo cancellation using optimized recursive prefiltering," *Proceedings of the 1993 International Symposium on Circuits and Systems*, Chicago IL, May 1993, pp. 463-466.

[2-40] K. Steiglitz and L.E. McBride, "A technique for identification of linear systems," *IEEE Transactions on Automatic Control*, vol. AC-10, pp. 461-464, 1965.

[2-41] K. Glover, "All optimal Hankel norm approximations of linear multivariable systems and their L_∞ error bounds," *International Journal of Control*, vol. 39, pp. 1115-1193, 1984.

[2-42] B. Beliczynski, I. Kale and G.D. Cain, "Approximation of FIR by IIR digital filters: an algorithm based on balanced model reduction," *IEEE Transactions on Signal Processing*, vol. 40, pp. 532-541, March 1992.

[2-43] T.W. Parks and C.S. Burrus, *Digital Filter Design.* New York: Wiley, 1987.

[2-44] I.M.Y. Mareels, R.R. Bitmead, M. Gevers, C.R. Johnson, Jr., R.L. Kosut and M.A. Poubelle, "How exciting can a signal really be?" *System and Control Letters*, vol. 8, pp. 197-204, January 1988.

[2-45] G.A. Williamson and C.R. Johnson, Jr., "Some effects of parameterization change in system identification," *Proceedings of the 1992 American Control Conference*, Chicago IL, June 1992, pp. 1268-1269.

[2-46] D.F. Marshall, W.K. Jenkins, Jr. and J.J. Murphy, "The use of orthogonal transforms for improving performance of adaptive filters," *IEEE Transactions on Circuits and Systems*, vol. CAS-36, pp. 474-484, April 1989.

[2-47] E.R. Ferrara, Jr., "Frequency-domain adaptive filtering," in *Adaptive Filters*, C.R.N. Cowan and P.M. Grant, eds. Englewood Cliffs, NJ: Prentice-Hall, 1985, ch. 6, pp. 145-179.

[2-48] G.A. Williamson, "Tracking properties of vector space adaptive filters," *Proceedings of the 26th Asilomar Conference on Signals, Systems, and Computers*, Pacific Grove CA, October 1992, pp. 30-34.

[2-49] G.A. Williamson, "Tracking random walk systems with vector space adaptive filters," to appear in *IEEE Transactions on Circuits and Systems II*, 1995.

[2-50] A. Benveniste, "Design of adaptive algorithms for the tracking of time-varying systems," *International Journal of Adaptive Control and Signal Processing*, vol. 1, pp. 3-29, 1987.

Chapter 3

Order Statistics and Adaptive Filtering[†]

PETER M. CLARKSON and GEOFFREY A. WILLIAMSON

Introduction

Recent years have seen a surge of interest in the role of order statistics in digital signal processing. *Order statistic* (OS) operations form the basis for a powerful class of nonlinear digital filters. Several OS filters, most notably the median filter, have been widely applied in signal and image processing. Median filters tend to smooth data but they also have properties that differ sharply from linear smoothing filters. In particular, median filters remove sparse impulses from data. They also preserve edges in signals. In some cases, however, median filters have been found to provide inadequate smoothing of non-impulsive data. This has motivated the use of other OS filters such as the trimmed mean, that can provide a compromise between the smoothing capacity of linear filters and the impulse attenuation achieved by the median. In signal estimation problems, OS filters can be defined that meet optimality criteria for a much broader class of inputs than is possible for linear filters. Additionally, many OS filters are *robust* to occasional large deviations from an assumed noise density.

With the high level of interest in OS filters, a natural development has been the design of algorithms to exploit the powerful properties of OS operators in an adaptive filtering framework. In this chapter, we will give an overview of these developments, beginning with a brief introduction motivating the various adaptive forms, describing the algorithms and their properties, and pointing out some of the applications.

[†]This work was supported by the NATIONAL SCIENCE FOUNDATION under grant nos. MIP-9102620 and MIP-9220769.

3.1 Median and Order Statistic Filters

As a preliminary, we will discuss median and other OS filters and some of their properties. The use of the median filter dates to the work of Tukey [3-1], who used a running median operation to smooth time series data:

$$z(n) = \text{Med}\{x(n)\ x(n-1), ..., x(n-N+1)\} \triangleq \text{Med}\{x(n)\}_N\ , \qquad (3.1\text{-}1)$$

where the operation $\text{Med}\{\ \}_N$ denotes the sample median of the N samples[1] and where $\{x(n)\}$ is taken to be an infinite two-sided sequence. Early applications of the median filter included speech analysis, where the filter was used to smooth raw 'pitch period' estimates [3-2], and image enhancement problems, where the median was used to smooth data corrupted by 'salt-and-pepper noise' [3-3].

The median filter defined by Eq. (3.1-1) is a highly nonlinear operator. This greatly complicates any analysis of the filter characteristics. Although some interesting general properties of median filters have been observed [3-4], two key characteristics of the filter are often cited as motivating the many applications of the filter:

i) edge preservation

ii) impulse suppression.

The first of these properties arises from the view of "edges" as monotonic signal regions. It follows intuitively from the definition (3.1-1) that a monotonic signal is invariant to median filtering:

Property 1 If $x(n) \leq x(n-1) \leq ... \leq x(n-N+1)$, then $z(n) = x(n-(N-1)/2)$ for all N. ∎

Hence a monotonic sequence is preserved by median filtering.[2] Since N is finite, the requirement for a monotonic sequence can be relaxed somewhat. A more useful property follows from the definition of *locally monotonic* (LOMO) sequences [3-5]. A LOMO(L) sequence is one for which any L point subsequence $\{x(n),\ x(n-1),\ ..., x(n-L+1)\}$ is monotonic. Thus, LOMO sequences consist entirely of monotonic subsequences, either non-increasing or non-decreasing, punctuated by constant regions. The following property is often used to justify the "edge-preservation" observed in applications of the median filter:

Property 2 Any LOMO(L) sequence is invariant to a median filter of length $N \leq 2L-2$. ∎

Note that, by contrast, a moving average filter defined by

$$z(n) = [x(n) + x(n-1) + \cdots + x(n-N+1)]/N, \qquad (3.1\text{-}2)$$

[1] We choose N odd throughout this chapter. This restriction allows slight simplification in the definitions of some OS estimators. The case of N even does not differ in any meaningful way.

[2] Apart from a shift (delay) of $(N+1)/2$ samples.

converts edges into ramps of width N [3-6].

We can give an intuitive interpretation to the impulse rejection property as follows.

Property 3 Consider a bounded sequence $\{x(n)\}$ such that $\max |x(n)| \le R$. Suppose that this sequence is corrupted by occasional high amplitude noise (impulsive noise) $\zeta(n)$, such that $|\zeta(n)| > R$. Then the median filter of Eq. (3.1-1) has the following property: $|z(n)| \le R$ for all n, provided that no more than $(N-1)/2$ impulses arise in any N sample interval. ∎

Thus, the impulses are entirely suppressed by the median filter.

We can illustrate these properties via a numerical example. Figure 3.1-1 illustrates the effects of average and median smoothing on a simple sequence. The synthetically constructed sequence is shown in Fig. 3.1-1a). The first part of the sequence (from $n = 0$ to $n = 400$) is a LOMO(25) sequence as defined above. The latter portion of the sequence (from $n = 401$ to 650) comprises constant regions with occasional sparse impulse-like values. Figures 3.1-1b) and c) show the resulting sequence after application of average and median filtering, respectively (with $N = 13$ in both cases).[3] As expected, the median filter preserves the signal edges and completely eliminates the impulses from the signal. In contrast, the average filter achieves neither of these, smearing the edges and reducing the amplitudes, but not eliminating the impulses. The conditions underlying the deterministic properties illustrated in this example are rarely precisely satisfied in practice. However, the tendency of the median filter to approximately preserve signal edges and largely remove impulses has been widely observed.

As we have already indicated, median filters act as smoothing filters for noisy data. However, as measured by the reduction of noise variance achieved by the filter, the smoothing obtained by a median filter may not be as great as that achieved by the average. For example, if equal length median and average filters are applied to zero-mean Gaussian white noise, the variance of the output sequence is greater by some 57% for the median filtered sequence [3-6]. This observation led several investigators to develop generalizations of median filters based on linear combinations of order statistics [3-7], [3-8]. We can define an *order statistic* (OS) filter as follows: An N-point non-recursive OS filter with input $\{x(n)\}$ produces output $\{z(n)\}$ according to the relation (see Fig. 3.1-2)

$$z(n) = \boldsymbol{a}^T \boldsymbol{x}_{(n)} \triangleq OS_{\boldsymbol{a}}\{\boldsymbol{x}(n)\}_N, \tag{3.1-3}$$

where the components $x_{(n)}(i)$ of the vector $\boldsymbol{x}_{(n)}$ are the last N input values $x(n), \ldots, x(n-N+1)$ ranked as $x_{(1)}(n) \le x_{(2)}(n) \le \cdots \le x_{(N)}(n)$, where $\boldsymbol{a}$ is a vector

[3] Since the sequence in this example is *not* infinite two-sided, an initialization effect occurs. In particular, in order to obtain $z(0)$, we need to assume values at $x(-1)$, $x(-2)$, ..., $x(-12)$. There are various ways such initial conditions can be constructed [3-6]. In this example we use "constant extension" i.e., $x(-12) = x(-11) = \cdots = x(0)$.

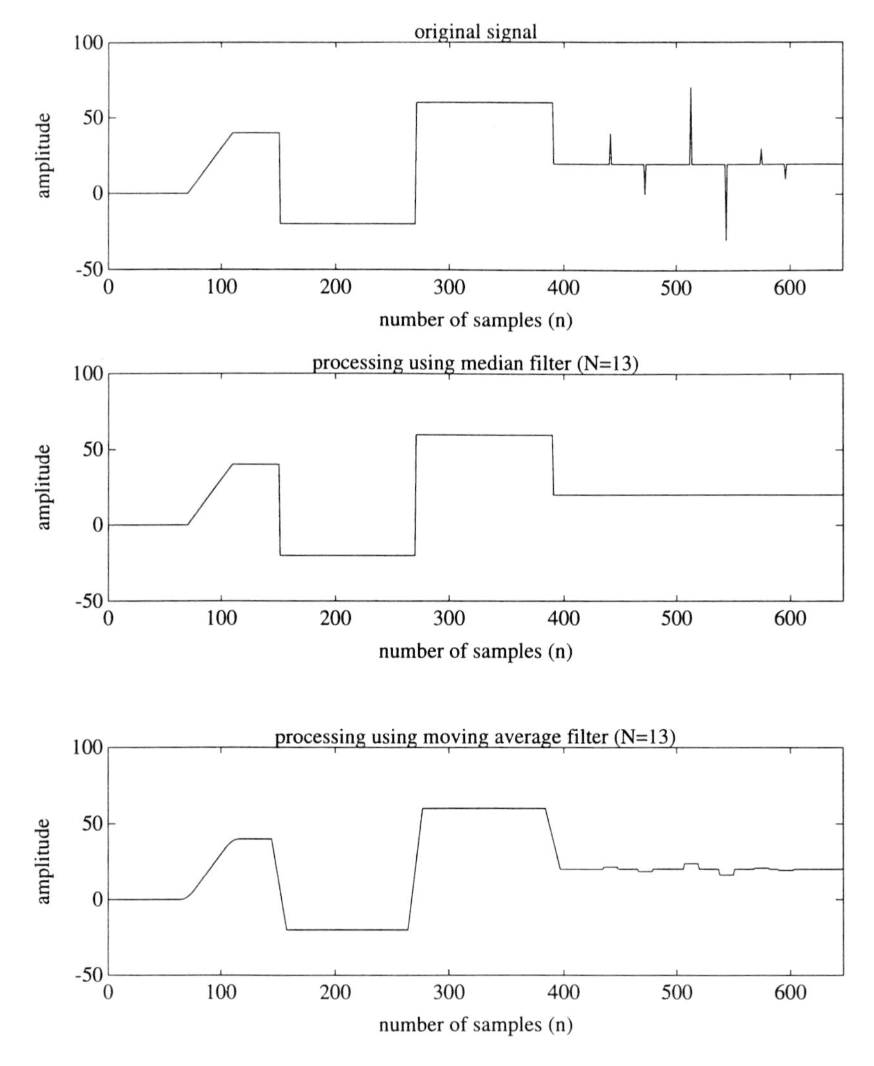

Figure 3.1-1: Average and median filtering applied to a synthetic sequence. a) Synthetic sequence consisting of LOMO region (from $n = 0$ to $n = 400$) and constant region with impulses (from $n = 401$ to $n = 650$. b) Median filter ($N = 13$) applied to a). c) Average filter ($N = 13$) applied to a).

weighting these ranked values, and where $\sum_{i=1}^{N} a_{(i)} = 1$ and $a(i) = a(N-i)$, $i = 1, 2, ..., (N-1)/2$.

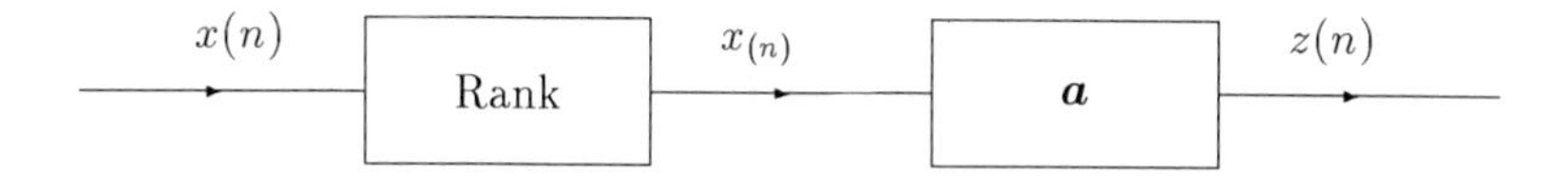

Figure 3.1-2: An order statistic filter – input data are algebraically ranked and then weighted.

The class of OS filters includes the median, for which

$$\boldsymbol{a} = [\underbrace{0 \cdots 0}_{\frac{N-1}{2}} \ 1 \ \underbrace{0 \cdots 0}_{\frac{N-1}{2}}]^T. \tag{3.1-4}$$

Equations (3.1-3) and (3.1-4) provide an alternative notation to Eq. (3.1-1). The moving average (3.1-2) is also an OS filter, with

$$\boldsymbol{a} = [\frac{1}{N} \ \frac{1}{N} \ \cdots \ \frac{1}{N}]^T. \tag{3.1-5}$$

This is the only linear member of the class and as we have seen corresponds to a simple *finite impulse response* (FIR) low-pass filter with impulse response given by Eq. (3.1-5). Another important OS filter is the trimmed mean for which

$$\boldsymbol{a} = [\underbrace{0 \cdots 0}_{M} \ \frac{1}{N-2M} \cdots \frac{1}{N-2M} \ \underbrace{0 \cdots 0}_{M}]^T. \tag{3.1-6}$$

The trimmed mean corresponds to ranking the data and censoring extreme values, and may be viewed as an intermediate between the median and the average filters [3-8]. A final example is provided by the outer mean filter for which

$$\boldsymbol{a} = [\frac{1}{2} \ 0 \ 0 \ \cdots \ 0 \ \frac{1}{2}]^T. \tag{3.1-7}$$

An obvious but often overlooked fact is that OS filters in general do *not* share the edge preservation and impulse suppression properties of the median [3-9]. OS filters *are* translation invariant and do preserve linear trends in data. Some OS filters, including notably the trimmed mean, may be preferred to the median because they can approximate the edge preservation and impulse suppression properties of median filters while simultaneously providing improved smoothing of non-impulsive noise.

It is in the context of smoothing, as measured by the efficiency of the operator, that the power of OS filters can be best appreciated. In particular, consider the following scenario: an unknown, locally constant signal s is embedded in zero-mean stationary noise $\{\eta(n)\}$, producing a measurement sequence $\{x(n)\}$ where

$$x(n) = s + \eta(n), \tag{3.1-8}$$

in some neighborhood of n. Consider the problem of estimating s from the observations. This problem may be approached via the application of OS filters to the noisy measurements $x(n)$. We can construct a *minimum variance unbiased* (MVUB) OS filter as the set of weights $\boldsymbol{a}^*$ that minimizes

$$J = E\{(z(n) - s)^2\}; \quad \text{subject to} \quad E\{z(n)\} = s, \tag{3.1-9}$$

where $z(n)$ is the OS filter output of Eq. (3.1-3). Combining Eqs. (3.1-3) and (3.1-8) we may write

$$E\{z(n)\} = \boldsymbol{a}^T \mathbf{1} s + \boldsymbol{a}^T E\{\boldsymbol{\eta}_{(n)}\}\ , \tag{3.1-10}$$

where $\mathbf{1}^T = [1\ 1 \cdots\ 1]$, and $\boldsymbol{\eta}_{(n)}^T = [\eta_{(n)}\ \eta_{(n-1)}\ \cdots\ \eta_{(n-N+1)}]$ is the vector of ranked samples. From Eq. (3.1-10) we observe that the unbiasedness constraint may be expressed in terms of two constraints:

$$\boldsymbol{a}^T \mathbf{1} = 1 \text{ (scaling constraint)}, \tag{3.1-11}$$

and

$$\boldsymbol{a}^T E\{\boldsymbol{\eta}_{(n)}\} = 0 \text{ (orthogonality constraint)}. \tag{3.1-12}$$

More compactly we write

$$C^T \boldsymbol{a} = \mathcal{F}, \tag{3.1-13}$$

where $C = [\mathbf{1}\ E\{\boldsymbol{\eta}_{(n)}\}]$, and $\mathcal{F} = [1\ 0]^T$. Since the constraint $\boldsymbol{a}^T\mathbf{1} = 1$ gives $z(n) = s + \boldsymbol{a}^T\boldsymbol{\eta}_{(n)}$, it follows directly from Eq. (3.1-9) that the minimization can equivalently be expressed as:

$$\text{Minimize } E\{(\boldsymbol{a}^T \boldsymbol{\eta}_{(n)})^2\}, \text{ subject to } C^T \boldsymbol{a} = \mathcal{F}\ . \tag{3.1-14}$$

This constrained minimization problem may be solved using Lagrange multipliers to give

$$\boldsymbol{a}^* = R_{(\eta)}^{-1} C (C^T R_{(\eta)}^{-1} C)^{-1} \mathcal{F}\ , \tag{3.1-15}$$

where $R_{(\eta)}$ is the $(N \times N)$ autocorrelation matrix of ordered noise measurements with elements $R_{(\eta)}(i,j) = E\{\eta_{(i)}(n)\eta_{(j)}(n)\}$.[4]

[4]Interestingly, the solution exactly parallels that for the constrained minimum variance linear beamformer [3-11].

When the noise sequence $\{\eta(n)\}$ consists of independent, identically distributed (iid) samples drawn from a symmetric density, one may show [3-10] that the MVUB OS filter $\boldsymbol{a}^*$ reduces to

$$\boldsymbol{a}^* = \frac{R_{(\eta)}^{-1}\mathbf{1}}{\mathbf{1}^T R_{(\eta)}^{-1}\mathbf{1}}. \tag{3.1-16}$$

For symmetric iid observations, the MVUB OS estimator $\boldsymbol{a}^*$ is the median when the noise distribution is Laplacian, the average when it is Gaussian, and the outer mean when it is uniform. For iid errors, it can easily be shown that the variance of this MVUB OS estimator is less than or equal to the output variance corresponding to the *best linear unbiased estimator*. In fact, for many distributions, the MVUB OS operator corresponds to the *uniformly* minimum variance unbiased estimator among *all* possible estimators [3-12].

3.2 Adaptive Filters and Order Statistics

In filter design, adaptation of the filter parameters provides a means for responding to data and system fluctuations as well as a mechanism for training in the face of unknown system characteristics.[5] These considerations apply equally to linear and to OS filters, and with the increasing interest in OS filters, the development of adaptive forms is a natural step.

It is important to differentiate between two forms of adaptation where OS operations are concerned:

A. **Order Statistic Adaptive Filters (OSAF)** – in which the nonlinear OS filtering operation (3.1-3) is embedded into the iterative update procedure for a *linear* filter (see Fig. 3.2-1a)). That is, where OS filtering forms a part of the update procedure in a conventional adaptive filter.

B. **Adaptive Order Statistic Filters (AOSF)** – in which the coefficients of the nonlinear OS filter (3.1-3) are iteratively updated (see Fig. 3.2-1b)).

In spite of the similar descriptions, these are very different classes of filter, having application to quite different problems. We discuss each of these separately below.

A. Order Statistic Adaptive Filters

Order statistic adaptive filters are a class of adaptive filters which employ OS operations to achieve performance enhancements over conventional adaptive filter designs. These improvements are obtained by embedding the OS operation in

[5] A detailed discussion of the motivation for adaptive solutions and the mechanisms of adaptation can be found in Chapter 2.

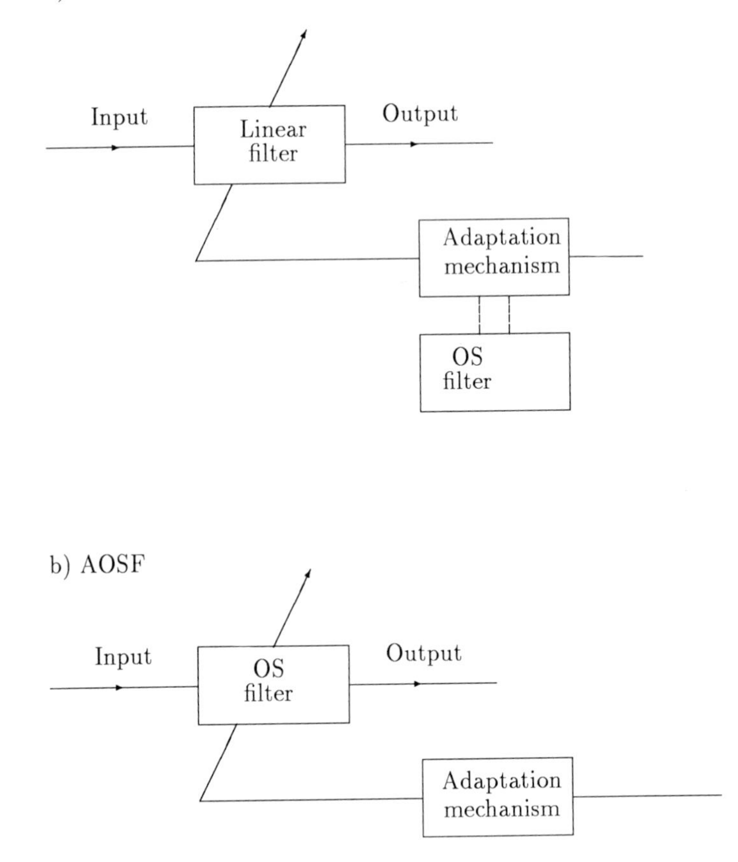

Figure 3.2-1: An overview of adaptation and order statistics in digital filter design. a) OSAF in which an adaptive filter is updated using a mechanism dependent on an OS operator. b) AOSF in which the coefficients of an OS filter are adaptively updated.

the update for the adaptive filter, with the objective of exploiting the powerful estimation properties of OS operators. While many adaptive algorithms could potentially benefit from the application of OS operators, it is the well-known *least mean squares* (LMS) algorithm [3-13] that has been used in the studies reported to date. LMS is an iterative update formula for a finite impulse response (FIR) linear filter which attempts to minimize the quadratic performance index

$$J = E\{e(n)^2\}, \tag{3.2-1}$$

where

$$e(n) = d(n) - y(n), \tag{3.2-2}$$

is the error between some desired, or ideal signal $d(n)$, and the filter output $y(n)$ given by

$$y(n) = \boldsymbol{f}^T(n)\boldsymbol{x}(n). \tag{3.2-3}$$

Here $\boldsymbol{f}(n) = [f_0(n)\ f_1(n) \cdots f_{L-1}(n)]^T$ is the vector of impulse response coefficients corresponding to time-index (update) n, and the data vector $\boldsymbol{x}(n) = [x(n)\ x(n-1)\ \cdots\ x(n-L+1)]^T$ (see Fig. 3.2-2).

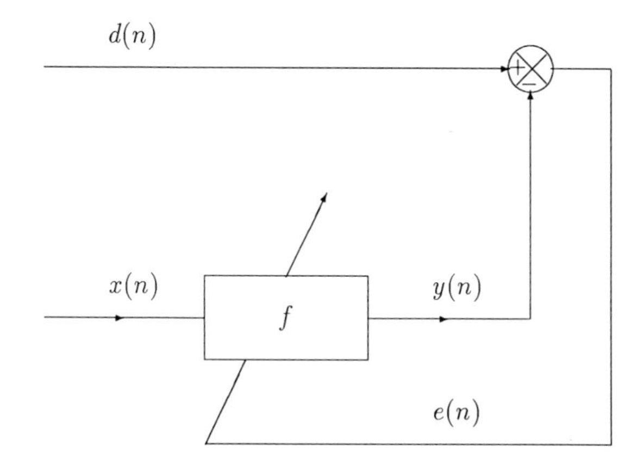

Figure 3.2-2: The LMS adaptive filter.

As detailed in Chapter 2, LMS is a steepest descent based algorithm, and is obtained by approximating the gradient of the mean-square error performance surface with an instantaneous data-based estimate. For an L-point non-recursive filter, this approach yields an iterative update equation for the filter coefficients of the form

$$f_i(n+1) = f_i(n) + \mu e(n)x(n-i)\ ; \qquad i = 0, 1, ..., L-1 \tag{3.2-4}$$

where μ is the adaptation step-size parameter. The LMS algorithm is subject to slow and generally non-uniform convergence, but the algorithm is well known for its simplicity and reliability. The use of the instantaneous estimate admits noise as well as the true gradient into the filter coefficient update, however, and this is reflected in an increase in the mean-square error over and above the minimum mean-squared error associated with the filter minimizing (3.2-1).

Given a sequence of instantaneous noisy gradient measurements, one may ask what is the optimal way to combine this data to get the best estimate of the true gradient? In most cases, the LMS algorithm is *not* the most efficient estimate. In fact, the optimality of the estimation procedure depends on the distribution of the raw gradient measurements, and thus indirectly on the data itself.

An improved estimate can be formed by replacing the instantaneous gradient estimate by the estimate obtained by applying an OS filter to the N most recent observations. Thus, an *order statistic LMS* (OSLMS) adaptive filter can be defined using an update equation [3-14]

$$f_i(n+1) = f_i(n) + \mu OS_{\boldsymbol{a}}\{e(n-i)x(n-i)\}_N \; ; \qquad i = 0, 1, ..., L-1 \qquad (3.2\text{-}5)$$

where the operation $OS_{\boldsymbol{a}}\{e(n)x(n-i)\}_N$ denotes the application of the OS operator to the N most recent gradient estimates for coefficient i. The filter output is given by Eq. (3.2-3), and together Eqs. (3.2-3) and (3.2-5) represent the class of OSLMS algorithms. It is important to realize that the adaptation is still intended to drive $\boldsymbol{f}(n)$ towards the least squares optimal filter $\boldsymbol{f}^*$ minimizing J from (3.2-1). The linear form of the *filtering operation* at each iteration is retained; the difference is the application of the nonlinear OS smoothing operation to the *gradient vector estimate.*

The incorporation of the OS operator into the LMS update generates a computational overhead for the algorithm. However, efficient algorithms facilitate the sorting of the data vector in $O(N)$ operations per coefficient with a comparable storage requirement [3-8]. In most studies values of $N \ll L$ have been employed, and overall computation remains $O(L)$ as in LMS.

1. The Median LMS Algorithm

The choice of OS filter employed by OSLMS should be conditioned on the nature of the input signals. If the gradient estimates are perturbed by impulsive noise, as may arise in a noisy communication channel for example, then the raw instantaneous estimate used by LMS will be subject to significant error whenever an impulse occurs. This was recognized by Clarkson and Haweel [3-15] who argued that replacing the instantaneous gradient with the median of the N most recent values would yield an improved gradient estimate. This observation leads to the *median LMS* (MLMS) algorithm with update (3.2-5) where $\boldsymbol{a}$ is given by Eq. (3.1-4). An alternative notation for the MLMS algorithm has the form

$$f_i(n+1) = f_i(n) + \mu \text{Med}\{e(n)x(n-i)\}_N \; ; \qquad i = 0, 1, ..., L-1 \; , \qquad (3.2\text{-}6)$$

where the operation $\text{Med}\{e(n)x(n-i)\}_N$ denotes the application of the median to the N most recent gradient estimates:

$$\text{Med}\{e(n)x(n-i)\}_N = \text{median}\{e(n)x(n-i), ..., e(n-N+1)x(n-i-N+1)\} \,. \tag{3.2-7}$$

Intuitively, the median operation provides the LMS update with some protection from the impact of sparse impulses in the input or desired signal. As a numerical example consider the behavior of the LMS and MLMS algorithms with input signals:

$$\begin{aligned} d(n) &= \boldsymbol{h}^T \boldsymbol{w}(n) + \eta(n) \\ x(n) &= w(n) + \zeta(n) \,, \end{aligned} \tag{3.2-8}$$

where $w(n)$ is a zero-mean, Gaussian iid sequence, $\zeta(n)$and $\eta(n)$ are sparse impulsive interferences, $\boldsymbol{h} = [h(0) \cdots h(L-1)]^T$ is the impulse response of a linear time-invariant system, and $\boldsymbol{w}(n) = [w(n) \;\; \cdots \;\; w(n-L+1)]^T$ (see Fig. 3.2-3).

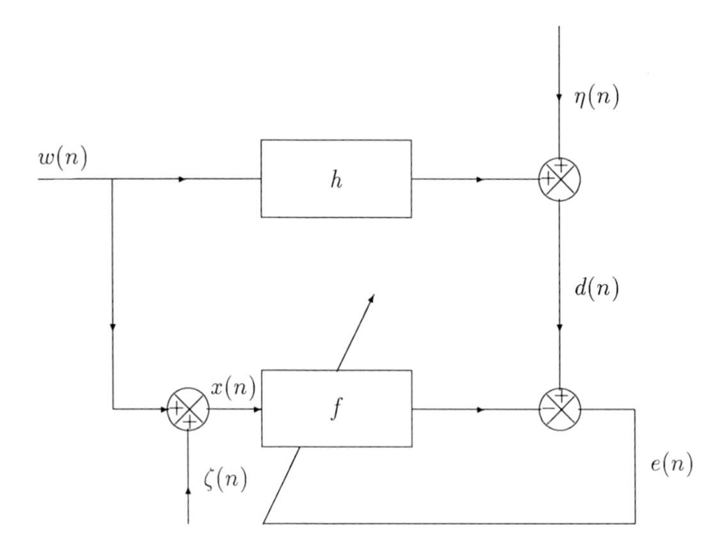

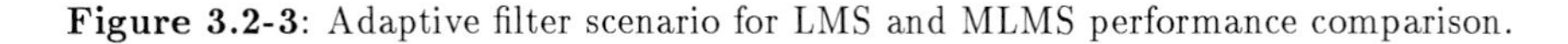

Figure 3.2-3: Adaptive filter scenario for LMS and MLMS performance comparison.

The objective of this experiment is to identify an unknown system $\boldsymbol{h}$, given only input and output measurements that are corrupted by sparse impulsive interference. Let us assume for simplicity that $\boldsymbol{h} = [-1 \; 0 \cdots 0]^T$ and that the filter coefficients are set to zero. We consider the behavior of the adaptive filter for two cases: a) $\zeta(n) \equiv 0$ (zero input noise) and b) $\eta(n) \equiv 0$ (zero output noise), in order to separately examine the impact of impulsive disturbance on the input and output of the unknown system $\boldsymbol{h}$. Referring to Fig. 3.2-3, in this simulation the input $w(n)$ is a zero-mean iid sequence with samples uniformly distributed over $[-1, 1]$. The disturbances $\eta(n)$, $\zeta(n)$ are 'sparse' sequences with non-zero values (impulses) occurring at random intervals with a mean separation of 100 samples. The impulse amplitudes are Gaussianly distributed with mean zero and standard deviation equal to 8. Figure 3.2-4a) shows the coefficient error $v_0(n) = f_0(n) - f_0^*(n)$ for $n = 0, 1, ..., 3999$ for case a) using the MLMS algorithm. For reference, the coefficient track obtained using conventional LMS is also included. The plots in this example were obtained using $L = 5$, and $\mu = 0.01$ in both cases. For MLMS, the median window length N in this trial was set at 3. The influence of the sparse impulsive noise is to perturb the coefficient solution for LMS. For MLMS, the coefficient track remains smooth and largely unaffected by impulses. Figure 3.2-4b) shows the corresponding results for case b) using a similar sparse impulsive disturbance. Here, the performance of LMS is even more seriously degraded, fluctuating wildly, and in fact it can be shown [3-16] that the mean convergence in this case is to a solution which is biased away from $\boldsymbol{h}$. MLMS, by contrast, again exhibits smooth convergence and has greatly reduced bias.

The MLMS algorithm has been the subject of several studies. The potential performance improvements obtainable by the algorithm have been convincingly demonstrated in many simulations (see, for example, [3-15], [3-14], [3-16], [3-17]). The algorithm has also been applied to the problem of *adaptive differential pulse code modulation* (ADPCM) coding over noisy channels [3-17]. Typically, in an ADPCM system the channel is subject to random bit errors which result in an impulsive-like interference at the receiver. Conventional ADPCM systems use "coefficient leakage" in the adaptive update to allow the deleterious effects of impulses to gradually decay from the filter response. A leaky version of the LMS filter of Eq. (3.2-4) takes the form

$$f_i(n+1) = \gamma f_i(n) + \mu e(n) x(n-i) \; ; \qquad i = 0, 1, ..., L-1 \tag{3.2-9}$$

where $0 < \gamma < 1$.[6] Obviously, the use of leakage introduces bias into the adaptive filter resulting in performance degradation in noise-free channels. As an alternative, an ADPCM system incorporating median protection into the predictor update has been proposed [3-17]. This MLMS-ADPCM system was compared to the conventional LMS-ADPCM in the context of a 32 kbps system. It was found that in an ideal (error-free) channel, the performance of the MLMS-ADPCM is comparable to the conventional LMS-ADPCM (without leakage). In a degraded channel, the MLMS outperforms LMS and appears to obviate the need for coefficient leakage in the predictor update.

Analytic performance predictions for MLMS have only proved possible for very

[6]The actual update used in an ADPCM system is considerably more complicated than this due to the use of input power normalization and 'sign-only' update terms; see for example [3-18] for a detailed discussion.

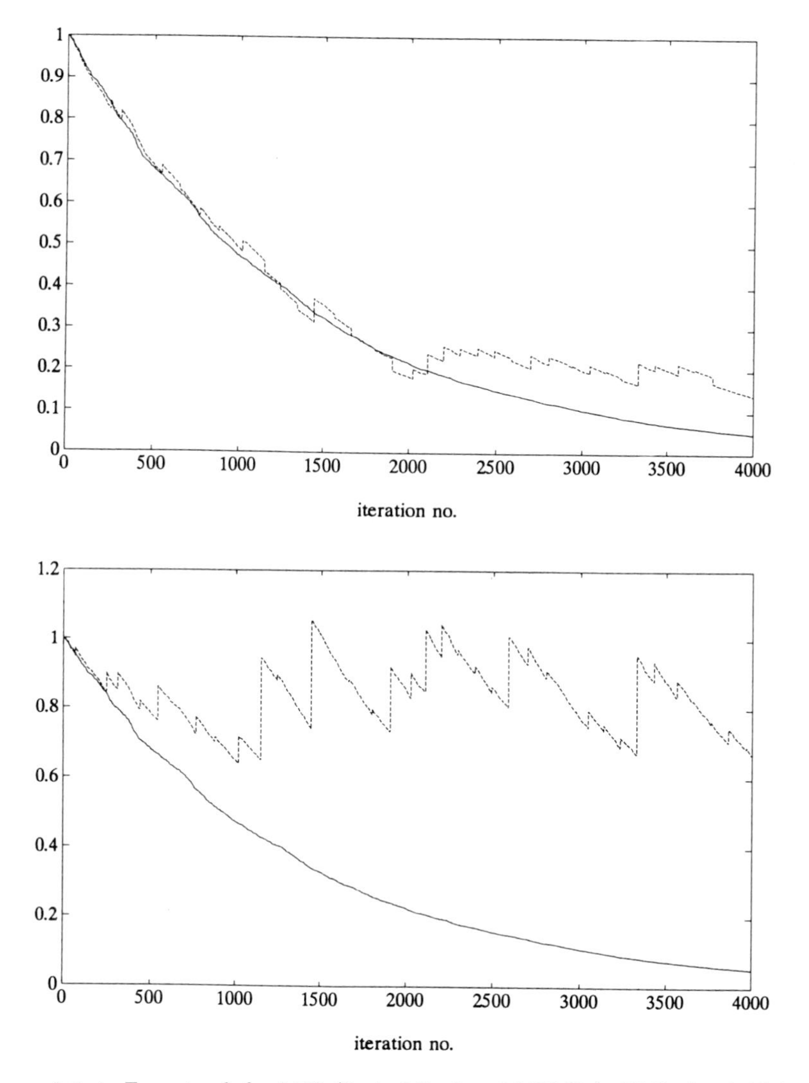

Figure 3.2-4: Error track for LMS (dashed line) and MLMS (solid line). a) $\zeta(n) = 0$, b) $\eta(n) = 0$.

restrictive inputs, and subject to a number of simplifying assumptions. For the case of zero-mean iid inputs, exponential convergence in the mean [3-16] and local stability of the filter coefficients [3-19] have been established for the MLMS algorithm.[7] Moreover, when the inputs are corrupted by sparse impulsive interferences, the rate of convergence for MLMS has been shown to have little dependence on the impulsive interference [3-16]. This contrasts sharply with LMS, which is degraded by the impulses. The steady-state behavior of the LMS and MLMS algorithms has also been compared for such inputs by examining the average deviation of the parameter estimates from their optimal values caused by the arrival of an impulse [3-16]. This analysis has also shown a significant performance advantage for MLMS. However, there are also a few pathological cases where instability can occur in the MLMS algorithm. One example identified in [3-16] involves an MLMS filter with $N = 3$, and filter length $L = 3$, with an input of the form

$$\begin{aligned} x(n) &= \ \ 3 \ \ ; n = 3m, \quad m \text{ integer} \\ &= -1 \ \ ; \text{otherwise.} \end{aligned} \tag{3.2-10}$$

This sequence can be viewed as a d.c. signal with a periodic impulse. Williamson *et al.* [3-16] showed that for particular filter coefficient vectors, the MLMS algorithm diverges with such inputs. Sethares and Bucklew [3-19] confirmed the local instability of the MLMS for such periodic inputs. This behavior can be contrasted with that of LMS, for which no such pathological behavior arises. It must be emphasized that such examples are rare, with all cases identified to date taking the form of highly artificial synthetic sequences as in the above example.[8]

2. OSLMS Algorithms

As we have indicated, MLMS is only a single member of the general class of OSLMS algorithms represented by Eq. (3.2-5). Other OSLMS filters investigated include the *average LMS* and the *trimmed mean LMS*. Simulations of OSLMS using the framework of Fig. 3.2-3 have been conducted [3-22], [3-23] using iid Gaussian sequences corrupted by unit variance iid data generated using a generalized exponential density:

$$f(x) = k_1 e^{-k_2|x|^\beta}\ ; \qquad 0 < \beta, \tag{3.2-11}$$

where

$$k_1 = (\beta k_2^{1/\beta})/2\Gamma(1/\beta), \quad k_2 = [\gamma(3/\beta)/\gamma(1/\beta)]^{\beta/2}, \tag{3.2-12}$$

and where Γ is the ordinary gamma function. As β in (3.2-11) increases, so the resulting density $f(x)$ varies from highly impulsive in nature ($\beta < 1$) to Laplacian

[7] The *rate* of convergence remains similar to LMS, however. Recently, an attempt to provide an increased convergence rate has been made via the development of a "split-path" median LMS algorithm [3-20].

[8] It is interesting to note that precisely the same example produces divergence in the well-known *signed-regressor* algorithm [3-21].

($\beta = 1$) to Gaussian ($\beta = 2$) to uniform ($\beta \to \infty$). Figure 3.2-5 shows the effect of varying β on the steady-state performance of several OSLMS algorithms. In these simulations [3-23], the OS window length was set to $N = 7$ and $L = 3$. Steady-state error is evaluated using the mean-square coefficient error $E\{\boldsymbol{v}^T\boldsymbol{v}\}$, where $\boldsymbol{v} = \boldsymbol{f}^* - \boldsymbol{f}_k$. Figure 3.2-5, shows post convergence values of J_{err} obtained by averaging over an "ensemble" of 1000 runs of 1000 iterations each. In these

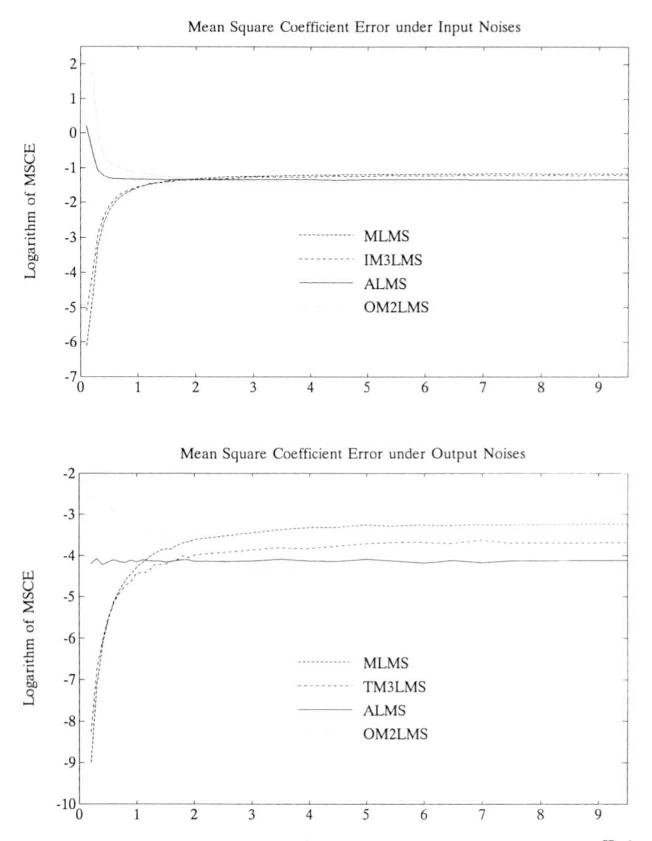

Figure 3.2-5: OSLMS steady-state performance – mean-square coefficient error a) $\zeta(n) = 0 = 0$; b)$\eta(n) = 0$. (Reprinted with permission ©IEEE 1993.)

simulations TM3LMS denotes $\boldsymbol{a} = [0\ 0\ \frac{1}{3}\ \frac{1}{3}\ \frac{1}{3}\ 0\ 0]^T$, and OM2LMS denotes the outer mean. One can observe from the diagram that for close to Gaussian noise ($\beta \approx 2$), the OSLMS algorithms all perform comparably. As the impulsivity increases (β decreases), the center weighted OSLMS algorithms perform best. On the other hand, as β increases, the OSLMS algorithms with outer weighting in $\boldsymbol{a}$ become slightly superior.[9]

Generally, the choice of a particular OS operator should be conditioned on the statistics of the gradient estimate $e(n)\boldsymbol{x}(n)$. For Gaussian or nearly Gaussian distributions, the simple averaging operation represents the optimal smoothing. For highly impulsive noise the median is appropriate, while for distributions between these extremes other order statistic filters, such as the trimmed mean, may yield superior performance. The ultimate objective of OSLMS filter design is to find estimates which are optimal with respect to the input data distribution and to define corresponding adaptive algorithms. In most instances, however, this distribution is not available *a priori* and can only be coarsely estimated using local data-based measures. Consequently, we seek methods which can produce close-to-optimal results using these crude estimates but which are *robust* to model imperfections.

Analytic performance predictions are even more difficult to obtain for general OSLMS algorithms than for MLMS. Fu *et al.* [3-23], [3-24] examined the asymptotic behavior of OSLMS algorithms applied to iid inputs. They showed that relative to LMS, in non-Gaussian environments, OSLMS can reduce the bias on filter coefficient estimates. They also demonstrated that OSLMS algorithms can reduce the excess mean-square error in steady state operation.

Ideally, the OSLMS filter should be capable of producing performance gains for all input distributions. The ultimate OSLMS filter would be one whose OS operator $\boldsymbol{a}$ is itself adapted so as to reach an optimal form with respect to the local density of the observed gradient estimates. Some work in this direction appears in [3-14], where the OS operator is adapted in response to local data-based impulse detectors. This adaptation mechanism is similar to the type of OS filter selection scheme devised by Restrepo and Bovik [3-25]. According to this approach a measure of the 'tail-behavior' γ, say, is determined using the gradient estimates defined within the local N-point window. The parameter γ reflects the impulsivity of the data in the window. The OS operator acts as a trimmed mean with the degree of trimming determined by the value of γ. Suitable tail statistics include the sample kurtosis and similar measures.[10] Haweel and Clarkson achieved good results by switching between mean and trimmed mean filters according to a simple threshold rule:

$$\begin{aligned} \text{Trimming factor } M &= M_1\ ; \quad \gamma \geq T \\ &= 0\ ; \quad \gamma < T, \end{aligned} \tag{3.2-13}$$

where M_1 and T are chosen in an empirical manner.

---

[9]This illustrates an important point about OS filtering in general – the effect of outliers on mismatched operators will likely impact performance far more than the impact of "inliers".

[10]Estimating the impulsivity of a distribution via data-based measures is discussed in greater detail in Section 3.4.

B. Adaptive Order Statistic Filters

Let us return to the constant signal model of Eq. (3.1-8). As we have seen, the MVUB solution of Eqs. (3.1.15) and (3.1.16) depends on the correlation matrix $R_{(\eta)}$. Bovik *et al.* [3-7] determined the optimal filter vector using known statistics for the observed noise. On the other hand, it is often necessary to construct OS filters *without* prior knowledge of the noise distribution. One approach is simply to form block estimates from the data which are then used to construct the OS filter [3-25]. Another possibility is to apply an iterative update to the OS filter vector. We refer to such a system as an *adaptive order statistic filter* (AOSF) (see Fig. 3.2-6).

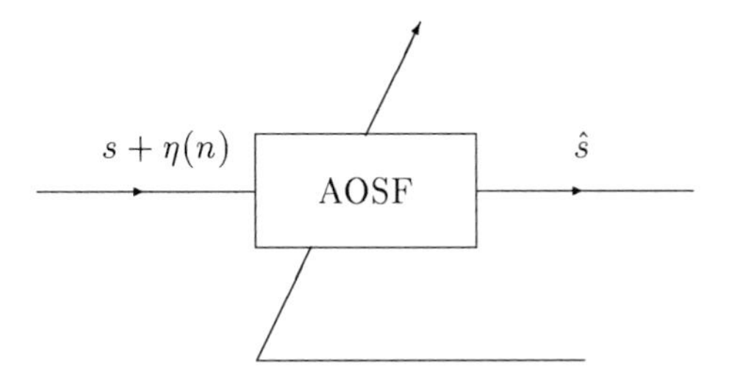

Figure 3.2-6: Estimating a constant signal using an adaptive order statistic filter.

Pitas and Venetsanopoulos [3-26] proposed a gradient based AOSF algorithm intended to achieve the constrained minimization of Eq. (3.1-9) using an LMS type update. Crucially, however, their algorithm employed only the constraint $\boldsymbol{a}^T\mathbf{1} = 1$, and forced satisfaction of the constraint by a simple scaling of the filter coefficients. The resulting filter vector does not converge to $\boldsymbol{a}^*$ in general, and the filter suffers from suboptimal and potentially poor performance. A gradient based algorithm that *does* satisfy both constraints was proposed by Williamson and Clarkson [3-10]. This AOSF algorithm satisfies both scaling and symmetry constraints by using a constrained minimization procedure analogous to that of Frost [3-11], but applied to ordered data. In this algorithm, the LMS update of Eq. (3.2-4) is modified to

$$\boldsymbol{a}(n+1) = \boldsymbol{P}[\boldsymbol{a}(n) \ + \ \mu\boldsymbol{x}_{(n)}e(n)] + \boldsymbol{F} \ , \tag{3.2-14}$$

where $\boldsymbol{F}$ and $\boldsymbol{P}$ are defined by

$$\boldsymbol{P} = \boldsymbol{I} - \boldsymbol{C}(\boldsymbol{C}^T\boldsymbol{C})^{-1}\boldsymbol{C}^T \tag{3.2-15}$$

$$\boldsymbol{F} = \boldsymbol{C}(\boldsymbol{C}^T\boldsymbol{C})^{-1}\mathcal{F} \ . \tag{3.2-16}$$

Equation (3.2-14) projects the LMS update $\boldsymbol{a}(n)+\mu\boldsymbol{x}_{(n)}e(n)$ into the space satisfying the constraint $\boldsymbol{C}^T\boldsymbol{a} = \mathcal{F}$ [3-11]. Note that the form of $\boldsymbol{C} = [\mathbf{1}\boldsymbol{m}_{(\eta)}]$ suggests a requirement for prior knowledge of the ordered mean $\boldsymbol{m}_{(\eta)}$. However, when $\eta(n)$ is zero-mean symmetric then $\boldsymbol{a}^T \boldsymbol{m}_{(\eta)} = 0$ for any *symmetric* $\boldsymbol{a}$.[11] Hence for such densities we may replace the orthogonality constraint by a simple symmetry constraint on $\boldsymbol{a}$ [3-10].

Much of the motivation for these algorithms has arisen from applications in image processing [3-25], [3-26]. More generally, one can motivate the study of such estimators via several important problems which share the common framework of signal modeling in the context of Eq. (3.1-8).

1. Doubly Adaptive OSLMS Algorithm

In an LMS type gradient based adaptive algorithm, the gradient at any iteration n can be modeled via Eq. (3.1-8) where s is the unknown gradient and $\eta(n)$ are the errors (noise) implicit in the use of a data-based estimation procedure. In using this model, it is assumed that the step-size is sufficiently small that the gradient is constant over short intervals. The problem is complicated somewhat by the fact that the samples $\eta(n)$ may not be independent or drawn from a symmetric density. Moreover, the slowly changing nature of the gradient, as well as the unknown noise statistics, motivates an iterative solution for $\boldsymbol{a}$. Nevertheless, one may in principle apply the AOSF mechanism of Eqs. (3.2-14) to iteratively update the OS operator $\boldsymbol{a}$ employed by the OSLMS algorithm. So far, as we described in the previous section, an adaptive update for the OS operator in OSLMS in the form of a simple threshold based switching scheme driven by a local measure of data impulsivity has been used. It is apparent that the use of an adaptive MVUB OS filter has potential for further improving OSLMS performance.

2. Array Processing

A classical problem in array processing is the minimization of array output power subject to a constraint in a specified 'look-direction' (the so-called *minimum variance distortionless response* (MVDR) processor) [3-27]. For a linear array of M equispaced elements, the received signals, although generally treated as complex, have a similar structure to that of Eq. (3.1-8). In particular, for a monochromatic plane wave incident from a known direction and embedded in iid noise, the received signals output from a steering (signal-alignment) array have the form

$$x_i = s + v_i \; ; \qquad i = 1, 2, ..., M, \tag{3.2-17}$$

where s is a constant (broadside) signal, and where the v_i's are steered noise terms. The linear processor applies a set of complex weights to the steered signals. For iid inputs, the optimal (minimum variance) weighting corresponds to the sample average of the steered signals. The MVUB OS constrained operator applies a

[11] By symmetry of a vector $\boldsymbol{a}$ we mean simply that $a_i = a_{N-i}$ for $i = 1, 2, ..., N-1$.

similar set of M weights to the *ranked* steered signals. In principle, the optimal OS operator can improve performance compared to the linear MVDR processor. In the linear case, adaptation is used to compensate for slowly varying noise statistics, and a similar motivation suggests the need for an iterative *MV order statistic* (MVOS) operator.[12]

3. AOSF Filters in Cardiac Arrhythmia Detection

Another application of the AOSF filter has been identified in the area of biomedical signal analysis. An *automatic implantable cardiovertor defibrillator* (AICD) is a device that is surgically implanted into the hearts of patients identified to be at high risk for sudden cardiac death. These devices monitor the electric activity of the patient's heart, automatically diagnose the presence of potentially fatal arrhythmias, and apply appropriate therapy once an arrhythmia is detected. Diagnosis relies crucially on accurate and timely estimates of important parameters of the electrogram such as the heart rate, its regularity, and other morphological features of the cardiac activation cycle. One of the most basic and widely used parameters is rate [3-28]. Estimation schemes for rate use a simple moving average applied to the raw observations. These raw values are typically obtained using a threshold calculated from the local peak amplitude of the electrogram to locate the depolarizations [3-29]. These observations contain minor beat-to-beat fluctuations due to respiratory variations which can be smoothed by averaging over a number of observations. However, even in well-organized rhythms, larger errors (outliers) can occur due to a variety of reasons including missed beats and double counted depolarizations due to amplitude fluctuations, as well as premature beats due to ectopic foci or abberant conduction.

Empirical procedures for adapting the depolarization threshold have been proposed to reduce the occurrence of outliers [3-31]. However, as we have already indicated, the occurrence of even a single outlier can be sufficient to severely degrade an estimate obtained by averaging. Moreover, even when the raw values are not corrupted by outliers, in most cases the observations are not Gaussianly distributed and averaging algorithms have only limited statistical efficiency.

We can use the interval between successive depolarizations $\tau(i)$, say, to generate raw rate observations $x(i) = 1/\tau(i)$. Over a short time interval, these values can be viewed as random variables $v(i)$; $i = 1, 2, ..., N$ drawn from a single distribution $f(v)$, with the rate r corresponding to the mean of the observations $r = E\{v(i)\}$. The measured values $x(i)$ can be modeled as the rate variables $v(i)$ corrupted by

[12]The MVDR array processor is the spatial equivalent of the *minimum variance spectral estimator*. An MVOS spectral estimator has potential advantages in most noise environments although issues related to the estimation of the moments of ordered data from finite sequences need to be addressed in order to develop effective solutions.

occasional large errors (outliers) $\zeta(i)$ as

$$x(i) = v(i) + \zeta(i) \; ; \quad i = 1, 2, ..., N, \tag{3.2-18}$$

where the outliers can be modeled via a 'sparse impulsive' distribution:

$$\begin{aligned} \text{prob}(\zeta(i) \neq 0) = & \; c(i) \\ \text{prob}(\zeta(i) = 0) = & \; 1 - c(i), \end{aligned} \tag{3.2-19}$$

where $c(i) \ll 1$. A common rate estimate is the *sample average*. However, we may also contemplate alternative estimators derived from nonlinear filters such as OS filters. In principle, we may apply an MVUB OS estimator to each sequence of N observations and use the resulting rate estimates to detect arrhythmias. In practice, the error distribution is not available *a priori* and one must apply data-driven (adaptive) algorithms. Moreover, heart rate is time-varying, exhibiting beat-to-beat fluctuations, gradual increases or decreases, and even rapid step-like transitions corresponding to onset and termination of paroxysmal tachycardias. In addition to the need for efficiency and robustness, a rate estimation scheme must be responsive to track such changes as they occur. Thus, for practical use we need an efficient, robust, *adaptive* estimator.

The AICD application illustrates both the strengths and weaknesses of the AOSF filter. The MVUB algorithm has been applied to this problem [3-32]. This algorithm produces adaptive updates of the MVUB estimator and *does* produce efficient estimates. On the other hand, the filter suffers from two limitations that restrict its practical utility. Firstly, we note that gradient based adaptation is often associated with slow convergence. Secondly, the AOSF can take the form of *any* OS operator and, as we discuss below in Section 3.4, not all OS operators are robust. Thus, in some applications, notably where rapid adaptation and robustness to outliers are required we must sacrifice the optimality of the MVUB AOSF. We will describe the construction of such algorithms in Section 3.4 below after first discussing the issue of robustness in greater detail.

3.3 OS Filters and Robustness

As we have indicated, while OS filters may be formulated that are optimal with respect to a particular distribution, not all OS filters are robust to deviations from that distribution [3-30]. For example, as we have repeatedly stressed, the average and the outer mean can be significantly degraded by a single outlier. We may characterize the robustness of OS estimators in a variety of ways. Two useful measures are the *breakdown point* [3-33] and the *influence function* [3-34].

A. *Breakdown point* – The breakdown point of an estimator is the smallest fraction of the observations that must be replaced by unbounded outliers before the estimation error can become unbounded. For example, the breakdown point of the average, applied to N observations, is equal to $1/N$ because it is sufficient to replace a single observation by one unbounded value. On the other hand, the sample median possesses the best possible breakdown point, namely $(N+1)/2N$. We have to replace more than half of the observations by outliers in order for the median to lie among the outliers.

B. *Influence function* – Let $\hat{s} = T_N(x(1), ..., x(N))$ be an estimator of a parameter s with asymptotic value

$$\lim_{N\to\infty} \{T_N(x(1), x(2), ..., x(N))\} \to T(G), \tag{3.3-1}$$

where $G(x)$ is the distribution of the iid observations $x(i)$. The *influence function* (IF) of T at F is defined by

$$IF(x; T, F) = \lim_{\delta\to 0} \left\{ \frac{T((1-\delta)F + \delta\Delta_x) - T(F)}{\delta} \right\}, \tag{3.3-2}$$

where Δ_x is a point mass at x. The influence function thus describes the effect on the estimate $T(F)$ of an infinitesimal contamination at point x. For example, for zero-mean symmetric distributions, the mean and median have influence functions:

$$IF_{mean}(x; T, \Phi) = x \tag{3.3-3}$$

$$IF_{med}(x; T, \Phi) = \frac{\text{sgn}(x)}{2f(0)}, \tag{3.3-4}$$

where sgn($\cdot$) denotes the sign operator. These functions are shown in Fig. 3.3-1 [3-34]. The diagram clearly illustrates the distinction between the two estimators – for the mean, the IF is unbounded – there is no limit to the influence of a single observation, while for the median the IF is strictly limited.

A useful measure derived from the IF is the *gross error sensitivity* [3-34]

$$\gamma^* = \sup_x |IF(x; T, F)|. \tag{3.3-5}$$

Estimates for which γ^* is finite are called *β-robust.* It is immediately apparent that the median is β-robust, while the mean is not.

In general, for outlier corrupted data a compromise between efficiency and robustness is provided by the trimmed mean operator. The breakdown point of the trimmed mean is $(M+1)/N$. This dependence on the trimming parameter M allows us to generate a continuous range of breakdown values between the extremes of the mean $(M=0)$ and the median $(M=(N-1)/2)$. Similarly, the IF of the trimmed mean lies between the extremes of mean and median. For data that is outlier corrupted, we may argue that restricting the OS operator to take the form of a trimmed mean with M adjusted adaptively, provides good, though not optimal efficiency, while giving good resistance to outliers.

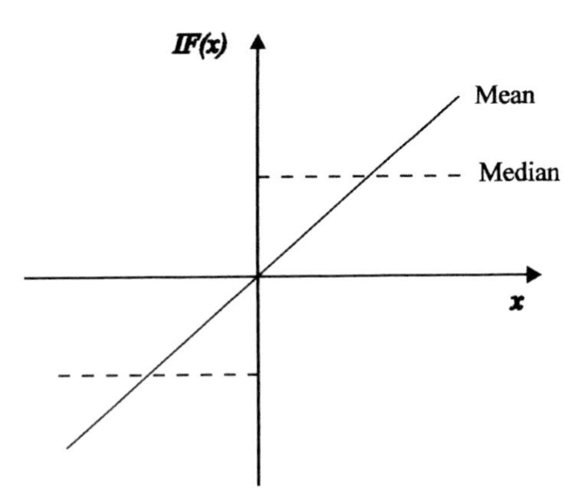

Figure 3.3-1: Influence functions for the mean and median for a zero-mean Gaussian sequence.

3.4 Rapid Adaptation – An Ad Hoc Estimator

The AOSF of Williamson and Clarkson can provide efficient adaptive signal estimates. However, like all gradient based adaptive algorithms, convergence of this algorithm typically requires many iterations. In many applications, significant changes can occur over relatively short segments of data. Gradient based parameter estimators are not well suited to such environments. Moreover, the MVUB OS operator produced by the AOSF estimator can take the form of any OS filter, and is therefore not necessarily robust. Finally, while the OS filters described above form a useful class of estimators for *symmetrically distributed* outliers, observation suggests that in many applications the outlier corruption can be significantly asymmetric. In summary, we seek an algorithm adapting over a much shorter interval than is possible with a gradient based update, constrained to be drawn from the subset of *robust* OS filters, and capable of producing effective estimates in the presence of asymmetric outliers. In order to achieve these objectives it is necessary to sacrifice the global optimality of the MVUB solution and accept a suboptimal solution. An algorithm that achieves each of these objectives has been developed [3-35], and will be described in some detail below.

A. A Two-Stage Adaptive OS Estimator

The adaptive procedure described in this section comprises a two-stage estimator with outlier detection and removal followed by adaptive trimmed mean filtering (see Fig. 3.4-1) [3-35].

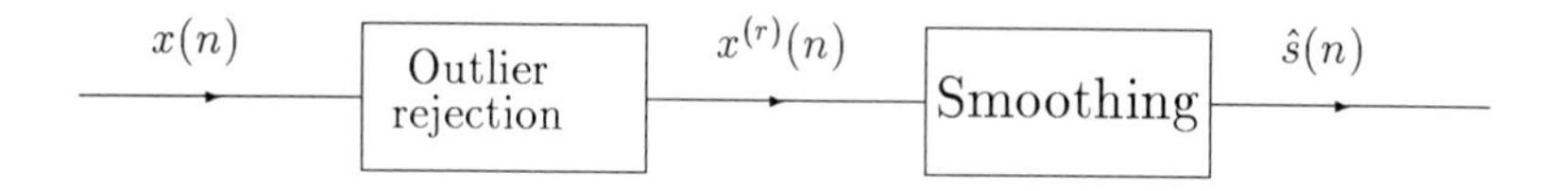

Figure 3.4-1: A two-stage OS estimator.

In the first stage, a highly robust outlier detection mechanism is applied to the observations. This both removes outliers from the data, and produces a residual set of observations that are approximately symmetric. A suitable outlier detection mechanism can be derived from the *median absolute deviation* (MAD), suggested by Hampel [3-36] as a robust estimator of the 'scale' of a set of observations as an alternative to the standard deviation which is notoriously non-robust. The MAD is the median of the absolute distances from the sample median of the observations and as such is the natural counterpart of the median for the estimation of scale [3-34]. For samples $x(i)$ for $i = 1, 2, ..., N$, the MAD is defined by

$$MAD = \text{ median } |x(i) - x_{med}| \ ; \quad i = 1, 2, ..., N, \tag{3.4-1}$$

where x_{med} is the sample median of the observations $x(i)$ for $i = 1, 2, ..., N$.

In the first stage of the estimator, an observation $x(i)$ such that

$$|x(i) - x_{med}| \geq K \times MAD, \tag{3.4-2}$$

is discarded as an outlier. Here, K is a constant chosen empirically in the range $4.0 - 5.5$.[13]

In the second stage of processing, the residual observations are subjected to a trimmed mean filtering operation, as defined by Eqs. (3.1-3), (3.1-6). The trimming factor M is adapted on the basis of a measure of the tail of the distribution of the residual observations (see also the discussion of Section 3.2). The approach taken in [3-35] is to use the measure Q_1 suggested by Hogg [3-37], and more recently adopted by Restrepo and Bovik [3-25]:

$$Q_1 = \frac{\overline{U}(.2) - \overline{L}(.2)}{\overline{U}(.5) - \overline{L}(.5)}, \tag{3.4-3}$$

[13]Sometimes the MAD of Eq. (3.4-1) includes a normalizing factor to produce a scale estimate that is consistent with the standard deviation when applied to Gaussian observations. Here, we absorb this factor into the constant K.

where

$$\overline{U}(\alpha) = \frac{1}{N\alpha} \sum_{i=N-[N\alpha]+1}^{N} x_{(i)} + (1 - \frac{[N\alpha]}{N\alpha})x_{(N-[N\alpha])}, \tag{3.4-4}$$

$$\overline{L}(\alpha) = \frac{1}{N\alpha} \sum_{i=1}^{[N\alpha]} x_{(i)} + (1 - \frac{[N\alpha]}{N\alpha})x_{(N-[N\alpha]+1)}. \tag{3.4-5}$$

$\overline{U}(\alpha)$ is the average of the $N\alpha$ largest order statistics, and $\overline{L}(\alpha)$ is defined similarly for the smallest order statistics, and where in both cases, $[\cdot]$ denotes the integer part of the argument. Hogg [3-37] argues that $[\overline{U}(.2)-\overline{L}(.2)]$ gives a good estimator of scale for Gaussian observations, while $[\overline{U}(.5) - \overline{L}(.5)]$ provides a good estimate for Laplacian data. On this basis, the ratio defined by Eq. (3.4-3) should provide a good indication of the tail behavior. One may examine this claim by computing values of Q_1 for various distributions. Figure 3.4-2 shows values of Q_1 obtained using data distributed according to the generalized exponential density of Eqs. (3.2-11), (3.2-12).

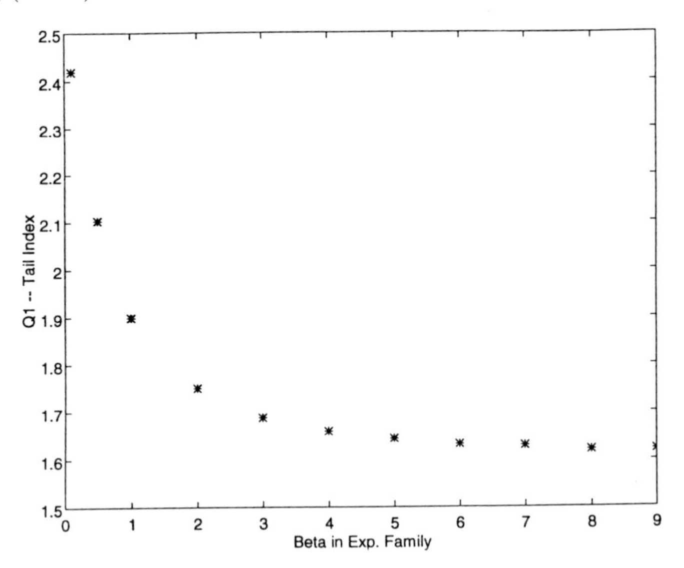

Figure 3.4-2: Tail parameter Q_1 for various distributions from the generalized exponential density parameterized by β. Each point is computed by averaging raw Q_1 values over an 'ensemble' of 1000 trials. $N = 20$ in each case.

It will be recalled from the earlier discussion that as β increases, so the generalized exponential density becomes less impulsive. Referring to Fig. 3.4-2 the decreasing values of Q_1 obtained as β increases is precisely what we would expect. Logically, the trimming factor M in Eq. (3.1-6) is then determined by comparing the value Q_1 against a set of predetermined thresholds. When the distribution is more impulsive, the tails are relatively heavy, Q_1 is higher, and the trimming factor is chosen to give a filter that is closer to the median; conversely when the distribution is closer to Gaussian, Q_1 is lower, and the estimator is more like the average. The algorithm of Fan and Clarkson [3-35] uses a threshold rule depending on N that is derived from a similar scheme suggested by Hogg [3-37]:

$$\begin{aligned}
10 < N \leq 20\,; &\quad M = [\tfrac{1}{8}N];\ Q_1 \leq 1.84 \\
; &\quad M = [\tfrac{2}{8}N];\ Q_1 > 1.84 \\
20 < N \leq 35\,; &\quad M = [\tfrac{3}{32}N];\ Q_1 \leq 1.81 \\
; &\quad M = [\tfrac{6}{32}N];\ 1.81 < Q_1 \leq 1.87 \\
; &\quad M = [\tfrac{9}{32}N];\ 1.87 < Q_1 \\
35 < N \quad ; &\quad M = [0\ N];\ Q_1 \leq 1.75 \\
; &\quad M = [\tfrac{3}{32}N];\ 1.75 < Q_1 \leq 1.81 \\
; &\quad M = [\tfrac{6}{32}N];\ 1.81 < Q_1 \leq 1.87 \\
; &\quad M = [\tfrac{9}{32}N];\ 1.87 < Q_1 \leq 1.91 \\
; &\quad M = [\tfrac{3}{8}N];\ 1.91 < Q_1\ .
\end{aligned}$$

Here, $[N \times \alpha]$ denotes the integer part of $N \times \alpha$. This is also called the α trimmed mean operator since the trimming factor M is determined by α. Note that, in order for the estimator to respond quickly, the estimation is based on observations made over a relatively short interval. However, if the number of observations is very small (10 or less, say) even the MAD scheme may fail to identify the outliers. For such short data segments it is often best to revert to the sample median.

For larger N, the threshold values can be optimized for a specific application by an empirical process using a 'training set'. In any case, as pointed out by Hogg, in most situations the performance of the algorithm is *not* critically dependent on the precise values for the thresholds [3-37].

1. Properties of Two-Stage Estimators

One may ask what are the benefits of a two-stage estimator such as that described above compared to, say, simple outlier rejection? It is known that while subjective or objective *ad hoc* outlier rejection can help prevent disaster due to extreme errors, outlier rejection alone is associated with limited efficiency [3-34]. As Hampel *et al.* point out, the basic weakness of all hard rejection rules is the inability to cope with contamination which lies in the "region of doubt". Augmenting the preliminary outlier rejection with a highly efficient estimator provides the protection of the outlier rejection mechanism coupled with the efficiency of a mean or trimmed mean estimator. This was recognized in [3-34] where a similar MAD rejection scheme was employed in conjunction with the sample mean, providing good efficiency for a range of distributions, coupled with a high breakdown point. Hampel *et al.* report that a simple scheme in which outlier rejection is followed by the arithmetic mean of the residual observations performs about as well as any *fixed* rejection rule. Combining the threshold rejection with an *adaptive* trimmed mean has two further advantages: i) adaptation facilitates response to changes in the noise statistics, and ii) use of the trimmed mean allows a less aggressive threshold to be set, thereby increasing efficiency.

The properties of such adaptive two-stage estimators are extremely complex due to the nonlinear nature of both stages of the algorithm, and the interaction between the stages. Nevertheless, using the algorithm of Fan and Clarkson as an example, some simple observations can be made:

i). *Equivariance*

The composite operation comprising the two-stage algorithm denoted by $T\{x_1, x_2, ..., x_N\}$ satisfies

$$T\{cx_1+d, ..., cx_N+d\} = cT\{x_1, ..., x_N\}+d\,; \qquad c > 0,\ -\infty < d < \infty. \quad (3.4\text{-}6)$$

This is important, ensuring that simple scaling of the observations or the addition of a d.c. offset cannot adversely impact the estimation procedure.

ii). *Robustness*

The two-stage operator $T\{x_1, x_2, ..., x_N\}$ achieves the highest possible breakdown point – $(N+1)/2N$. As measured by the breakdown point, it is as robust as the sample median.

For the usual adaptive trimmed mean estimator, the breakdown point is $(M+1)/N$, the robustness changes with the trimming factor M. In many cases there exists a trade-off between robustness and efficiency. For the two-stage robust adaptive estimator described above, the first stage uses the MAD to remove the distant outliers. Since it never "breaks down" unless the sample median is an unbounded outlier, so the two-stage operator $T\{x_1, x_2, ..., x_N\}$ achieves the same breakdown point as the sample median, which is the highest possible. This follows because in stage two only bounded samples remain (unless breakdown has occurred in the first stage).

It is intuitively obvious that the limiting effect of the outlier rejection mechanism results in a finite influence function. Hence, the two-stage estimator is β-robust.

iii). *Symmetry*

The MAD scheme also improves the symmetry of the observations by removing the (generally asymmetric) outliers. This is important because for symmetric distributions, Q_1 is independent of the parameter s.

This may be viewed as a direct deduction from ii) above. If the underlying distribution $f(v)$ is symmetric, then up to $(N+1)/2N$ distant one-sided outliers will be removed.

The restriction to (inner) trimmed mean filtering may produce some loss of efficiency if the observations are drawn from a short-tailed distribution. However, simulation studies described below have demonstrated that this effect is not significant, whereas residual outliers not removed by the initial detector would produce serious errors in the resulting estimates.

2. Application to Cardiac Rate Estimation

In this section we compare the performance of a number of estimators, both adaptive and non-adaptive, in the context of cardiac rate estimation. The algorithms to be examined include the simple mean, the median, trimmed mean, and the adaptive two-stage estimator discussed above. For the purposes of this comparison, both simulated data with known distribution, and experimental data recorded during surgical procedures in an electrophysiology laboratory is available [3-35]. Three kinds of simulated data corresponding to Laplacian, Gaussian, and uniform distributions have been analyzed. In each case, the data generated were iid, and were additively corrupted by impulse noise according to the model (3.2-19). Two forms of impulse noise were used corresponding to one and two-sided noise. The two-sided noise was generated according to the model:

$$\begin{aligned} \text{prob}(\zeta_k = A) &= a \\ \text{prob}(\zeta_k = B) &= b \\ \text{prob}(\zeta_k = 0) &= 1 - a - b, \end{aligned} \tag{3.4-7}$$

where $a, b \ll 1$. Single-sided noise can be generated by setting a equal to zero in (3.4-7). Both one and two-sided models can be used to generate asymmetrically distributed outliers, which is the usual case in experimental data. The performance of this two-stage estimation scheme has been compared with that of simple averaging, the median, the Hodges–Lehmann (H–L) estimate [3-38], [3-39][14] and fixed trimmed mean filtering. We may compare the mean-squared error for each estimator calculated using

$$J = \frac{1}{L} \sum_{i=1}^{L} (r - \hat{r}_i)^2, \tag{3.4-8}$$

[14]The estimate is the median of $(x_i + x_j)/2$ for any $i \leq j$. Bickel [3-40] recommends the H–L estimate in situations where the degree of contamination and type of distribution is not known with great precision.

over an 'ensemble' of $L = 1000$ repetitions [3-35]. Table 3.4-1 shows a typical set of results obtained by assuming that observations of normal heart rate are uniformly distributed from 65 beats/min to 75 beats/min, and are corrupted by one-sided impulse noise at 35 beats/min with probability 0.1, which may be caused by undersensing the depolarizations, for example. Table 3.4-2 shows results from the same data corrupted by two-sided impulses at 35 beats/min and 120 beats/min which may be caused by oversensing. In each case the mean-squared error for the adaptive algorithm is close to that achieved by the fixed optimum for the particular distribution, but *without* prior knowledge of the distribution.

N	*Avg*	*Med*	*H–L*	α-*Trim*	*TSO*
10	24.09	2.94	6.51	6.29	2.94
15	20.19	2.70	4.49	3.42	.92
20	18.19	1.59	2.81	1.74	.73
25	17.08	1.36	1.68	1.23	.49
30	16.01	1.20	2.19	1.14	.45
35	15.85	1.11	1.19	.99	.34
40	15.31	1.00	.91	.81	.25

Table 3.4-1: Mean-squared estimation errors obtained for x_k distributed as $f(x) = 0.1\delta(x-35) + 0.9\ \text{uniform}(65, 75)$. *Avg* = sample average, *Med* = sample median, *H–L* = Hodges–Lehmann estimate, α-*Trim* = α trimmed mean with $\alpha = .25$, *TSO* = two-stage operator.

N	*Avg*	*Med*	*H–L*	α-*Trim*	*TSO*
10	22.73	2.36	5.45	5.34	2.36
15	13.22	1.87	1.45	1.16	1.01
20	10.81	1.38	.85	.97	.70
25	8.39	1.14	.64	.76	.51
30	7.39	.96	.55	.65	.42
35	6.27	.82	.44	.53	.35
40	5.63	.73	.39	.51	.25

Table 3.4-2: Mean-squared estimation errors obtained for x_k distributed as $f(x) = 0.05\delta(x-35) + 0.9\text{uniform}(70-5, 70+5) + 0.05\delta(x-120)$. *Avg* = sample average, *Med* = sample median, *H–L* = Hodges–Lehmann estimate, α-*Trim* = α trimmed mean with $\alpha = .25$, *TSO* = two-stage operator.

We observe that the outlier removal process significantly improves the accuracy of the estimation, whichever estimate is applied to the residual observations. This is because the outlier removal estimate is as robust as the sample median. Hence any contamination is decreased. For the adaptive two-stage estimator, the estimate is not only as robust as the sample median, but also almost as efficient as the optimal estimate.

Twenty-five recordings derived from the electrophysiology lab at the University of Chicago Hospital have also been analyzed. This set included fourteen episodes of normal sinus rhythm and eleven episodes of ventricular tachycardias. These recordings were derived from 1 cm. bipolar electrodes on catheters in the right atrium and ventricle. The recordings were digitized at 1000 samples per second. All episodes were truncated to 10 seconds duration. For these experimental data, the true rates are unknown, and, except for highly organized rhythms, objective comparisons are difficult. However, one may observe the impact of outlier rejection and estimation qualitatively from Tables 3.4-3 and 3.4-4, and from Fig. 3.4-3.

Case	Avg	Med	H–L	α-$Trim$	TSO
1	242.1	244.0	246.0	246.2	246.3
2	222.7	242.0	239.0	240.0	242.9
3	273.3	270.0	271.5	270.3	269.7
4	222.2	268.0	216.5	236.4	281.8
5	286.9	286.0	286.0	285.5	284.7
6	336.5	348.0	342.5	346.7	345.5
7	206.8	245.5	226.5	235.4	250.6
8	268.3	267.0	267.0	267.1	267.3
9	245.1	248.0	248.5	248.6	249.3
10	261.5	256.0	262.5	259.1	261.5
11	213.1	212.0	212.5	212.4	212.1

Table 3.4-3: Ventricular tachycardia from ventricular channel. Avg = sample average, Med = sample median, H–L = Hodges–Lehmann estimate, α-$Trim$ = α trimmed mean with $\alpha = .25$, TSO = two-stage operator.

From Tables 3.4-3 and 3.4-4, some significant differences between the results obtained from the various algorithms may be observed. In Table 3.4-3 we may cite Cases 4 and 7 as examples showing significant variation, Case 2 as showing some variation, and the others as exhibiting little variation. For sinus rhythm, as shown in Table 3.4-4, usually no significant change is observed.

Fig. 3.4-3a) shows a histogram of raw rate intervals from episode 7 of ventricular tachycardia. The observed raw rate values contain a significant number of outliers due to amplitude fluctuations in the data. Fig. 3.4-3b) shows the corresponding histogram after processing to remove outliers as described above (K in Eq. (3.4-2) was set to 4 for this example). The outliers have been removed, and the result of

Case	Avg	Med	$H–L$	α-$Trim$	TSO
1	85.6	86.0	85.5	85.7	85.6
2	87.3	87.0	87.0	87.1	86.9
3	76.0	87.0	86.5	86.5	87.0
4	86.5	87.0	86.5	86.6	86.5
5	80.8	84.5	84.0	84.3	84.4
6	99.1	99.0	99.0	99.0	98.6
7	96.4	96.0	96.5	96.3	96.4
8	97.3	97.0	97.5	97.0	97.2
9	104.6	105.0	104.5	104.8	104.6
10	109.1	109.0	109.0	109.3	109.3
11	75.2	73.0	73.0	71.3	71.3
12	87.9	85.0	85.0	84.9	84.1
13	72.5	73.0	72.5	72.6	72.6
14	74.0	74.5	74.3	74.8	74.3

Table 3.4-4: Sinus rhythm from ventricular channel. Avg = sample average, Med = sample median, $H–L$ = Hodges–Lehmann estimate, α-$Trim$ = α trimmed mean with $\alpha = .25$, TSO = two-stage operator.

the two-stage processing is a final rate estimate of 250.6. This differs significantly from the rate estimate of 206.8 obtained using the simple average.

These results confirm that for normal sinus rhythm, simple averaging works well. On the other hand, for arrhythmias, even regular arrhythmias such as ventricular tachycardia, it is important to protect the rate estimate by incorporating outlier rejection into the estimation procedure.

Summary

Order statistic (OS) filters have a number of interesting properties that contrast with those of linear filters, and serve to augment the digital filter design tools available to the system designer. The median filter provides data smoothing coupled with preservation of signal edges and rejection of impulsive noise, properties that cannot be obtained from linear filters. OS filters can be designed that provide optimal smoothing matched to the distribution of the noise. These estimators provide the uniformly minimum variance unbiased estimates from among *all* estimators in many cases.

Our motivation for the coupling of adaptation and OS filter design is intimately connected with the need for robustness in signal analysis. However, we have been careful to point out that not all OS filters are robust to distributional deviations.

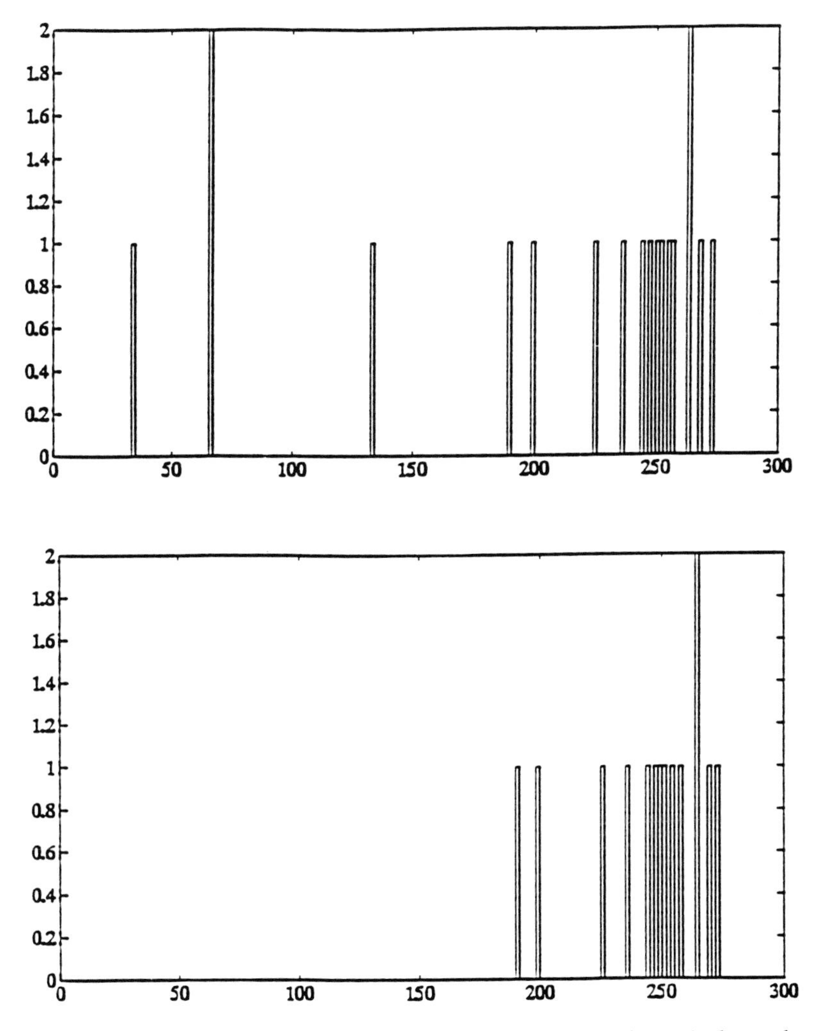

Figure 3.4-3: Histograms of raw rate intervals from an episode of ventricular tachycardia. a) Raw rate intervals. b) Intervals after outlier rejection.

We have seen that OS operations can enhance the performance of conventional adaptive filters subjected to non-Gaussian noise. This class of OSLMS algorithms has performed well in simulation studies and has started to have a significant impact in applications including the design of ADPCM systems for noisy channels. Current work on the application of the OSLMS filters to equalization problems shows great promise [3-41].

The same considerations that motivate the use of adaptation in linear filter design are also relevant to OS filter design. Persuing the analogy further, consideration of gradient based LMS algorithms leads to a class of gradient-descent *adaptive order statistic filters* (AOSFs) designed to provide an iterative update towards the MVUB OS operator. While potentially optimal, such algorithms are limited by the slow convergence associated with gradient descent adaptation algorithms and in some cases by a lack of robustness. In applications where rapid adaptation and distributional robustness are essential, it is often necessary to sacrifice the optimality of the MVUB OS operator in favor of an *ad hoc* approach. This may be as simple as an OS filter whose form is conditioned by a local data-based distribution measure. In other cases, the form of the data can motivate a more complicated multi-stage procedure designed to respond to several characteristics of signal or noise (asymmetry of the noise distribution, rapid changes in data or noise distribution etc.) The two-stage algorithm described in Section 3.4 is one example of a successful approach and one could envisage many other situations where a similar algorithm could be applied.

References

[3-1] J.W. Tukey, *Exploratory Data Analysis*, Addison-Wesley, 1971.

[3-2] L.R. Rabiner and R.W. Schaffer, *Digital Processing of Speech Signals*, Prentice-Hall, 1978.

[3-3] R.F.Gonzalez and P. Wintz, *Digital Image Processing*, 2nd edn., Addison-Wesley, 1987.

[3-4] N.C. Gallagher and G.L. Wise, "A theoretical analysis of the properties of median filters," *IEEE Trans. Acoust., Speech, Signal Processing*, vol. ASSP-29, pp. 1136-1141, 1981.

[3-5] S.G. Tyan, "Median filtering: deterministic properties," *Topics in Applied Physics*, vol. 43, pp. 197-217, Springer-Verlag, 1981.

[3-6] B.I. Justusson, "Median filtering: statistical properties," *Topics in Applied Physics*, vol. 43, pp. 161-182, Springer-Verlag, 1981.

[3-7] A.C. Bovik, T.S. Huang and D.C. Munson, "A generalization of median filtering using linear combinations of order statistics," *IEEE Trans. Acoust., Speech, Signal Processing*, vol. ASSP-31, pp. 1342-1349, 1983.

[3-8] J.B. Bednar and T.L. Watt, "Alpha-trimmed means and their relationship to median filters," *IEEE Trans. Acoust., Speech, Signal Processing*, vol. ASSP-32, pp. 145-153, 1984.

[3-9] H.G. Longbotham and A.C. Bovik, "Theory of order statistic filters and their relationship to linear filters," *IEEE Trans. Acoust., Speech, Signal Processing*, vol. ASSP-37, pp. 275-287, 1989.

[3-10] G.A. Williamson and P.M. Clarkson, "On signal recovery with adaptive order statistic filters," *IEEE Trans. Signal Processing*, vol. SP-40, pp. 2623-2626, 1992.

[3-11] O.L. Frost, III, "An algorithm for linearly constrained adaptive array processing," *Proc. IEEE*, vol. 60, pp. 926-935, 1972.

[3-12] E.L. Lehmann, *Theory of Point Estimation*, Wiley, 1983.

[3-13] B. Widrow and S.D. Stearns, *Adaptive Signal Processing*, Prentice-Hall, 1985.

[3-14] T.I. Haweel and P.M. Clarkson, "A class of order statistic LMS algorithms," *IEEE Trans. Signal Processing*, vol. SP-40, pp. 44-53, 1992.

[3-15] P.M. Clarkson and T.I. Haweel, "A median LMS algorithm," *Proc. of the IEE Electr. Letts.*, vol. 25, pp. 520-522, 1989.

[3-16] G.A. Williamson, P.M. Clarkson, and W.A. Sethares, "Performance characteristics of the median LMS adaptive filter," *IEEE Trans. Acoust. Speech, Signal Processing*, vol. 41, pp. 667-680, 1993.

[3-17] M. Givens and P.M. Clarkson, "The application of the median LMS algorithm to ADPCM systems," *Proc. of the IEEE Int. Conf. Acoust., Speech and Signal Processing*, pp. 3665-3668, 1991.

[3-18] J.D. Gibson, "Adaptive prediction in speech differential encoding systems," *Proc. IEEE*, vol. 68, pp. 488-525, 1980.

[3-19] W.A. Sethares and J.A. Bucklew, "Local stability of the median LMS filter," *IEEE Trans. Signal Processing*, vol. SP-42, Nov. 1994.

[3-20] K.F. Wan and P.C. Ching, "A fast convergence median LMS algorithm," *Proc. IEEE ISCAS*, vol. 2, pp. 377-390, 1994.

[3-21] W.A. Sethares, B.D.O. Anderson, and C.R. Johnson, Jr., "Adaptive algorithms with filtered regressor and filtered error," *Mathematics of Control and Signal Processing*, vol. 2, pp. 381-403, 1989.

[3-22] T.I. Haweel and P.M. Clarkson, "Analysis and generalization of a median adaptive filter," *Proc. of the IEEE Int. Conf. Acoust., Speech and Signal Processing*, pp. 1269-1272, 1990.

[3-23] Y. Fu and G.A. Williamson, "Asymptotic behavior of order statistic least mean square (OSLMS) algorithms in non-Gaussian environments," *Proc. of the IEEE Int. Conf. Acoust., Speech and Signal Processing*, pp. 747-750, 1993.

[3-24] Y. Fu, G.A. Williamson and P.M. Clarkson, "Adaptive algorithms for non-Gaussian noise environments: the order statistic least mean square algorithms," *IEEE Trans. Signal Processing*, vol. SP-42, Nov. 1994.

[3-25] A. Restrepo and A. Bovik, "Adaptive trimmed mean filters for image restoration," *IEEE Trans. Acoust., Speech, Signal Processing*, vol. ASSP-36, pp. 1326-1337, 1988.

[3-26] I. Pitas and A.N. Venetsanopoulos, "Adaptive filters based on order statistics," *IEEE Trans. Signal Processing*, vol. SP-39, pp. 518-522, 1991.

[3-27] S. Haykin (ed.), *Array Signal Processing*, Prentice-Hall, 1985.

[3-28] S.C. Vlay, "Clinical recognition of cardiac arrhythmias," in *Manual of Cardiac Arrhythmias*, S.C. Vlay (ed.), Little, Brown and Company, 1988.

[3-29] K.L.Ripley, T.E. Bump and R.C. Arzbaecher, "Evaluation of techniques for recognition of ventricular arrhythmias by implanted devices," *IEEE Trans. Biomed. Eng.*, vol. BME-36, pp. 618-624,1989.

[3-30] H.A. David, *Order Statistics*, Wiley, 1981.

[3-31] A. Polikaitis and R.C. Arzbaecher, "Validation of an adaptive software trigger and arrhythmia diagnostic algorithm," to appear in *Journal of Electrocardiography*, 1993.

[3-32] P.M. Clarkson. Q. Fan, G.A. Williamson and R. Arzbaecher, "Robust adaptive parameter estimation in arrhythmia detection," *J. Electrocardiology*, vol. 25(S), pp. 207-211, 1993.

[3-33] F.R. Hample, "A general qualitative definition of robustness," *Ann. Math. Stat.*, vol. 42, pp. 1887-96, 1971.

[3-34] F.R. Hampel, E.M. Ronchetti, P.J. Rousseeuw and W.A.Stahel, *Robust Statistics: The Approach Based on Influence Functions*, Wiley, 1986.

[3-35] Q. Fan and P.M. Clarkson, "A robust adaptive estimator of rate for cardiac arrhythmia detection," *IEEE ISCAS Proceedings*, vol. 1, pp. 822-825, 1993.

[3-36] D.F. Andrews, P.J. Bickel, F.R. Hampel, P.J. Huber, W.H. Rogers and J.W. Tukey, *Robust Estimates of Location: Survey and Advances*, Princeton University Press, Princeton, NJ.

[3-37] R.V. Hogg, "Adaptive robust procedures: a partial review and some suggestions for future applications and theory," *J. Amer. Statist. Ass.*, vol. 69, pp. 909-927, 1974.

[3-38] J.L. Hodges Jr. and E.L. Lehmann, "Efficiency of some nonparametric competitors of the t-test," *Ann. Math. Stat.*, vol. 27, pp. 324-325, 1956.

[3-39] J.L. Hodges Jr. and E.L. Lehmann, "Estimates of location based on rank tests," *Ann. Math. Stat.*, vol. 34, pp. 598-611, 1963.

[3-40] P.J. Bickel, "On some robust estimates of location," *Ann. Math. Stat.*, vol. 36, pp. 847-858, 1965.

[3-41] Y. Fu, "A class of adaptive algorithms for non-Gaussian noise environments: order statistic least mean squared (OSLMS) algorithms," Ph.D. Thesis, ECE Department, Illinois Institute of Technology, 1994.

Chapter 4

Multi-Layer Perceptron Neural Networks with Applications to Speech Recognition

ERIC R. BUHRKE and JOSEPH L. LoCICERO

Introduction

Artificial neural networks were originally created by neurobiologists as a model of biological processes [4-1], [4-2]. Since their inception, the scope of neural nets has been expanded and they are now used by engineers and physicists in many different applications [4-3], [4-4], [4-5]. Examples include echo cancellation in long-distance telephony, handwritten zip-code recognition,[1] text-to-speech conversion, and automatic speech recognition. Artificial neural networks are attractive for pattern recognition problems because they are nonlinear processors with powerful theoretical representation properties which are capable of self-adapting and learning.

The analogy between biological and artificial neural networks comes from a theory of the brain and thought [4-6] sometimes referred to as cognition. The brain can solve many difficult pattern recognition problems relatively easily. It is hypothesized that this ability comes from the cooperation between large numbers of biological neurons connected together operating in parallel [4-6]. Artificial neural networks model this cooperative structure with large arrays of very simple interconnected processors called *artificial neurons.* Artificial neurons are crude models of biological neurons that only account for a few of their basic characteristics. An artificial neuron consists of a series of weighted inputs, a summer, a nonlinearity,

[1]See also Section 2.6 in Chapter 2.

and an activation threshold. The inputs come from other artificial neurons or from external sources. The artificial neuron produces a positive output if the weighted sum of the inputs is greater than the activation threshold Θ. The output of an artificial neuron can be expressed in terms of the neural nonlinearity $\sigma(\cdot)$ as

$$y = \sigma[\langle \mathbf{x}, \mathbf{w} \rangle + \Theta], \tag{4-1}$$

where $\langle \mathbf{x}\ \mathbf{w} \rangle$ is the inner product of the input vector $\mathbf{x}$ and the weight vector $\mathbf{w}$. The nonlinearity $\sigma(\cdot)$ can be either chosen as a hard nonlinearity, such as a limiter, or a soft nonlinearity such as a hyperbolic tangent. An artificial neuron is depicted in Fig. 4-1.

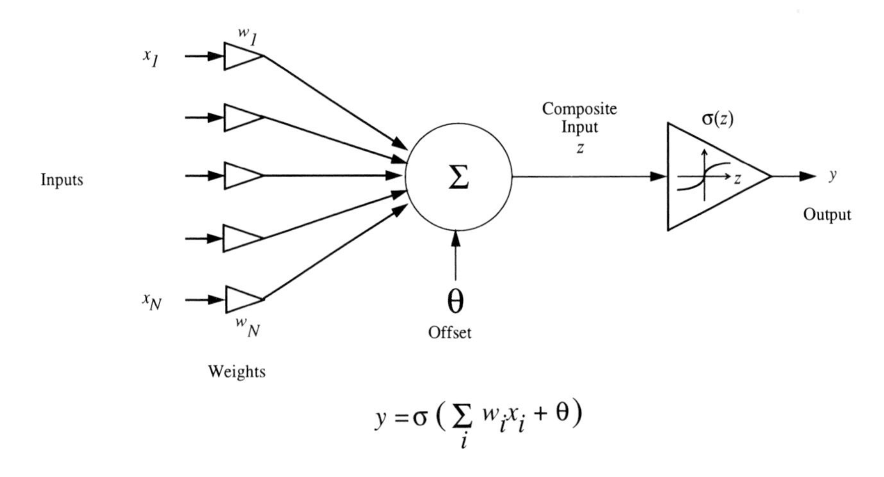

Figure 4-1: Elements of an artificial neuron with a soft nonlinearity.

Neural networks can be divided into two categories, depending on how their neurons are connected. These categories are *feed-forward neural networks* and *feedback neural networks.* Feedback neural networks are used for optimization and associative memories [4-7]. To do justice to feedback neural networks and their applications, an entire chapter would be required. Consequently, they will not be considered further here.

A feed-forward neural network contains many neurons that are grouped into sets called *layers* [4-8]. The feed-forward neural network is a hierarchy of layers where the network input is applied at the bottom layer and the network output emits from the top layer. Layers which do not receive or emit external signals are by convention called *hidden layers*. This chapter considers feed-forward neural networks in detail by examining the most popular feed-forward neural network, namely the *multi-layer perceptron* [4-8].

4.1 The Multi-Layer Perceptron

A. Brief History

The perceptron was proposed by Rosenblatt in 1958 [4-2] based on characteristics of the visual systems in animals. Later Widrow modified the perceptron and created the adaptive linear element or ADELINE [4-9] that laid the foundations for adaptive filtering [4-10]. In 1969 Minsky and Papert [4-11] pointed out that a single perceptron could only linearly bisect a decision space, and thus can only solve linearly separable problems. This result caused the perceptron to be all but ignored until the work of Rumelhart and McClelland in 1986 [4-8]. Rumelhart and McClelland derived an approximation rule for designing multi-layer perceptrons that circumvented some of the limitations pointed out by Minsky and Papert. This technique is known as the back-propagation algorithm. It iteratively finds the network weights by attempting to minimize the output mean-squared error via gradient descent. Following the derivation of the back-propagation algorithm, applications such as those noted above appeared with increasing frequency.

B. Fully Interconnected Neurons

A multi-layer perceptron is a layered neural network. Each layer contains a fixed number of neurons. The neurons in each layer are typically fully interconnected to the neurons of the previous layer. This implies that the output of all neurons in the previous layer are connected to all neurons in the next layer. A special constant neuron is associated with each layer. The constant neuron receives no inputs and always has an output of unity. It is used to implement the activation threshold. Thus the activation threshold can be treated as an additional weight.

The neurons in a feed-forward neural net receive their input only from neurons in the previous layer. The sum of these weighted inputs is applied to a nonlinearity $\sigma(\cdot)$ which serves to limit the signal. We shall denote the output of artificial neuron j in layer k as y_j^k, where k ranges from 0 at the network input to N at the network output for an N-layer perceptron. Analogously, the sum of the weighted inputs, called the composite input, is given by z_j^k. Thus

$$y_j^k = \sigma(z_j^k) , \tag{4.1-1}$$

where

$$z_j^k = \sum_i w_{ji}^k y_i^{k-1} , \tag{4.1-2}$$

and w_{ji}^k is the weight to destination neuron j in layer k from source neuron i in layer $k-1$. Finally, we shall use L_k to indicate the number of neurons in layer k. Thus the sum in Eq. (4.1-2) will have L_{k-1} terms. An example of a two-layer perceptron with zero activation thresholds is given in Fig. 4.1-1. Note that the inputs x_j are identical to y_j^0.

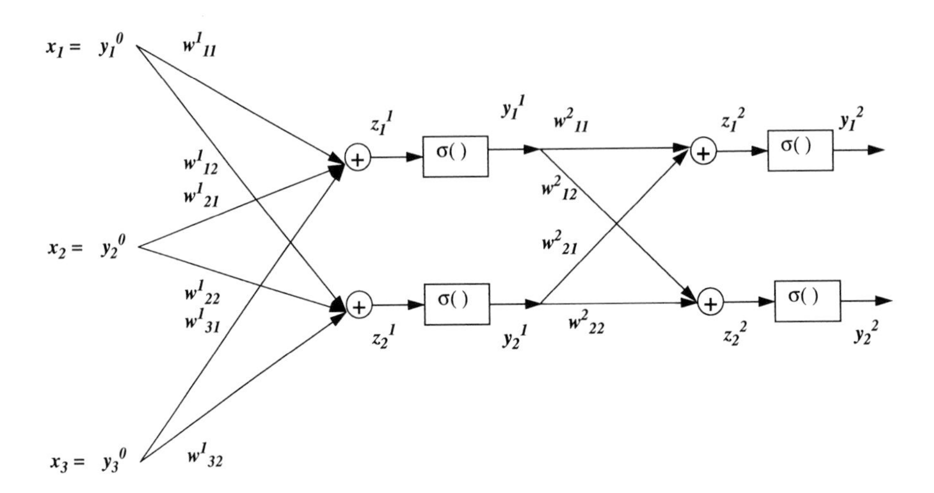

Figure 4.1-1: An example of a two-layer perceptron with zero activation thresholds.

The multi-layer perceptron has been used to solve a number of difficult problems. A famous application of the multi-layer perceptron is the NETtalk text-to-speech system of Sejnowski and Rosenberg [4-12]. The network was trained to convert raw text into parameters from which a speech waveform could be synthesized.

The successful application of a multi-layer perceptron requires the consideration of three design issues. These are: specification of the network architecture, approximation of the neural weights, and generalization ability of the network. These aspects are discussed later in this chapter.

C. Function of the Perceptron

Two distinct computations produce the output of an artificial neuron and thus define the functional characteristics of a single-layer perceptron. The first computation forms the composite input from the weighted sum of neural inputs. Following this, the second computation forms the neural output by applying the composite input to a nonlinearity.

The composite input can be visualized geometrically. Specific pictorial examples are given in the next section. The weighted sum forming the composite input is a hyperplane. This hyperplane bisects the neural input space. The weight vector of the neuron is the vector normal to this hyperplane and can be arbitrarily oriented to point into or out of the hyperplane. The location of an input point is reflected in the magnitude and sign of the composite input. The sign of the composite input reflects which side of the hyperplane the input is on. The magnitude of the composite input gives information about the distance of the input to the hyperplane. If the hyperplane can dichotomize two pattern classes, then it is possible to discriminate them with a single neuron.

After the composite input is computed, it is passed through a nonlinearity sometimes referred to as a sigmoid function. The nonlinearity forms the neural output. If the nonlinearity is a hard limiter, such as a step function, then the neural output is a binary quantity that specifies where the input lies with respect to the hyperplane. On the other hand, a soft limiter nonlinearity, such as a hyperbolic tangent, retains some distance information. Thus neurons with soft nonlinearities provide location and distance information.

The weighted sum of the composite input and the neural nonlinearity combine to form a neuron. This neuron is only capable of implementing a decision based on the hyperplane. A single neuron, sometimes referred to as a single-layer perceptron, is thus limited in its abilities.

Single-layer perceptrons perform linear discriminant analysis. Therefore single-layer perceptrons can only solve linearly separable problems exactly. The fraction of the total number of problems that are linearly separable F_s is the number of linear dichotomies of n points in d dimensions divided by the total number of problems (2^{n-1}) [4-13], that is

$$F_s = \frac{1}{2^{n-1}} \sum_{k=0}^{d-1} \binom{n-1}{k} . \tag{4.1-3}$$

This fraction of linearly separable problems F_s approaches zero as the dimensionality d and number of data points n becomes large. Therefore single-layer perceptrons can only solve a small fraction of problems. Minsky and Papert [4-11] point out that this is a serious drawback of the single-layer perceptron.

Having established the operation of a neuron and the function of single-layer perceptron, it is natural to consider solving more complicated problems using tiers of single-layer perceptrons or multi-layer perceptrons. Examples are given in the next section that show how a two-layer binary neural network allows the formation of disjoint decision regions.

It is well known that any Boolean function can be implemented with two layers of binary logic gates. Therefore a two-layer perceptron can implement any Boolean function. However, this is only possible when there are enough hidden layer neurons to cover all the necessary minterms of the Boolean function [4-14]. For complicated Boolean functions, the number of hidden layer neurons can be quite large. If there are I input neurons to a multi-layer perceptron, then an upper bound on the number of hidden layer neurons is 2^{I-1}.

The functionality of a multi-layer perceptron is evidenced by the power of a three-layer network. Kolmogorov showed that, if a sufficient number of hidden layer neurons is used, such a perceptron can represent any continuous mapping function of several variables by the superposition of functions of a single variable [4-15], [4-16]. The key elements are the number of hidden layer neurons and the neural nonlinearity function. These items will be examined in the context of examples in the next section.

4.2 Signal Classification Design Examples

A great deal can be learned about neural network architecture and operation by doing some design examples using only perceptrons. In this section we shall study the neural network solution to several classification problems. The first network architecture example is a single neuron and the single-layer perceptron. Following this the minimal network architecture is found for three more complicated nonlinearly separable problems. These pathological problems lead to a formal discussion, in the next section, of the number of layers and the number of neurons per layer. The number of input and output neurons are assumed to be design constraints in all examples.

A. Binary Exclusive-OR (XOR)

The binary XOR problem, discussed in [4-8], classifies the points $(-1, -1)$ and $(1, 1)$ as members of class A and the points $(1, -1)$ and $(-1, 1)$ as members of class B. This problem is similar to the design of an XOR logic gate. The input decision space is shown in Fig. 4.2-1.

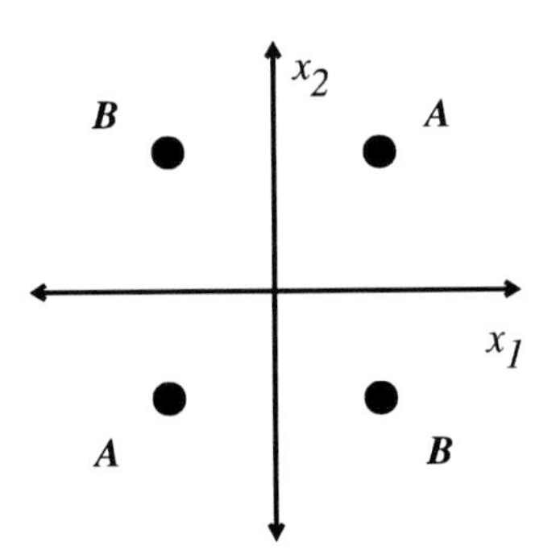

Figure 4.2-1: Binary XOR problem.

We shall solve this problem with a two-layer network of binary neurons. These binary neurons have outputs that are either 1, when the composite input is greater than zero, or -1, when the composite input is less than zero.

The patterns are not linearly separable and a multilayer network must be used. If we use the two hyperplanes, as shown in Fig. 4.2-2, such that

$$z_1 = x_1 - x_2 + 1 , \tag{4.2-1}$$

$$z_2 = x_1 - x_2 - 1 , \tag{4.2-2}$$

where x_1 and x_2 are the inputs to the neural network, then we can map the input space into the hidden unit space where class B is mapped to the points $(-1, -1)$ and $(+1, +1)$ and class A is mapped to the point $(-1, 1)$ as shown in Fig. 4.2-3. This is seen by using the hard nonlinearities

$$y_1 = \operatorname{sgn}(z_1) , \tag{4.2-3}$$

and

$$y_2 = \operatorname{sgn}(z_2) , \tag{4.2-4}$$

where $\operatorname{sgn}(\cdot)$ is the signum function, defined as

$$\begin{aligned} \operatorname{sgn}(\alpha) &= +1, \qquad \alpha > 0, \\ \operatorname{sgn}(\alpha) &= -1, \qquad \alpha < 0 . \end{aligned} \tag{4.2-5}$$

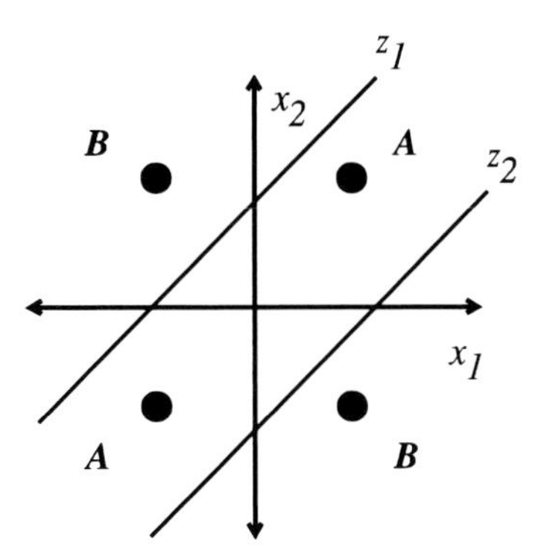

Figure 4.2-2: Hidden layer hyperplanes in the binary XOR problem.

This mapping yields a linearly separable set of points at the output of the hidden layer that can be dichotomized with the hyperplane

$$z_{out} = y_1 - y_2 - 1 . \tag{4.2-6}$$

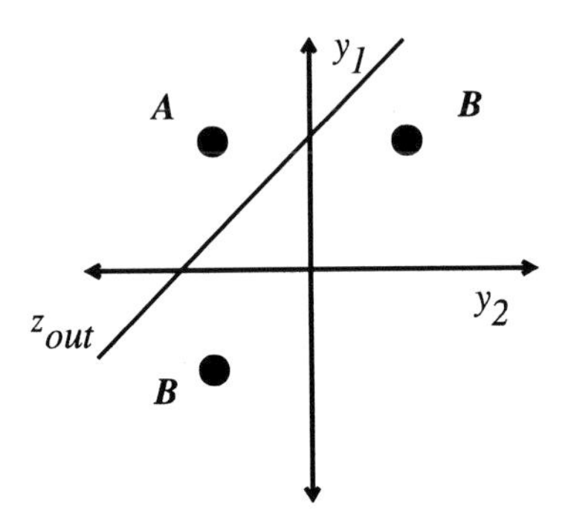

Figure 4.2-3: Decision space for the binary XOR problem.

If the final output is

$$y_{out} = \mathrm{sgn}(z_{out}) \,, \tag{4.2-7}$$

then class A is represented by $y_{out} = +1$ and class B is represented by $y_{out} = -1$. Note that the coefficients of the hyperplanes need not be specified exactly because a binary output is desired and the signum function used. A neural network that solves the binary XOR problem has two layers with two neurons in the hidden layer. It is depicted in Fig. 4.2-4.

The binary XOR example showed that adding a hidden layer gave the network the additional capacity to form disjoint decision regions.

B. Hypercube and Hypersphere

Consider the pattern classes where all elements of class A lie within the hypercube $-1 < x_i < 1, i = 1, \cdots, M$, and elements of class B lie outside the hypercube. This is shown in Fig. 4.2-5 for $M = 2$.

We shall solve the problem of distinguishing between these two classes with a two-layer network and binary neurons with output $[-1, +1]$. This problem is not linearly separable. Therefore hyperplanes must be found that form disjoint decision regions. For this problem all hyperplanes are adjacent to a face of the hypercube. For convenience, orient the hyperplanes so that the positive normal vector of the hyperplane points to the interior of the hypercube. This is shown for the two-dimensional case in Fig. 4.2-6 where the arrows indicate the positive direction.

The output of each neuron is positive only when the input is on the "cube-side" of the hyperplane. Hyperplanes chosen in this manner isolate the decision class A. The output of all the neurons will be positive simultaneously when the input point is in the hypercube. The output layer is simply the

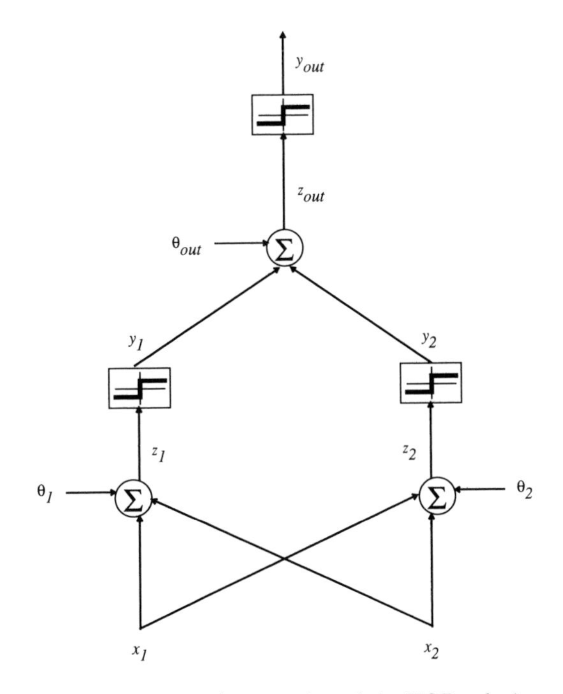

Figure 4.2-4: Implementation of the XOR solution.

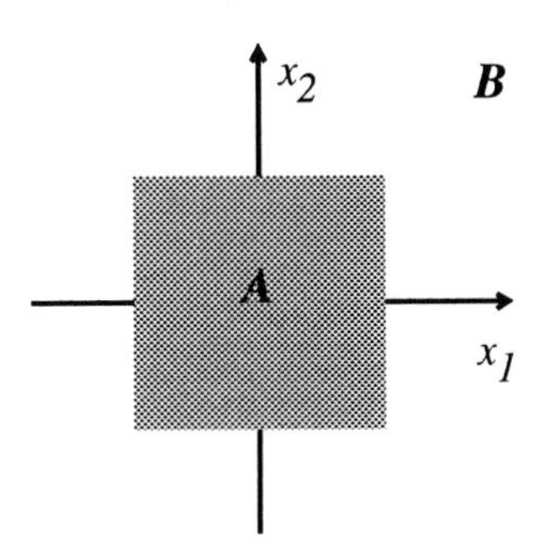

Figure 4.2-5: Hypercube problem.

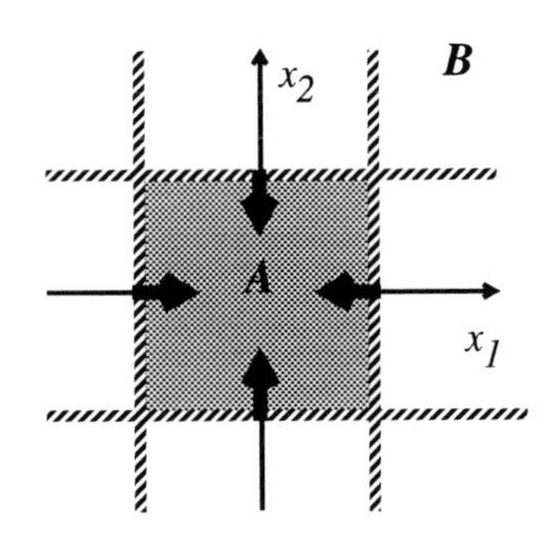

Figure 4.2-6: Hyperplanes for the hypercube problem.

AND of the outputs of the hidden layer. The output of the hidden layer is then linearly separable and can be dichotomized with a single neuron. Here the minimal network architecture is a two-layer network with $2M$ neurons in the hidden layer. When $M = 2$, the network simplifies to Fig. 4.2-7.

The classification problem becomes more complex when the input pattern classes are no longer binary. In the example just discussed, regions of space were associated with classes. These regions were isolated with the first layer of neurons. The second layer of neurons discriminates between the outputs of the first hidden layer with an AND operation or an OR operation using the first layer outputs [4-14].

Consider a spherical decision region shown in two dimensions in Fig. 4-2.8. The decision region A is inside a unit sphere described in M-dimensions as $x_1^2 + x_2^2 + \cdots + x_M^2 = 1$. The decision region B is outside the unit sphere. This problem is nonlinearly separable. Proceeding as in the previous examples we try to isolate the decision region A with hyperplanes. This is shown in Fig. 4-2.9.

Unfortunately we can only approximate the surface of the sphere. Any finite number of hyperplanes will only form a polyhedral region around the sphere. The sphere cannot be exactly represented by hyperplanes, and there is an error associated with approximating the decision surface, and classifying points in the space as belonging to region A or B. If it were possible to have an infinite number of hyperplanes then we could choose each one to correspond to a tangent plane of the sphere. Analogous to the hypercube example, we choose the positive orientation of each hyperplane to point towards the center of the sphere. The neuron of the second layer sums the output from each hidden layer neuron. A positive value of the neural sum classifies a point inside the sphere while a negative value classifies a point as outside.

The hypersphere poses a problem that requires an infinite number of hidden neurons for an exact solution. Any finite number of neurons only approximates

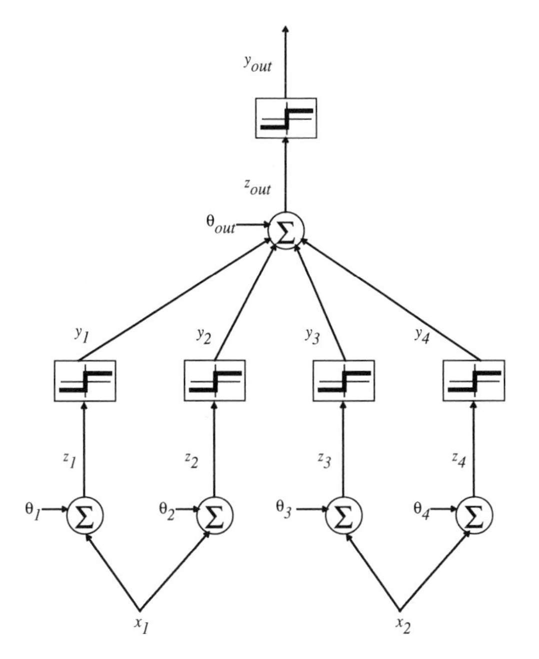

Figure 4.2-7: Implementation of a two-dimensional hypercube solution.

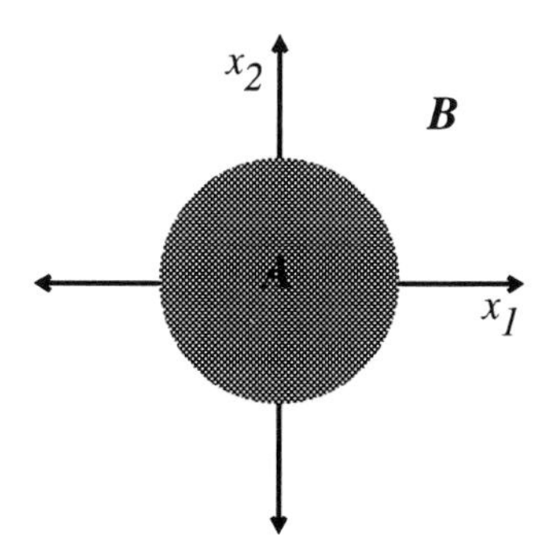

Figure 4.2-8: Hypersphere problem.

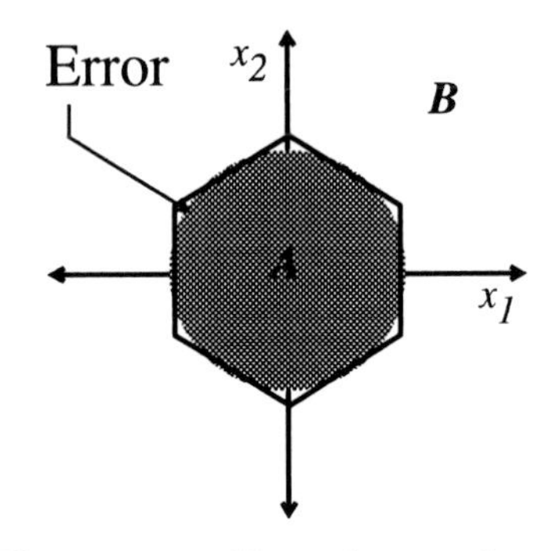

Figure 4.2-9: Hyperplanes and error.

the decision surface. It is infeasible to have an infinite number of hidden neurons. Therefore our neural model can only approximate curved decision regions.

The hypersphere problem described here can be solved with a single higher-order neuron [4-17]. A single neuron with a composite input $z = x_1^2 + x_2^2 + \cdots + x_M^2 - 1$ can solve this problem exactly.

C. Region Exclusive-OR

The final example is the distributed XOR problem. The binary XOR problem solved above is modified to have continuous regions. We decide class A when $(x_1 > 0) \cap (x_2 > 0)$ or $(x_1 < 0) \cap (x_2 < 0)$. We decide that the input is a member of class B if $(x_1 > 0) \cap (x_2 < 0)$ or $(x_1 < 0) \cap (x_2 > 0)$. Here we use neurons with smooth continuous sigmoids such as the hyperbolic tangent or error function. To minimize the amount of error induced by the network, the first layer maps the data along the same axes. If the nonlinearities were hard, each quadrant of space would then be mapped to a single point. Instead, the quadrants are mapped to a compact region with tails extending towards the axes. A second hidden layer maps these compact regions to class decisions. There is a finite amount of error from the tails of the decision regions. With larger weights and more hidden layers the error is reduced.

In this example, in contrast to the binary problem, the network no longer maps the regions defined by the weights to a single point but maps the input region to an output region. The output region is more compact with tails that are determined by the slope of the nonlinearity. The finite slope of the sigmoid can cause tails in the classification regions. These tails introduce error because of their overlap. It is necessary to have many layers in the network when soft nonlinearities are used [4-18].

4.3 Perceptron Architecture and Learning

In this section we discuss factors influencing the choice of the neural network architecture and the all-important learning algorithm called *back-propagation.* The term neural network architecture includes the number of layers in the network and the number of neurons in each layer [4-19]. The network architecture is set *a priori* and is not adjusted during learning.

This section also contains an in-depth discussion of the two constraints limiting the network architecture. First, the network must be large enough to solve the problem [4-20]. This implies that there are enough layers in the network and enough neurons in each layer. Second, the network must be small enough to be reliably trained with a limited size data set [4-21]. If either of these constraints is violated by the network architecture then the network will perform poorly.

A. The Number of Hidden Layers

The literature contains many theoretical discussions about the number of layers required by a multi-layer perceptron. Lippman [4-22], [4-23] argued that a perceptron with a single hidden layer can solve all convex problems and some non-convex problems. Longstaff and Cross [4-24] showed that it is possible to approximate any classification region with at least two layers of hidden neurons through superposition of convex regions. A theorem by Cybenko [4-25] extends this by proving that with enough neurons in a single hidden layer, coupled with an input and an output layer, it is possible to approximate any function to any desired degree of accuracy. Unfortunately, this theorem places no limit on the number of neurons in the hidden layer. A key to this proof is a representation theorem by Kolmogorov noted above [4-15], [4-16]. This theorem proves that it is possible to represent any function of several variables by the superposition of functions of one variable. These theorems provide the rigorous mathematical result that a sufficiently large neural network is theoretically capable of implementing any functional mapping.

B. The Number of Neurons per Layer

Once the number of network layers has been chosen, it is necessary to specify the number of neurons in each layer. Each neuron is a hyperplane that dichotomizes its input. These hyperplanes can divide the input space into disjoint decision regions. If this division is optimal, then the second layer can classify the points using AND or OR operations on the outputs of the first layer. The number of neurons in each layer is therefore identical to the number of hyperplanes necessary to segment the input into distinct decision regions. Unless the structure of the data is known *a priori*, the number of neurons needed in each layer is difficult to predict. Because it is difficult to choose the number of neurons in each layer, many researchers have conducted problem-specific experiments. In these experiments the number of input and output neurons was fixed and the number of hidden layer neurons was varied. These experiments all have similar trends. The network performance, such as the probability of correct classification, is usually poor for small networks and better for larger networks. If the number of hidden layer neurons continues to increase, there is no substantial increase in network performance [4-26], [4-27]. This behavior is easily explained. The network performance is poor when there are not enough neurons to solve the problem. The performance increases until enough hidden units are present to solve the problem after which only small incremental improvements are impossible.

Several theoretical results about the number of neurons per layer have been published. Most of these examine the number of linear threshold functions necessary to separate a set of points in a multi-dimensional space. Cover [4-13] provides a count of the number of dichotomies implementable by a single threshold unit. A threshold unit with T weights has a probability of one-half of implementing a random dichotomy on $2(T+1)$ random vectors. Nilsson [4-28] derived an upper bound on the number of hyperplanes necessary to separate P points in an M-dimensional space as simply $P-1$. Baum [4-29] refined these bounds and proved that a neural network with one hidden layer containing P/M units can completely dichotomize an M-dimensional space. These theoretical bounds provide the minimal number of neurons needed to categorize a training data set composed of linearly independent random vectors. In most practical problems, the numbers of neurons required is much less than the theoretical upper bounds because the data has structure and does not consist of totally random vectors.

C. Neural Network Training

Before a neural network, such as a multi-layer perceptron with a predetermined architecture, can operate properly, it must be trained. One dictionary defines learning as "the acquisition of knowledge or skill through experience or study". Learning as it applies to neural networks is slightly different. Neural network learning is the optimization of the neural weights through repeated exposure to examples, frequently termed *exemplars*.

A formal description of learning is given by Tishby, Levin, and Solla [4-30]. In their discussion a space, $\mathcal{S}$, is constructed containing all real functions. A neural network is only able to exactly realize a small set of functions in the space $\mathcal{S}$ because the capabilities of the neural network are limited by its architecture, and all other functions must be approximated. We shall denote the set of ϵ-admissible functions by Ω_ϵ, that is, the set of all functions that can be realized by the neural network with an approximation error less than ϵ. The elements of Ω_ϵ are denoted by ω. A probability density function, $f_\Omega(\omega)$, determines the probability that the neural network can realize a particular function from Ω_ϵ. Before learning, in the absence of any *a priori* knowledge, $f_\Omega(\omega)$ is flat because all realizable functions are equally probable. As learning progresses $f_\Omega(\omega)$ develops peaks around likely functions. Learning causes some of the functions in Ω_ϵ to have a low probability because they do not satisfy the constraints imposed by the training data. Other functions in Ω_ϵ have their probabilities increased because learning makes them more likely. If the learning algorithm were ideal, and there were enough exemplars in the training data set to rule out all invalid functions, then $f_\Omega(\omega)$ would develop into an array of impulse functions centered at the correct networks.

Unfortunately learning algorithms are not ideal and the size of the training data sets is limited. Learning is further complicated by the fact that different weights do not always yield functionally distinct networks. The result is that learning is a difficult problem to which sub-optimal solutions are found [4-20].

There are two general classes of learning algorithms used to train neural networks. The first class is entitled *supervised learning*. Supervised learning requires a known training set of desired functional inputs and associated outputs. The matched functional pairs describe the target system, and the network is "taught" by the desired output "teacher" to mimic the function. The second class is entitled *unsupervised learning* and involves unsupervised or self-organizing methods of training. In unsupervised learning the system adapts itself to the data. While supervised learning requires a "teacher", unsupervised learning is analogous to "independent study". Unsupervised techniques do not require the *a priori* knowledge of a desired response.

1. Supervised Learning via Back-propagation

The classical supervised learning algorithm for the multi-layer perceptron was derived independently by several authors [4-31], [4-32], [4-33]. This learning algorithm is referred to as *back-propagation* or the *generalized delta rule*, and is currently the most popular learning algorithm for a perceptron network. Key to the back-propagation algorithm is the use of a smooth monotonic nonlinearity with a continuous first derivative as part of the neuron model.

Back-propagation is a supervised gradient descent algorithm [4-34]. The desired neural output is either a pattern class or a functional relationship. For each exemplar in the training set, the actual network output is compared to the desired output and a mean-squared difference is calculated. The mean-squared difference is minimized with a gradient descent algorithm. The back-propagation algorithm minimizes the mean-squared error by adjusting the weights proportional to the error gradient. The mean-squared error at time n can be written as

$$E(n) = \frac{1}{2} \sum_{i=1}^{L_N} [y_i^N(n) - d_i(n)]^2 , \tag{4.3-1}$$

where $d_i(n)$ is the desired output at time n, $y_i^N(n)$ is the output of neuron j in the output layer N at time n, and the constant $1/2$ is included for mathematical convenience. The adjustment to each network weight, w_{ji}^k, is proportional to the gradient of the performance measure, thus

$$w_{ji}^k(n+1) = w_{ji}^k(n) + \alpha \frac{\partial E(n)}{\partial w_{ji}^k(n)} , \tag{4.3-2}$$

which is usually written as

$$w_{ji}^k(n+1) = w_{ji}^k(n) + \alpha \delta_j^k y_i^{k-1} , \tag{4.3-3}$$

where $w_{ji}^k(n)$ are the weights to destination node j of layer k, from source neuron i in layer $k-1$, and α is the learning or adaptation constant. The term δ_j^k represents the component of the derivative of an upper layer and is derived below.

The partial derivative of a neural output with respect to one of its weights is found from Eqs. (4.1-1) and (4.1-2) to be

$$\frac{\partial y_j^k}{\partial w_{ji}^k} = \frac{\partial \sigma(z_j^k)}{\partial z_j^k} y_i^{k-1} . \tag{4.3-4}$$

Equation (4.3-4) can be simplified by choosing the nonlinearity, $\sigma(\cdot)$, to be the function

$$\sigma(z) = \frac{1}{1 + e^{-z}} . \tag{4.3-5}$$

This function is very convenient because its derivative can be evaluated as the simple expression

$$\frac{d\sigma(z)}{dz} = \frac{e^{-z}}{(1 + e^{-z})^2} = \sigma(z)[1 - \sigma(z)] . \tag{4.3-6}$$

The derivative of the output of a neuron, y_j^k, with respect to one if its weights, w_{ji}^k, is thus

$$\frac{\partial y_j^k}{\partial w_{ji}^k} = y_j^k (1 - y_j^k) y_i^{k-1} , \tag{4.3-7}$$

where we have used the relationship $y_j^k = \sigma(z_j^k)$ from Eq. (4.1-1). The derivative of the composite signal with respect to the layer input is obtained from Eq. (4.1-2) as

$$\frac{\partial z_j^k}{\partial y_i^{k-1}} = w_{ji}^k . \tag{4.3-8}$$

Using the chain rule we can write the derivative of the mean-squared error with respect to weight w_{ji}^k as

$$\frac{\partial E}{\partial w_{ji}^k} = \frac{\partial E}{\partial y_j^k} y_j^k (1 - y_j^k) y_i^{k-1} . \tag{4.3-9}$$

The derivative of the mean-squared error with respect to the input of layer k is thus

$$\frac{\partial E}{\partial y_i^{k-1}} = \sum_{j=1}^{L_k} \frac{\partial E}{\partial z_j^k} w_{ji}^k \, . \tag{4.3-10}$$

The summation in Eq. (4.3-10) stems from the fact that every output neuron is connected to all input neurons, that is, a fully interconnected multi-layer perceptron.

The above derivatives can be rewritten using the δ notation. Comparing Eqs. (4.3-2) and (4.3-3) at the output layer we have

$$\frac{\partial E}{\partial w_{ji}^N} = \delta_j^N y_i^{N-1} \, , \tag{4.3-11}$$

where Eqs. (4.3-9) and (4.3-1) allow us to find δ_j^k at the output layer as

$$\delta_j^N = y_j^N (1 - y_j^N)(y_j^N - d_j) \, . \tag{4.3-12}$$

For other layers, the derivative of the mean-squared error with respect to the weights is given as

$$\frac{\partial E}{\partial w_{ji}^k} = \delta_j^k y_i^{k-1} \, , \tag{4.3-13}$$

where δ_j^k can be computed recursively with the aid of Eqs. (4.3-9) and (4.3-10) as

$$\delta_j^k = y_j^k (1 - y_j^k) \sum_i \delta_i^{k+1} w_{ji}^{k+1} \, . \tag{4.3-14}$$

The weight perturbations are calculated with Eq. (4.3-3) where each δ is computed with Eq. (4.3-12) for the output layer and Eq. (4.3-14) for all other layers. The recursive relationship requires computation of the δ values at the output layer first, and then calculation of the δ values at the next lower layer. The name back-propagation results from this backward δ propagation.

(a) Stopping and Adaptation

Implementation of the back-propagation algorithm requires the consideration of two heuristic factors. These are the stopping criterion and the choice of the adaptation constant. The specification of these factors is problem dependent.

Popular stopping criteria depend on aspects of the learning algorithm. Learning can be stopped after a fixed number of iterations. The number of iterations is typically chosen as a large value to prevent premature halting of the learning process. Another stopping criterion examines the value of the mean-squared error. If the mean-squared error is less than a specified tolerance, or does not decrease significantly in a preset time interval, then the learning is halted. These techniques result in premature stopping if the error surface is relatively flat.

The adaptation constant controls the rate of convergence. A small adaptation constant results in long training times while a large adaptation constant can cause the learning algorithm to be unstable. Typical values for the adaptation constant range from 0.1 to 0.5 [4-8], but have been set as large as 7.0 [4-35] and as small as 0.05 [4-26]. The actual value of the adaptation constant is highly dependent on the gradient of the mean-squared error; it is problem specific; and its choice is somewhat arbitrary.

(b) Convergence Speed-Up

It has been shown experimentally that back-propagation converges extremely slowly. The convergence is approximately exponential [4-36], [4-37]. It has also been proven that there is no algorithm that can find the *optimal* value for the weights in a reasonable amount of time [4-38], [4-39]. Therefore many heuristics for increasing the speed of back-propagation have been developed. Only a few of these are discussed here.

A popular technique for speeding the convergence of back-propagation [4-8], adds a momentum term to the update algorithm in Eq. (4.3-3). The momentum term modifies the weight update equation so that it becomes

$$w_{ji}^k(n+1) = \beta w_{ji}^k(n) + (1-\beta)w_{ji}^k(n-1) + \alpha\delta_j^k y_i^{k-1} , \qquad (4.3\text{-}15)$$

where β is a momentum coefficient. The momentum term is an attempt to smooth the descent when the gradient is large, and increase the convergence when the gradient is small [4-40]. A time-varying momentum coefficient has been used [4-41] to increase its effectiveness.

Convergence of the back-propagation algorithm slows as the weights become large. This is because the derivative of the nonlinearity is nearly zero for large input values. To avoid this problem, a decay term for the weights has been suggested to decrease their magnitudes and keep them in the active region [4-20], [4-42]. The decay term also reduces unnecessary weights to zero.

(c) Gradient Descent

In a general optimization problem, the error surface is not smooth, but has many suboptimal local minima and many globally optimal minima. Since back-propagation is gradient based, optimization always proceeds toward the nearest minimum. It is highly probable that back-propagation will produce a sub-optimal solution. To minimize the likelihood that the final solution is a local minimum, the training procedure must be run several times, each time with different initial conditions [4-41]. Different initial conditions begin the optimization at different locations on the error surface. After several training runs have converged, the best is chosen. Although this repetitive procedure is computationally expensive, it has proven to give better performance than just a single training run [4-38].

A reason for getting stuck in a local minimum is the lack of cooperation between the neurons during learning. It is possible that several neurons form identical representations while leaving other areas unrepresented. The lack of cooperation can be mediated by beginning with a very large network and training it till convergence. After convergence with this technique, we can eliminate all weights that have changed very little, and collapse neurons that have outputs which are highly correlated with other neurons [4-43], [4-44].

Back-propagation is a gradient descent algorithm on a complicated nonlinear error surface. With this type of error surface, gradient descent algorithms are inefficient because each successive gradient estimate is nearly perpendicular to the previous estimate. Gradient estimates of this nature cause the optimization to zig-zag down long surfaces [4-45]. More efficient methods of optimization exist that make better use of the gradient. Conjugate gradient algorithms and variable metric methods are optimization procedures that make efficient use of the gradient information [4-43]. It has been shown that if the error function is a quadratic form, then these algorithms will converge in G passes, where G is the number of weights in the network. Both of these algorithms accumulate information about the second derivative matrix, or the Hessian, and use it in the minimization to achieve the maximum amount of minimization at every step of the procedure. These algorithms achieve appreciably quicker convergence results with little extra overhead when compared with back-propagation [4-40], [4-46].

2. Unsupervised Learning

This subsection addresses the issue of unsupervised learning. Unsupervised learning lets the network find its own structure in the input data. Other related methodologies used for supervised learning are vector quantization and clustering [4-47]. These techniques are not explored in this discussion.

A simple method of unsupervised learning makes use of the algorithms available for supervised learning. This method essentially folds the network over on itself, forcing the input of the network to equal the output of the network. With the input forced to equal the output, the network finds its own representation of the input data at the hidden layer. This technique has been used by Saund [4-48] to form nonlinear regression surfaces, and by Elman and Zipser [4-49] for speech feature extraction.

Hebbian learning is another technique used for unsupervised learning [4-6]. This learning modification is based on Hebb's law which states that the synapse efficacies are modified in proportion to the strength of the signals reaching the synapse. If two neurons fired in synchronism, then the efficacy is strengthened. Stated in another way, the strength of the neural weights between two neurons is increased in proportion to the correlation between the two neural outputs.

A brief analysis of a linear network where the weights are modified via a Hebb rule reveals some of the properties of Hebbian learning. Assume that the network is a simple linear transform with column vector $\mathbf{x}$ being the input, column vector $\mathbf{y}$ being the output, and $\mathbf{W}$ the linear transformation matrix. Assume that the input vector $\mathbf{x}$ is drawn from a stationary distribution of training data and all samples are independent and identically distributed. The Hebbian update rule is given by

$$\mathbf{W}_{k+1} = \mathbf{W}_k + \eta \mathbf{x}_k \mathbf{y}_k^T = \mathbf{W}_k + \eta \mathbf{x}_k \mathbf{x}_k^T \mathbf{W}_k , \qquad (4.3\text{-}16)$$

where η is a small, positive update constant. Taking the expected value of $\mathbf{W}_k$, assuming $\mathbf{W}_{k-1}$ is known, we arrive at

$$E(\mathbf{W}_k | \mathbf{W}_{k-1}) = (\mathbf{I} + \eta \mathbf{R}_{xx}) \mathbf{W}_{k-1} , \qquad (4.3\text{-}17)$$

where $\mathbf{R}_{xx}$ is the covariance matrix of the input $\mathbf{x}$, and $\mathbf{I}$ is the identity matrix. Assuming independent inputs at each iteration, Eq. (4.3-17) can be extrapolated to a known $\mathbf{W}$ at time zero as

$$E(\mathbf{W}_k | \mathbf{W}_0) = (\mathbf{I} + \eta \mathbf{R}_{xx})^k \mathbf{W}_0 . \qquad (4.3\text{-}18)$$

Observe that the initial configuration of the network greatly influences its final state. Since the magnitude of the largest eigenvalue of $\mathbf{I} + \eta \mathbf{R}_{xx}$ is greater than unity, the network weights diverge as time goes to infinity. Network stability is achieved only if the largest eigenvalue of the matrix $\mathbf{I} + \eta \mathbf{R}_{xx}$ is no greater than unity which will not be the case. Therefore it is necessary to normalize the Hebbian learning rule. In biological systems, this normalization may take the form of a saturation due to the physical constraints of the system.

Networks using Hebbian learning have been created by Linsker [4-50]. These networks produce feature detecting cells that perform analogous to those in the visual cortex. Oja [4-51] has also shown that a Hebbian training rule will perform a principal component analysis on the input signal.

D. Generalization Ability

This section is concluded with a discussion of the ability of neural networks to generalize. A neural network generalizes well when it achieves acceptable performance by accurately classifying both the training data and the independent testing data. Generalization ability is influenced by many factors including training set size, estimation of the network performance, the initial values of the network weights [4-20], the order of presentation of the training exemplars [4-52], and overlearning [4-12], [4-53]. Here we only consider the most important factors, namely the training set size and the estimation of the network performance. The generalization ability of the network is highly dependent on both of these factors.

1. Size of the Training Set

Many empirical studies have investigated the relationship between the size of the training set and the ability of the networks to generalize. It has been suggested that the minimal number of exemplars required to achieve good generalization varies in proportion to the number of weights in the network and is inversely proportional to the desired performance [4-29], [4-37]. This relationship is highly problem dependent. It is commonly accepted that the more data available for training the better the network generalizes [4-20], [4-52], [4-54].

Theoretical studies concerning the number of training exemplars required for good generalization yield expressions that are consistent with the empirical studies. Lower bounds on the number of exemplars needed to achieve good generalization have been derived. Blumer *et al.* [4-55] have shown that it is possible to achieve a misclassification probability of P_ϵ on independent data by using Q exemplars, where Q is bounded by

$$Q \geq 8\frac{V}{P_\epsilon} \log\left(8\frac{V}{P_\epsilon}\right) , \tag{4.3-19}$$

where V is the Vapnik–Chervonenkis [4-56] measure of the network complexity. The bound in Eq. (4.3-19) was refined by Baum and Haussler [4-57] by assuming that the training exemplars all come from a single probability distribution. To achieve a misclassification probability of P_ϵ on independent testing patterns drawn from the same distribution as the training patterns, the number of training exemplars Q can then be bounded by

$$Q \geq O\left(\frac{N_w}{P_\epsilon} \log \frac{N_T}{P_\epsilon}\right) , \tag{4.3-20}$$

where N_w is the total number of weights in the network and N_T is the total number of neurons. When the logarithm term is neglected the result of Eq. (4.3-20) is consistent with the empirical results.

2. Estimation of Generalization Performance

The design of a practical neural network requires an estimate of its final performance. Unfortunately the final performance is difficult to estimate because all estimates are biased to the characteristics of the training data. A better estimate of the network performance can be obtained by using the pattern recognition techniques discussed below.

The first pattern recognition technique to be discussed is the *substitution method* [4-58]. This technique trains and tests the network on the same data set. The substitution method provides the least biased and most consistent estimate possible. This low bias and high consistency is possible because the substitution method makes use of the entire training set. Unfortunately the performance measure estimate using this technique is extremely optimistic. There is no way to determine whether the final network will produce accurate decisions with independent tests.

A second technique called the *hold-out method* [4-59], divides the data set into two halves. One half is used for testing and the other half is used for training. The estimate of the network weights has more bias and is less consistent than if the network was trained with the substitution method. The hold-out method however gives two measures of the network performance; one on the testing set and one on the training set. The performance estimate for the training data set is optimistic while the performance estimate for the testing data set is pessimistic. The testing data set performance estimate is poor because it is highly dependent on how the testing and training sets were partitioned. The final estimate of the network performance is either the average of the two estimates or just the training performance estimate.

The *leave-one-out method* [4-60] is a technique whose performance estimate is the median of these three techniques. Here the data set is divided into D partitions. Of these partitions, $D - 1$ are used for training and the remaining one is used to estimate the performance. This procedure is repeated D times, each with a different partition and the average performance measure is computed. The network parameters that give the median performance are selected as the final network configuration. This produces a less biased measure of the network performance while giving ample training data for the estimate of the neural weights. For large problems though, the number of training repetitions becomes prohibitive.

A final compromise technique is to train the neural network twice [4-61], first with the entire data set and second with a fraction of the data. The first training run provides an optimistic value of the performance measure and a good estimate of the network parameters. The second training run provides a poor estimate of the performance measure. The performance measures from the first and the second runs are averaged to produce the final performance measure. The final network weights are the ones found in the first training run.

4.4 Statistical Training of Multi-Layer Perceptrons

A. Complexity Issues

A key problem with the multi-layer perceptron is the specification of the interconnection weights, w_{ji}, from a set of training data. The noisy and incomplete training data set along with the nonlinear nature of the neural network prohibit an analytical solution. Thus it is necessary to resort to approximate techniques that solve the problem iteratively. This process is referred to as *fast learning*.

The training time for a neural network is dominated by three factors. These factors are the number of training patterns N_{tp}, the number of computations per pattern N_{cp}, and the required number of iterations needed to reach convergence N_{it}.

A temporal complexity measure, T_c, is the product of these three factors [4-62]

$$T_c = N_{tp} \cdot N_{cp} \cdot N_{it} . \tag{4.4-1}$$

The number of computations per iteration per pattern N_{cp} is fixed by design constraints and cannot be adjusted once learning has begun. The factor N_{it} has been investigated by many authors [4-40], [4-41], [4-62], and can be reduced by tuning the learning algorithm as discussed in Section 4.3. The purpose of this section is to investigate the term N_{tp}. In problems such as speech recognition, there is typically an abundance of training data. For example a training database with 500 exemplars of 0.5 seconds each may be blocked into 20 ms frames. This yields more than 10^4 frames. Thus in studying the complexity T_c, the term N_{tp} can be the dominant factor in the training time.

This section considers perceptron learning as a statistical optimization problem. The learning problem is initially discussed and formulated. Then an approximate learning technique is explained, and some experimental results using this algorithm presented.

B. Statistical Optimization

To formally restate the learning problem [4-30], consider a weight space $\mathcal{S}_W$ which is the domain of all possible network weights for a predefined topology. Associated with each point in this space is a neural network which implements a particular function. Because of the nature of the network [4-20] each function can be associated with a subset of points in $\mathcal{S}_W$.

We are given a finite set of training data Ω that partially describes the probability density of the network input $f_{\mathbf{x}}(\mathbf{x})$. Consider a performance measure Ψ that indicates how well a function satisfies the constraints imposed by Ω. The performance measure might be the mean-squared error at the network output, between the expected output and the actual output; or the probability of classification error. The learning problem finds the set of weights U that gives the optimum value of Ψ given Ω.

In the learning problem, the training data exemplars are assumed to be realizations from a stationary process, and it is assumed that learning minimizes the expected value of the performance measure. Thus the *probability density function* (pdf) of the network input ensemble is fixed and the network is trained by minimizing the expected value of Ψ with respect to the network weights. This problem is mathematically stated as

$$\left(\bar{\Psi}\right)_{OPT} = \min_{U} \left[E\left(\Psi[\mathbf{y}(U, \mathbf{x})] \right) \right] , \tag{4.4-2}$$

where $\mathbf{y}$ is the output of the neural network and the expectation is over the pdf of the network input $f_{\mathbf{x}}(\mathbf{x})$. We denote the network weights collectively as the set U.

Unfortunately the optimization specified in Eq. (4.4-2) is ill-posed [4-63] because $f_{\mathbf{x}}(\mathbf{x})$ is unknown. Additionally, there are many solutions to Eq. (4.4-2). Thus it is imperative to approximate Eq. (4.4-2). An equivalent optimization problem uses the pdf of the network output rather than the pdf of the network input in computing the expectation. This optimization problem is readily solved with a parametric form for the output density [4-64].

C. Approximate Solution

A parameterized form for the output density of a single neuron denoted as $f[y\,;\,\Gamma(U,\,X)]$ can be found [4-64]. Here $\Gamma(U,\,X)$ indicates that the neuron output density depends on the weights U, and appropriate input moments X. An analogous functional form can be assumed for the joint density of the outputs of a single layer in a neural network. This parametric form for the density of the neuron output implies that there is a mapping from the neuron input moments to the neuron output moments. The linear relationship in Eq. (4.1-2) between the neuron inputs y_i and the composite output z_j provides part of this mapping. It has been shown [4-65] that the probability density of the composite input can be approximated by a Gaussian distribution. Thus, we need only the mean and covariance at the neuron output to approximate the density of the composite input in the next layer neuron. The task is thus to transform the mean and covariance of z_j to the mean and covariance of y_j via $\sigma(z_j)$, following the relationship in Eq. (4.1-1).

For purposes of this development, define the neuron nonlinearity as the error function:

$$\sigma(z) = \frac{1}{2}\mathrm{erf}(z) = \frac{1}{2}\int_0^z \frac{2}{\sqrt{\pi}} e^{-z^2} dz \,. \tag{4.4-3}$$

This is a saturable function that monotonically increases with z from $-\frac{1}{2}$ to $+\frac{1}{2}$. The mean of the neural output (and input to the next layer) is given as

$$\eta_y = \int_{-\infty}^{\infty} \mathrm{erf}\left(\frac{\lambda}{\sqrt{2}}\right) \exp\left[-\frac{(\lambda - \eta_z)^2}{2\sigma_z^2}\right] d\lambda \,, \tag{4.4-4}$$

where η_z and σ_z^2 are the mean and variance, respectively, of the composite input z from Eq. (4.1-2). The expression in Eq. (4.4-4) can be evaluated in closed form as

$$\eta_y = \frac{1}{2}\mathrm{erf}\left(\frac{\eta_z}{\sqrt{2(\sigma_z^2 + 1)}}\right) \,. \tag{4.4-5}$$

In a similar manner, the variance and covariances at the neuron output can be related to the mean, variance, and covariance of the composite input. Convergent series, rather than closed form expressions are found for these statistics. The interested reader is referred to [4-65] for additional details.

By exploiting the layered structure of the neural network we can propagate the input moments of the neural network to the network output moments. This technique can be used to approximate the expected value of the performance measure as a function of the network weights and the moments of the network input. With this functional relationship between the expected value of the performance measure, the network weights, and the network input statistics, we can find the best network weights using standard optimization techniques [4-66].

By formulating the learning problem in the manner described above, we are able to reduce the training time because the training patterns are only used to compute the network input moments. Thus, there are several orders of magnitude less training data. When the reduction of the training data offsets the increase in the computational load, this technique is significantly faster than other training algorithms [4-33].

D. Experimental Results

Several experiments were conducted to compare the performance of back-propagation and statistical neural training introduced above. Three experiments are discussed in this section: a simple Gaussian discrimination problem with a large quantity of training data, the classical XOR problem, and a phonetic discrimination problem.

1. Gaussian Test with a Small Network

The first experiment is a detection problem where the neural network is trained to discriminate between two equally likely one-dimensional Gaussian distributions. This problem was chosen because of its simplicity. The two Gaussian distributions have unit variance and means of -1 and 2 respectively. The optimum Bayesian detector for this problem is a simple threshold device with the threshold set at 0.5.

A small neural network was used for discrimination. This neural network had a single input node, two hidden layers, each with two nodes, and a single output node. This network had 13 weights and is clearly too small for the Gaussian approximation on the composite input to be valid. Thus the performance of the network trained with the statistical neural network techniques will be impaired. The convergence time of the *statistical neural network* (SNN) training and *back-propagation* (BP) training was examined. Initially we assumed that an infinite amount of training data was available (convergence after 100,000 exemplars). The BP learning algorithm was not optimized and no momentum was used in adapting the weights. The learning adaptation constant was set at 0.01 to avoid unstable training. The *mean-squared error* (MSE) at the network output was computed as the average of the squared difference between the expected output and the actual output. The squared error for the BP algorithm was averaged over 1000 presentations. Convergence occurred when the change in averaged squared error between the current iteration and the previous iteration was less than 10^{-5}. This convergence criterion was used for both algorithms. The results of this experiment, shown in Table 4.4-1, demonstrate that both training techniques achieve rather similar performance. The MSE for both cases is close, as is the *variance* (Var) of the MSE. The network trained with BP achieves slightly less MSE than the network trained with the statistical neural network technique, but the statistical neural network training resulted in a lower MSE variance. The network trained with BP required 208.3 s of *central processing unit* (CPU) time on a 0.75 MFLOP Sun 4 workstation, while SNN required 29.19 s. The training time was thus reduced by a factor of seven, or by more than 86% when the SNN is compared to back-propagation.

	Time	MSE	Var
BP	208.30 s	0.101	0.0496
SNN	29.19 s	0.114	0.0435

Table 4.4-1: Training times and performances of BP and SNN.

2. Exclusive-OR Problem

The previous experiment compared the performance of BP and SNN training techniques on a problem with a large number of training exemplars. For this problem the temporal complexity measure T_c in Eq. (4.4-1) was dominated by N_{tp}. Statistical neural network training techniques had a definite advantage in this case. The next experiment was chosen to remove this advantage.

What follows is the classic XOR problem discussed in Section 4.2, i.e., class A $\{(+1, +1), (-1, -1)\}$ and class B $\{(+1, -1), (-1, +1)\}$. However, there are only four training exemplars. A small neural network was used to solve this problem. There were two input nodes, two hidden nodes, and two output nodes. The network was trained to have an output of $(+0.5, -0.5)$ when the input was from class A and $(-0.5, +0.5)$ when the input was from class B.

The BP learning algorithm updated the weights after every exemplar. The algorithm used an adaptation constant of 1.0, and no momentum. The algorithm converged when the change in the squared error for the four patterns was less then 10^{-5}. The SNN learning used the same convergence criterion as BP with an adaptation constant of 0.3. The training times for this experiment measured in seconds of CPU time on a Sun 4 workstation are given in Table 4.4-2.

Back-propagation is the faster algorithm for this experiment: it converges about 20% faster than the SNN algorithm. There are several reasons for this. First the SNN algorithm was formulated for large neural networks. In this small neural network the Gaussian assumption is violated and hinders the performance of the

Training times (in seconds)	
Back-propagation	1.95
SNN gradient	2.41

Table 4.4-2: Training times for the XOR problem.

algorithm. Additionally, the SNN training technique has a clear advantage when there is a large amount of training data. This problem has few training exemplars, thus the larger number of computations of the SNN training approach are a disadvantage. Despite these disadvantages, the SNN approach does well on this problem, achieving performance comparable to BP.

3. Fricative Discrimination

In this experiment, an integral part of a large vocabulary speech recognizer was built. The objective was to discriminate between the three unvoiced fricatives *F*, *S*, and *SH*, and band-limited white background noise. Six neural network classifiers were designed to discriminate between the pairs: *S/SH, S/F, S/noise, SH/F, SH/noise,* and *F/noise*. For these initial tests, the speech samples were from a single male speaker, collected in a typical office environment. The speech was preprocessed using a filter bank with four filters spaced linearly between 100 Hz and 3300 Hz. The output of the filter bank was averaged over a 20 ms window and sampled to produce a set of four features at a rate of 50 Hz.

The complete neural network consisted of six two-layer perceptron neural networks. Each two-layer perceptron has four inputs, twelve neurons in the hidden layer, and two output neurons. The neural networks were trained so that their outputs were either $(0.4, -0.4)$ or $(-0.4, 0.4)$. One complete network was trained using the BP algorithm and another with the SNN learning algorithm described above. The same set of initial weights was used for all networks. Both the BP algorithm and the SNN learning algorithm used an adaptation constant of 0.01 and no momentum. The training was aborted when the value for the averaged squared error for an epoch through the database was less than 0.01 or when 1000 passes through the training data set were completed. These values were chosen to avoid excessive training times. The training times in seconds of CPU time on a Sun 4 workstation and the recognition performance are shown in Table 4.4-3.

The results in Table 4.4-3 demonstrate the training time advantage of the SNN learning technique. The total training time was reduced by more than 75% when compared to BP. Thus SNN learning significantly reduces the training time while still achieving a performance that is comparable to back-propagation. Both algorithms give questionable performances on the *S/F* recognition. This is expected because of the difficulty of this discrimination due to the acoustic similarity of the phonemes.

Network	Back-propagation		Statistical	
	Time	% Correct	Time	% Correct
S/SH	57.1	100.0	36.1	99.8
S/F	363.9	59.9	46.1	65.4
S/noise	391.5	98.7	71.2	94.2
SH/F	199.3	100.0	41.5	99.8
SH/noise	183.2	100.0	66.2	99.8
F/noise	357.5	96.8	122.2	91.7

Table 4.4-3: Training times, in seconds, and recognition performance for the fricative discriminator networks.

4.5 Combining Multi-Layer Perceptrons for Speech Recognition

A *time delay neural network* (TDNN) is a modified multi-layer perceptron used to classify time-dependent patterns [4-68]. A TDNN used for speech recognition is designed by adjusting the neural weights with a learning algorithm that minimizes a performance measure over a set of training data. The typical design procedure requires specifying a network architecture and training it with all the available data. This technique can result in extremely long training times and poor generalization. The end result is disappointing performance especially with complex problems such as speech recognition. The network fails because of its inability to generalize and because there are many local minima in the cost function [4-20]. A further disadvantage of this design procedure is that it prohibits the inclusion of any *a priori* knowledge about the problem during training [4-67].

A more reasonable approach to the design of large neural networks is a divide and conquer technique [4-68], i.e., design the network modularly by judiciously dividing the design problem into several smaller sub-problems. These sub-problems are solved with smaller sub-networks that converge faster and require less data for training, thereby achieving better generalization. These sub-problems can be chosen prudently so that prior knowledge can be incorporated into the network design. The use of prior knowledge improves the overall performance by tuning the neural network to specific aspects of the problem [4-20]. When approaching the neural network design modularly, it is vital to have a method for combining the separately designed sub-networks. The task of combining sub-networks was considered in [4-68], where special training methodologies were used to form a "neural glue". In this section, a method of combining the sub-networks is detailed. This technique is termed *network fusion*, based on the theory of data fusion [4-68]. It can be shown that when the sub-networks are independent, this method of combining is Bayesian optimal.

These ideas are illustrated in Fig. 4.5-1. The large neural network is decomposed into several smaller sub-networks. These sub-networks are trained on the same input data, or a partition of this input data. After training, the sub-networks are combined into a single larger network using network fusion.

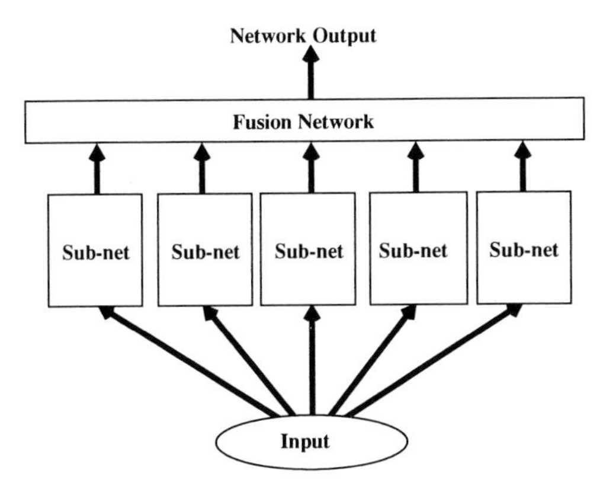

Figure 4.5-1: The network fusion procedure.

In this section a network fusion rule is derived by considering the log-likelihood ratio of each sub-network output. It is shown that this rule can be implemented with a neural network, and formulas for the neuron weights and offsets are given. Following this, the fusion rule is examined and an iterative procedure defined for fine tuning and performance improvements. The network fusion procedure is then illustrated with several experiments.

A. The Network Fusion Rule

Let each of the N_S sub-networks to be combined into a single network to produce a single decision v_i, $i = 1, \cdots, N_S$, where $v_i \in [-1, +1]$. For simplicity, assume the sub-network training corresponds to a -1 output for class A training patterns, and a $+1$ for class B training patterns. This can be achieved by post-processing the sub-networks output such that $u_i = -1$ when $v_i \leq 0$ and $u_i = +1$ when $v_i > 0$. The $\{u_i\}$ results by passing the $\{v_i\}$ through a hard limiter centered around zero.

Now form two hypotheses, H_0 and H_1, where H_0 is the hypothesis that the pattern has been drawn from class A and H_1 is the class B hypothesis. Also assume that there are highly reliable estimates of the probabilities $P(u_i = -1|H_0)$, $P(u_i = -1|H_1)$, $P(u_i = +1|H_0)$, and $P(u_i = +1|H_1)$ for $i = 1, \cdots, N$. We wish to optimally combine the hard-limited outputs into a single global decision that minimizes the total probability of error. This is achieved with a Bayesian detector, and the likelihood ratio [4-70] is given as

$$\frac{P(u_1, u_2, \cdots, u_N|H_1)}{P(u_1, u_2, \cdots, u_N|H_0)} \lessgtr \frac{P_0}{P_1}, \tag{4.5-1}$$

where H_0 corresponds to less than and H_1 corresponds to greater than P_0/P_1. In this classification problem there is no bias toward either class A or B, thus the *a priori* class probability $P_0 = P_1$.

If all sub-networks are trained independently, it is reasonable to make the simplifying assumption that their outputs are independent. Additionally the decision capabilities of the Bayesian detector remain unchanged if a monotonic function is applied to both sides of Eq. (4.5-1). Taking the logarithm and using the independence assumption gives the log-likelihood ratio

$$\sum_{i=1}^{N_S} \log\left(\frac{P(u_i|H_1)}{P(u_i|H_0)}\right) = \lambda \lessgtr 0, \tag{4.5-2}$$

where class A is chosen if the sum of the individual log-likelihoods is less than zero and class B if they are greater than zero.

Since the variables $\{u_i\}$ are discrete, we can rewrite Eq. (4.5-2) using a linear interpolation. Any individual log-likelihood ratio in the sum in Eq. (4.5-2) is one of two values because u_i is a binary variable. A linear interpolation between one point of a pair $(u_i = -1, \log[P(u_i = -1|H_1)/P(u_i = -1|H_0)])$ and another point of the pair $(u_i = +1, \log[P(u_i = +1|H_1)/P(u_i = +1|H_0)])$ takes the form

$$\log\left(\frac{P(u_i|H_1)}{P(u_i|H_0)}\right) = w_i u_i + \theta_i, \tag{4.5-3}$$

where

$$w_i = \frac{1}{2}\log\left(\frac{P(u_i = -1|H_0)P(u_i = +1|H_1)}{P(u_i = -1|H_1)P(u_i = +1|H_0)}\right), \tag{4.5-4}$$

and

$$\theta_i = \frac{1}{2}\log\left(\frac{P(u_i = -1|H_1)P(u_i = +1|H_1)}{P(u_i = -1|H_0)P(u_i = +1|H_0)}\right). \tag{4.5-5}$$

The total log-likelihood ratio linear interpolation is then

$$\lambda = \sum_{i=1}^{N_S} w_i u_i + \Theta, \tag{4.5-6}$$

where $\Theta = \sum_{i=1}^{N_S} \theta_i$.

The decision rule based on Eq. (4.5-6) decides in favor of hypothesis H_0 when $\lambda > 0$ and hypothesis H_1 when $\lambda < 0$. This has the form of a single neuron with a hard limiter $\sigma(\cdot)$ as a nonlinearity, that is

$$y = \sigma\left(\sum_{i=1}^{N_S} w_i u_i + \Theta\right). \tag{4.5-7}$$

When y in Eq. (4.5-7) is less than zero we decide H_0; when y is greater than zero we decide H_1. Thus a single neuron is the optimal decision function in this case. The next subsections examine the network fusion rule when several of the assumptions made in its derivation are relaxed.

1. Soft Nonlinearities

The presence of a hard limiter before the network fusion rule results in a loss of information. If we relax the hard limiter condition, and follow the work of Gish [4-71], it can be shown that the output, y, of a well-trained sub-network is given approximately as

$$y \approx P(H_1|x) - P(H_0|x). \tag{4.5-8}$$

If the input pattern x is a clear member of H_1 then the magnitude of the output of the neural network will be nearly unity. However, if there is some ambiguity, the output of the sub-network will be close to zero. This can be used to improve the network performance. After the sub-networks have been combined using network fusion, the hard nonlinearities can be removed leaving only the network outputs $\{y_i\}$. The log-likelihood ratios are then scaled by the difference between the *a posteriori* class probabilities given in Eq. (4.5-8). This decreases the contribution to the total decision of those sub-networks making inferior decisions, and allows the fusion neuron to decide using only the sub-networks with "confident" decisions.

2. Insufficient Training Data

The effect of insufficient training data manifests itself as noisy estimates of the probabilities. In [4-56] a bound on the error between the noisiest estimate of a probability $\bar{p}$ and its true value p was derived as

$$P(|p - \bar{p}| > \delta_p) = 1 - \epsilon_p \leq g(F)\exp(-\gamma F \delta_p^2), \tag{4.5-9}$$

where δ_p and ϵ_p are small positive numbers, F is the number of terms used in the relative frequency estimate of p, $g(F)$ is a polynomial function of F, and γ is a constant.

Omiting the algebraic details found in [4-65] for the sake of brevity, the following bound on the difference between the true value of the fusion weights w_i and the estimated value $\bar{w}_i$ can be derived

$$| w_i - \bar{w}_i | \leq \frac{\Gamma_t}{\sqrt{\gamma F}} \left[\log \left(\frac{g(F)}{1 - \epsilon_p} \right) \right]^{1/2} , \qquad (4.5\text{-}10)$$

where Γ_t is a function of the true values of the probabilities. The difference in Eq. (4.5-10) tends to zero as $1/\sqrt{F}$. Thus, for sufficiently large training data sets, the network fusion procedure is relatively insensitive to F. A similar result can be derived for the fusion offsets, θ_i.

3. Self-Organizing Sub-network Improvement

The independence of the sub-network outputs was a key assumption made in the derivation of the fusion rule. In the general case, the sub-networks *are dependent* and the final network performance can be improved by taking advantage of this fact and making the cross-connections between the sub-networks non-zero. A self-organizing iterative algorithm has been proposed for this purpose [4-65]; an outline of this algorithm follows.

To determine the sub-network interconnections, first design all sub-networks individually. Second, fix the connections of each sub-network and increase the average magnitude of the output of all sub-networks in the ensemble using the sub-network interconnections. Third, perform network fusion on the resulting ensemble of sub-networks.

Experience with this technique has shown that maximizing the second stage of the self-organizing algorithm can be performed extremely efficiently. If a single line search is conducted in the direction of the initial gradient, the maximum is attained quickly. Continued line searches do not improve the performance further.

B. Experimental Results

1. Gaussian Random Process Discrimination

The network fusion technique is demonstrated with a Gaussian discrimination problem. This problem was selected because it is readily solved analytically and limiting theoretical performances can therefore be calculated. During training the neural network was allowed to see only part of the input data. Two three-element Gaussian random vectors were used. Both distributions have identity covariance matrices. Class A has a mean vector of $(-1, -1, -1)$ and class B has a mean vector of $(+1, +1, +1)$. One thousand points were generated from each distribution. The limiting Bayesian classification probability is 0.958, while the optimal Bayesian classifier [4-70] probability is 0.957 with the empirical set of data.

Three two-layer sub-networks were trained independently. Each sub-network has two inputs, four hidden layer nodes, and one output. The sub-networks were trained to classify the problem using only two inputs. Thus the relationship between all the input variables could not be exploited fully by any of the sub-networks. The performance of the sub-networks is given in Table 4.5-1 where network X_iX_j denotes the network that used only inputs i and j; and y denotes the generic output variable. The best sub-network has a performance, i.e., the probability of making

Network	$P(y > 0 \mid A)$	$P(y < 0 \mid B)$	$P(\text{correct})$
X_1X_2	0.909	0.825	0.867
X_1X_3	0.844	0.851	0.848
X_2X_3	0.902	0.905	0.904

Table 4.5-1: Empirical network probabilities for the original sub-networks in the Gaussian discrimination problem.

a correct decision, $P(\text{correct})$, which is 94% of the Bayesian rate given above. Note that $P(\text{correct})$ is just the average of $P(y > 0 \mid A)$ and $P(y < 0 \mid B)$.

The sub-networks were combined using the network fusion technique discussed above and achieved a $P(\text{correct})$ of 0.907. By replacing the hard nonlinearities by soft nonlinearities, the performance increased to 0.909. The ensemble of the three sub-networks was trained using the self-organizing improvement algorithm where only the interconnection weights of the second layer were optimized. The fusion was then repeated and the performance increased substantially to 0.933 with soft nonlinearities. Finally, global training was performed on the second and third layers of the neural network using a supervised conjugate gradient algorithm. This further increased the performance to 0.939. These results are summarized in Table 4.5-2.

Type	$P(\text{correct})$
Bayes limit	0.958
Bayes detector	0.957
Hard-nonlinearities	0.907
Soft-nonlinearities	0.909
After iteration hard-nonlinearities	0.933
After iteration soft-nonlinearities	0.933
After iteration and global training	0.939

Table 4.5-2: Classification probabilities.

2. Speech Recognition

The database used for this experiment contained three vocabulary entries, the Spanish words *uno*, *dos*, and *tres*, that were recorded in Madrid, Spain over conventional telephone lines. More than 1000 speakers contributed to the database. All were unfamiliar with speech recognition technology. The signal-to-noise ratio of the recorded speech ranged from −17.0 dB to 46.0 dB. The speech was band-limited from 300 Hz to 3.4 kHz and sampled at 8.0 kHz. It was manually verified and endpointed. Originally there were 3044 speech files in the database, of which 1685 were male speakers, and 1359 female speakers. It was initially split into a training set of 1850 files and a testing set of 1194 files. Verification resulted in files being discarded because they were too long, too short, or they had a very low signal-to-noise ratio. The final database contained 1069 training files and 593 testing files.

The speech was pre-processed using filter bank techniques [4-71]. The entire speech utterance was energy normalized and passed through a bank of twenty band-pass filters. The filters were spaced to approximate the Mel scale [4-73] and the bandwidths of the filters were chosen to achieve an all-pass composite spectrum. The output of each filter was rectified by squaring and low-pass filtering to produce twenty features at a 50 Hz rate.

The goal was to build a time delay neural network with a 420 ms sliding window to recognize the above vocabulary. When the utterance was in the window, the network produced a positive pulse identifying the proper vocabulary item. Otherwise the network output was negative. Thus the time delay neural network was a word spotting system. A word is recognized correctly when the index of the maximum positive output of the neural network is the same as that of the speech utterance in the window.

Each vocabulary word was separated into three distinct sub-word acoustic events. These corresponded to roughly an *onset*, an extended *center*, and a terminating trailing *end*. Several files were segmented by hand and used as prototypes for the segmentation of the entire training database. After the database was segmented, it was examined further to remove all excessively short and long utterances.

Single-layer perceptron networks were then trained to recognize the vocabulary word using only data from a specific acoustic event. Training took place using statistical neural network techniques as explained in Section 4.4. The performance, i.e., probability of error, of these sub-networks is shown in the upper half of Table 4.5-3. The column labeled "Errors" contains the number of misclassification errors, while the column labeled "Misses" gives the number of no decisions made by the network. Table 4.5-3 shows that these sub-networks achieve poor-to-moderate performance.

The *onset* sub-network accepts three frames of data corresponding to 60 ms. The *center* sub-network receives eight frames of data, representing 160 ms. There are seven data frames, over 140 ms, applied to the *end* sub-network.

Sub-network performance			
Network errors	Errors	Misses	Error probability
onset	267	30	0.236
center	120	42	0.107
end	289	16	0.252
Fused sub-network performance			
onsetF	179	138	0.175
centerF	99	44	0.089
endF	164	98	0.155

Table 4.5-3: Performances of the sub-networks.

Nine copies of the *onset* sub-network were spaced one frame apart and combined together with the fusion procedure to form another sub-network *onsetF*. This fusion enabled us to spread the detector over an area of the time window to help the network in processing time-varying speech. This is similar to the slaved weights technique in [4-68]. As with the *onset* sub-networks, six copies of the *center* sub-network spaced one frame apart were fused together to produce the sub-network *centerF*. Three copies of the sub-network *end*, spaced one frame apart, were also fused to produce *endF*. The performances of these sub-networks are given in the lower half of Table 4.5-3. The performance increase is significant using network fusion and a spreading technique. The error probability is decreased by 38% for the *end* sub-network.

The final step was to combine the feature detecting sub-networks into one large network, again using network fusion techniques. The network *onsetF* was placed at zero lags. The network *centerF* was placed at three lags while the network *endF* was placed at twelve lags. Each lag corresponds to 20 ms for the 50 Hz feature rate. The performance of the composite network is given in Table 4.5-4, for the training and testing databases. The scores given in these tables are the overall error rates for the Spanish words *uno, dos,* and *tres*. Observe that by fusing the sub-networks, the error decreases by almost 70% over the best performance of any individual sub-network. Table 4.5-4 also confirms the conjecture that the fusion method provides good generalization, as predicted by Eq. (4.5-10). The entire neural network was designed in two hours using network fusion techniques on a 2 MFLOP workstation.

Summary

This chapter has provided a broad introduction to artificial neural networks. We have focused on multi-layer perceptrons, and have seen how they can be used

Two layer performance			
Database	Errors	Misses	Error Probability
Training	31	92	0.029
Testing	16	48	0.027

Table 4.5-4: Final performance of the word spotting network.

in applications that require signal classification, decision, and detection. A brief history was provided, and a mathematical model of a basic neural processing element presented. The artificial neuron was seen to consist of a linear portion, that is, the weighted sum of inputs, and a nonlinear portion often referred to as a sigmoid function, $\sigma(\cdot)$.

A series of design examples demonstrated the power of a multi-layer perceptron in bisecting the decision space with several hyperplanes. The perceptron architecture was discussed, including the number of hidden layers and the number of neurons per layer needed to solve a particular design problem. Perceptron training was covered in detail, specifically the all-important back-propagation algorithm traditionally used to compute the neuron weighting coefficient values during the learning phase of a neural network design.

The last two sections of this chapter dealt with the concepts of statistical training and network fusion. Here learning complexity was explored and an approximate solution to find the neuron weights explained. This solution related the network weights to the input statistics, and permitted an order of magnitude reduction in training time compared to the training time required by back-propagation.

Network fusion was introduced as an effective method of designing a large neural network by judiciously dividing the design problem into several smaller sub-problems. The sub-problems were solved with smaller sub-networks that converge faster and require less data for training, thereby achieving better generalization. The network fusion rule was based on Bayesian estimation, and can be realized with a single neuron. Experimental results applied to speech recognition, specifically word spotting, clearly demonstrate the value of both statistical training and network fusion design.

References

[4-1] W.H. Pitts and W.S. McCulloch, "A logical calculus of ideas immanent in nervous activity," *Bull. Math. Biophy.*, vol. 5, pp. 115-133, Jan. 1943.

[4-2] F. Rosenblatt, "The perceptron: A probabilistic model for information storage and organization in the brain," *Psychological Review,* vol. 65, pp. 386-408, May 1958.

[4-3] Y. LeCun, I. Guyon, L.D. Jackel, D. Henderson, B. Boser, R.E. Howard, J.S. Denker, W. Hubbard and H.P. Graf, "Handwritten digit recognition: Applications of neural network chips and automatic learning," *IEEE Commun. Mag.,* vol. 11, pp. 41-46, Nov. 1989.

[4-4] H. Sompolinsky, "Statistical mechanics of neural networks," *Physics Today,* vol. 41, pp. 70-80, Dec. 1988.

[4-5] R.P. Gorman and T.J. Sejnowski, "Analysis of hidden units in a layered network trained to classify sonar targets," *Neural Networks*, vol. 1, pp. 75-89, 1988.

[4-6] D.O. Hebb, *The Organization of Behavior: A Neurophysiological Theory.* New York, NY: John Wiley and Sons, Inc., 1949.

[4-7] J.J. Hopfield, "Neural networks and physical systems with emergent collective computational abilities," *Proc. Natl. Acad. Sci. USA*, vol. 79, pp. 2554-2558, April 1982.

[4-8] D. Rumelhart, J. McClelland and the PDP Research Group, *Parallel Distributed Processing.* Cambridge, MA: Bradford Books, 1986.

[4-9] B. Widrow and M.E. Hoff, "Adaptive switching circuits," in *Proc. IRE WESCON Convention*, pp. 96-104, 1960.

[4-10] B. Widrow and S.D. Stearns, *Adaptive Signal Processing.* Englewood-Cliffs, NJ: Prentice-Hall, Inc., 1985.

[4-11] M.L. Minsky and S. Papert, *Perceptrons.* Cambridge, MA: MIT Press, 1969.

[4-12] T.E. Sejnowski and C.R. Rosenberg, "Parallel networks that learn to pronounce English text," *Complex Systems*, vol. 1, pp. 145-168, Jan. 1987.

[4-13] T.M. Cover, "Geometrical and statistical properties of systems of linear inequalities with applications in pattern recognition," *IEEE Trans. Electronic Computers*, vol. 14, pp. 326-334, June 1965.

[4-14] J. Makhoul, R. Schwartz and A. El-Jaroudi, "Classification capabilities of two-layer neural nets," in *Proc. IEEE Internat. Conf. Acoust., Speech, Signal Processing*, vol. 1, pp. 635-638, May 1989.

[4-15] A.N. Kolmogorov, "On the representation of continuous functions and many variables by superposition of continuous functions of one variable and addition," *Doklady Akademii Nauk SSSR*, vol. 144, pp. 679-681, 1963.

[4-16] G.G. Lorentz, "The 13-th problem of Hilbert," in *Proc. Symp. Pure Mathematics*, vol. 28, pp. 419-430, May 1975.

[4-17] C.L. Giles and T. Maxwell, "Learning, invariance and generalization in high-order neural networks," *Applied Optics*, vol. 26, pp. 4972-4976, Nov. 1987.

[4-18] A. Weiland and R. Leighton, "Geometric analysis of neural network capabilities," in *Proc. IEEE Internat. Conf. Neural Networks*, vol. III, pp. 385-392, June 1987.

[4-19] R.C. Eberhart, "Standardization of neural network terminology," *IEEE Trans. Neural Networks,* vol. 1, pp. 244-245, June 1990.

[4-20] J. Denker, D. Schwartz, B. Wittner, S. Solla, R. Howard, L. Jackel, and J. Hopfield, "Large automatic learning, rule extraction and generalization," *Complex Systems*, vol. 1, pp. 877-922, May 1987.

[4-21] D.H. Foley, "Considerations of sample and feature size," *IEEE Trans. Info. Th.*, vol. 18, pp. 618-626, Sept. 1972.

[4-22] R.P. Lippmann, "An introduction to computing with neural nets," *IEEE Acoust., Speech, Signal Processing Mag.*, vol. 4, pp. 4-22, April 1987.

[4-23] W.Y. Huang and R.P. Lippmann, "Neural net and traditional classifiers," in *Neural Information Processing Systems*, D.Z. Anderson (ed.). New York, NY: American Institute of Physics, pp. 387-396, 1988.

[4-24] I.D. Longstaff and J.F. Cross, "A pattern recognition approach to understanding the multilayer perceptron," Royal Signals and Radar Establishment Memorandum No. 3936, July 1986.

[4-25] G. Cybenko, "Approximations by superpositions of a sigmoidal function," *Mathematics of Control, Signals and Systems*, vol. 2, pp. 303-314, March 1989.

[4-26] M.D. Bedworth and J.S. Bridle, "Experiments with the back propagation algorithm: A systematic look at a small problem," Royal Signals and Radar Establishment Memorandum No. 4049, June 1987.

[4-27] D.J. Burr, "Experiments on neural net recognition of spoken and written text," *IEEE Trans. Acoust., Speech, Signal Processing,* vol. 36, pp. 1162-1168, July 1988.

[4-28] N.J. Nilsson, *Learning Machines,* New York, NY.: McGraw-Hill Book Co., 1965.

[4-29] E.B. Baum, "On the capabilities of multilayer perceptrons," *Journal of Complexity*, vol. 4, pp. 193-215, Mar. 1988.

[4-30] N. Tishby, E. Levin and S. Solla, "Consistent inference of probabilities in layered networks: Predictions and generalization," in *Proc. IEEE Internat. Conf. Neural Networks*, vol. II, pp. 403-409, June 1989.

[4-31] P.J. Werbos, "Backpropagation: Past and future," in *Proc. IEEE Internat. Conf. Neural Networks,* vol. I, pp. 343-353, June 1987.

[4-32] D.B. Parker, "Optimal algorithms for adaptive networks: Second order back propagation, second order direct propagation and second order Hebbian learning," in *Proc. IEEE Internat. Conf. Neural Networks,* vol. II, pp. 593-600, June 1987.

[4-33] D.E. Rumelhart, G.E. Hinton and R.J. Williams, "Learning internal representations by error propagation," in *Parallel Distributed Processing: volume 1,* D.E. Rumelhart and J.L. McClelland (eds.). Cambridge, MA: Bradford Books, pp. 318-362, 1986.

[4-34] W. Press, B. Flannery, S. Teukolsky and W. Vetterling, *Numerical Recipes in C.* Cambridge, England: Cambridge University Press, 1989.

[4-35] A. Von Lehmen, E.G. Paek, P.F. Liao, A. Marrakchi and J.S. Patel, "Factors influencing learning by backpropagation," in *Proc. IEEE Internat. Conf. Neural Networks*, vol. I, pp. 335-341, June 1988.

[4-36] S. Ahmad, *A Study of Scaling and Generalization in Neural Networks,* MS Thesis. Urbana, IL: University of Illinois, 1988.

[4-37] S. Ahmad and G. Tesauro, "Scaling and generalization in neural networks: A case study," in *Advances in Neural Information Processing Systems I*, D.S. Touretzky (ed.). San Mateo, CA: Morgan Kaufnam Publishers, pp. 161-168, 1989.

[4-38] J.F. Kolen, "Faster learning through a probabilistic approximation algorithm", in *Proc. IEEE Internat. Conf. Neural Networks,* vol. I, pp. 449-454, June 1988.

[4-39] S. Judd, "Learning in networks is hard," in *Proc. IEEE Internat. Conf. Neural Networks,* vol. II, pp. 685-692, June 1987.

[4-40] R.L. Watrous, "Learning algorithms for connectionist networks: Applied gradient methods of nonlinear optimization," in *Proc. IEEE Internat. Conf. Neural Networks,* vol. II, pp. 619-628, June 1987.

[4-41] T.P. Vogl, J.K. Mangis, A.K. Rigler, W.T. Zink and D.L. Alkon, "Accelerating the convergence of the back-propagation method," *Biol. Cybern.*, vol. 59, pp. 257-263, March 1989.

[4-42] D.C. Plaut, S.J. Nowlan and G.E. Hinton, "Experiments on learning by back propagation," Technical Report CMU-CS-86-126, Carnegie-Mellon Univ., Pittsburgh, PA, 1986.

[4-43] J. Sietsma and R.J.F. Dow, "Neural net pruning - Why and how," in *Proc. Internat. Conf. Neural Networks,* vol. I, pp. 325-333, June 1988.

[4-44] Q. Xue, Y. Hu and P. Milenkovic, "Analyses of the hidden units of the multilayer perceptron and its application in acoustic-to-articulatory mapping," in *Proc. IEEE Internat. Conf. Acoust., Speech, Signal Processing*, vol. 2, pp. 869-872, April 1990.

[4-45] D.A. Pierre, *Optimization Theory with Applications.* New York, NY: John Wiley and Sons, Inc., 1969.

[4-46] E.D. Dahl, "Accelerated learning using the generalized delta rule," in *Proc. IEEE Internat. Conf. Neural Networks,* vol. II, pp. 523-530, June 1987.

[4-47] A.K. Jain and R.C. Dubes, *Algorithms for Clustering Data.* Englewood Cliffs, NJ: Prentice-Hall, Inc., 1988.

[4-48] E. Saund, "Dimensionality reduction using connectionist networks," *IEEE Trans. Patt. Anal. Mach. Intell.*, vol. 11, pp. 304-314, Mar. 1989.

[4-49] J.L. Elman and D. Zipser, "Learning the hidden structure of speech, " *J. Acoust. Soc. Am.*, vol. 83, pp. 1615-1626, April 1988.

[4-50] R. Linsker, "Self-organization in the perceptual network," *Computer*, vol. 21, pp. 105-117, Mar. 1988.

[4-51] E. Oja, "A simplified neuron model as a principal component analyzer," *J. Math. Biology*, vol. 15, pp. 267-273, Mar. 1982.

[4-52] G. Josin, "Neural-space generalization of a topological transformation," *Biol. Cybern.*, vol. 59, pp. 283-290, Mar. 1988.

[4-53] J. Oglesby and J. Mason, "Optimization of neural models for speaker identification," in *Proc. IEEE Internat. Conf. Acoust., Speech, Signal Processing,* vol. 1, pp. 261-264, April 1990.

[4-54] T. Maxwell, C.L. Giles and T.C. Lee, "Generalization in neural networks: The contiguity problem," in *Proc. IEEE Internat. Conf. Neural Networks,* vol. II, pp. 41-46, June 1987.

[4-55] A. Blumer, A. Ehrenfeucht, D. Haussler and M. Warmuth, "Classifying learnable geometric concepts with the Vapnik-Chervonenkis dimension", in *Proc.* 18th *ACM Symp. on the Theory of Computing*, vol. 1, pp. 273-282, 1986.

[4-56] V.N. Vapnik and A.Y. Chervonenkis, "On the uniform convergence of relative frequencies of events to their probabilities," *Theory of Probability and its Applications*, vol. 17, pp. 264-280, March 1971.

[4-57] E.B. Baum and D. Haussler, "What size net gives valid generalization?" in *Advances in Neural Information Processing Systems I*, D.S. Touretzky (ed.). San Mateo, CA: Morgan Kaufnam Publishers, pp. 81-91, 1989.

[4-58] G. Toussaint, "Bibliography on estimation of misclassification," *IEEE Trans. Info. Th.*, vol. 20, pp. 472-479, July 1974.

[4-59] W.H. Highleyman, "The design and analysis of pattern recognition experiments," *Bell Syst. Tech. J.*, vol. 41, pp. 723-744, March 1962.

[4-60] M. Hills, "Allocation rules and their error rates," *J. Roy. Statist. Soc.*, series B, vol. 28, pp. 1-31, Jan. 1966.

[4-61] F. Mosteller, "The jackknife," *Rev. Int. Statist. Inst.*, vol. 39, pp. 363-368, March 1971.

[4-62] P. Haffner, A. Waibel, H. Sawai and K. Shikano, "Fast back-propagation learning methods for large phonemic neural networks," in *Proc. Eurospeech 1989*, vol. 1, pp. 553-556, 1989.

[4-63] Y.Z. Tsypkin, *Adaptation and Learning in Automatic Systems.* New York, NY: Academic Press, 1971.

[4-64] K. Fukunaga, *Introduction to Statistical Pattern Recognition.* New York, NY: Academic Press, 1972.

[4-65] E.R. Buhrke, *Time Delay Neural Networks for Speech Recognition*, Ph.D. Thesis. Chicago, IL: Dept. of Electrical and Computer Engineering, Illinois Institute of Technology, 1991.

[4-66] F.A. Lootsma (ed.), *Numerical Methods for Nonlinear Optimization.* New York, NY: Academic Press., 1972.

[4-67] E.W. Felten, O. Martin, S.W. Otto and J. Hutchison, "Multi-scale training of a large backpropagation net," *Biol. Cybern.*, vol. 62, pp. 503-509, May 1990.

[4-68] A. Waibel, T. Hanazawa, G. Hinton, K. Shikano and K.J. Lang, "Phoneme recognition using time-delay neural networks," *IEEE Trans. Acoust., Speech, Signal Processing*, vol. 37, pp. 362-339, Mar. 1989.

[4-69] Z. Chair and P.K. Varshney, "Optimal data fusion in multiple sensor detection systems," *IEEE Trans. Aerospace Electronic Systems,* vol. AES-22, pp. 98-101, Jan. 1986.

[4-70] H.L. Van Trees, *Detection, Estimation and Modulation Theory.* New York, NY: John Wiley and Sons, Inc., 1968.

[4-71] H. Gish, "A probabilistic approach to the understanding and training of neural network classifiers," in *Proc. Internat. Conf. Acoust., Speech, Signal Processing*, vol. 3, pp. 1362-1364, April 1990.

[4-72] L.R. Rabiner and R.W. Schafer, *Digital Processing of Speech Signals.* Englewood Cliffs, NJ: Prentice-Hall, Inc., 1978.

[4-73] J.R. Deller, Jr., J. Proakis and J. Hansen, *Discrete Time Processing of Speech Signals.* New York, NY: Macmillan Publishing Co., 1993.

Chapter 5

Auditory Localization Using Spectral Information†

DIBYENDU NANDY and JEZEKIEL BEN-ARIE

Introduction

Human hearing involves the perception and interpretation of sensory input to the ears. Such perception may be classified as *auditory events* as opposed to *sound events* [5-1]. A sound event is the physical phenomenon of mechanical vibration which can characterize a sound source, while an auditory event is the brain's perception of an auditory phenomenon which may or may not be due to a sound event. The brain's auditory perception of the world around us is based on the cause–effect relationships that exist between most sound events and auditory events. The brain perceives sound sources by associating them with auditory events. Auditory function is carried out by the auditory system which culminates in the auditory cortex in the brain. The complex electromechanical and electrochemical signal processing scheme that constitutes our auditory system thus enables us to perceive sound events as auditory events and interpret them according to the visual world around us. We are able to follow and segregate sound streams by their frequency content, timbre and pitch. This ability is considerably enhanced by the functional capability of the human auditory system to localize and track a sound source. Auditory localization is the phenomenon by which we perceive sound sources to be at specific locations relative to our pose. Auditory localization is an aspect of the interpretation of auditory events which enables the brain to build an *auditory scene* of the immediate environment and associate it with the visual picture that is acquired simultaneously by the human visual system.

†This work was supported by The Whitaker Foundation

The human auditory system has been shown to possess remarkable abilities in localizing and tracking sound sources [5-1]–[5-4]. This phenomenon has been the subject of intensive multi-disciplinary research for several decades. Several theories have been suggested to explain the neural signal processing that could take place in the process of localization. Such modeling has been restricted by the complexity of the auditory system. Many tantalizing riddles still remain unexplained by most of these models. Such phenomena include localization in the medial saggital plane, which is the ability that humans have to localize sounds that are at different elevations but are equidistant from either ear, or the *cocktail party* effect, a phenomenon where a listener is simultaneously able to localize and separate a number of conversations, and to attend at will to any one of them. Various researchers [5-5]–[5-9] have attempted to explain the localization process on the basis of various phenomenological and computational models. The basic premise of all such models is that localization is the result of processing two primary acoustic cues in the auditory cortex. These are the relative *interaural time differences* (ITDs) and *interaural amplitude differences* (IADs) at the two ears. This approach was first formulated by Lord Rayleigh in his classic *duplex theory* of sound localization [5-10]. A model to extract ITD cues was proposed by Jefferess [5-5]. The Jefferess model suggests that the time-frequency analysis in the cochlea that is provided by the basilar membrane is processed by a cross-correlation network. Most models that are based on ITD cues use modifications of this cross-correlation model [5-9], [5-11]. Other models include count comparison models [5-7] and models based on the equalization and cancellation theory of sound localization [5-6].

While these (ITD and IAD) cues certainly have a role in the ability to localize, models which use only ITDs and IADs have been able to explain the localization phenomenon only partially. Researchers have shown that the use of only ITD and IAD cues in auditory stimuli degrades the ability to localize to a large extent [5-3]. Furthermore, such stimuli result in sensations of the sound source being perceived to be located intracranially (on an axis passing through the two ears), rather than a genuine sensation of an extracranially located sound source, or true localization [5-1], [5-12]. Even more significant, the use of only ITD and IAD cues does not explain the ability of humans to localize sounds in the vertical medial plane [5-13]–[5-15], where ITD and IAD cues are negligible. Furthermore, it is well known that the ability of auditory nerve fibers to follow temporal fluctuations in auditory stimulation (known as *phase-locked activity*) degrades with increasing frequencies above about 1.3 kHz [5-16], [5-2]. Above these frequencies, it is observed that auditory nerve fibers show steady increase in firing rates, at their particular center frequencies, but are unable to lock onto the phase of the incoming signal. Traditional spatial or temporal correlation based localization models utilize this phase-locked activity to extract the ITD cues which play a major part in localization. Hence, localization ability over a large range of perceptible audio frequencies is unexplained by these traditional models.

In many models dating back to the 19th century, researchers had suggested that the external ears or pinnae play an important role in localization by spectrally modifying the sound signal perceived at the eardrum according to the direction of its origin. In 1901, Angell and Fite showed that localization is possible with only one ear [5-17], although the accuracy of binaural localization is superior. This effect was restudied in detail by Butler and his colleagues [5-18]–[5-20] and it was shown that monaural localization is also possible both in azimuth (horizontal plane) and elevation (vertical plane). Like binaural localization in the vertical medial plane, the phenomenon of monaural localization cannot be explained by only ITD and IAD cues, which evidently are binaural cues. Psychophysical research has indicated that for a genuine spatial sensation of sound, additional high-frequency cues are essential [5-14], [5-15], [5-18]. Such cues are included in the form of highly direction-dependent filtering by the pinnae, head and torso [5-12], [5-21], [5-22]. Pinnae cues are primarily responsible for the outside-the-head sensation or *externalization* [5-12].

The importance of pinnae cues was first demonstrated by Batteau [5-23]. He used sounds recorded from microphones mounted in the pinnae of an artificial human head and then fed the recording to the headphone of a listener sitting in an isolated room, who was able to localize sounds with excellent accuracy. Batteau's procedure was a precursor to modern day design of realistic 3-D acoustic displays which can present waveforms over headphones to emulate those due to free-field stimulation by natural sources. The obvious assumption has been that if one is able to set up the same waveform in the ear canal as due to free-field stimulation, then one can simulate localization of such synthetic sources. The modifications imposed on the free-field source are experimentally estimated by identifying the transfer function of the contributing effects of the pinnae, head and torso. The effects of spectral modification by the pinnae, head and torso, has been called the *head related transfer function* (HRTF). Experimental determination of the HRTFs from all source directions about the head have been made by several researchers [5-21], [5-22], [5-24], [5-25], [5-26]. Such measurements indicate the directional dependence of the HRTFs, i.e., the frequency-dependent characteristics depending on the location of the free-field sound source. The use of such HRTFs to synthesize 3-D sound [5-21], [5-22], [5-27], [5-28], in particular the Convolvotron developed at the NASA Ames Center by Wenzel *et al.* have been quite successful. While not perfected fully (externalization is not always successful [5-29]), this approach indicates that the auditory system is able to identify spectral auditory cues for purposes of localization.

There is evidence that the anteroventral cochlear nucleus (AVCN) extracts an estimate of a dilated version of the short-time amplitude spectrum of perceived auditory signals [5-30]. Yang *et al.* [5-30] refer to such a representation as the auditory spectrum. The physiological estimation of such an auditory spectrum suggests that frequency cues could be extracted in the peripheral auditory system for the purpose of localization. The role of HRTFs in modeling the localization process has been hitherto overlooked despite substantial experimental evidence that points to its relevance in spatial hearing. Yang's model of auditory signal representation in the cochlear nucleus implies that localization information embedded as the HRTF in the auditory signal spectrum is available for further processing.

In Section 5.1, we consider a localization model based on extracting HRTF information as spectral localization cues. The auditory representation suggested by Yang *et al.* [5-30], allows a means of extracting HRTF ratio patterns, which are invariant to incoming signal spectra. Such ratio patterns may be associated with the directions of the incoming sound signal. Thus classifying a given HRTF ratio pattern can yield an estimate of the localization. We consider the neural processing that occurs in the cochlear nucleus and the *superior olivary complex* (SOC) and assess its functionality in carrying out such pattern recognition.

In Section 5.2, we further consider matching of such HRTF ratio patterns. We define a novel measure, the *discriminative matching measure* (DMM) by which we can estimate the "goodness" of a match. In the following sections, we consider several solutions to the matching problem from a neural signal processing viewpoint. We consider *correlation*-based approaches in Section 5.3. In Section 5.4 we develop a signal processing approach based on optimizing the DMM. A nonlinear approach based on the *back-propagation* algorithm is described in Section 5.5. Finally, in Section 5.6 we compare the results of simulations that have been conducted using these methods. The chapter concludes with a discussion of our experimental results and the inferences we draw from our model.

5.1 A Localization Model Based on HRTFs

A. Physiology of Binaural Processing

A schematic diagram of the *afferent auditory nervous system* adapted from [5-2] is depicted in Fig. 5.1-1. The complex nature of the system is immediately apparent. The afferent fibers of the acoustic nerve leave the *cochlea* and travel to the *cochlear nucleus*. Nerve fibers from the left cochlea are indicated in white and those from the right cochlea are indicated in black. Afferent nerves from the cochlear nucleus either travel *ipsilaterally* (on the same side) or *contralaterally* (along the opposite side). Some of these fibers may bypass intermediate neuronal nuclei. Generally nerves leaving the cochlear nucleus synapse with the *superior olivary complex* (SOC). This is the first stage where binaural processing may occur, strictly because this is where the first set of ipsilateral and contralateral nerves synapse together. Nerves from this stage lead to the *lateral lemiscus*, the *inferior colliculus*, the *medial geniculate body* and finally the *auditory cortex*. Current research has still not indicated the nature of electrophysiological activity in the higher levels of the auditory system of man in sufficient detail for us to model the processing at those

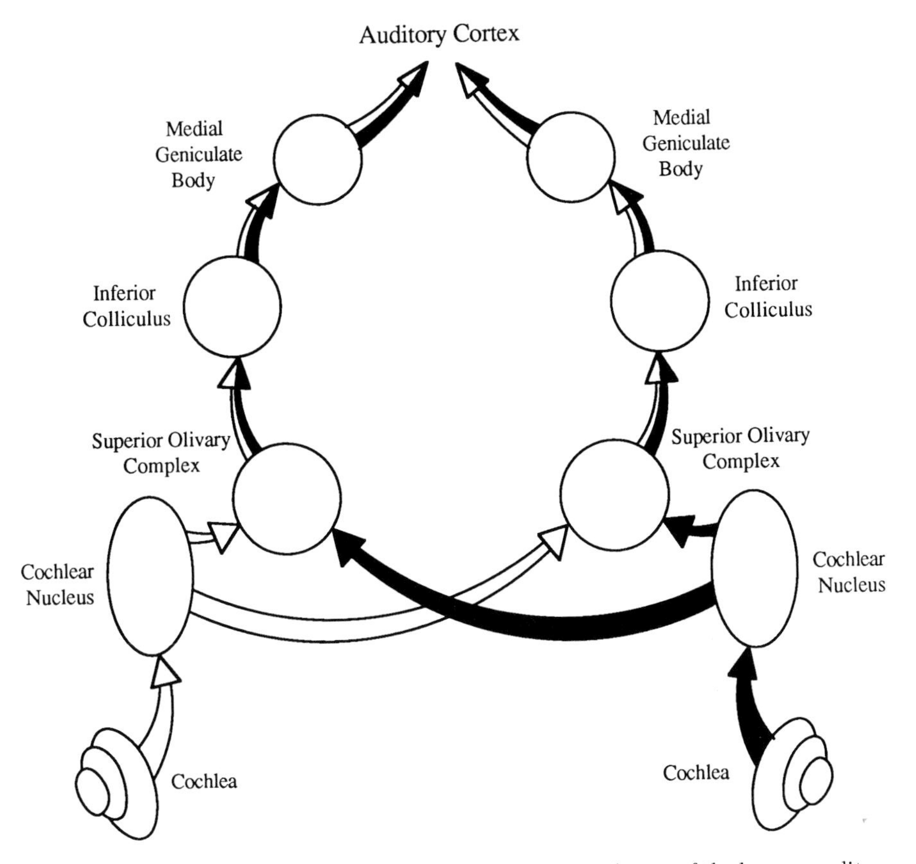

Figure 5.1-1: A schematic diagram of the afferent nerve pathways of the human auditory system.

levels. It is believed however that the binaural processing required for localization can and does occur to a large extent within the SOC.

Neural processing in the cochlea primarily serves to transduce the mechanical vibrations of the cochlea to firings of the auditory nerve fibers. In the process, several filtering functions occur, which can be modeled by a velocity coupling or a derivative stage, followed by half-wave rectification and low-pass filtering, as discussed in [5-30]. These filters model the manner in which the vibration of the basilar membrane is transmitted through the cochlear fluid to the *cilia* (hair) of the *inner hair cells* (IHCs), and how the nerve fibers innervating these IHCs respond to the motion of the cilia attached to these IHCs. The point of note is that nerve fibers are arranged so as to be excited by frequencies in decreasing order from the base to the apex of the cochlea. Thus each nerve fiber is excited by its particular or *characteristic frequency* (CF), depending on its location in the array of fibers innervating the IHCs. The cochlear nucleus processes the received stimulus further before transmitting it to the SOC. There is evidence to indicate that the *tonotopic* (place-frequency) ordering is maintained in the three different parts of cochlear nucleus: the *anteroventral cochlear nucleus* (AVCN), the *posteroventral cochlear nucleus* (PVCN) and the *dorsal cochlear nucleus* (DCN) [5-2]. Neurons in the *lateral superior olive* (LSO) and the *medial superior olive* (MSO), both of which are primary nuclei of the SOC, receive inputs from both ipsilateral and contralateral cochlear afferent nerves. Electrophysiological studies indicate that these nuclei respond to binaural cues, but in different ways. The majority of the LSO units are *most sensitive to mid-to-high frequency stimuli*, having CFs greater than 3 kHz. They are excited by stimulation of the ipsilateral ear and inhibited by stimulation of the contralateral ear [5-31], [5-32]. This response characteristic is termed E/I: the ipsilateral stimulus excites and the contralateral one inhibits. *It is important to note that medium-to-high-frequency response of this neural structure supports our assumption that high-frequency cues have a significant role to play in localization.*

The MSO units on the other hand are reported to have an E/E response and are excited by stimuli from either ipsilateral or contralateral stimuli [5-32]. These MSO units have been characterized to be sensitive to the stimulus's *interaural phase difference* (IPD) [5-33], [5-34] for binaural tonal stimuli with low CFs ($<$ 1 kHz) (the ITD manifests itself as a phase difference rather than a simple time difference for continuous tones).

B. HRTF Based Modeling

The geometry of the pinnae, head and torso affect the incident acoustic signal by modifying its phase and amplitude spectrum. Differences in the acoustic path lengths at all frequencies lead to ITDs between the perceived auditory signals at the two ears. The shadowing and multiple reflection effects of the head, torso and pinnae lead to varying IADs over the perceived auditory bandwidth. ITDs and IADs are present as modifications imposed on the amplitude and phase spectrum of the signal impinging on the eardrums. The effects of ITDs, IADs and the head-related filtering are captured by modeling the transfer characteristics of the incoming signal path as a linear filter from the source to the ear canal. The direction-dependent linear filter thus obtained is called the *head related transfer function* (HRTF). The HRTF is determined as a finite impulse response model which is transformed to the frequency domain using the *discrete Fourier transform* (DFT). We note that the HRTF is a complete representation since it implicitly includes ITDs and IADs in the phase and amplitude spectrum respectively. As outlined by Wightman and Kistler [5-21] and others [5-27], the HRTF depends upon the directional parameters of the azimuth and elevation of the sound source relative to the head location.

Models have been proposed to explain the binaural processing by means of which ITD cues can be extracted at low frequencies (less than 1 kHz) and there is sufficient physiological evidence to support such models (the response of the MSO units). At high frequencies (above 1 kHz) the peripheral auditory system is unable to extract ITD cues [5-2]. The inability of the peripheral auditory system to extract ITD cues at frequencies above 1 kHz is contrary to the human ability to localize broadband sound sources above this range as shown by Musicant and Butler [5-19]. Musicant and Butler also reported that humans are unable to accurately localize narrowband (bandwidths of 1 kHz or less) noise sources, centered above 4 kHz. Similarly the ability of humans to localize sounds monaurally [5-19], though limited, cannot be explained on the basis of binaural cues like ITDs and IADs in the input stimuli. This leads us to conclude that spectral cues are very essential for localization of high-frequency sounds and that the extraction of such cues must be explained within the framework of a localization model. In our model we consider the localization of broadband high frequency sounds which can provide such frequency cues. We consider signals that have non-zero spectral components for a sufficiently broad band of frequencies in the audible frequency range. Using the index k as the discrete frequency variable, we define

$$X_I(k,i) = H_I(k,i)X(k,i) \ , \tag{5.1-1}$$

and

$$X_C(k,i) = H_C(k,i)X(k,i) \ , \tag{5.1-2}$$

where $X(k,i)$ represents the DFT of the free-field sound source (from direction i), $X_I(k,i)$ and $X_C(k,i)$ are the DFTs of simulated sound at the ipsilateral and contralateral eardrums respectively, from direction i, and $H_I(k,i)$ and $H_C(k,i)$ are the HRTFs of the ipsilateral and contralateral ears associated with direction i.

Yang *et al.* proposed that the AVCN is capable of extracting a dilated version of the short-time amplitude spectrum [5-30] of the incoming stimulus, namely, the *auditory spectrum*:

$$X(n;s) = c \mid X(n;a^s k_0) \mid \ . \tag{5.1-3}$$

Here $X(n;s)$ represents the time-frequency response, evoked in the nerve fiber, displaced by s from the location of a base frequency (k_0), in the AVCN, for a short-time frame (indexed by n). The spatial variable s is inherently discrete, by virtue of being an index to the finite number of nerve fibers which innervate the IHCs in the cochlea. c is a scaling constant, and $X(n;a^s k_0)$ is a dilated estimate of the input signal spectrum $X(k)$ with a being a dilation constant, such that $0 < a < 1$. Thus, in Eq. (5.1-1) and Eq. (5.1-2), $k = a^s k_0$. This dilation models the nonlinear frequency scale of the basilar membrane in the mid-to-high frequencies. While the scale is more or less linear below 500 Hz, it becomes gradually logarithmic with higher frequencies [5-30]. Yang *et al.* model the basilar membrane response as an affine wavelet transform, with individual cochlear bandpass filters forming the wavelet basis. The individual filters are finite impulse response models and are defined to be dilated versions of each other. Note that the logarithmic scale does not affect the psychophysical arguments and the mathematical formulations presented in the following sections. The outputs of the basilar membrane filters are processed by a lateral inhibition network, postulated to exist in the AVCN, to give rise to the dilated short-time auditory spectral representation.

Hereafter, the auditory spectrum is considered for a particular time frame indexed by n and in the interests of readability, the dependence on n is suppressed in the notation. We introduce instead, the variable i in Eq. (5.1-3), corresponding to the direction of the sound source in Eq. (5.1-1) and Eq. (5.1-2). It is easy to show that the ratio of the ipsilateral to the contralateral auditory spectrum is dependent only on the ratio of the HRTFs, when such a ratio can be defined, i.e., when the relevant frequency component exists in the sound source. A neural compressive non-linearity, whose response can be modeled as a logarithm, occurring at a previous stage followed by LSO units with an E/I response yields a difference of logarithms, or equivalently, a ratio estimate of the ipsilateral $X_I(s,i)$ and contralateral $X_C(s,i)$ auditory spectra. The detected E/I response of the LSO units suggests such an operation. Thus, this E/I response can be modeled as the ratio:

$$h(s,i) = \frac{X_I(s,i)}{X_C(s,i)} = \frac{| X_I(a^s k_0, i) |}{| X_C(a^s k_0, i) |} . \tag{5.1-4}$$

Equation (5.1-4) indicates that the equivalent response of the LSO units is a spatial pattern corresponding to the ratio of the ipsilateral and contralateral auditory spectra. From Eq. (5.1-3) and Eq. (5.1-4), and noting that the HRTF amplitude is non-negative, we have

$$h(s,i) = \frac{| H_I(a^s k_0, i) X(a^s k_0, i) |}{| H_C(a^s k_0, i) X(a^s k_0, i) |} = \frac{| H_I(a^s k_0, i) |}{| H_C(a^s k_0, i) |} . \tag{5.1-5}$$

Based on this model, we believe that the auditory system can extract direction-dependent HRTF ratio patterns in real time from incoming auditory signal spectra.

The spatial pattern $h(s,i)$ can now be considered for matching over a broad band of the space-frequency variable s, whose extent depends on the bandwidth of

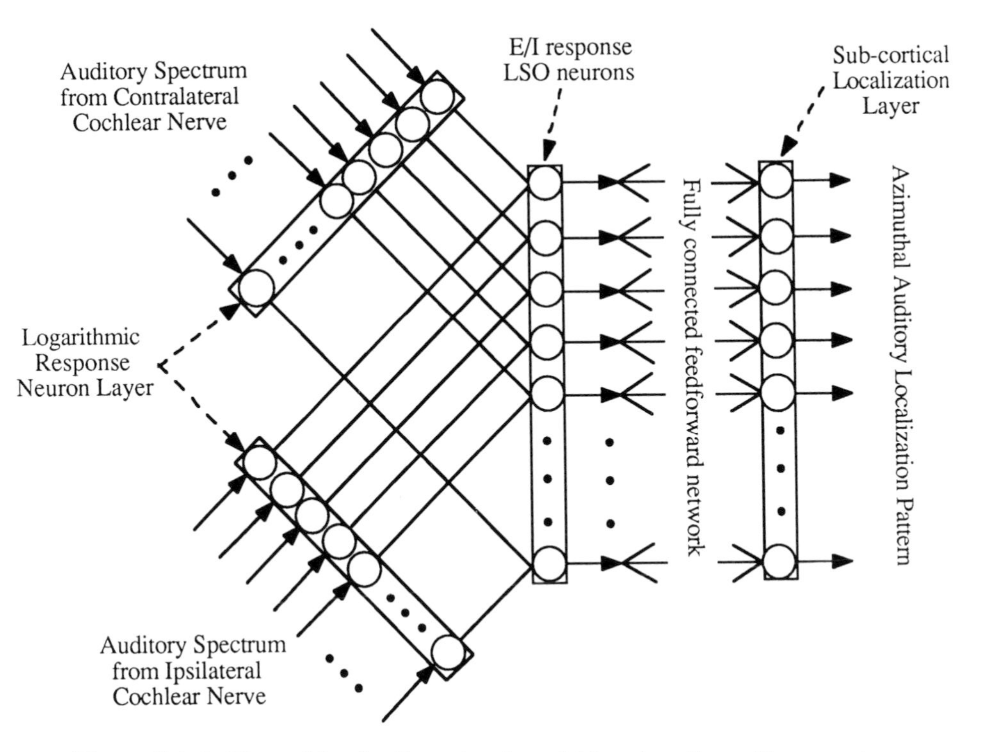

Figure 5.1-2: Binaural localization network model based on the auditory spectrum.

the source signal spectrum. We define the spatial pattern vector $\boldsymbol{h}(i)$ in terms of $h(s,i)$ as

$$\boldsymbol{h}(i) = [h(1,i), h(2,i), \cdots, h(s,i), \cdots]^T . \tag{5.1-6}$$

s represents the spatial index of the nerve fibers innervating the cochlea and extends over the mid-to-high-frequency range, which is the effective range of response of the LSO units [5-2]. If the spatial pattern vector $\boldsymbol{h}(i)$ in Eq. (5.1-5) has different characteristics for every direction i, a feedforward network which projects such incoming spectral ratios onto optimally generated localization filters can estimate the direction from which a (detected) sound is emitted. Figure 5.1-2 indicates how the activity of the ipsilateral and contralateral nerve carry auditory spectral information [5-30] from the cochlear nucleus to the SOC, specifically to the E/I LSO units. Further processing at the next subcortical layer would be responsible for the process of ratio pattern recognition, as described above and in Fig. 5.1-2. The output of such a network is a pattern which should peak in a group of nodes which correspond to the estimated sound source direction.

5.2 Template Matching and a Matching Measure

In the previous section, we have shown the independence of the HRTF ratio to the incoming broadband signal spectrum and have suggested a simple network which could use such ratios to localize the source. We now consider the kind of filters that enable us to recognize these HRTF ratio patterns. Due to the invariance of the HRTF ratios, the problem reduces to one of recognizing HRTF ratio patterns where each pattern relates to a particular direction. For the recognition of such patterns three methods are investigated: *normalized correlation* (NC), classification using a *back-propagation-trained neural network*, and a novel optimal template matching scheme based upon *expansion matching* (EXM) [5-35], [5-36]. We compare the effectiveness of each of these matching methods for incorporation into the overall model for localization. Since the chief purpose of this discussion is the evaluation of the template matching capabilities of each method, the problem is studied from a signal processing standpoint.

Template matching from the perspective of localization is formulated as follows. Given the ratio of the HRTFs of the ipsilateral and contralateral ear for a sound source emanating from a particular direction, this incoming HRTF ratio vector is to be matched to a standard library of HRTF ratio templates each representing a particular direction, and the best match is ascertained. It is to be noted that HRTFs and consequently their ratios vary in a smooth manner from direction to direction. These HRTFs are primarily the result of imposed modification on the spectrum by the pinnae. The general shape of the pinnae in humans as well as other animals suggest that they have evolved to optimize certain criteria of hearing and localization while being constrained by smoothness requirements imposed by acoustics. The possible relationships between these criteria and constraints and the resulting HRTFs are explored by Rao and Ben-Arie [5-37]. This nature of HRTFs constrains us to view the HRTF ratio map as a continuous, smooth function of azimuth and elevation. The HRTF ratio map is the ratio of the HRTFs plotted for different azimuths and elevations on an imaginary sphere which includes all possible directions around the head. In the present case, we examine a subset of this ratio map consisting of a discrete number of HRTF ratio vectors varying in the azimuthal directions. The same approach could be extended to include variations due to elevation changes as well. Our formulation of the problem is quite general, since we actually address the problem of matching an arbitrary input pattern to a set of stored template patterns.

Mathematically, the problem can be formulated as follows. Given an input vector $\boldsymbol{s}$ of dimension M, which represents the spectral ratio pattern of the incoming sound, and a set of M-dimensional template vectors $\boldsymbol{h}(i)$; $i = 1, \cdots, N_h$, which represent the stored HRTF ratio patterns, we seek to determine which of the N_h templates matches the input $\boldsymbol{s}$ best. We quantify the "goodness" of the matching by a *discriminative matching measure* (DMM). If the input vector $\boldsymbol{s}$ is actually closest to the lth template $\boldsymbol{h}(l)$, and the matching scores (result of whichever matching method was used) of the input vector $\boldsymbol{s}$ with the ith template $\boldsymbol{h}(i)$ is given by $c(i)$,

$$c(i) = \boldsymbol{s}^T \boldsymbol{h}(i) \; ; \quad i = 1, \cdots, N_h \; , \tag{5.2-1}$$

where the template $\boldsymbol{h}(i)$ represents the appropriate filter vector of the particular matching method used.

We define the DMM as the ratio

$$DMM = \frac{c^2(l)}{\frac{1}{N_h - 1} \sum_{i=1}^{N_h} [\tilde{\psi}(l) - \tilde{\psi}(i)]^2 c^2(i)} \; , \tag{5.2-2}$$

where $\tilde{\psi}(i)$; $i = 1, \cdots N_h$, is the normalized azimuthal angle of the template for direction i. Thus a matching score that peaks for a filter template that is azimuthally "closer" to the direction of the incoming signal than one which is azimuthally "farther" deserves a correspondingly higher DMM score. The squaring operation penalizes incorrect scores as the square of the azimuthal error. The term discriminative is used here to denote that the ratio is between the response from the *true directional filter* and all other responses from the remaining *off-directional filters*.

This definition of DMM is somewhat related to the definition of *discriminative signal-to-noise ratio* (DSNR) that is used in *expansion matching* (EXM) of templates [5-35].[1] However the DSNR emphasizes peak sharpness in the context of

[1] See Chapter 8 for a more complete discussion on expansion matching.

linear shift-invariant filtering, whereas the DMM signifies the ability of the matching method to distinguish one amongst a set of templates, with the added feature that responses closer to the actual desired response are more desirable.

In the following sections, we compare three matching methods in terms of the DMM that they yield over a wide variety of test inputs. The DMM measures the matching performance of each of these methods.

5.3 Normalized Correlation Matching

Correlation is one of the most widely used matching techniques, primarily due to its simplicity of implementation and ease of analysis. Basically, correlation involves projecting the input vector $\boldsymbol{s}$ onto the set of template vectors $\boldsymbol{h}(i)$; $i = 1, \cdots, N_h$, to obtain the matching scores $c_{cor}(i)$

$$c_{cor}(i) = \boldsymbol{s}^T \boldsymbol{h}(i) \; ; \quad i = 1, \cdots, N_h \, . \tag{5.3-1}$$

Ben-Arie and Rao have studied the correlation matching scheme (also called matched filtering in the case of shift-invariant matching) [5-35]. Correlation in general yields broad matching peaks and also many spurious responses, thus hampering good detection. Furthermore, correlation is affected by changes in the input amplitude (energy).

To alleviate to some extent the amplitude sensitivity of correlation matching it has been suggested [5-38] that the vector and template be normalized first before applying the correlation matching. This involves additional computational complexity. Thus, we define vector $\tilde{\boldsymbol{s}}$ to be a normalized input vector as

$$\tilde{\boldsymbol{s}} = \frac{\boldsymbol{s}}{\sqrt{\boldsymbol{s}^T \boldsymbol{s}}} \, . \tag{5.3-2}$$

Similarly, $\tilde{\boldsymbol{h}}(i)$ are the normalized versions of the template vectors $\boldsymbol{h}(i)$; $i = 1, \cdots, N_h$. The normalized correlation matching scores $c_{ncor}(i)$ can be defined as

$$c_{ncor}(i) = \tilde{\boldsymbol{s}}^T \tilde{\boldsymbol{h}}(i) \; ; \quad i = 1, \cdots, N_h \, . \tag{5.3-3}$$

Since the normalization eliminates the amplitude sensitivity of the projection approach, normalized correlation can be expected to perform more robustly than the direct correlation scheme. The normalized correlation matching scores are bounded by

$$| \, c_{ncor}(i) \, | \leq 1 \; ; \quad i = 1, \cdots, N_h \, , \tag{5.3-4}$$

and equality is attained only when the input vector matches the corresponding template vector exactly (within a constant scaling). The problem with this technique is that it yields comparably high scores even for non-matching templates, i.e., the difference between the matching and non-matching scores could be very small, implying a very low level of discrimination. In other words, the results have low DMM.

5.4 Optimal DMM Matching

Normalized correlation suffers from the problem of low DMM, though it has less amplitude sensitivity than the direct correlation approach. In this section, we formulate a set of optimal templates that, apart from being normalized, yield maximal possible DMM when correlated with the input vector. In our formulation, we also include additive noise and thus use expectations of random variables to define the DMM in a slightly different manner.

We assume that the normalized input vector $\tilde{\boldsymbol{s}}(l)$ is a noisy version of the lth normalized template vector $\tilde{\boldsymbol{h}}(l)$, i.e.

$$\tilde{\boldsymbol{s}}(l) = \tilde{\boldsymbol{h}}(l) + \boldsymbol{\lambda} \,, \tag{5.4-1}$$

where $\boldsymbol{\lambda}$ is a random noise vector. The DMM for matching the lth template can be redefined for random signals as

$$DMM_l = \frac{\left(\boldsymbol{E}[\boldsymbol{c}_{opt}(l)]\right)^2}{\frac{1}{N_h - 1}\sum_{i=1}^{N_h}[\tilde{\psi}(l) - \tilde{\psi}(i)]^2 \boldsymbol{E}[\boldsymbol{c}^2_{opt}(i)]} \;; \quad \boldsymbol{c}_{opt}(i) = \boldsymbol{\theta}^T(l)\tilde{\boldsymbol{s}}(i) \,. \tag{5.4-2}$$

$\boldsymbol{E}[\;]$ denotes the statistical expectation operator and $\boldsymbol{\theta}(l)$ is the desired optimal template for matching the lth pattern. Equation (5.4-2) reduces to the definition of Eq. (5.2-2) for the case of deterministic values. In order to maximize the DMM, it is sufficient to constrain the numerator of Eq. (5.4-2) and minimize the denominator. This would yield the optimal template vector $\boldsymbol{\theta}(l)$, which when projected onto the normalized incoming vector $\tilde{\boldsymbol{s}}(l)$, would yield the desired value (on the average), and for any other input vector $\tilde{\boldsymbol{s}}(i); i \neq l$, the response would be minimized on the average. In other words, to obtain $\boldsymbol{\theta}(l)$ we need to minimize

$$\frac{1}{N_h - 1}\sum_{i=1}^{N_h}[\tilde{\psi}(l) - \tilde{\psi}(i)]^2 \boldsymbol{E}[\boldsymbol{c}^2_{opt}(i)], \tag{5.4-3}$$

subject to the constraint

$$\boldsymbol{E}[\boldsymbol{c}_{opt}(l)] = u_l \,, \tag{5.4-4}$$

where u_l is the desired matching score for the lth template. Using vector notation, the minimization can be rewritten as

$$\begin{aligned}
&\frac{1}{N_h - 1}\sum_{i=1}^{N_h}\;[\tilde{\psi}(l) - \tilde{\psi}(i)]^2 \boldsymbol{E}[\boldsymbol{c}^2_{opt}(i)] \\
&\quad = \frac{1}{N_h - 1}\sum_{i=1}^{N_h}[\tilde{\psi}(l) - \tilde{\psi}(i)]^2 \boldsymbol{E}\left[(\boldsymbol{\theta}^T(l)\tilde{\boldsymbol{s}}(i))^2\right] \\
&\quad = \frac{1}{N_h - 1}\sum_{i=1}^{N_h}[\tilde{\psi}(l) - \tilde{\psi}(i)]^2 \left(\boldsymbol{\theta}^T(l)\boldsymbol{E}[\tilde{\boldsymbol{s}}(i)\tilde{\boldsymbol{s}}^T(i)]\boldsymbol{\theta}(l)\right) \to \min
\end{aligned} \tag{5.4-5}$$

subject to the constraint

$$\boldsymbol{E}[\boldsymbol{\theta}^T(l)\tilde{\mathbf{s}}(l)]\boldsymbol{\theta}^T(l)\boldsymbol{E}[\tilde{\mathbf{s}}(l)] = u_l \ . \tag{5.4-6}$$

We can now use the method of Lagrange multipliers and rewrite the objective function to be minimized as

$$\begin{aligned} J(\boldsymbol{\theta}(l)) = \frac{1}{N_h - 1} & \sum_{i=1}^{N_h} [\tilde{\psi}(l) - \tilde{\psi}(i)]^2 \left(\boldsymbol{\theta}^T(l)\boldsymbol{E}[\tilde{\mathbf{s}}(i)\tilde{\mathbf{s}}^T(i)]\boldsymbol{\theta}(l)\right) \\ + & \ \eta_l \left(u_l - \boldsymbol{E}[\boldsymbol{\theta}^T(l)\tilde{\mathbf{s}}(l)]\right) , \end{aligned} \tag{5.4-7}$$

where η_l is the Lagrange multiplier chosen to satisfy the constraint of Eq. (5.4-6).

If the noise $\boldsymbol{\lambda}$ of Eq. (5.4-1) is assumed to be zero mean and uncorrelated with the optimal vector $\boldsymbol{\theta}(l)$, the following identities are obtained.

$$\boldsymbol{\theta}^T(l)\boldsymbol{E}[\tilde{\mathbf{s}}(l)] = \boldsymbol{\theta}^T(l)\tilde{\boldsymbol{h}}(l) = u_l \ , \tag{5.4-8}$$

and

$$\begin{aligned} \boldsymbol{E}[\tilde{\mathbf{s}}(i)\tilde{\mathbf{s}}^T(i)] &= \boldsymbol{E}\left[(\tilde{\boldsymbol{h}}(i) + \boldsymbol{\lambda})(\tilde{\boldsymbol{h}}(i) + \boldsymbol{\lambda})^T\right] \\ &= \tilde{\boldsymbol{h}}(i)\tilde{\boldsymbol{h}}^T(i) + \boldsymbol{E}[\boldsymbol{\lambda}\boldsymbol{\lambda}^T] \\ &= [\boldsymbol{R}_{h_i}] + [\boldsymbol{R}_\lambda] \ , \end{aligned} \tag{5.4-9}$$

where $[\boldsymbol{R}_{h_i}]$ and $[\boldsymbol{R}_\lambda]$ are defined as the autocorrelation matrices of the ith template and the noise respectively.

Now the objective function Eq. (5.4-7) can be rewritten as

$$\begin{aligned} J(\boldsymbol{\theta}(l)) = \frac{1}{N_h - 1} & \sum_{i=1}^{N_h} [\tilde{\psi}(l) - \tilde{\psi}(i)]^2 \boldsymbol{\theta}^T(l)\left[[\boldsymbol{R}_{h_i}] + [\boldsymbol{R}_\lambda]\right]\boldsymbol{\theta}(l) \\ + & \ \eta_l \left(u_l - \boldsymbol{\theta}^T(l)\tilde{\boldsymbol{h}}(l)\right) . \end{aligned} \tag{5.4-10}$$

Taking the partial derivative with respect to $\boldsymbol{\theta}(l)$ and equating to zero, we obtain

$$\begin{aligned} \frac{\partial J(\boldsymbol{\theta}(l))}{\partial \boldsymbol{\theta}(l)} &= \frac{2}{N_h - 1}\sum_{i=1}^{N_h} [\tilde{\psi}(l) - \tilde{\psi}(i)]^2 \left[[\boldsymbol{R}_{h_i}] + [\boldsymbol{R}_\lambda]\right]\boldsymbol{\theta}(l) - \eta_l \tilde{\boldsymbol{h}}(l) \\ &= 0 \ , \end{aligned} \tag{5.4-11}$$

which can be simplified to yield

$$\begin{aligned} \boldsymbol{\theta}(l) &= \frac{\eta_l}{2}\left[\frac{1}{N_h - 1}\sum_{i=1}^{N_h} [\tilde{\psi}(l) - \tilde{\psi}(i)]^2 \left([\boldsymbol{R}_{h_i}] + [\boldsymbol{R}_\lambda]\right)\right]^{-1} \tilde{\boldsymbol{h}}(l) \\ &= \frac{\eta_l}{2}\left[[\boldsymbol{R}_{H_l}] + \alpha_l[\boldsymbol{R}_\lambda]\right]^{-1} \tilde{\boldsymbol{h}}(l) \ , \end{aligned} \tag{5.4-12}$$

where $[\boldsymbol{R}_{H_l}]$ is a weighted average of $\boldsymbol{R}_{h_i}$ as shown in Eq. (5.4-13) and α_l is a weighting constant on the noise autocorrelation matrix $\boldsymbol{R}_\lambda$ as indicated in Eq. (5.4-14):

$$[\boldsymbol{R}_{H_l}] = \frac{1}{N_h - 1} \sum_{i=1}^{N_h} [\tilde{\psi}(l) - \tilde{\psi}(i)]^2 \boldsymbol{R}_{h_i} \, . \tag{5.4-13}$$

$$\alpha_l = \frac{1}{N_h - 1} \sum_{i=1}^{N_h} [\tilde{\psi}(l) - \tilde{\psi}(i)]^2 \, . \tag{5.4-14}$$

We also need to satisfy the constraint given by Eq. (5.4-8) or

$$\boldsymbol{\theta}(l)^T \tilde{\boldsymbol{h}}(l) = \frac{\eta_l}{2} \tilde{\boldsymbol{h}}^T(l) \, [[\boldsymbol{R}_{H_l}] + \alpha_l [\boldsymbol{R}_\lambda]]^{-1} \, \tilde{\boldsymbol{h}}(l) = u_l \, , \tag{5.4-15}$$

which yields

$$\eta_l = 2 \frac{u_l}{\tilde{\boldsymbol{h}}^T(l) [[\boldsymbol{R}_{H_l}] + \alpha_l [\boldsymbol{R}_\lambda]]^{-1} \tilde{\boldsymbol{h}}(l)} \, . \tag{5.4-16}$$

From Eq. (5.4-12) and Eq. (5.4-16) the final solution for $\boldsymbol{\theta}(l)$ is

$$\boldsymbol{\theta}(l) = \frac{u_l [[\boldsymbol{R}_{H_l}] + \alpha_l [\boldsymbol{R}_\lambda]]^{-1} \tilde{\boldsymbol{h}}(l)}{\tilde{\boldsymbol{h}}^T(l) [[\boldsymbol{R}_{H_l}] + \alpha_l [\boldsymbol{R}_\lambda]]^{-1} \tilde{\boldsymbol{h}}(l)} \, . \tag{5.4-17}$$

Thus, Eq. (5.4-17) yields the optimal DMM template to be correlated with the input vector. This template can be made optimal in the presence of any given noise by including the statistics of the noise $[\boldsymbol{R}_\lambda]$ in the above formulation.

Experimental results in Section 5.6 confirm that this formulation is indeed optimal in the sense of DMM. In the case of additive white noise of increasing power, i.e., lower SNR, it is easy to show that the optimal DMM filter converges to the NC solution in the same manner as the EXM filter developed in [5-35].

5.5 Matching Using Back-propagation Neural Networks

In order to compare the linear matching methods discussed above (i.e., NC and DMM optimization) with a non-linear method, we also examine a pattern recognition method using neural networks. Neural networks are generally non-linear in nature and have been found to robustly recognize patterns.[2] Neural networks are also credited with the capability of generalization from a relatively small set of training data.

The training of a neural network consists of presenting a set of training patterns or exemplars to the network and demanding a corresponding set of desired outputs

[2] Refer to Chapter 4 for an extensive discussion of feedforward neural networks and their applications.

from the network. A learning algorithm modifies the weights of the network in accordance with the error between the actual and desired outputs. While many such algorithms exist, we use the *back-propagation* algorithm mainly because it is simple and, in general, as effective as other more sophisticated algorithms. The topic of back-propagation is well developed [5-39], [5-40] and is discussed in Chapter 4. We do not present any details here other than noting that the algorithm performs a gradient descent minimization on the mean-squared error surface in the weight space of the neural network.

A. Network Model

The back-propagation network used in our model is a three-layered network consisting of an input layer, an intermediate processing layer and an output layer. The input layer serves as a distribution layer for the network. Each succeeding layer is fully connected to the previous layer.

The network defines a non-linear mapping from the input vector space to the output vector space. The mapping is a function of the number of layers, the weight matrix associated with each layer and the nonlinearity parameters of each neuron. Each layer imposes its own mapping on its input vector. The mapping imposed by each layer of neurons define decision surfaces in the weight space for a pattern recognition problem such as the one we are dealing with.

In our experiments, the number of input and output nodes is fixed by the choice of vector length and number of HRTFs, respectively. Networks with even one hidden layer are seen to be quite efficient in extracting the relevant features of the HRTF ratio vectors corresponding to different directions and can thus classify such vectors very well. The number of hidden nodes is also chosen empirically. The HRTF ratio templates show on the order of 10 to 15 significant features, which can be detected by visual inspection. The network tends to extract these features in the hidden nodes and hence we use a network with 12 hidden nodes. We have found that this architecture converges faster than either significantly larger or smaller sized networks.

B. Error Back-propagation Training

The assignment of weights in the network is done by iteratively converging to a desired (good local) minimum on the mean-squared error surface of the network mapping function. The network is initialized with small random weights. The error associated with each output node is defined as the difference between the node target output and the achieved node output. Errors associated with hidden layer nodes are determined by applying the inverse mapping of each layer on the error vector associated with that layer (*back-propagating* the error associated with each node of the next layer through the nonlinearity). The use of a sigmoidal nonlinearity which is continuous allows an analytical derivation of this inverse mapping [5-39].

The network is trained in a batch mode. The training data set includes the original template patterns along with noisy versions of the same. The target pattern for each direction is a delta function indicating the desired direction with all other outputs ideally at zero. The inclusion of training patterns corrupted by additive noise with variance σ_λ^2 (SNRs of 10 dB and 20 dB were used) ensures robustness and better generalization capabilities [5-41].

5.6 Experiments

A. Experimental Setup

In our simulations, the HRTF ratio matching problem is treated objectively as a pattern classification problem. We direct our efforts towards trying to identify the best matching approach comparing the methods of: *normalized correlation* (NC), *DMM optimization* (ODMM) and *back-propagation*. In our first set of experiments we analyze the discriminative ability and robustness to various distortions.

To simplify the problem, we first attempt to discriminate twelve directions 30° apart, by using the HRTF ratio vectors corresponding to these twelve directions. This set of experiments is conducted in three parts. In the first part, we attempt to quantify the performance of each method, when the input patterns are distorted by noise. This experiment aims at measuring robustness of the matching methods with variations or distortions in the actual shape of the HRTFs. Such variations may be due to spectral estimation errors in the cochlear nucleus. In the second part, we investigate the performance of each method with HRTF ratio vectors, which have some contiguous parts of the vector missing. This kind of distortion can be likened to the HRTF ratio patterns that would result if an input signal has exciting frequency components over only a part of the auditory bandwidth that is analyzed by the LSO units. Thus, such a scheme gives a measure of the robustness of the matching methods for conditions under which certain narrow bands of frequency are unavailable in the short-time auditory spectrum, as in regular speech. In this scheme, we fill in the missing frequency band by linearly interpolating from the data points at the boundaries of the band under consideration. This procedure follows from the principles of *gestalt psychology* by supposing that the human auditory cortex performs interpolations to generate an estimate of missing data akin to the phenomenon of subjective contouring in the human visual cortex [5-42], [5-4]. In the third part, we consider localization acuity for narrowband, high-frequency ratio patterns. In this case, we select the band of frequencies with a rectangular window whose boundaries are smoothed by Gaussians. This experiment corresponds to the case where the auditory system is excited by a few tones or excited over a narrow range of audible frequencies.

In our second set of experiments, we tackle the more realistic case with a large number of directions to discriminate from. We consider the case of discriminating between 360 different directions by means of the associated HRTF ratio patterns. The ratio patterns for the 360 directions are generated by a cubic spline interpolation between the HRTFs measured by psychophysical experiments [5-21]. This satisfies our constraint that the HRTFs change smoothly along the dimension of azimuth. Such a setup can be better visualized as an image or map. Fig. 5.6-1 shows the HRTF ratios in a 3-D isometric projection of the original 12 HRTF ratio patterns from which the data is interpolated for intermediate directions.

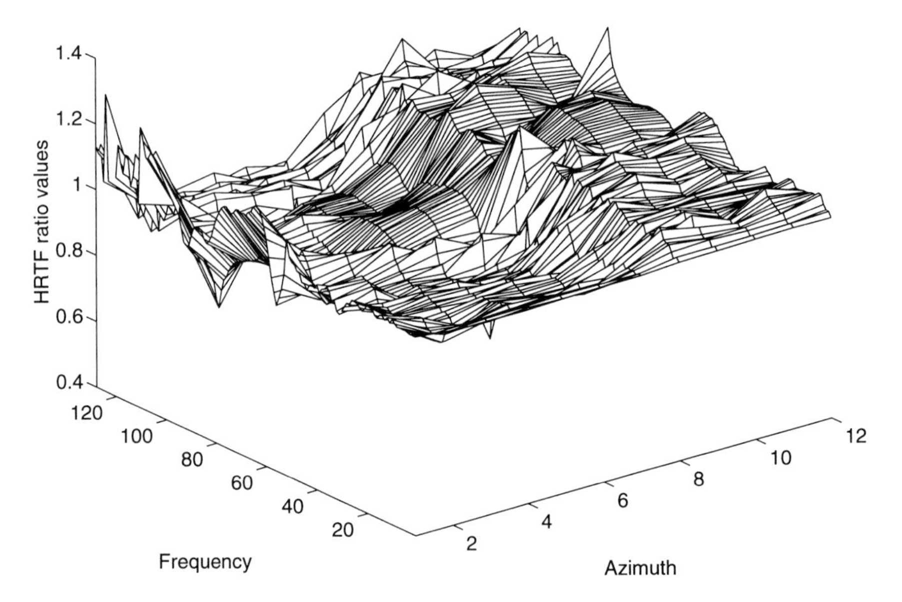

Figure 5.6-1: A 3-D relief map of the HRTF ratio patterns generated from data determined by Wenzel *et al.* [5-27]. The data is shown for 12 azimuthal directions 30° apart. Each pattern is the ratio of the ipsilateral and contralateral HRTFs corresponding to the left ear.

In this study we compare the performance of the two linear matching methods, namely, *optimal DMM* (ODMM) matching and *normalized correlation* (NC) matching and their robustness to additive noise.

B. Experimental Results

For our first set of experiments, matching tests are performed on HRTF ratio patterns by the three methods explained above. The resulting outputs are compared by measuring the DMM for each test, and averaging these DMM values over all the test samples. Results for the three parts of our experiments are shown in Figs. 5.6.2 through 5.6-4.

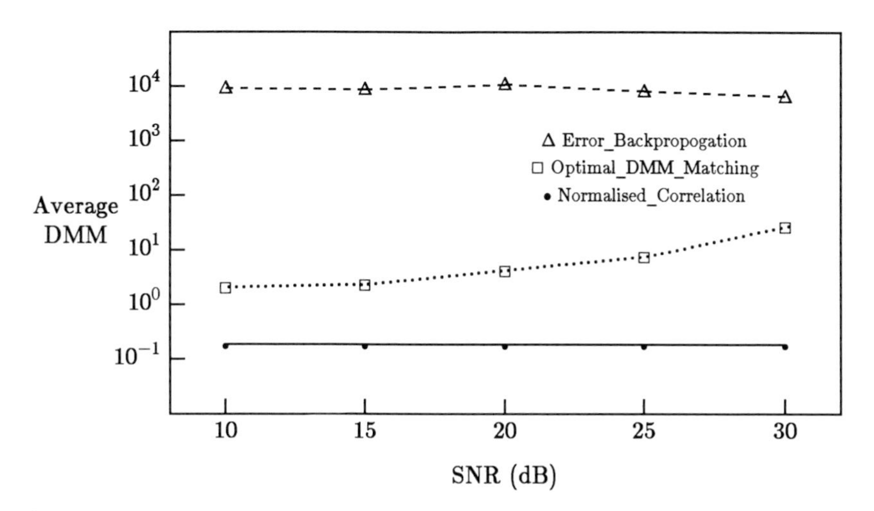

Figure 5.6-2: Average DMM results obtained by matching with noisy input HRTF ratio vectors. The level of the additive noise is given by the SNR. The back-propagation network is seen to give the best results followed by the ODMM method.

Figure 5.6-2 indicates the results of matching noisy HRTF ratio templates. The ODMM matching method is seen to be better than NC for all SNRs. The back-propagation gives superior DMM outputs in comparison to the linear methods. Improved performance is observed with higher values of SNR for the ODMM matching method.

Figure 5.6-3 shows that ODMM matching is seen to outperform other linear methods (in the sense of DMM) for broadband excitations. Back-propagation gives better average DMM results. The performance of the ODMM matching method is seen to decrease as the exciting bandwidth is reduced.

Figure 5.6-4 indicates that the ODMM method performs better than NC, while back-propagation is seen to be better than both in localizing narrowband excitation.

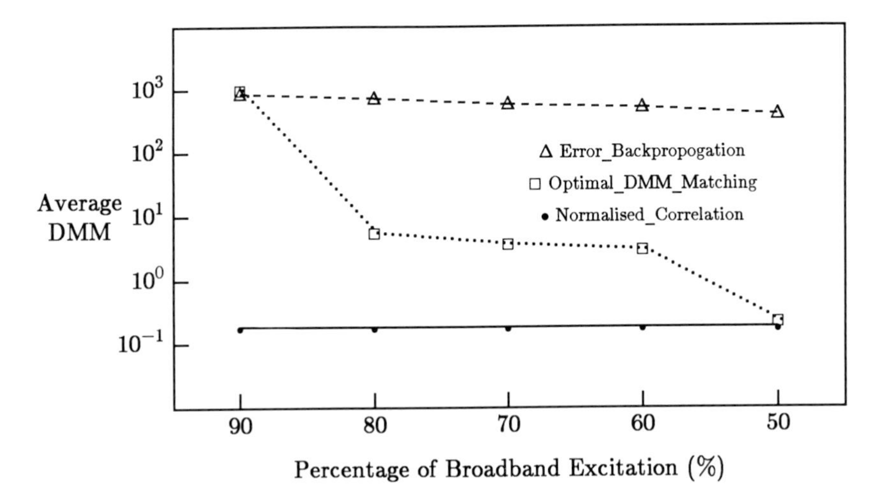

Figure 5.6-3: Average DMM results obtained by matching with broadband excitation. The abscissa indicates the percentage of broadband excitation. The back-propagation network shows robustness to reduced data. The ODMM method shows a gradual degradation which is similar to observed degradation of human localization performance as the bandwidth of the input signal is reduced.

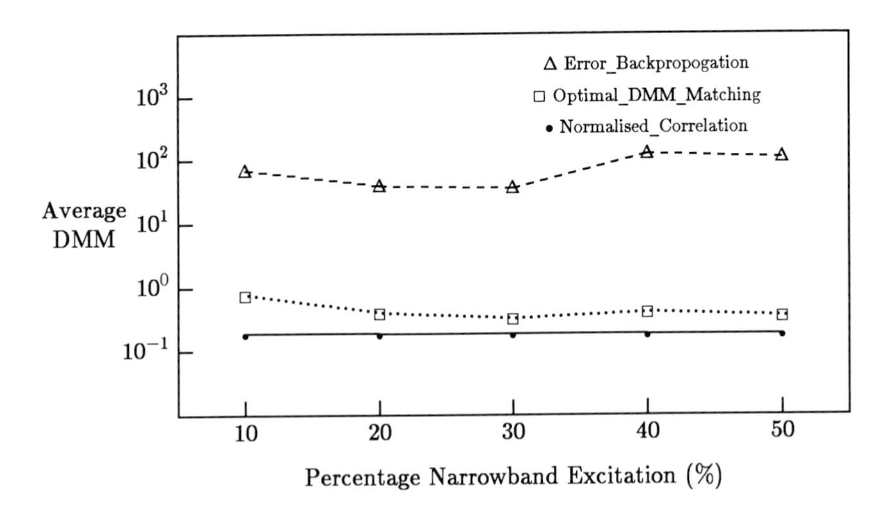

Figure 5.6-4: Average DMM results obtained by matching with narrowband excitation. The abscissa indicates the percentage of excitation. The performance of the back-propagation is still the best. The performance of the optimal DMM method has drops gradually.

However the values obtained for the DMM are lower for all the three methods, as compared to the performance with broadband excitation. As we indicated in Section 5.1, humans too are unable to correctly localize narrowband excitation [5-19].

We now discuss our second experiment using a large number of HRTF ratio patterns. Figure 5.6.5 shows the result of matching noisy HRTF ratios in the HRTF ratio map using the ODMM filters and Fig. 5.6-6 show the results with NC.

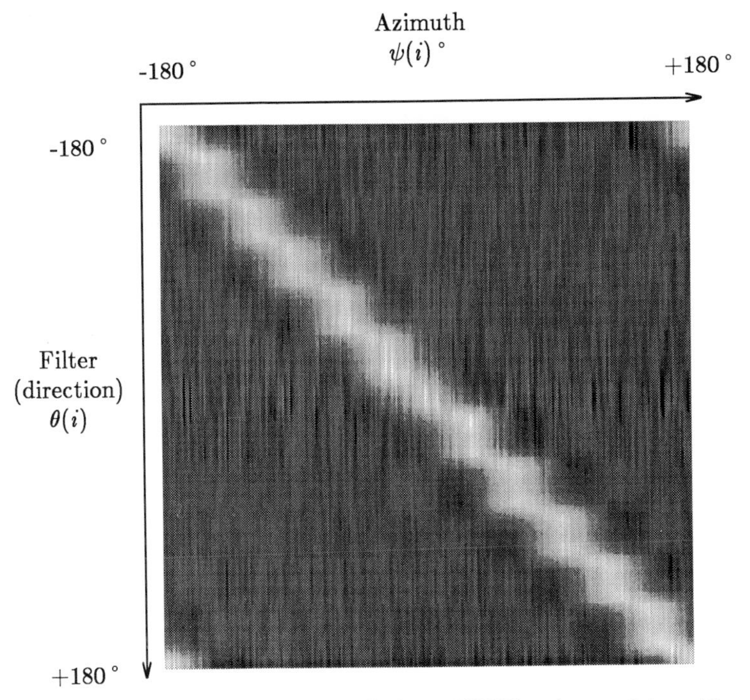

Figure 5.6-5: ODMM matching response for input HRTF ratio templates with additive noise (SNR = 20 dB). The response is fairly well localized along the diagonal with a sharp drop-off.

The resulting image is defined in terms of the θ-ψ axes. Coefficients along the θ axis indicate the results of matching an incoming HRTF ratio $\boldsymbol{h}(l)$ corresponding to direction $\psi(l)$ with the various directional filters $\theta(i)$. Thus a good match for $\boldsymbol{h}(l)$ with good discrimination would be indicated by a peak value at or near the point $(\psi(l), \theta(l))$ with all other points $(\psi(l), \theta(i))$; $i \neq l$, being minimized.

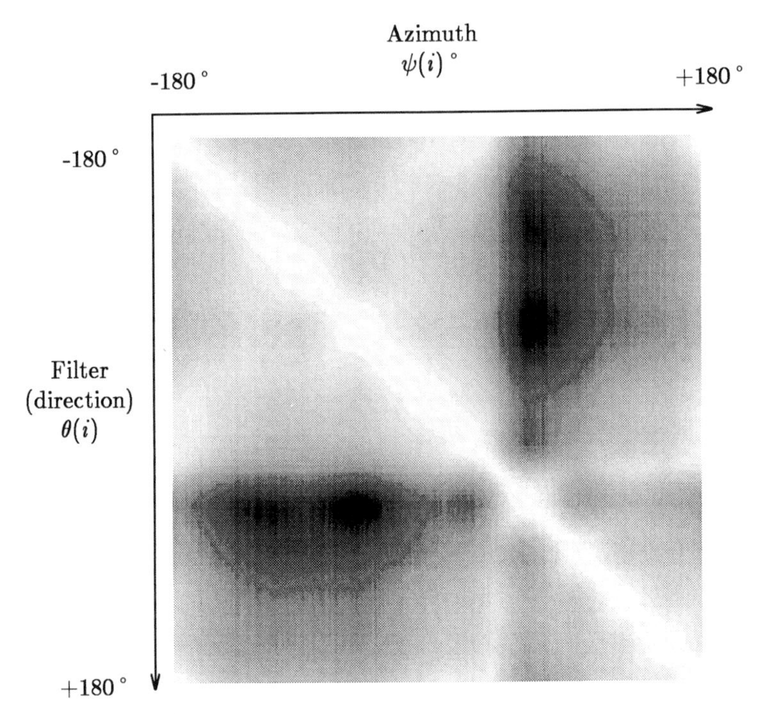

Figure 5.6-6: NC matching response for input HRTF ratio templates with additive noise (SNR = 20 dB). Note the relatively large regions of white which indicate a relatively flat response with poor localization.

Ideal localization would result in having all the peaks along the diagonal with the response minimized elsewhere.

The results for the ODMM matching method indicates superior discrimination as compared to the results for the NC method, which has a much broader peak and higher average response energy at all locations. This is more evident in Fig. 5.6-7 where typical response profiles of both the methods to the ratio pattern at $\psi(l) = -74°$ are shown.

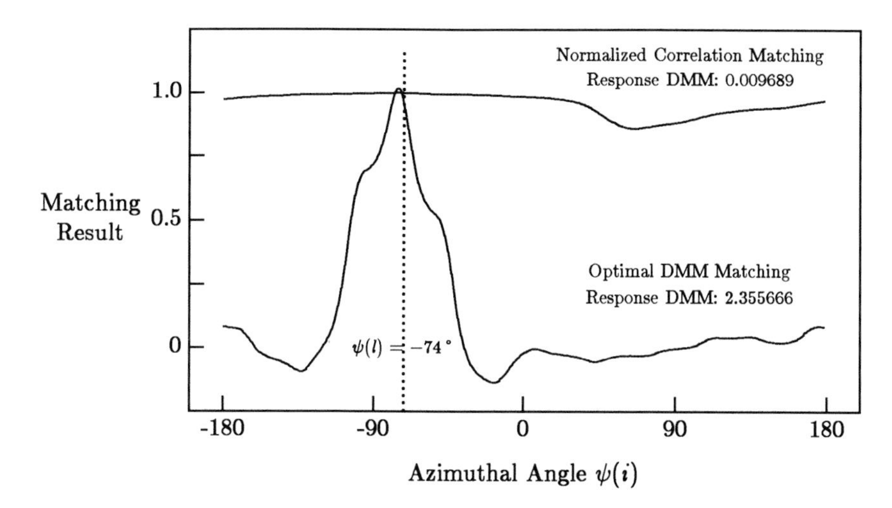

Figure 5.6-7: Typical response profiles drawn from Fig. 5.6-5 and Fig. 5.6-6 for $\psi(l) = -74°$. Note the well-localized response of the ODMM method as compared to the relatively flat response obtained by the NC method.

The true discrimination of the ODMM method as compared to the NC method is evident. We see that ODMM matching is much better and generates a sharp peak in the correct direction, while the correlation method generates a very broad and flat peak which hampers correct localization. Matching using both the ODMM method and the NC method was carried out over a large range of SNRs. The average DMM over all 360 directions for these noise values is shown in Fig. 5.6-8. The ODMM matching method is seen to be better by a factor of $10 \sim 1000$ depending on the SNR of the input vector.

Summary

We have modeled localization in a novel manner using spectral information extracted by the cochlear nucleus. This model is able to explain localization in the mid-to-high-frequency range, using directional frequency cues that are encoded in

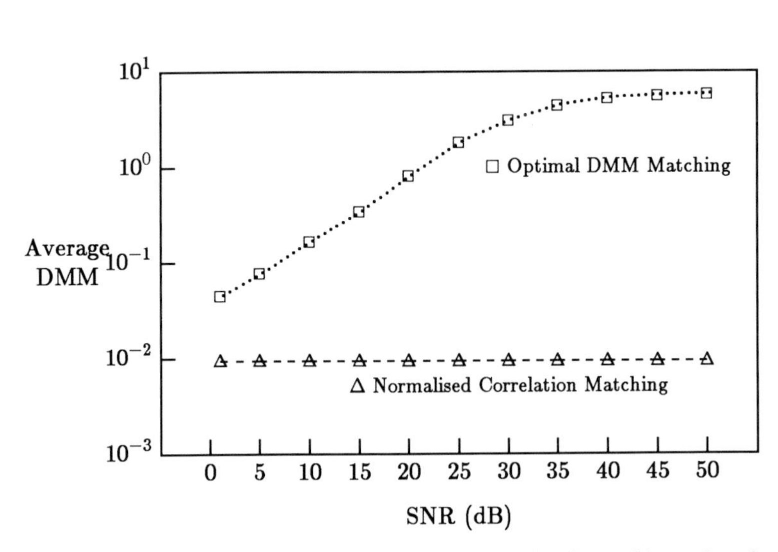

Figure 5.6-8: Comparison of ODMM filter matching and NC matching using the DMM averaged over several test inputs. As is evident the ODMM method gives a much higher DMM on average than does NC Matching.

the *head related transfer functions* (HRTFs). From our experiments, we believe that localization is better modeled using HRTF ratio templates, than by following the traditional models using only *interaural time* and *amplitude differences* (ITDs and IADs). However, we emphasize that this model applies *only* to broadband medium and high-frequency sound sources (frequencies above 1 kHz), that propagate towards the head in the horizontal azimuthal plane.

An extension of this model to include elevation dependencies can be easily realized, whereas the traditional localization models have not been able to explain the ability to localize sounds at different elevations. The spectral modifications imposed on incoming signal spectra, due to elevation dependencies of the HRTFs, gives us strong reason to believe that this model will be able to explain localization in the medial saggital plane for mid and high-frequency sources.

Both optimal matching and back-propagation based feedforward networks offer viable alternatives, as models for pattern matching in the auditory nervous system that depend on HRTF frequency cues.

In extending our experiments to the more practical case of an improved resolution HRTF map, we observe that the *optimal DMM* (ODMM) matching method provides a better solution to the localization problem using frequency domain cues. This implies that the auditory localization model could generate good results on matching the HRTF ratio employing the ODMM method. The model is linear and optimizes the DMM criteria (Section 5.2), which is formulated on the principle of obtaining good localization while maintaining a relatively smooth and gradual fall-off in response away from the peak. We believe this to be a factor that is driven by evolutionary needs and by constraints imposed by the environment.

References

[5-1] J. Blauert, *Spatial Hearing: The Psychophysics of Human Sound Localization*, MIT Press, Cambridge, MA, pp. 37-50, 1983.

[5-2] W.A. Yost and D.W. Nielsen, *Fundamentals of Hearing: An Introduction*, CBS College Publishing, NY, pp. 96-111, 151-161, 1985.

[5-3] A.W. Mills, "Auditory localization," in *Foundations of Modern Auditory Theory Vol. 2*, J. Tobias (ed.), Academic Press, NY, pp. 303-347, 1972.

[5-4] A.S. Bregman, *Auditory Scene Analysis: The Perceptual Organization of Sound*, The MIT Press, Cambridge, pp. 1-44, 595-618, 1990.

[5-5] L. Jefferess, "A place theory of sound localization," *J. Comp. Physio. Psych.* vol. 61, pp. 468-486, 1948.

[5-6] N.I. Durlach, "Binaural signal detection: equalization and cancellation theory," in *Foundations of Modern Auditory Theory Vol. 2*, J. Tobias (ed.), Academic Press, NY, 1972.

[5-7] H.S. Colburn and N.I. Durlach, "Models of binaural interaction," in *Handbook of Perception IV: Hearing*, E.C. Carterette and M.P. Friedman (eds.), Academic Press, NY, pp. 467-515, 1978.

[5-8] J. Blauert, "Modeling of interaural time and intensity difference discrimination," *Psychophysical, Physiological, and Behavioural Studies in Hearing*, G. van den Brink and F. Bilsen (eds.), Delft U. Press, Delft, The Netherlands, pp. 421-424, 1980.

[5-9] R.F. Lyon, "Computational models of binaural localization and separation," *Proc. IEEE ICASSP*, Boston, MA, pp. 1148-1151, 1983.

[5-10] Lord Rayleigh (J.W. Strutt), "On our perception of sound direction," *Phil. Mag.*, vol. 13, pp. 214-232, 1907.

[5-11] S.A. Shamma, N. Shen and P. Gopalaswamy, "Stereausis: binaural processing without neural delays," *J. Acoust. Soc. Am.*, vol. 86, pp. 989-1005, 1989.

[5-12] G. Plenge, "On the difference between localization and laterization," *J. Acoust. Soc. Am.*, vol. 56, pp. 944-951, 1953.

[5-13] J. Blauert, "Sound Localization in the medial plane," *Acustica*, vol. 22, pp. 205-213, 1969.

[5-14] R.A. Butler and K. Belendiuk, "Spectral cues utilized in the localization of sound in the medial saggital plane," *J. Acoust. Soc. Am.*, vol. 61, pp. 1264-1269, 1977.

[5-15] S.R. Oldfield and S.P.A. Parker, "Acuity of sound localization : a topography of sound space: pinnae cues absent," *Perception*, vol. 13, pp. 600-617, 1984.

[5-16] J. Zwislocki and R. Feldman, "Just noticeable differences in dichotic phase," *J. Acoust. Soc. Am.*, vol. 28, pp. 860-864, 1956.

[5-17] J.R. Angell and W. Fite, "The monaural localization of sound," *Psych. Rev.*, vol. 8, pp. 225-243, 1901.

[5-18] R.A. Butler, E.T. Ley and W.D. Neff, "Apparent distance of sounds recorded in echoic and anechoic chambers," *J. Exp. Psych., Human Perception and Performance*, vol. 6, pp. 745-750, 1980.

[5-19] A.D. Musicant and R.A. Butler, "Influence of monaural spectral cues on binaural localization," *J. Acoust. Soc. Am.*, vol. 77, pp. 202-208, 1985.

[5-20] R.A. Butler and C.C. Helwig, "The spatial attributes of stimulus frequency and their role in sound localization," *American J. of Otolaryngol.*, vol. 4, pp. 165-173, 1983.

[5-21] F.L. Wightman and D.J. Kistler, "Headphone simulation of free-field listening. I: stimulus synthesis," *J. Acoust. Soc. Am.*, vol. 85, pp. 858-867, 1989.

[5-22] F.L. Wightman and D.J. Kistler, "Headphone simulation of free-field listening. II: psychophysical validation," *J. Acoust. Soc. Am.*, vol. 85, pp. 868-878, 1989.

[5-23] D.W. Batteau, "The role of pinnae in human localization," *Proc. of the Royal Society*, London, England, pp. 158-180, 1967.

[5-24] M.B. Gardner, "Some monaural and binaural facets of the medial plane localization," *J. Acoust. Soc. Am.*, vol. 54, pp. 1489-1495, 1973.

[5-25] R. Flannery and R.A. Butler, "Spectral cues provided by the pinnae for monaural localization in the horizontal plane," *Perception and Physics*, vol. 29, pp. 438-444, 1981.

[5-26] S. Mehrgardt and V. Mellert, "Transformation characteristics of the external human ear," *J. Acoust. Soc. Am.*, vol. 61, pp. 1567-1576, 1977.

[5-27] E.M. Wenzel, "Three-dimensional virtual acoustic displays," *NASA Technical Memorandum 103835*, July 1991.

[5-28] T.J. Doll, "Development of three-dimensional audio signals," *SAE Aerospace Technology Conference and Exposition*, CA, 1986.

[5-29] D.R. Begault and E.M. Wenzel, "Headphone localization of speech stimuli," *Proceedings of the Human Factors Society*, San Francisco, CA., Sept. 1991.

[5-30] X. Yang, K. Wang and S.A. Shamma, "Auditory representations of acoustic signals," *IEEE Trans. Information Theory*, vol. IT-38, pp. 824-839, 1992.

[5-31] J. Bodreau and C. Tsuchitani, "Binaural interaction in the superior olive *S*-segment," *J. Neurophysiol.*, vol. 21, pp. 422-454, 1968.

[5-32] J. Guinan Jr., B. Norris and S. Guinan, "Single auditory units in the superior olivary complex II: locations of unit categories and tonotopic organization," *Int. J. Neurosci.*, vol. 4, pp. 147-166, 1972.

[5-33] J. Goldberg and P. Brown, "Functional organization of the dog superior olivary complex: an anatomical and electrophysiological study," *J. Neurophysiol.*, vol. 31, pp. 639-656, 1968.

[5-34] T. Yin and C. Chan, "Interaural time sensitivity in medial superior olive of cat," *J. Neurophysiol.*, vol. 64, pp. 465-488, 1990.

[5-35] J. Ben-Arie and K.R. Rao, "A novel approach for template matching by non-orthogonal image expansion," *IEEE Trans. Circts. and Systs. for Video Technology*, vol. 3, pp. 71-84, 1993.

[5-36] D. Nandy, K.R. Rao and J. Ben-Arie, *Optimal Template Matching for Auditory Localization*, Technical Report, IIT-ECE-TR-015, ECE Dept. Illinois Institute of Technology, May 1993.

[5-37] K.R. Rao and J. Ben-Arie, *On the Design of an Optimal Set of Head Related Transfer Functions*, Technical Report, IIT-ECE-TR-014, ECE Dept. Illinois Institute of Technology, May 1993.

[5-38] A.C. Rosenfeld and A. Kak, *Digital Picture Processing*, Academic Press, NY, 1982.

[5-39] D.E. Rumelhart, G.E. Hinton and R.J. Williams, "Learning internal representations by error propagation," in *Parallel Distributed Processing: Explorations in the Microstructure of Cognition : Foundations Vol. 1*, D.E. Rumelhart and J.L. McClelland (eds.) MIT Press, Cambridge, MA, 1986.

[5-40] P.J. Werbos, *Beyond Regression: New Tools for Prediction and Analysis in the Behavioural Sciences*, Doctoral Dissertation, Applied Mathematics, Harvard University, Boston, MA, November 1974.

[5-41] L. Holmström and P. Koistinen, "Using additive noise in back-propagation training," *IEEE Trans. Neural Nets.*, vol. 3, pp. 24-38, 1992.

[5-42] M.D. Levine, "Spatial and frequency domain processing," *Vision in Man and Machine*, McGraw-Hill, pp. 226-227, 1985.

Chapter 6

Signal Processing by Projection Methods: Applications to Color Matching, Resolution Enhancement, and Blind Deconvolution

HENRY STARK and YONGYI YANG

Introduction

The method of mathematical projections is a powerful technique for solving source recovery and signal synthesis problems in the physical and mathematical sciences. In its most common realization, the projection algorithm has an iterative structure in which operators, called projectors, modify an estimate of the function to be recovered so that what emerges is a new and better estimate that satisfies known constraints imposed upon the solution. A very nice property of projections is that the method is entirely neutral with respect to human aesthetic. It merely produces a recovered signal (or image) that satisfies known constraints imposed by the physics of the problem. Unlike maximum likelihood it doesn't seek to optimize an *objective* criterion that is *subjectively* imposed (i.e., the result of the human aesthetic). While projection methods date back to the 1960s, when a group of Russian mathematicians [6-1] invented the method of *projections onto convex sets* (POCS), the application of such methods to solving engineering problems is much more recent owing to the development of inexpensive powerful workstations and, in part, to the delay associated with the percolation of new mathematical ideas into the engineering community.

In this chapter, we shall introduce the reader to the basic ideas of sets and projections and illustrate how abstract mathematical ideas can be forged into concrete algorithms for solving specific engineering problems. We shall distinguish between two types of constraints: convex and non-convex and illustrate some pitfalls that materialize when non-convex constraints have to be imposed.

To illustrate how useful projection methods are, we shall apply them to three relatively sophisticated engineering problems: color matching, resolution enhancement and blind deconvolution. In color matching we ask: (1) what is the smallest adjustment required of a given spectrum to produce color matching in a color-sensitive visual system? (2) What combination of basis colors will produce a match to a particular set of measurements? and (3) What is a parsimonious mixture of colors that when added to a given spectrum will produce a predetermined color match?

The second problem we shall discuss is resolution enhancement. Here we ask how can a series of low-resolution images be combined into a single high-resolution image in a straightforward fashion. We shall demonstrate that a natural vehicle for this synthesis is POCS.

The final problem that we solve by projection methods is more difficult than the first two. It involves the new idea of projection of pairs of functions and addresses the perplexing problem of trying to recover a source signal (e.g., a cluster of stars) from data which has been distorted by an unknown distorting function. Because both the original and distorting functions are unknown, the recovery of the original signal is known as "blind deconvolution".

The utility of POCS is also displayed in other places in this book. In Chapter 7 it is used to recover compressed images without artifacts, while in Chapter 10 it is used to recover a probability density function from observed data.

The intent of this chapter is to introduce the reader to the extraordinary power and versatility of projection methods so that with little additional help, he or she will be able to consider these methods for solving their own engineering problems, whatever they may be.

6.1 The Method of Projections Onto Convex Sets

POCS is an iterative method for finding points (e.g., vectors or functions) that lie at the intersection of convex sets. It turns out that for a significant number of practical problems in optics and image processing, finding a solution to the problem is equivalent to finding a point at the intersection of convex sets. We begin with some definitions that we need to understand the POCS algorithm.

Convex set C: A set C is convex if for any two points $x_1, x_2 \in C$, the point $x_3 = \sigma x_1 + (1-\sigma)x_2$ also lies in C for any value of the parameter σ in the range $0 \leq \sigma \leq 1$. If one takes any two points in a convex set and draws the line segment connecting them, all the points on the line segment will also be in the set. Figure 6.1-1a) shows a convex set; Fig. 6.1-1b) shows a non-convex set.

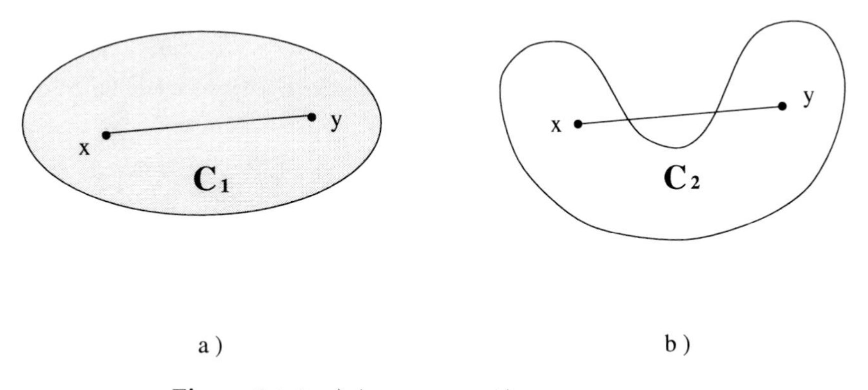

Figure 6.1-1: a) A convex set; b) a non-convex set.

Closed set C: A set C is closed if it contains all of its limit points. For example, the set [0,1], i.e., the set of all points in the interval zero to one inclusive, is closed. The set (0,1) which leaves out the end points is open. In both cases zero and one are limit points but only in the former case are they included in the set. A closed convex set is a set which is both convex and closed.

Projection onto the set C: Given any point x, its projection onto C is the point in C, say z, which is closest to the original point x. If x is already in C then its projection is merely itself. For a closed convex set, the projection is always unique. The word closest implies a distance measure. For n-dimensional Euclidean vectors $\mathbf{x} = (x_1, x_2, ..., x_n), \mathbf{y} = (y_1, y_2, ..., y_n)$, etc., a reasonable distance measure between $\mathbf{x}$ and $\mathbf{y}$ is

$$d = \left[\sum_{i=1}^{n}(x_i - y_i)^2\right]^{1/2}. \tag{6.1-1}$$

Projection operator or projector onto C: A function, map or rule P is called a projection operator or projector if it assigns to each point x a point $Px \in C$ which is closest to the original point. Projectors satisfy the relation $P^2x = Px$.

Hilbert space $\mathcal{H}$: A Hilbert space is an inner product space with certain desirable convergence properties (completeness) [6-2]. For our purpose, the Hilbert space under consideration is the set of square integrable functions with a norm defined as

$$\| f \| \triangleq \left[\int_{-\infty}^{\infty} | f(x) |^2 \; dx \right]^{1/2} . \tag{6.1-2}$$

The inner product of two functions $f(x)$ and $g(x)$ is defined as

$$\langle f, g \rangle \triangleq \int_{-\infty}^{\infty} f(x)\overline{g(x)}\, dx, \tag{6.1-3}$$

where the overbar denotes conjugation and the distance d between two functions f and g is given by

$$d(f,g) = \|f - g\| = \langle f - g,\; f - g \rangle^{\frac{1}{2}} . \tag{6.1-4}$$

The space so defined is a Hilbert space and is denoted by $\mathcal{L}_2$.

Consider a closed convex set $C \subset \mathcal{H}$; for any $f \in \mathcal{H}$, the projection Pf of f onto C is the element in C closest to f. As already stated, if C is closed and convex, Pf exists and is uniquely determined by f and C from the minimality criterion [6-3]:

$$\|f - Pf\| = \min_{g \in C} \|f - g\|. \tag{6.1-5}$$

This rule, which assigns to every $f \in \mathcal{H}$ its nearest neighbor in C, defines (in general) the nonlinear projection operator $P : \mathcal{H} \to C$ without ambiguity.

The basic idea of POCS is as follows: Every known property of the unknown $f \in \mathcal{H}$ will restrict f to lie in a closed convex set C_i in $\mathcal{H}$. Thus, for m known properties, there are m closed convex sets $C_i, i = 1, 2, ..., m$ and $f \in C_0 \triangleq \cap_{i=1}^{m} C_i$. Then the problem is to find a point in C_0 given the sets C_i and the projection operators P_i projecting onto $C_i, i = 1, 2, ..., m$.

Consider now the sequence $\{f_k\}$ generated by the recursion relations

$$f_{k+1} = P_m P_{m-1} \cdots P_1 f_k; \quad k = 0, 1, 2, \ldots . \tag{6.1-6}$$

The convergence of this sequences to an $f \in C_0$ has been demonstrated by Opial [6-4] and by Gubin *et al.* [6-1]. Indeed they showed that a more general recursion relation will also converge to an $f \in C_0$, this recursion relation being

$$f_{k+1} = T_m T_{m-1} \cdots T_1 f_k; \quad k = 0, 1, 2, \ldots . \tag{6.1-7}$$

In Eq. (6.1-7) the T_i's are called *relaxed projectors* although they are not, generally, true projectors like the P_i's. The relation between T_i and P_i is

$$T_i \triangleq I + \mu_i(P_i - I), \tag{6.1-8}$$

where I is the identity operator, and μ_i is a relaxation parameter that, when properly adjusted, can increase the rate of convergence. For Eq. (6.1-7) to converge to an $f \in C_0$, μ_i must be constrained to the range $0 < \mu_i < 2$.

In their work, Gubin and Opial show that the sequence $\{f_k\}$ converges *weakly* to a point in C_0. Weak convergence is defined as follows. A sequence $\{f_k\}$ converges weakly to a point, say f^*, if

$$\lim_{k \to \infty} \langle f_k, g \rangle = \langle f^*, g \rangle \tag{6.1-9}$$

for every $g \in \mathcal{H}$. This is sometimes called inner product convergence. Strong convergence, i.e.,

$$\lim_{k \to \infty} \|f_k - f^*\| = 0 \tag{6.1-10}$$

always implies weak convergence but not the other way around. In a finite dimensional space, weak convergence is equivalent to strong convergence. Equation (6.1-6) has a simple graphical interpretation which is shown in Fig. 6.1-2a). The solution region, i.e., the intersection of the constraint sets, is approached by alternating projections between the sets. Note that, because of the convexity of the sets, a solution is always reached in the limit so long as the intersection set is non-empty. If the sets are non-convex, topological anomalies called traps can prevent a correct solution from being reached, as illustrated in Fig. 6.1-2b).

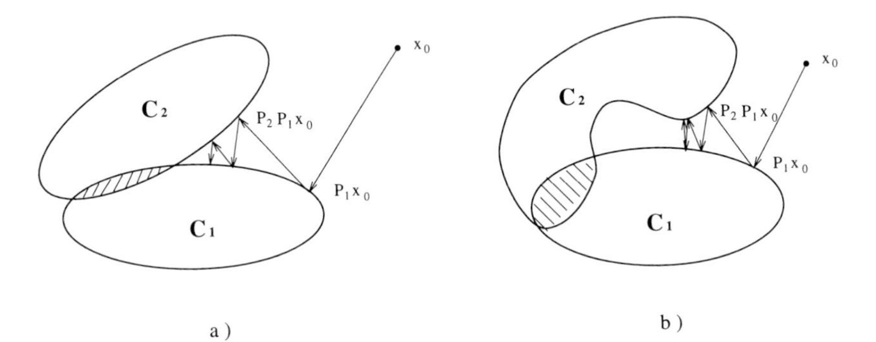

Figure 6.1-2: a) When all the constraint sets are convex, the iteration always proceeds to a solution; b) non-convex sets can create traps which prevent a solution from being reached.

As stated above, the relaxed projectors T_i in Eq. (6.1-7) are generally not true projectors. In Fig. 6.1-3 the action of the relaxed projectors for $\mu > 1$ is shown; as can be inferred from Fig. 6.1-3 faster convergence is possible for some set configuration by choosing $\mu > 1$.

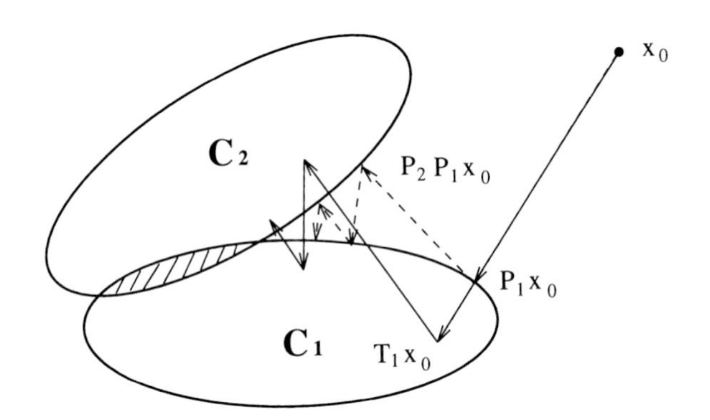

Figure 6.1-3: The action of relaxed projections. Faster convergence is possible using relaxed projections.

We conclude this section by giving a simple example using Euclidean vectors. Suppose we are seeking a vector $x_s = (x_1^{(s)}, x_2^{(s)})$ such that x_s has simultaneous membership in

$$C_1 = \left\{ x : x_1^2 + x_2^2 \leq 4 \right\} \tag{6.1-11}$$

and

$$C_2 = \{ x : x_1 = 2 \} . \tag{6.1-12}$$

The projection onto C_1 of any point x_0 is

$$P_1 x_0 = \begin{cases} 2\frac{x_0}{\|x_0\|} & \text{if } x_0 \notin C_1 \\ x_0 & \text{if } x_0 \in C_1, \end{cases} \tag{6.1-13}$$

while the projection of any point $x_0 = (x_1^{(0)}, x_2^{(0)})$ onto C_2 is

$$P_2 x_0 = (2, x_2^{(0)}). \tag{6.1-14}$$

Consider the iteration $x_{k+1} = P_2 P_1 x_k$ with $x_k = (x_1^{(k)}, x_2^{(k)})$. From the above we have

$$x_{k+1} = \left(2, \frac{2}{\sqrt{4 + (x_2^{(k)})^2}} x_2^{(k)} \right) . \tag{6.1-15}$$

Therefore, the sequence of iterates x_k generated by the algorithm will converge to $x^* = (2, 0)$, as shown in Fig. 6.1-4. Note that there is a unique solution to this problem. This is not always the case however: in most problems there are many solutions.

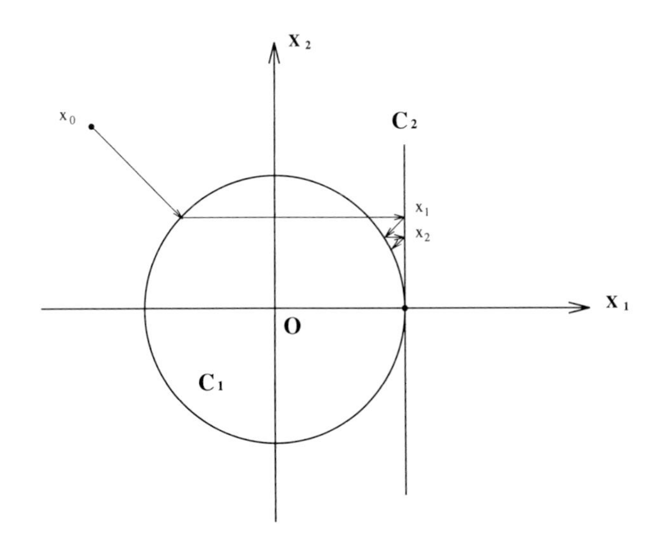

Figure 6.1-4: An example of how POCS finds the vector (x_1, x_2) that satisfies $x_1^2+x_2^2 \leq 4$ and $x_1 = 2$.

6.2 Color Matching Problems

Color matching refers to the fact that a human observer will perceive two spectral distributions as being the same if they evoke the same responses at the retinal sensors (cones) in the eye. Since the mapping from the spectral space to the response space involves a many-to-three dimensionality reduction, for most lights (spectral power distributions) there is a family of other possible lights that would appear the same to a human. In a more general setting one can conceive of a generalized color sensitive visual system (GCSVS) that perceives two spectra to be identical if the responses of all n of its sensors are the same for both. For example in one version of LANDSAT each frame consists of four images in four separate bands (two in the visible and two in the infrared) [6-5]. Then presumably two points would appear the same in their spectra if they evoke the same response in all four bands; here $n = 4$.

Assume that we have a system with n sensors, the ith one having spectral sensitivity $S_i(\lambda), i = 1, 2, ..., n$, where λ is the wavelength. Let $W(\lambda)$ be a given color spectrum. The response of the ith sensor to $W(\lambda)$ is given by

$$\alpha_i = \int_\Lambda S_i(\lambda)W(\lambda)d\lambda, \tag{6.2-1}$$

where Λ is the support of the integrand. Now consider two spectra $W^{(1)}(\lambda)$ and $W^{(2)}(\lambda)$ and denote

$$\alpha_i^{(1)} = \int_\Lambda S_i(\lambda)W^{(1)}(\lambda)d\lambda \tag{6.2-2}$$

$$\alpha_i^{(2)} = \int_\Lambda S_i(\lambda) W^{(2)}(\lambda) d\lambda. \tag{6.2-3}$$

If $\alpha_i^{(1)} = \alpha_i^{(2)}, i = 1, 2, .., n$, then $W^{(1)}(\lambda)$ and $W^{(2)}(\lambda)$ are perceived as being the same; they are said to be *metamers*, and the color matching condition is satisfied.

Problem 1 We are given a spectrum $W'(\lambda)$ and wish to find the *closest* $W(\lambda)$, say $W^*(\lambda)$, that evokes a prescribed response $\alpha_i^*, i = 1, 2, ..., n$. By closest we mean that the normed distance $\|W' - W^*\|$ is the smallest among all metamers that provoke responses $\boldsymbol{\alpha}^* = (\alpha_1^*, \alpha_2^*, ..., \alpha_n^*)$. (Note that here all quantities are real. The asterisk is an identification mark, not conjugation.)

To solve this problem, we define the set

$$C_c \triangleq \left\{ W(\lambda) : \int_\Lambda S_i(\lambda) W(\lambda) d\lambda = \alpha_i^*, i = 1, 2, .., n \right\}. \tag{6.2-4}$$

Clearly

$$C_c = \cap_{i=1}^n C_i, \tag{6.2-5}$$

where

$$C_i \triangleq \left\{ W(\lambda) : \int_\Lambda S_i(\lambda) W(\lambda) d\lambda = \alpha_i^* \right\}. \tag{6.2-6}$$

The set C_i is convex, i.e., the convex combination of any two elements in the set is also in the set, since if $W^{(1)}(\lambda)$ and $W^{(2)}(\lambda) \in C_i$ then $W^{(3)}(\lambda) \triangleq \sigma W^{(1)}(\lambda) + (1 - \sigma) W^{(2)}(\lambda) \in C_i$. To see this we use the notation introduced in Eq. (6.1-3)

$$\langle S_i, W \rangle \triangleq \int_\Lambda S_i(\lambda) W(\lambda) d\lambda, \tag{6.2-7}$$

then

$$\begin{aligned} \langle S_i, W^{(3)} \rangle &= \sigma \langle S_i, W^{(1)} \rangle + (1 - \sigma) \langle S_i, W^{(2)} \rangle \\ &= \sigma \alpha_i^* + (1 - \sigma) \alpha_i^* \\ &= \alpha_i^*, \qquad \text{for } 0 < \sigma < 1. \end{aligned}$$

Thus C_i is convex. Moreover we see that $W^{(3)}(\lambda)$ is in the set for any $-\infty < \sigma < \infty$. Hence C_i is not only convex, it is also a *linear variety*[1] (LV). To show that C_i is closed, consider a sequence $W^{(j)} \in C_i$, $j = 1, 2, ...$, with limit point W, i.e.,

$$\lim_{j \to \infty} \|W^{(j)} - W\| = 0. \tag{6.2-8}$$

By applying the Schwartz inequality, we have

$$| \langle S_i, W^{(j)} - W \rangle | \leq \|S_i\| \, \|W^{(j)} - W\| \to 0 \tag{6.2-9}$$

[1] A linear variety is a convex set for which the parameter σ is allowed the range $-\infty < \sigma < \infty$.

because W is a limit point for $\{W^{(j)}\}$ and $S_i \in \mathcal{L}_2$. Thus

$$\langle\ S_i, W^{(j)}\ \rangle \quad \rightarrow \quad \langle\ S_i, W\ \rangle\,. \tag{6.2-10}$$

The projection of the point $W'(\lambda)$ onto C_i is easily computed using the method of Lagrange multipliers. The method of Lagrange multipliers is a common technique for solving for the extreme point(s) of a function when side conditions or constraints are applied [6-6], which appears in several chapters in this book. The solution is obtained by writing

$$J(W) \triangleq \|W' - W\|^2 + \gamma(\ \langle\ S_i,\ W\ \rangle\ - \alpha_i^*), \tag{6.2-11}$$

where γ is the Lagrange multiplier, and computing

$$\frac{dJ}{dW} = 0 \tag{6.2-12}$$

which yields

$$W(\lambda) = W'(\lambda) - \frac{\gamma}{2} S_i(\lambda). \tag{6.2-13}$$

Finally, applying the constraint $\langle\ S_i, W\ \rangle\ = \alpha_i^*$ yields as the projection

$$W^*(\lambda) \triangleq P_i W'(\lambda) = W'(\lambda) + \frac{\alpha_i^* - \langle\ S_i, W'\ \rangle}{\|S_i\|^2} S_i(\lambda). \tag{6.2-14}$$

Thus applying P_i to $W'(\lambda)$ yields a spectrum that evokes the appropriate response α_i^* in the ith sensor. However for color matching we must evoke the responses $\alpha_1^*, \alpha_2^*, ..., \alpha_n^*$ simultaneously.

From the fundamental theorem of POCS (see Eq. (6.1-6)) this is accomplished by the algorithm

$$W_{k+1}(\lambda) = P_n P_{n-1} ... P_1 W_k(\lambda), \quad W_0 = W'(\lambda). \tag{6.2-15}$$

Finally, how do we know that W_k converges to a spectrum, say $W_c(\lambda)$, that is closest to the original $W'(\lambda)$? Had we projected $W'(\lambda)$ directly onto the set C_c, then by the definition of a projection we would have been guaranteed that the resulting spectrum would have been the closest. The iterative algorithm that we furnish in Eq. (6.2-15) is, however, not a projection onto C_c and therefore should, in general, not converge to the same point as a one-step projection onto C_c. Fortunately, the following theorem states that the sequence $\{W_k(\lambda)\}$ does indeed converge strongly to the projection onto C_c.

Theorem 6.2-1 If each set C_i in $C_c = \cap_{i=1}^n C_i$ is a linear variety and if the intersection C_c of these linear varieties is non-empty, the sequence $W_{k+1}(\lambda) = P_n P_{n-1} ... P_1 W_k(\lambda)$ with W_0 as the initial point converges strongly to $P_c W_0$ for every $W_0 \in \mathcal{H}$ [6-3]. ∎

In other words, the same point is reached either by direct projection onto C_c or by iteration of the alternating projections onto the C_i's. Figure 6.2-1 illustrates the theorem using Euclidean vectors in the two-dimensional case. Thus Eq. (6.2-15) with projectors $P_i, i = 1, 2, ..., n$, as given in Eq. (6.2-14) does produce a color-matched spectrum that is closest to the original spectrum $W'(\lambda)$.

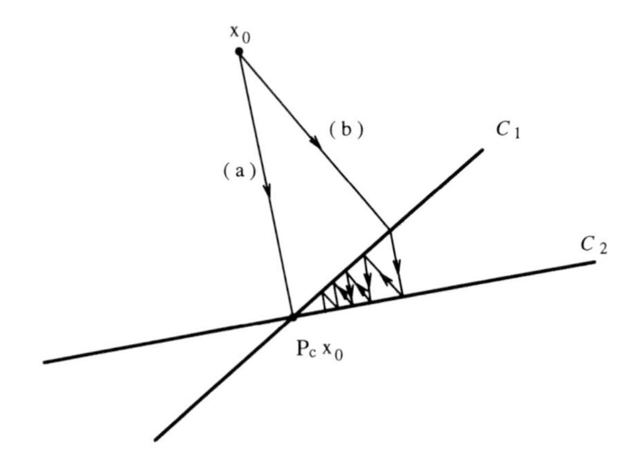

Figure 6.2-1: When constraint sets are all linear varieties, the projection of a point onto the intersection can be reached by alternating orthogonal projections.

Problem 2 The solution to the previous problem finds the closest color-matching spectrum to a given spectrum but does not describe what combination of basis colors $R_i(\lambda)$ need to be added to accomplish this task. We use the term *basis* loosely here meaning only those colors that are available to the user for generating or modifying spectra. As a first step in solving this problem we consider the following: assume that we have available m basis colors with spectra $R_j(\lambda), j = 1, 2, ..., m$ and we wish to create a spectrum $W^*(\lambda)$ that evokes the prescribed column vector responses $\boldsymbol{\alpha}^* = (\alpha_1^*, \alpha_2^*, ..., \alpha_m^*)^T$ by some linear combination of the basis color, i.e.,

$$W^*(\lambda) = \sum_{j=1}^{m} \beta_j^* R_j(\lambda). \tag{6.2-16}$$

Thus finding $\beta_j^*, j = 1, 2, ..., m$, is equivalent to finding $W^*(\lambda)$. We introduce the following notation

$$\boldsymbol{\beta} \triangleq (\beta_1, \beta_2, ..., \beta_m)^T$$

$$\begin{aligned} a_{ij} &\triangleq \langle\, S_i, R_j \,\rangle \triangleq \int_\Lambda S_i(\lambda) R_j(\lambda) d\lambda \\ \mathbf{a}_i &\triangleq (a_{i1}, a_{i1}, ..., a_{im})^T, \end{aligned}$$

for $i = 1, 2, ..., n; \quad j = 1, 2, ..., m$. To determine an appropriate set of constraints, we need to be somewhat more specific about our underlying assumptions. We assume that the basis vectors represent colored lights such as those generated in the experiments by Land and McCann [6-7] in which illuminating projectors with sharp cut-off band-pass filters are used. In their experiment, the flux for each projector/filter combination was independently controlled by a variable transformer. Thus the β_j's are proportional to the transformer power, and are essentially non-negative. However, in other situations, the basis color vectors might have positive or negative coefficients such as, for example, when the zeroth basis vector is taken to be a mean vector. In the latter case, the appropriate constraints would require that the net radiance or reflectance be everywhere non-negative. However in this problem we require that $\beta_j \geq 0, j = 1, 2, ..., m$. Consider now the following sets

$$C_i \triangleq \left\{ \boldsymbol{\beta} : \langle\, S_i, \sum_{j=1}^{m} \beta_j R_j \,\rangle = \alpha_i^* \right\} \tag{6.2-17}$$

for $i = 1, 2, ..., n$ and

$$C_{n+1} \triangleq \{\, \boldsymbol{\beta} : \beta_j \geq 0, j = 1, 2, ..., m \,\}. \tag{6.2-18}$$

These sets are convex and closed; also C_{n+1} is not an LV. The sets in Eq. (6.2-17) can be rewritten as

$$C_i \triangleq \left\{ \boldsymbol{\beta} : \boldsymbol{\beta}^T \cdot \mathbf{a}_i = \alpha_i^* \right\} \quad \text{for } i = 1, 2, ..., n. \tag{6.2-19}$$

With P_i projecting onto C_i, for $i = 1, 2, ..., n+1$ the algorithm

$$\boldsymbol{\beta}_{k+1} = P_{n+1} P_n ... P_1 \boldsymbol{\beta}_k, \quad \boldsymbol{\beta}_0 \text{ arbitrary} \tag{6.2-20}$$

converges to a $\boldsymbol{\beta}^*$ such that $W^*(\lambda)$ computed as in Eq. (6.2-16) evokes the prescribed responses $\boldsymbol{\alpha}^*$. Thus all that is needed is to find the projectors $P_1, P_2, ..., P_{n+1}$. Without going through the details it is directly shown that given an arbitrary $\boldsymbol{\beta}$, its projection onto $C_i, 1 \leq i \leq n$, is given by [6-28]

$$\boldsymbol{\beta}^* \triangleq P_i \boldsymbol{\beta} = \boldsymbol{\beta} + \frac{\alpha_i^* - \boldsymbol{\beta}^T \cdot \mathbf{a}_i}{\|\mathbf{a}_i\|^2} \mathbf{a}_i \tag{6.2-21}$$

and when $i = n + 1$:

$$\boldsymbol{\beta}^* \triangleq P_{n+1} \boldsymbol{\beta} = \boldsymbol{\beta}^+, \tag{6.2-22}$$

where $\boldsymbol{\beta}^+$ is the *rectification* of $\boldsymbol{\beta}$, i.e., the components of $\boldsymbol{\beta}^+$ are the same as the non-negative components of $\boldsymbol{\beta}$ and zero otherwise.

There is no guarantee that Eq. (6.2-20) will converge unless it is known *a priori* that a linear combination of basis colors $R_j(\lambda)$ can actually achieve a match. Assuming this to be the case, the algorithm in Eq. (6.2-20) will converge to a spectrum that satisfies the color matching responses. Note that in this case the one-step projection onto $\cap_{i=1}^{n+1} C_i$ would typically not yield the same point as the convergent point in Eq. (6.2-20) because, as stated earlier, C_{n+1} is not a linear variety.

Problem 3 We are given a color spectrum $W'(\lambda)$ that fails the color match test. The problem is to add the smallest (or at least the near-smallest) amount of basis colors $R_j(\lambda), j = 1, 2, ..., m$, to evoke the color match vector $\boldsymbol{\alpha}^* = (\alpha_1^*, \alpha_2^*, ..., \alpha_m^*)^T$. This problem is similar to Problem 1 except that the solution includes the actual synthesis of the spectrum, i.e., it requires finding the proportion of basis colors to be added for a color match. The synthesized spectrum has the form

$$W(\lambda) = \beta_0 W'(\lambda) + \sum_{j=1}^{m} \beta_j R_j(\lambda). \tag{6.2-23}$$

A mathematical statement of the problem is to

$$\begin{aligned}
&\text{minimize} \quad \textstyle\sum_{j=1}^{m} \beta_j \\
&\text{subject to} \\
&\quad 1.\ \ \langle \beta_0 W' + \textstyle\sum_{j=1}^{m} \beta_j R_j, S_i \rangle = \alpha_i^* \quad \text{for } i = 1, 2, ..., n, && (6.2\text{-}24) \\
&\quad 2.\ \ \beta_j \geq 0 \ \text{ for } j = 0, 1, 2, ..., m. && (6.2\text{-}25)
\end{aligned}$$

A direct solution to this problem using POCS may not be possible; an indirect method is given below. Define

$$\begin{aligned}
a_{i0} &\triangleq \langle S_i, W' \rangle \\
a_{ij} &\triangleq \langle S_i, R_j \rangle \\
\mathbf{a}_i &\triangleq (a_{i0}, a_{i1}, ..., a_{im})^T \\
\boldsymbol{\beta} &\triangleq (\beta_0, \beta_1, ...\beta_m)^T \\
\hat{\boldsymbol{\beta}} &\triangleq (\beta_1, \beta_2, ..., \beta_m)^T,
\end{aligned}$$

where $i = 1, 2, ..., n; j = 1, 2, ..., m$. Note that the vector $\mathbf{a}_i$ has been augmented by the component a_{i0}. The solution must lie in the following constraint sets:

$$\begin{aligned}
C_i &\triangleq \left\{ \boldsymbol{\beta} : \boldsymbol{\beta}^T \cdot \mathbf{a}_i = \alpha_i^* \right\} \quad \text{for } i = 1, 2, ..., n \\
C_{n+1} &\triangleq \left\{ \boldsymbol{\beta} : \sum_{j=1}^{m} \beta_j \leq E \right\} \\
C_{n+2} &\triangleq \{ \boldsymbol{\beta} : \beta_j \geq 0, j = 0, 1, ..., m \}.
\end{aligned} \tag{6.2-26}$$

Some remarks are in order: the sets C_i, $i = 1, 2, ..., n$, are similar to those in Eq. (6.2-17) except the augmented constraint vector $\mathbf{a}_i$ is used here; the conditions $\boldsymbol{\beta}^T \cdot \mathbf{a}_i = \alpha_i^*$ enforce the matching constraints. The set C_{n+1}, with bound E chosen *a priori*, enforces the parsimony on the use of basis color; a way to approximate the minimization of $\sum_{j=1}^{m} \beta_j$ is to reduce E until convergence of the algorithm becomes problematic. This estimate of E_{min} can be done in trial runs. Also with respect to C_{n+1}: the constraint is on the sum involving only $\beta_1, \beta_2, .., \beta_m$ and not β_0; it is assumed that adjusting the level (i.e., β_0 of the initial spectrum $W'(\lambda)$) involves no cost. If it did we would merely modify C_{n+1} to $C_{n+1} \triangleq \left\{ \boldsymbol{\beta} : \sum_{j=0}^{m} \beta_j \leq E \right\}$. Finally C_{n+2} ensures that the synthesis is physically realizable. The projections onto $C_i, i = 1, 2, ..., n$ and $i = n + 2$ are the same as in Problem 2. The projection onto C_{n+1} is computed as follows.

Consider an arbitrary vector $\boldsymbol{\beta}'$ whose projection onto C_{n+1} we seek. Write $\boldsymbol{\beta}' = (\beta_0, \hat{\boldsymbol{\beta}}'^T)^T$; then the projection, say $\boldsymbol{\beta}^*$, will have the form $(\beta_0, \hat{\boldsymbol{\beta}}^{*T})^T$ since there are no constraints on β_0. Hence we can compute the projection as follows. Define the m-component unit vector $\mathbf{u} = (1, 1, ..., 1)^T$, and let

$$J(\hat{\boldsymbol{\beta}}) \triangleq \|\hat{\boldsymbol{\beta}}' - \hat{\boldsymbol{\beta}}\|^2 + \gamma(\mathbf{u}^T \cdot \hat{\boldsymbol{\beta}} - E) \tag{6.2-27}$$

and compute

$$\frac{dJ(\hat{\boldsymbol{\beta}})}{d\hat{\boldsymbol{\beta}}} = 0. \tag{6.2-28}$$

This yields

$$\hat{\boldsymbol{\beta}} = \hat{\boldsymbol{\beta}}' - \frac{\gamma}{2}\mathbf{u}. \tag{6.2-29}$$

Now use the fact that the projection for convex sets occurs at the boundary. Hence

$$\mathbf{u}^T \cdot \hat{\boldsymbol{\beta}} = \mathbf{u}^T \cdot \hat{\boldsymbol{\beta}}' - \frac{\gamma}{2} m = E \tag{6.2-30}$$

or

$$\frac{\gamma}{2} = \frac{1}{m}(\mathbf{u}^T \cdot \hat{\boldsymbol{\beta}}' - E) = \frac{1}{m}(\sum_{j=1}^{m} \beta_j - E) \tag{6.2-31}$$

and the projection onto C_{n+1} is

$$\boldsymbol{\beta}^* = P_{n+1}\boldsymbol{\beta}' = (\beta_0, \beta_1 - \kappa, \beta_2 - \kappa, ..., \beta_m - \kappa)^T, \tag{6.2-32}$$

where

$$\kappa = \frac{1}{m}(\sum_{j=1}^{m} \beta_j - E). \tag{6.2-33}$$

The algorithm

$$\boldsymbol{\beta}_{k+1} = P_{n+2}P_{n+1}...P_1\boldsymbol{\beta}_k, \quad \boldsymbol{\beta}_0 \text{ arbitrary} \tag{6.2-34}$$

will yield a solution near the optimum solution to Problem 3.

A. A Numerical Example

In the following we assume that we have a system with four sensors of which the spectral sensitivity functions are defined separately as follows:

$$\begin{aligned} S_1(\lambda) &= \exp[-4(\lambda-1)^2] \cdot I_\Lambda \\ S_2(\lambda) &= \exp[-4(\lambda-2)^2] \cdot I_\Lambda \\ S_3(\lambda) &= \exp[-4(\lambda-3)^2] \cdot I_\Lambda \end{aligned}$$

and

$$S_4(\lambda) = \begin{cases} \lambda - 1.5 & \text{if } 1.5 \leq \lambda < 2.5 \\ 3.5 - \lambda & \text{if } 2.5 \leq \lambda \leq 3.5 \\ 0 & \text{otherwise,} \end{cases} \tag{6.2-35}$$

where $\Lambda = \{\lambda : \lambda \in [0, 3.5]\}$ and I_Λ is given by

$$I_\Lambda \triangleq = \begin{cases} 1, & \text{if } \lambda \in \Lambda \\ 0, & \text{if } \lambda \notin \Lambda. \end{cases} \tag{6.2-36}$$

The graphs of these functions are shown in Fig. 6.2-2.

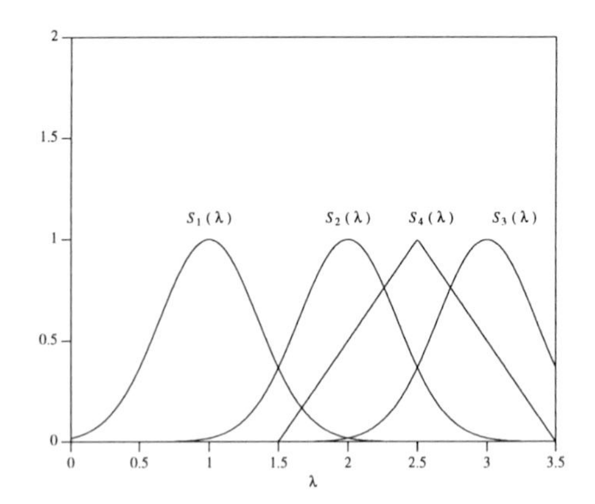

Figure 6.2-2: Four spectral sensitivity functions of a fictitious visual system.

For numerical convenience and ease of graphing we used an arbitrary set of wavelengths which do not have a direct physical meaning. Assume that we are given a measurement of a color source with a spectral density defined by

$$W(\lambda) = \begin{cases} 1 & \text{if } 0 \leq \lambda < 1 \\ \exp(1-\lambda) + \frac{1}{2}[1+\cos(\lambda-2)\pi] & \text{if } 1 \leq \lambda < 3 \\ \exp(1-\lambda) & \text{if } 3 \leq \lambda < 3.5 \\ 0 & \text{otherwise.} \end{cases} \quad (6.2\text{-}37)$$

The plot of this function is shown in Fig. 6.2-3.

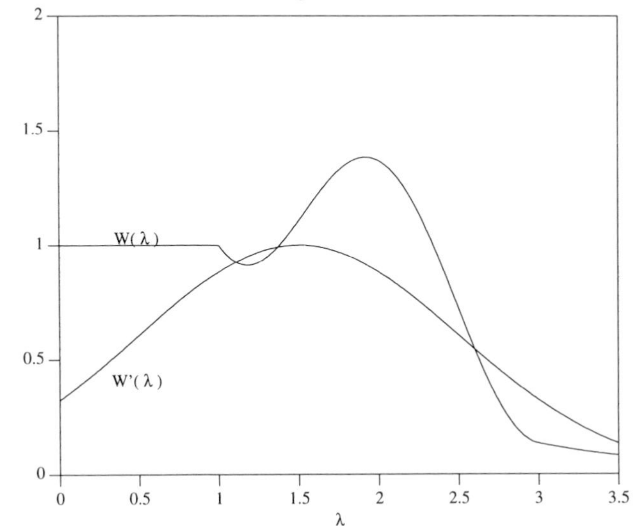

Figure 6.2-3: Spectra of fictitious color sources. $W(\lambda)$ produces a desired match. $W'(\lambda)$ does not.

The responses of the four sensors to this color source can be computed according to Eq. (6.2-1) as

$$\boldsymbol{\alpha}^* = (0.884552,\ 1.02886,\ 0.225009,\ 0.730532)^T. \quad (6.2\text{-}38)$$

Our objective then is to find a spectrum which will match this response in the settings of Problems 1, 2 and 3 accordingly.

Problem 1 Assume that we have a color source with the following spectral density

$$W'(\lambda) = \exp\left[-\frac{1}{2}(\lambda - \frac{3}{2})^2\right] \cdot I_\Lambda. \quad (6.2\text{-}39)$$

The plot of $W'(\lambda)$ is also shown in Fig. 6.2-3. The responses of the four sensors to this source is

$$\boldsymbol{\alpha} = (0.747151,\ 0.747677,\ 0.300817,\ 0.603613)^T. \quad (6.2\text{-}40)$$

Obviously this source fails the match. As mentioned before, our task here is to find the spectrum which has minimum deviation from the available source $W'(\lambda)$ and also matches the response in Eq. (6.2-38).

The algorithm defined in Eq. (6.2-15) has been implemented. The numerical solution after 46 iterations is plotted in Fig. 6.2-4.

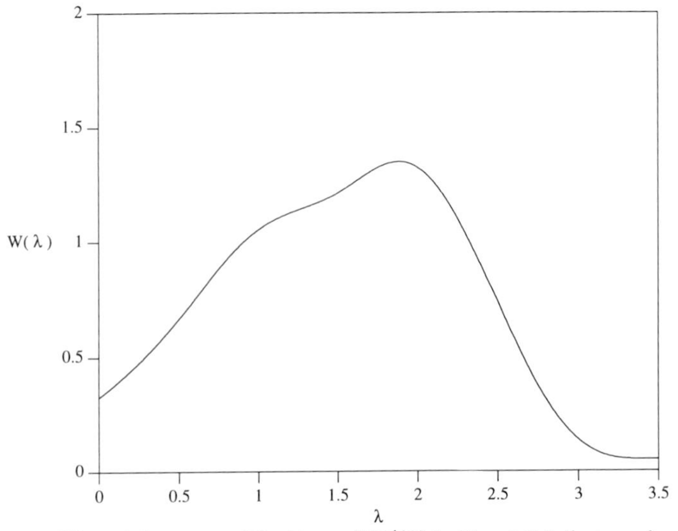

Figure 6.2-4: The minimum modifications of $W'(\lambda)$ in Fig. 6.2-3 that produces a match to $W(\lambda)$.

The responses of the four sensors to this solution are

$$\boldsymbol{\alpha}' = (0.884499,\ 1.02921,\ 0.225435,\ 0.730532)^T. \tag{6.2-41}$$

Of course, the accuracy of matching can be further improved by performing more iterations, as illustrated in the numerical realization of Problem 2 that follows.

Problem 2 Assume here that we have available six color sources as our basis colors. Their spectral densities are defined as

$$R_j(\lambda) = \text{tri}(\lambda - \frac{j}{2}) \tag{6.2-42}$$

for $j = 1, 2, ..., 6$, where tri$(\cdot)$ is a triangular-shaped function defined by

$$\text{tri}(x) = \begin{cases} 2x + 1 & \text{if } -\frac{1}{2} \leq x < 0 \\ 1 - 2x & \text{if } 0 \leq x \leq \frac{1}{2} \\ 0 & \text{otherwise.} \end{cases} \tag{6.2-43}$$

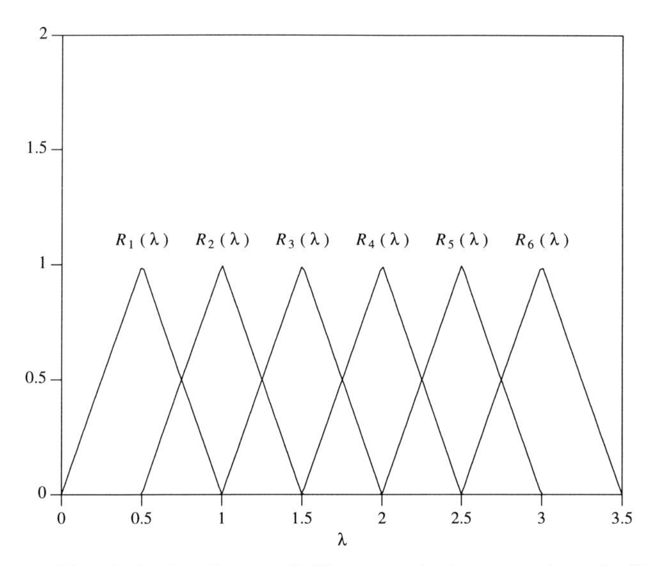

Figure 6.2-5: The six basis colors available to synthesize a spectrum in Problem 2.

All these functions are shown in Fig. 6.2-5.

Our objective here then is to find a suitable mixture of these basis colors that will match the required response in Eq. (6.2-38). The algorithm in Eq. (6.2-20) is implemented and the result after 1000 iterations (here more iterations are performed intentionally to illustrate that matching accuracy can be improved with more iterations, in contrast to the results of Problem 1) is

$$\boldsymbol{\beta} = (0.5300,\ 1.1715,\ 1.1655,\ 1.4243,\ 0.7177,\ 0.1074)^T. \tag{6.2-44}$$

The corresponding spectrum of this mixture can be computed by Eq. (6.2-16) and is shown in Fig. 6.2-6.

The response of the four sensors to this solution is

$$\boldsymbol{\alpha}' = (0.884551,\ 1.02886,\ 0.225014,\ 0.730532)^T. \tag{6.2-45}$$

The resulting spectrum almost matches exactly (to machine precision) the prescribed response in Eq. (6.2-38).

Problem 3 Assume that we have the same spectrum $W'(\lambda)$ available as in Problem 1 which fails the match. We have also the same basis colors available as in Problem 2. As discussed earlier, here we want to add the smallest (or at least the near-smallest) amount of basis colors to match the responses in Eq. (6.2-38).

To achieve this goal, the algorithm defined in Eq. (6.2-34) is implemented and tested for different values of E. As we said earlier, reducing the value of E reduces

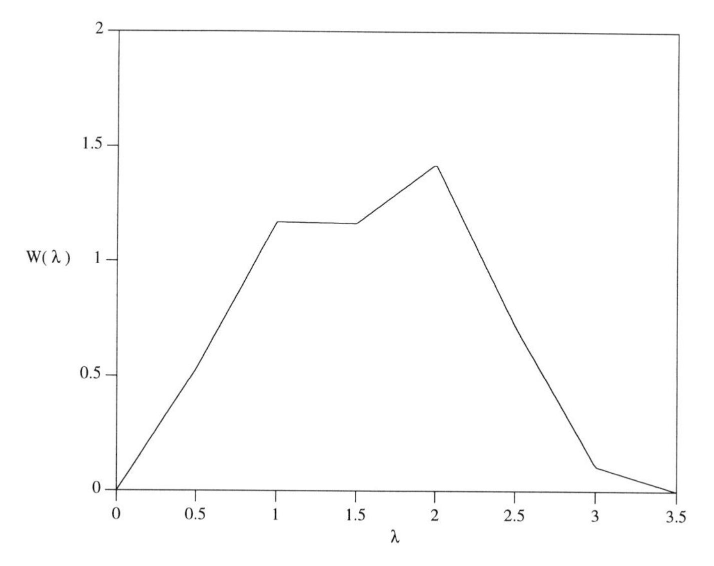

Figure 6.2-6: The spectrum synthesized from the basis colors in Fig. 6.2-5 that gives the correct match, i.e., matches the responses in Eq. (6.2-38).

the total volume of colors added to the existing spectrum to induce a match. The results of reducing E are listed in Table 6.2-1. We started with a value of $E = 3.5$ and achieved convergence very rapidly. As E was reduced from this value, convergence took longer, but the total volume of added colors diminished. Indeed for $E = 2.7$, color matching could be achieved without using $R_3(\lambda)$ and $R_6(\lambda)$. The corresponding spectral densities are shown in Figs. 6.2-7a) (for $E = 3.5$), 6.2-7b) (for $E = 3.0$) and 6.2-7c) (for $E = 2.7$). For E much less than 2.7, e.g., E =2.5, convergence is not possible indicating, therefore, that we are close to using the smallest volume of colors to achieve a match.

E	β_0	β_1	β_2	β_3	β_4	β_5	β_6	No. of iterations
3.5	0.4546	0.4460	0.8629	0.2604	1.2912	0.2940	0.	787
3.0	0.4845	0.4029	0.8842	0.1502	1.3134	0.2493	0.	1672
2.7	0.5198	0.1492	1.0134	0.	1.3430	0.1946	0.	7128
2.5	Non-convergence							——

Table 6.2-1: Attempting to match a color spectrum using the least volume of colors (Problem 3). Note that color $R_6(\lambda)$ is never needed. As the volume number E is decreased eventually $R_3(\lambda)$ is not needed. For E less than 2.7, matching is not possible.

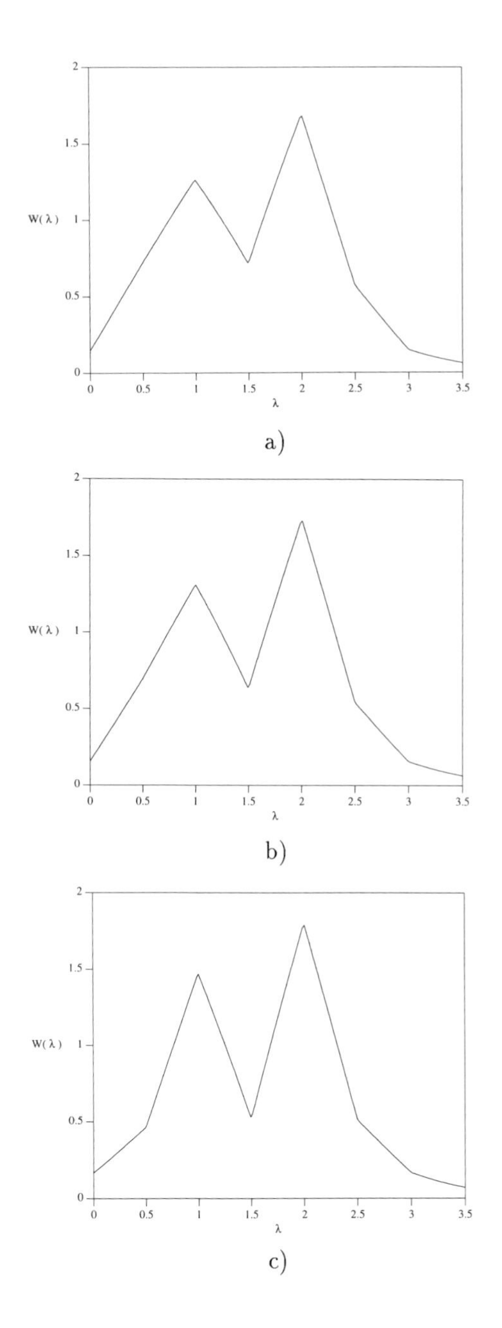

Figure 6.2-7: Modifying a spectrum using the colors in Fig. 6.2-5 to match the response in Eq. (6.2-38). a) $E = 3.5$; b) $E = 3.0$; c) $E = 2.7$.

B. Remarks

In Section 6.2 we showed how projection methods are useful in color matching under various constraint settings. In the first setting, the proposed algorithm converges to a spectrum which not only matches the measurement but also enjoys a minimum distance measure (which implies minimum adjustment physically) from an initial non-matching density. In the second setting, the proposed algorithm can find a suitable, i.e., a physically realizable ($\beta_j \geq 0$), mixture of the available color sources which will meet the measurement requirement. It is well known that standard matrix inversion methods do not guarantee physical realizability [6-8], [6-9]. In the last setting, our proposed algorithm can find a parsimonious mixture of available colors that, when added to a weighted non-matching spectrum, will match a given measurement requirement.

6.3 Resolution Enhancement

Resolution enhancement refers to the problem of recovering a high-resolution signal from low-resolution data. In image processing, the recovered high-resolution signal is usually an image and the low-resolution data is either a single blurred image, or a collection of blurred images, or even a sequence of numbers representing detector-readings from large-area-detector arrays. A great deal has been written on the subject of resolution enhancement. Several recent books are partially or wholly devoted to this subject [6-10]–[6-12]. One of the best examples of resolution enhancement, in which a set of low-resolution images is combined into a single high-resolution image, is side-looking radar, discussed by Goodman in [6-13].

In remote sensing and astronomy, images are often reconstructed from data obtained by fixed or scanning detector arrays superimposed upon the image field [6-14]. Since detectors are typically much larger than the blur spot of the imaging optics, a naive reconstruction of the image from the detector data would produce an image of lower resolution than that furnished by the imaging optics. The reduction of the detector size to match the blur spot of the imaging optics may not be technically feasible, and even if it were, the associated signal-to-noise ratio (SNR) of the detector output might be too low to be useful.

Frieden and Aumann [6-15] argued that, if one permits an overlap of successive scans and rescans of the same area from different directions, there might be enough information in the acquired data to permit a higher-resolution reconstruction, i.e., one more commensurate with the imaging optics. Frieden and Aumann illustrated such increased-resolution reconstructions by using a least-square algorithm that they called filtered localized projection.

In this section we review an approach, based on POCS, first described by Stark and Oskoui [6-16] and subsequently used by Wernick [6-17] to obtain high-resolution images in *positron-emission tomography* (PET). The high-resolution image recovery algorithm has the form of Eq. (6.1-6) where all the projectors P_i, $i = 1, ..., m$, project onto convex sets. Hence, provided that $C_o \triangleq \cap_{i=1}^{m} C_i$ is non-empty, convergence to a solution that satisfies all prior constraints is guaranteed. Exactly what these constraints, their associated sets, and their projectors are is discussed next.

A. Data Constraints

Consider a detector with a response function $\sigma(x, y)$ when it is centered at the origin and aligned with some convenient orthogonal coordinate system (x, y), for example, the rectangularly shaped detector of length a and width b in Fig. 6.3-1a). Now assume that the detector is superimposed upon the image $f(x, y)$, and let its response function when it is in position j be $\sigma_j(x, y)$. Then the detector output d_j is given by

$$d_j = \int_{-\infty}^{\infty} \int_{-\infty}^{\infty} f(x, y)\sigma_j(x, y)dxdy. \qquad (6.3\text{-}1)$$

The geometry is shown in Fig. 6.3-1b). The rectangular detector in Fig. 6.3-1 would, for example, have a response

$$\sigma_j(x, y) = \text{rect}\left[\frac{(x - x_j)\cos\theta_j + (y - y_j)\sin\theta_j}{a}\right] \times \text{rect}\left[\frac{(y - y_j)\cos\theta_j - (x - x_j)\sin\theta_j}{b}\right] \qquad (6.3\text{-}2)$$

when shifted to (x_j, y_j) and rotated, as shown, by an amount θ_j.

In the notation of Section 6.1, we rewrite Eq. (6.3-1) as $d_j = \langle f, \sigma_j \rangle$. The problem is then as follows: given readings $d_j, j = 1, 2, ..., K$, and prior constraints or properties π_{K+j} $(j = 1, 2, ..., N - K)$, how do we reconstruct an image whose resolution is higher than that which would result from a naive reconstruction from the detectors alone? N here is the total number of measurements and prior constraints.

In real-world situations, the processing is often done by computer. In this case the finite size of the pixels must be accounted for. One way to do this is to include in the detector response the fractional pixel areas included within the detector footprint. When d_j is obtained numerically, the detector output at the jth reading can be written as

$$d_j = \sum_l \sum_k f(l, k)\sigma_j(l, k), \qquad (6.3\text{-}3)$$

where $f(l, k)$ is the brightness of the pixel centered at (l, k) and $\sigma_j(l, k)$ is the detector response at position (l, k) and includes the fractional area of the pixel at (l, k) within the detector footprint when it is at the jth position. For example, suppose that the detector has a uniform, say, unity, response over its footprint;

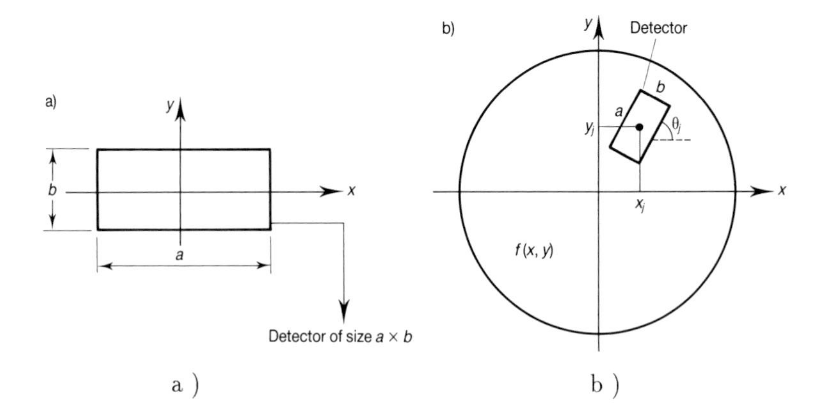

Figure 6.3-1: a) Footprint of a rectangular detector of length a and width b; b) footprint of the same detector rotated by an amount θ_j.

then

$$\sigma_j(l,k) = \begin{cases} 0, & \text{when pixel centered at } (l,k) \text{ lies wholly outside the footprint} \\ 1, & \text{when pixel centered at } (l,k) \text{ lies wholly within the footprint} \\ r_j, & \text{otherwise.} \end{cases} \tag{6.3-4}$$

In Eq. (6.3-4), r_j is the fractional area of the portion of the (l,k)th pixel that lies within the detector footprint. Equation (6.3-3) is recognized to be a discrete two-dimensional inner product. Therefore it, too, can be written as $d_j = \langle f, \sigma_j \rangle$.

We now define the sets C_j $(j = 1, ..., K)$ as

$$C_j = \{ f : \langle f, \sigma_j \rangle = d_j \}. \tag{6.3-5}$$

In words, C_j is the set of all image functions whose inner product with the detector function σ_j is d_j. We note that the K sets described in Eq. (6.3-5) are in effect K constraints on the image. Since f occurs linearly in the expression $\langle f, \sigma_j \rangle = d_j$, it is possible, in principle, to obtain enough equations, even with detectors larger than a pixel, so that a formal solution can be obtained by matrix inversion. For example, if the image has P^2 pixels and $K = P^2$ is the number of independent equations $\langle f, \sigma_j \rangle = d_j$ then, by matrix inversion of the P^2 linear equations, $f(x,y)$ can be found [6-18]. When P is large, the matrix inversion process is typically numerically unstable.

A close examination of the constraint set in Eq. (6.3-5) reveals that it has the same form as the inner product constraint of Eq. (6.2-6). Hence the set C_j is convex and closed and by analogy with Eq. (6.2-14) it has a projection given by

$$g(x,y) = P_j h = \begin{cases} h(x,y) & \text{if } \langle h, \sigma_j \rangle = d_j \\ h(x,y) + \frac{d_j - \langle h,\sigma_j \rangle}{\langle \sigma_j,\sigma_j \rangle}\sigma_j(x,y) & \text{otherwise.} \end{cases} \tag{6.3-6}$$

In Eq. (6.3-6), $h(x,y)$ is an arbitrary function in the $\mathcal{L}_2$ space of square-integrable functions, $\langle h, \sigma_j \rangle$ is an inner product, $g(x,y)$ is the projection of $h(x,y)$ onto C_j, P_j is the projector, and d_j is the measurement constraint and is assumed to be available. In POCS, Eq. (6.3-6) is used recursively. Thus if $f_n(x,y)$ represents the nth estimate of the correct but unknown image $f(x,y)$, then the improved estimate $f_{n+1}(x,y)$ would be obtained at the next step, from

$$f_{n+1}(x,y) = \begin{cases} f_n(x,y) & \text{if } \langle f_n, \sigma_j \rangle = d_j \\ f_n(x,y) + \frac{d_j - \langle f_n,\sigma_j \rangle}{\langle \sigma_j,\sigma_j \rangle}\sigma_j(x,y) & \text{otherwise.} \end{cases} \tag{6.3-7}$$

We note that $\langle \sigma_j, \sigma_j \rangle$ in Eq. (6.3-7) is not merely the area of the detector; it includes terms involving r_j^2 and therefore depends on the position of the detector array relative to the image-plane coordinates.

B. Prior Knowledge

In addition to the constraints imposed by the data, there are additional constraints that can be imposed from prior knowledge (all functions in this discussion are assumed to belong to L_2 and the usual L_2 norm and inner product are used throughout). The only prior knowledge constraint sets used in this discussion are as follows:

1. The amplitude constraint set C_A:

$$C_A = \{ g : \alpha \le g(x,y) \le \beta, \beta > \alpha \}. \tag{6.3-8}$$

The projection of an arbitrary function $h(x,y)$ onto C_A is

$$g = P_A h = \begin{cases} \alpha & \text{if } h(x,y) < \alpha \\ h(x,y) & \text{if } \alpha \le h(x,y) \le \beta \\ \beta & \text{if } h(x,y) > \beta. \end{cases} \tag{6.3-9}$$

2. The energy constraint set C_E:

$$C_E = \left\{ g : \|g\|^2 \le E \right\}, \tag{6.3-10}$$

where E is the maximum permissible energy in the reconstructed image.

The projection of the function $h(x, y)$ onto C_E is

$$g = P_E h = \begin{cases} h(x,y) & \text{if } \|h\|^2 \leq E \\ (E/E_h)^{1/2} h(x,y) & \text{if } \|h\|^2 > E, \end{cases} \tag{6.3-11}$$

where $E_h \triangleq \|h\|^2$.

3. The reference-image constraint set C_R:

$$C_R = \{ g : \|g - f_R\| \leq \epsilon_R \}, \tag{6.3-12}$$

where f_R is some previously known reference function, and ϵ_R is the permitted root mean square (rms) deviation from the reference and is known *a priori*. C_R is sometimes called the sphere constraint because the set includes all $\mathcal{L}_2$ functions that lie within a sphere of radius ϵ_R and centered at f_R. The projection onto C_R is [6-18]

$$g = P_R h = \begin{cases} h & \text{if } \|h - f_R\| \leq \epsilon_R \\ f_R + \epsilon_R \frac{h - f_R}{\|h - f_R\|} & \text{if } \|h - f_R\| > \epsilon_R. \end{cases} \tag{6.3-13}$$

Note that C_R is a generalization of C_E. By setting $f_R = 0, \epsilon_R = E^{1/2}$, and $E_h = \|h\|^2$, we obtain Eq. (6.3-11).

4. The bounded support-constraint set C_s:

$$C_s = \{ g : g(x, y) = 0 \text{ for } (x, y) \notin A \}. \tag{6.3-14}$$

In Eq. (6.3-14) A is some finite region in $R \times R$. The projection operator onto C_s is

$$g = P_s h = \begin{cases} h(x,y) & \text{if } (x,y) \in A \\ 0 & \text{otherwise.} \end{cases} \tag{6.3-15}$$

The derivations of all these projectors were given previously (see, e.g., [6-18], [6-19]). We do not repeat them here.

C. Numerical Results

As a numerical example, we attempt to reconstruct an image from low-resolution data gathered by sequentially rotating a low-resolution detector array, superimposed on the image field, to M equispaced angular positions (Fig. 6.3-2). If the image has P^2 pixels, and the detector array had D detectors then $MD \geq P^2$ for a unique recovery.

The image to be reconstructed is a 64×64 version of the widely used Shepp–Logan phantom [6-20] shown in Fig. 6.3-3a). Superimposed upon the image is a rectangular detector array consisting of 8 detectors along the horizontal and 16

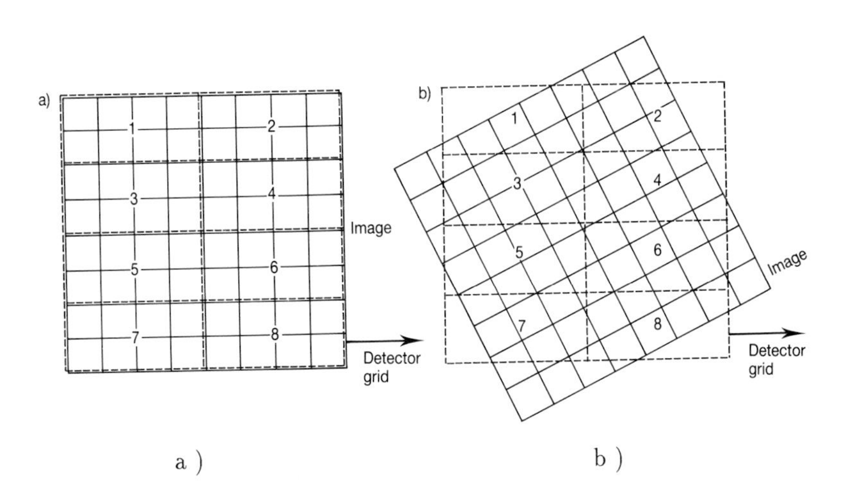

Figure 6.3-2: Obtaining additional information by rotating a low-resolution detector array on the image field: a) original position; b) rotated position.

detectors along the vertical. Each detector has dimensions of 8 pixels along the horizontal and 4 pixels in the vertical. As there are $64 \times 64 = 4096$ unknown brightnesses, a full data set would require a minimum of 32 angular displacements ($128 \times 32 = 4096$). Note that for each angular displacement we obtain approximately 128 equations.

The first experiment shows the feasibility of using POCS as an alternative to matrix inversion for reconstructing the image. Figure 6.3-3 shows the reconstruction of the Shepp–Logan phantom obtained by using POCS and only the data constraints given in Eq. (6.3-3) for $j = 1, 2, ..., 4096$; no prior knowledge about the image was assumed. Figures 6.3-3b), c), and d) show the reconstructions after 10, 50, and 100 iterations, respectively. The fine structure in the Shepp–Logan phantom becomes evident after 50 iterations and except for some noise, is reproduced exactly after 100 iterations. The algorithm used to get these results is given by

$$f_{n+1}(x,y) = P_K....P_1 f_n(x,y), \quad n = 0, 1, \cdots . \qquad (6.3\text{-}16)$$

The starting point for the recursion in Eq. (6.3-16) is the uniform ellipse shown in Fig. 6.3-4. In general the visual influence on the reconstruction of the starting point f_0 decreases with increasing number of iterations. In Fig. 6.3-5a) and b) the recursion of Eq. (6.3-4) is used with f_0 as in Fig 6.3-4 and with $f_0 = 0$, respectively. Both are clearly converging to the correct solution, but the reconstruction initialized by the more propitious f_0 is smoother than the reconstruction initialized by $f_0 = 0$. The rms error history vis-a-vis the original image for the two reconstructions starting with different initializations is shown in Fig. 6.3-6. Figure 6.3-7 shows the effect of limited-view reconstruction by POCS. Of a full angular range of 180° (divided into 32 view angles), we did not collect data in regions of angular diameters of 20°, 40°, 60°, and 90° in four separate experiments. Hence the numbers of available detector readings and the numbers of equations available for reconstruction are reduced from the full data set of 4096 to $(8 \times 16) \times 29 = 3712, (8 \times 16) \times 25 = 3200, (8 \times 16) \times 21 = 2688$, and $(8 \times 16) \times 15 = 1920$, respectively. To ameliorate the effect of insufficient data, we used prior knowledge associated with the sets C_A and C_R: the amplitude and reference-image constraint sets, respectively. In C_A we let $\alpha = 0$ and $\beta = 1$, and in C_R we let f_R be the image of Fig. 6.3-4, which we also picked as f_0; the parameter ϵ_R was set to $\epsilon_R = \|f_R - f_T\|$, where f_R is the true image shown in Fig. 6.3-3a). The results are shown in Fig. 6.3-7. As expected, the visibility decreases with increasing loss of data, but, even with 40° of missing view data, the fine structure in the Shepp–Logan reconstruction is still visible in Fig. 6.3-7b). The results should be contrasted with Fig. 6.3-8 which shows reconstructions that were attempted without using prior knowledge. In the latter, the fine structure is all but obliterated.

D. Remarks

In Section 6.3 we showed how to recover a high-resolution image from a sequence of low-resolution images using POCS. By using convex projections, we avoided matrix inversion of large matrices. Moreover, POCS enabled us to use readily available prior knowledge in a systematic fashion. A detailed discussion of this approach is furnished in [6-16] where comparison with another well-known method (least-squares) is furnished.[2] As stated earlier, Wernick successfully used this method to extract high-resolution images in PET, where the "wobble" notion of the PET machine conveniently furnished the extra data required to recover sub-detector-size impulse responses.

6.4 Generalized Projections

The last problem discussed in this chapter, blind deconvolution, is more difficult than the first two because the constraint sets are not convex. There are many important signal recovery problems for which constraints are non-convex. Recovering a signal from its magnitude, finding the antenna currents to produce a given far-field intensity, source recovery in correlation astronomy, are but a few of such problems. In this section we describe the method of generalized projections which is applicable for non-convex sets. When sets are non-convex (see Fig. 6.1-1b)) the

[2] Dr. Oskoui did much of the original research described here.

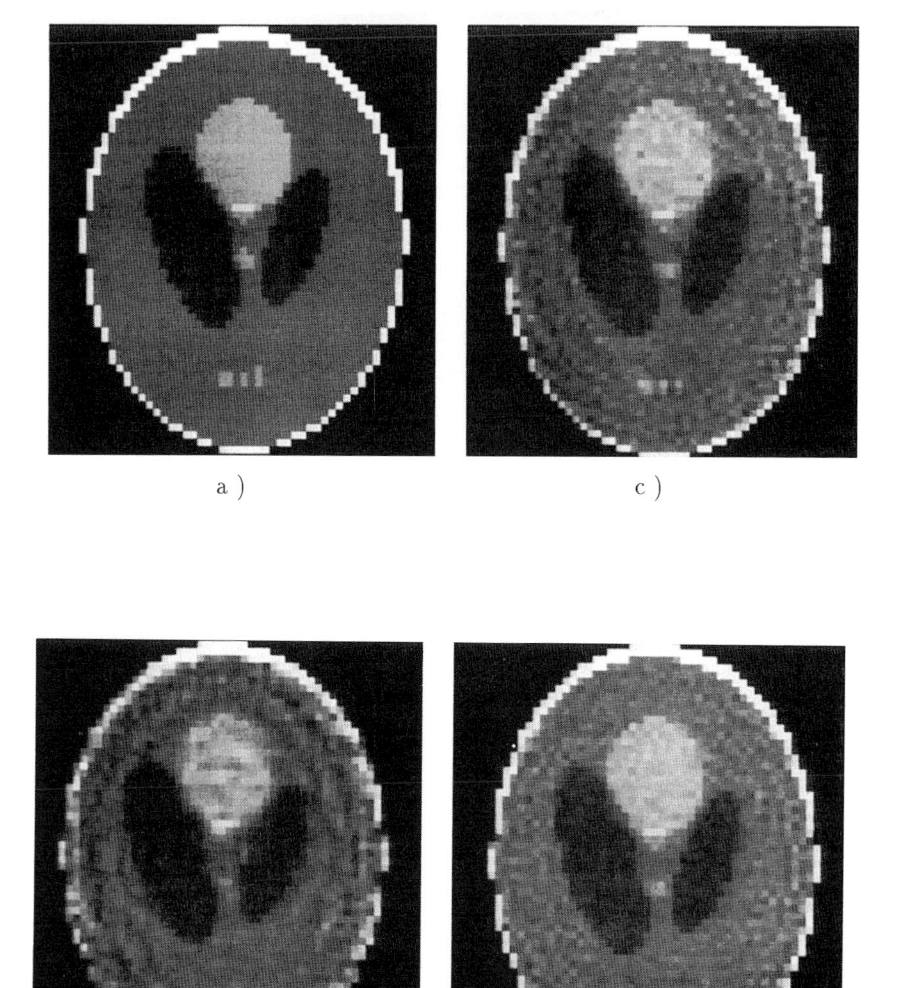

Figure 6.3-3: Reconstruction of the Shepp–Logan head phantom from low-resolution detectors using POCS: a) Original; b) after 10 iterations; c) after 50 iterations; d) after 100 iterations.

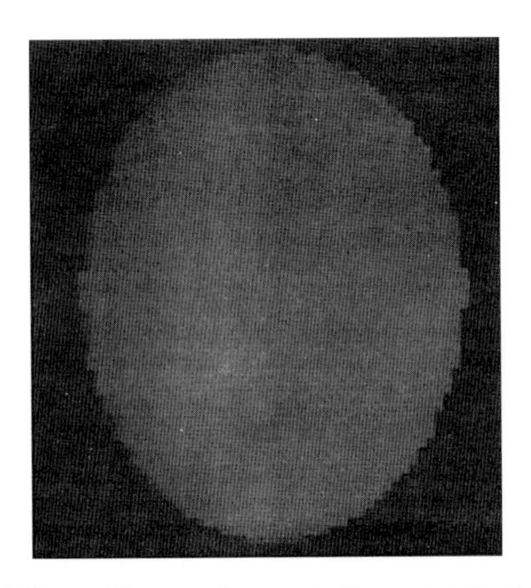

Figure 6.3-4: Uniform ellipse used as a starting point for POCS iteration.

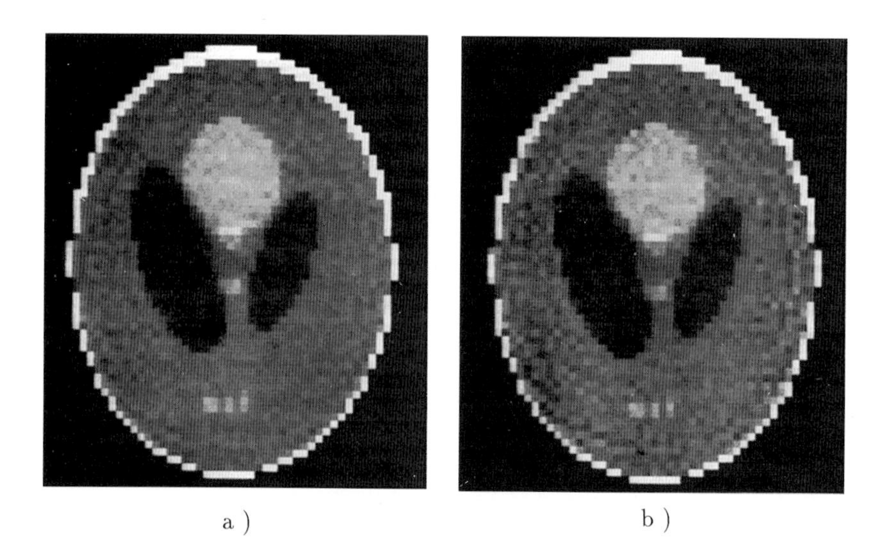

a) b)

Figure 6.3-5: Effect of starting point on reconstruction: a) using the uniform ellipse as a starting point; b) using the 0 image as a starting point.

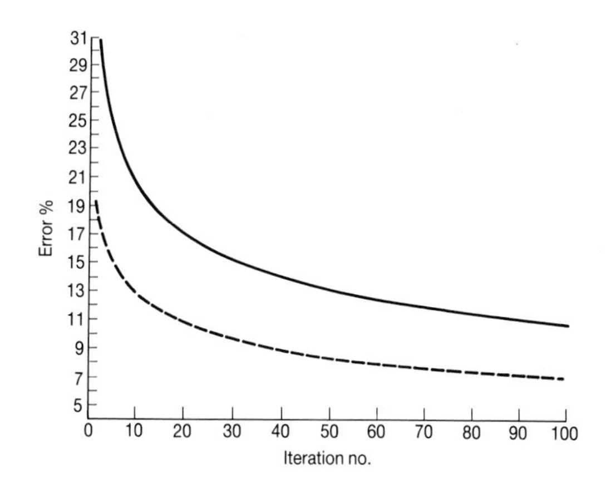

Figure 6.3-6: The error history (error versus number of iterations) for reconstructions using different starting points.

convergence properties enjoyed in the method of projections onto convex sets do not generally hold. Indeed in the case of non-convex sets many strange things can happen: there may be many points that satisfy the definition of a projection; projection operators may be expansive; convergence might take place, then again might not; if convergence does take place it might be to the wrong solution; and finally it is not even clear that projections always exist (i.e., there is no proof that they do). However, to be on the safe side we shall assume that all sets that appear in problems for which the method of generalized projections may be useful have at least one projection.

Thus in general, when sets are non-convex, it is unlikely that the algorithm

$$f_{n+1} = T_1 T_2 \ldots T_m f_n, \tag{6.4-1}$$

where $T_i = I + \mu_i(P_i - I)$, will converge to a point $f \in C_0 \triangleq \cap_{i=1}^m C_i$. However, in his Ph.D. thesis [6-21], Levi studied the non-convex case and showed that under certain circumstances, the iterative algorithm in Eq. (6.4-1) is useful in image recovery. We develop this idea in detail below.

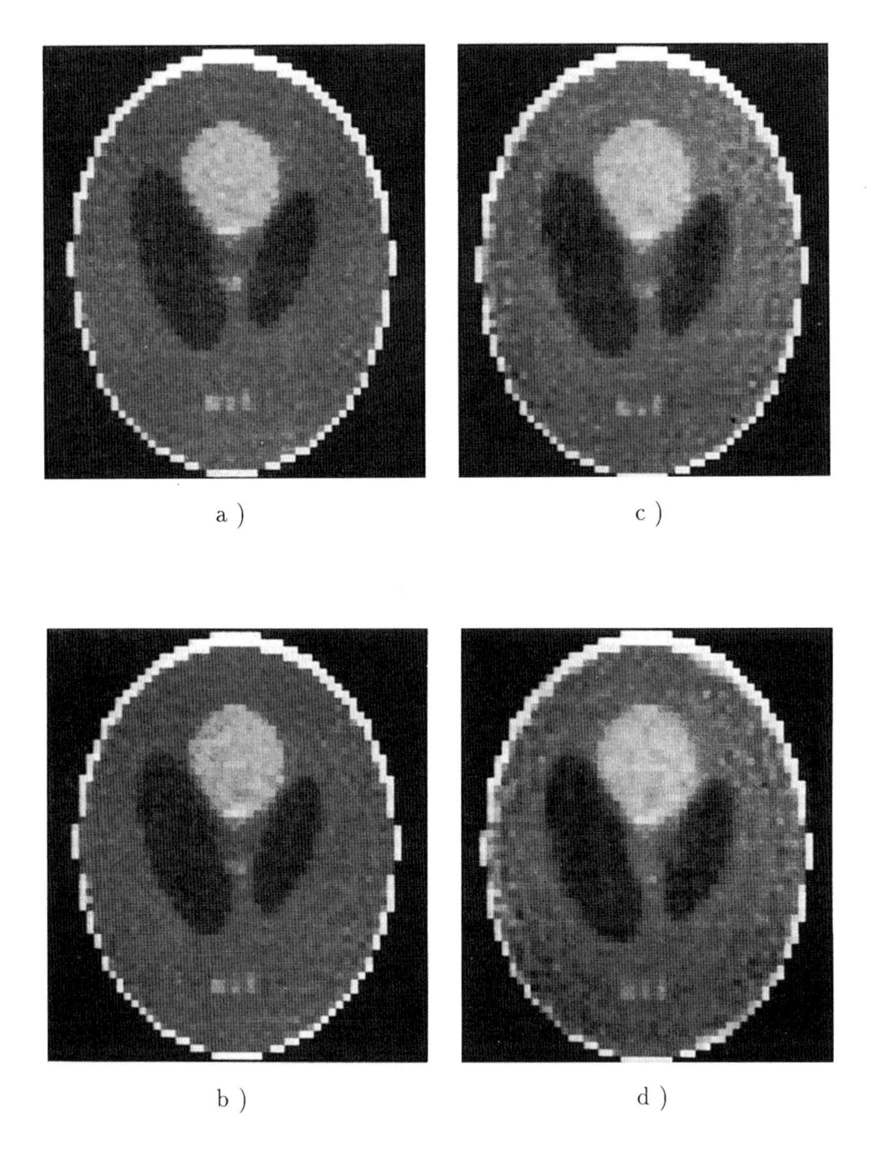

Figure 6.3-7: POCS reconstruction of the Shepp–Logan phantom in the limited view case: a) 20 degrees of missing data; b) 40 degrees of missing data; c) 60 degrees of missing data; d) 90 degrees of missing data. Prior knowledge about amplitude levels and nearness to a reference image were used.

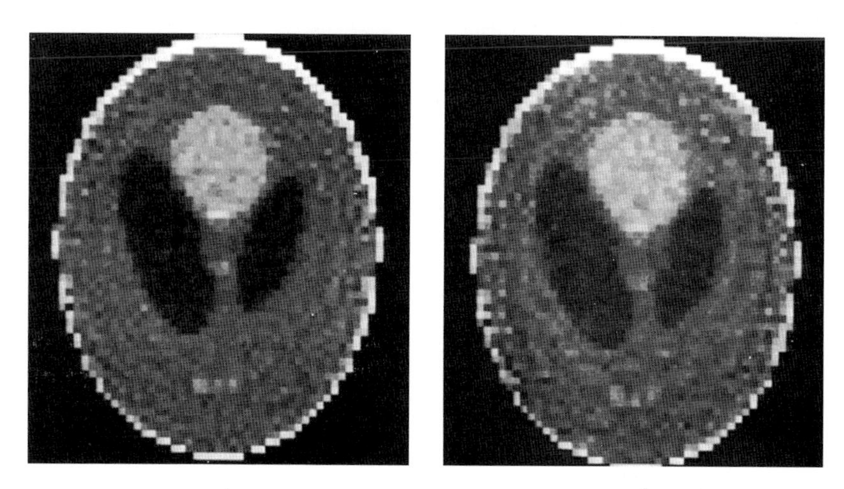

a) c)

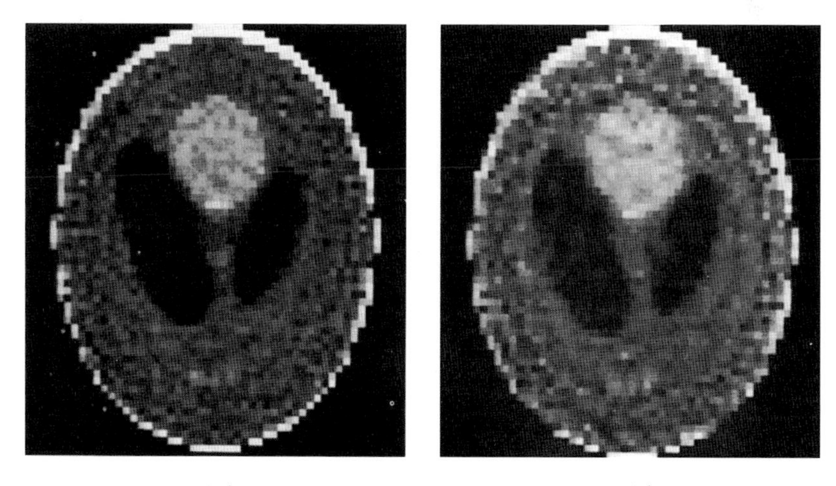

b) d)

Figure 6.3-8: POCS reconstruction as in Fig. 6.3-7 but no prior knowledge was used.

A. Fundamental Theorem of Generalized Projections

As in POCS, we define the distance of a point q from a set C, denoted by $d(q, C)$, as

$$d(q, C) = \min_{g \in C} \|g - q\|. \tag{6.4-2}$$

From this definition and the definition of a projection it is clear that

$$d(q, C) = \|Pq - q\|, \tag{6.4-3}$$

where P is the projector that projects onto C.

Definition 6.4-1 Let C_i for $i = 1, ..., m$ be any m sets with projectors $P_1, P_2, ..., P_m$ and relaxed projectors $T_1, T_2, ..., T_m$. Let f_n be the estimate of f at the nth iteration of the form

$$f_{n+1} = T_1 T_2 \ldots T_m f_n, \tag{6.4-4}$$

then the *summed distance error* (SDE) of f_n and the sets $C_i, i = 1, ..., m$, is defined as

$$J(f_n) = \sum_{i=1}^{m} \|P_i f_n - f_n\|. \tag{6.4-5}$$

Discussion Clearly $J(f_n)$ is the sum of the distances of f_n to each of the sets that the solution is known to belong to. Since $J(f_n) = 0$ if $f_n \in C_0$, and $J(f_n) \neq 0$ if $f_n \notin C_0$ the SDE can be used, unfortunately at some risk, as a measure of the performance of the algorithm in Eq. (6.4-4). Short of obtaining convergence, we would like to enjoy the property that

$$J(f_{n+1}) \leq J(f_n). \tag{6.4-6}$$

Unfortunately Eq. (6.4-6) does not hold in general; however under some circumstances it does hold and this is stated in the following fundamental theorem of generalized projections.

Theorem 6.4-1 (Fundamental Theorem of Generalized Projections) For $m = 2$, the recursion in Eq. (6.4-4) has the property that

$$J(f_{n+1}) \leq J(T_2 f_n) \leq J(f_n) \tag{6.4-7}$$

for every μ_1 and μ_2 that satisfy

$$0 \leq \mu_i \leq \frac{A_i^2 + A_i}{A_i^2 + A_i - \frac{1}{2}(A_i + B_i)}, \; i = 1, 2 \tag{6.4-8}$$

where

$$A_1 \triangleq \frac{\|P_1T_2f_n - T_2f_n\|}{\|P_2T_2f_n - T_2f_n\|} \tag{6.4-9}$$

$$A_2 \triangleq \frac{\|P_2f_n - f_n\|}{\|P_1f_n - f_n\|} \tag{6.4-10}$$

$$B_1 \triangleq \frac{\langle P_2T_2f_n - T_2f_n, \; P_1T_2f_n - T_2f_n \rangle}{\|P_2T_2f_n - T_2f_n\|^2} \tag{6.4-11}$$

$$B_2 \triangleq \frac{\langle P_1f_n - f_n, \; P_2f_n - f_n \rangle}{\|P_1f_n - f_n\|^2}. \tag{6.4-12}$$

∎

Some remarks are in order:

1. It is easily shown that $A_i \leq B_i$ and hence
$$A_i \geq \frac{1}{2}(A_i + B_i). \tag{6.4-13}$$

2. The value of $\mu_i = 1$ is always in the range given in Eq. (6.4-8). Hence the following iteration
$$f_{n+1} = P_1P_2f_n \tag{6.4-14}$$
has the set distance reduction property

3. If $A_i \to \infty$, the range on μ_i becomes $0 \leq \mu_i \leq 1$

4. The SDE property in Eq. (6.4-7) does not hold in general, for $m > 2$.

Examples illustrating remark 4 are easy to construct [6-22]. The proof of the Fundamental Theorem is furnished in [6-21].

B. Practical Aspects of the Generalized Projection Algorithm

A first point that arises is: if Eq. (6.4-4) does not posses the SDE decreasing property for $m > 2$, is it not overly restrictive in practice? The answer is no because the Fundamental Theorem does not restrict the complexity of the sets and therefore C_1 and C_2 can indeed include signals with multiple constraints. For example, the two sets

$$C_A \triangleq \{\, g : \; g(x) = 0 \;\text{ for }\; x \notin S_A \,\} \tag{6.4-15}$$

$$C_B \triangleq \{\, g : |\, g(x)\,| = P(x) \text{ for } x \in S_B \,\} \tag{6.4-16}$$

with projectors (here q is an arbitrary element in $\mathcal{L}_2$)

$$P_A q = \begin{cases} q, & x \in S_A \\ 0, & x \notin S_A \end{cases} \tag{6.4-17}$$

$$P_B q = \begin{cases} P(x), & q(x) > 0 \text{ and } x \in S_B \\ -P(x), & q(x) \leq 0 \text{ and } x \in S_B \\ q(x), & x \notin S_B \end{cases} \tag{6.4-18}$$

can be combined into the single set

$$C_1 = \left\{ g(x) : \begin{array}{l} g(x) = 0 \text{ for } x \notin S_A \text{ and} \\ \mid g(x) \mid = P(x), \text{ for } x \in S_A \cap S_B \end{array} \right\} \tag{6.4-19}$$

with projector

$$P_1 q = \begin{cases} 0 \;, & x \notin S_A \\ P(x), & q(x) > 0, x \in S_A \cap S_B \\ -P(x), & q(x) \leq 0, x \in S_A \cap S_B \\ q(x), & \text{otherwise.} \end{cases} \tag{6.4-20}$$

Traps and Tunnels

Algorithms involving non-convex sets are subject to pathological behavior called *traps* and *tunnels*. Traps never occur when only convex sets are involved but tunnels can occur. We define a trap as a fixed point of the composition $T_1, T_2, ..., T_M$ which is not a fixed point of one or more of the T_i's. When the algorithm is in a trap, $f_{n+1} = f_n$ but it is not true that $T_i f_n = f_n$ for all i. Thus for a trap

$$J(f_{n+1}) = J(f_n) \neq 0 \tag{6.4-21}$$

while for a solution $J(f_{n+1}) = J(f_n) = 0$. This property of traps and the solution enables one to determine when one is in a trap or a solution.

A tunnel is a region where

$$f_{n+1} = T_1 T_2 ... T_m f_n \simeq f_n \tag{6.4-22}$$

i.e., where the change from one iteration to the next is negligible, yet f_n is not a point in the solution set. Tunnels represent set morphologies in which the surfaces of two or more sets are nearly tangent. Tunnels are observed in POCS algorithms also. Traps are shown in Fig. 6.1-2b).

There are several other interesting properties of generalized projections which are discussed in [6-22]. However, the material in this section will help the reader understand why the method of generalized projections (traps notwithstanding) is a reasonable choice for solving the blind deconvolution problem discussed next.

6.5 Projection Based Blind Deconvolution

The blind deconvolution problem refers to a class of problems of the form

$$g(\mathbf{x}) = \int f(\mathbf{x}-\mathbf{y})h(\mathbf{y})d\mathbf{y} + n(\mathbf{x}), \tag{6.5-1}$$

where $g(\mathbf{x})$ is the observed data, $f(\mathbf{x})$ is the unknown source signal, $h(\mathbf{x})$ is the unknown blurring function (impulse response) of the system and $n(\mathbf{x})$ is noise. Recovering both $f(\mathbf{x})$ and $h(\mathbf{x})$ from $g(\mathbf{x})$ alone is a difficult problem.

With or without noise, Eq. (6.5-1) admits an infinite number of solutions if no other constraints are available for $f(\mathbf{x})$ and $h(\mathbf{x})$. However, once constraints are imposed, the set of feasible solutions to Eq. (6.5-1) can be greatly reduced. A number of algorithms have been proposed to solve the blind deconvolution problem. A commonly used technique is described in [6-23]. It is an iterative algorithm according to which the constraints based on *a priori* knowledge about both $f(\mathbf{x})$ and $h(\mathbf{x})$ are enforced. A difficulty with this algorithm is the lack of rigorous justification of its convergence, making it difficult to determine when to stop the iterations. Indeed Lane in [6-24] argues that the general iterative deconvolution loop of Ayers and Dainty [6-23] can diverge in the sense that new estimates can be both visually and computationally worse than the previous ones. In [6-25], an error function was minimized subject to a positivity constraint to obtain a solution by a tedious numerical process called *simulated annealing*. However for a large-scale problem this approach is not numerically efficient. In [6-26], a steepest decent algorithm is used to minimize a cost function directly. However, due to the non-convexity of the cost function used, avoiding suboptimal solutions is a problem. In his paper [6-24], Lane proposed an iterative algorithm based on conjugate gradient minimization in which he incorporated features that yield the stable convergence properties of Fienup's error reduction algorithm [6-27]. In this algorithm the constraints are incorporated into an error function and the problem is reformulated as an unconstrained minimization problem. Since the constraints are not enforced directly, it is not guaranteed that the convergent point will satisfy them. Furthermore, this algorithm is not scale-invariant. In other words, the final solution of this iterative algorithm is sensitive to scaling by a constant factor in a non-trivial fashion.

A. The Generalized Projection Based Algorithm

For the noiseless case, Eq. (6.5-1) becomes

$$g(\mathbf{x}) = \int f(\mathbf{x}-\mathbf{y})h(\mathbf{y})d\mathbf{y}. \tag{6.5-2}$$

In this problem, we assume that both $f(\mathbf{x})$ and $h(\mathbf{x})$ in Eq. (6.5-2) are of finite energy, in other words, both are square integrable functions. Let $\mathcal{L}_2$ be the space of square integrable functions. The space that we are working with is $\mathcal{H} = \{ (f,h), f \in \mathcal{L}_2 \text{ and } g \in \mathcal{L}_2 \}$ which carries a norm

$$\|(f,h)\| \triangleq \left[\int | f(\mathbf{x}) |^2 \, d\mathbf{x} + \int | h(\mathbf{x}) |^2 \, d\mathbf{x} \right]^{\frac{1}{2}} . \tag{6.5-3}$$

Define the set

$$C_g \triangleq \{ (u,v) \ : \ u * v = g \} . \tag{6.5-4}$$

as the set of all solution pairs, i.e., the set of all function pairs $u(\mathbf{x})$ and $v(\mathbf{x})$ such that

$$g(\mathbf{x}) = \int u(\mathbf{x} - \mathbf{y}) v(\mathbf{x}) d\mathbf{y}. \tag{6.5-5}$$

The symbol $*$ denotes convolution. We reserve the symbols f and h for the *true* source and *true* point source response; clearly, the pair $(f,h) \in C_g$. Even to readers familiar with the theory of alternating projections, the object defined in Eq. (6.5-4) may seem strange. It differs from other constraint sets in that its elements are two-tuples, i.e., function pairs; thus the constraint is simultaneously imposed on two functions. As is to be expected, the projection onto C_g is considerably more difficult to compute than that in the vast majority of ordinary constraint sets, i.e., those involving only a single function. In general, the set C_g contains infinitely many elements each of which is a feasible solution to the blind deconvolution problem if no further information is provided about either the signal $f(\mathbf{x})$ or the system $h(\mathbf{x})$ or both. Thus, we must use as much prior knowledge about f and h as possible in order to exclude spurious solutions in the set C_g and obtain a meaningful solution. Let C_f, C_h denote the constraint sets based on the prior knowledge about f and h, respectively, i.e.,

$$C_f \triangleq \{(u,v) \ : \ u \text{ satisfying prescribed prior knowledge about } f\} \tag{6.5-6}$$

$$C_h \triangleq \{(u,v) \ : v \text{ satisfying prescribed prior knowledge about } h\} . \tag{6.5-7}$$

Then an element in the set $C_0 \triangleq C_g \cap C_f \cap C_h$ will be a solution to Eq. (6.5-2) which also satisfies all the available prior knowledge. We expect the set C_0 to contain fewer elements than the set C_g, because its elements are constrained by physical reality. The specific definitions of the set C_f and set C_h are based on available prior knowledge, and they are problem dependent. Usually they are determined by our understanding of the problem. Examples of such sets are:

1. Sets based on the support region of the signal:

$$\begin{aligned} C_f &\triangleq \{ (u,v) : u(\mathbf{x}) = 0, \forall \mathbf{x} \notin \Omega_f \} \\ C_h &\triangleq \{ (u,v) : v(\mathbf{x}) = 0, \forall \mathbf{x} \notin \Omega_h \} , \end{aligned}$$

   where Ω_f, Ω_h denote the support of f and h, respectively.

2. Sets based on the intensity range. For example,

$$C_f \triangleq \{ (u,v) : u(\mathbf{x}) \in [f_{\min}, f_{\max}] \}, \tag{6.5-8}$$

where $f_{\min}, f_{\max}$ are determined by the specific problem.

3. Sets based on the physical properties of the system. For example, if the system is band-limited, then a constraint set can be defined as

$$C_h \triangleq \{ (u,v) : v \leftrightarrow V(\omega) = 0, \forall \omega \in \Omega_H \}, \tag{6.5-9}$$

where Ω_H is the support of the Fourier transform of h.

4. Sets based on experimental measurements. For example, if we have some gross measurement $\tilde{h}(\mathbf{x})$ about the system impulse response function $h(\mathbf{x})$, then we have

$$C_h \triangleq \left\{ (u,v) : \|v - \tilde{h}\| \leq E \right\}, \tag{6.5-10}$$

where E is some constant which can be determined by the measurement accuracy.

A brief remark is in order; the sets described in 1. through 4. have elements which are two-tuples but the constraint is applied to only one member of the function pair. In this sense, these sets are much more like the constraint sets used in alternating projections than the set in Eq. (6.5-5).

Let P_g, P_f and P_h denote the projection operators onto the sets C_g, C_f and C_h, respectively; then generalized projection theory guarantees that the recursion

$$(f,h)_{(k+1)} = P_h P_f P_g (f,h)_k; \; k = 0,1,2,... \tag{6.5-11}$$

where $(f,h)_k$ is the estimated recovery of (f,h) at the kth iteration and $(f,h)_0$ is our initial estimate, i.e., the starting point, will converge to a point in C_0 provided that the algorithm does not stagnate at traps [6-21]. Note: in [6-21] Levi showed that an algorithm like the one in Eq. (6.5-11), i.e., involving non-convex sets, has the summed-distance error convergence property if the number of constraint sets involved is at most two. However, as the reader can see, Eq. (6.5-11) involves *three* sets C_h, C_f and C_g. The statement surrounding Eq. (6.5-11) is still true because it is readily shown that if $C_h \cap C_f \triangleq C_{hf}$, then $P_{hf} = P_h P_f$ and thus, despite its appearance, Eq. (6.5-11) can be thought of as involving only two sets.

The projections P_f and P_h corresponding to the sets defined above are well known [6-18]. The projector P_g onto the set C_g can be derived as follows: consider any two-tuple element (f', h') in $\mathcal{H}$ but not in C_g; its projection onto C_g, denoted by the two-tuple $(\tilde{f}, \tilde{h})$, is solved by minimizing the quantity $\|(u,v) - (f',h')\|$ under the constraint that $(u,v) \in C_g$. This can be done by using the Lagrange multiplier technique. The solution is obtained by the two-step procedure:

1. Solve for $\tilde{F}(\omega) = \mathcal{F}[\tilde{f}(\mathbf{x})]$ from

$$| \tilde{F} |^4 - | \tilde{F} |^2 \tilde{F} F' + H' G^* \tilde{F} = | G |^2, \qquad (6.5\text{-}12)$$

where $\mathcal{F}$ is the Fourier transform operator, $F' = \mathcal{F}[f']$, $H' = \mathcal{F}[h']$, $G = \mathcal{F}[g]$ and G^* denotes the complex conjugate of G.

2. Solve for $\tilde{H}(\omega) = \mathcal{F}[\tilde{h}(\mathbf{x})]$ from

$$\tilde{H}(\omega) = G(\omega)/\tilde{F}(\omega). \qquad (6.5\text{-}13)$$

The projection is found by solving the implicit nonlinear equation (6.5-12). This can be done numerically although there might be more than one solution to Eq. (6.5-12) which is due to the fact that set C_g is non-convex. Another difficulty in finding the projection is that in the region where $\tilde{F}(\omega)$ is small, numerical inaccuracies pose a problem in the computation of $\tilde{H}(\omega)$ in Eq. (6.5-13).

B. Modifications of the Pure Projection Algorithm

In addition to the numerical difficulties encountered in solving Eqs. (6.5-12), traps present a danger of obtaining false solutions. Therefore, the generalized projection algorithm presented above may not be very practical to implement. Instead, consider the functional

$$J_{(u,v)} \triangleq \|g - u * v\|^2 = \int [g(\mathbf{x}) - (u * v)(\mathbf{x})]^2 d\mathbf{x} \geq 0. \qquad (6.5\text{-}14)$$

Any element (u, v) in the set C_g is the point where $J_{(u,v)}$ assumes its global minimum. The algorithm in Eq. (6.5-11) will converge to a feasible solution or stagnate at a trap (a morphological local minimum). At a feasible solution, the known constraints imposed by the sets C_f and C_h are satisfied and $J_{(u,v)} = 0$, since the solution has membership in C_g. However, to avoid problems associated with solving Eqs. (6.5-12) and (6.5-13) we propose an alternative algorithm which, while still subject to erroneous solutions associated with local minima of J, is significantly easier to implement. The algorithm involves the following steps:

1. Take an educated initial guess of (f, h) and set it as (f_0, h_0).

2. Solve for $f_{(k+1)} = \arg_{f \in C_f} \left\{ \min J_{(f,h_k)} \right\}$.

3. Solve for $h_{(k+1)} = \arg_{h \in C_h} \left\{ \min J_{(f_{(k+1)},h)} \right\}$.

4. Set $k = k + 1$; repeat step 2 and step 3 till convergence is achieved.

Theorem 6.5-1 The iteration defined above will converge to a local or global minimum of the functional $J_{(u,v)}$.

Proof: Note that, from steps 2 and 3, we have

$$J_{(f_{(k+1)},h_{(k+1)})} \leq J_{(f_{(k+1)},h_k)} \leq J_{(f_k,h_k)} \tag{6.5-15}$$

and that $J_{(u,v)}$ is non-negative for all $(u,v) \in \mathcal{H}$. Hence the non-increasing sequence either converges to zero or to a local minimum. ∎

Remarks

1. If $J_{(f_k,h_k)}$ converges to 0, then a global minimum is found and the resulting two-tuple (f,h) will also be a stationary point of the earlier generalized projection based algorithm in Eq. (6.5-11).
2. If more accurate information is available about f than about h, we can interchange the role of f with h in the algorithm.
3. In some earlier work [6-25], [6-26], a functional of similar nature is minimized to yield the solution. Due to the fact that the $J_{(u,v)}$ is a non-convex function of (u,v), the proposed minimization algorithms are not efficient numerically.

The functional $J_{(u,v)}$ is a quadratic function with respect to u if v is fixed and vice versa. Therefore the minimization in step 2 and step 3 can be done by a standard gradient projection algorithm. In step 2, the following iteration will yield the global minimum: for the kth update

$$f_k^{l+1} = P_f(f_k^l - \alpha_k \bigtriangledown J_{(f_k^l,h_k)}), \quad l = 0,1,2,\cdots \tag{6.5-16}$$

where $\bigtriangledown$ is the gradient operator, P_f is the projection operator onto C_f, $f_k^0 = f_{(k-1)}$ and α_k is a constant chosen to guarantee convergence [6-27]. In Eq. (6.5-16) f_k^l is the estimate of f during the kth cycle (held fixed as l varies) at the lth iteration. Equation (6.5-16) needs some discussion. First observe that Eq. (6.5-16) is a contraction mapping since P_f is non-expansive and for the proper value of α_k, the operation in parentheses is a contraction. Also the numerical realization of the functional $J_{(h,f)} = \|g - f * h\|^2$ when h is held fixed at, say, the vector value $\mathbf{h}_k$ becomes $J = \|\mathbf{g} - \mathbf{H_k} \cdot \mathbf{f}\|^2$ where $\mathbf{g}$ and $\mathbf{f}$ are vectors, and $\mathbf{H}_k$ is a circulant/Toeplitz matrix of the shifted $\mathbf{h}_k$'s. The gradient of J is then easily computed to be

$$\bigtriangledown J = 2\mathbf{H_k}^T\mathbf{H_k}\mathbf{f} \; - \; 2\mathbf{H_k}^T\mathbf{g}, \tag{6.5-17}$$

where T denotes the transpose. If λ_k denotes the largest eigenvalue of the matrix $\mathbf{H_k^T H_k}$ then it is well-known that Eq. (6.5-16) will lead to convergence, i.e., be a contraction mapping for $0 < \alpha_k < 1/\lambda_k$.

Another remark about Eq. (6.5-16) is relevant. We note that Eq. (6.5-16) interrupts the gradient search at each iteration by projecting onto the constraint set C_f. Why not let the gradient portion of the algorithm iterate until it finds an unconstrained global minimum and *then apply* the constraint? Logical as this procedure sounds, it could lead to the wrong answer as the following simple, but non-trivial example shows. Let the functional to be minimized be given by

$$J(x,y) = x^2 + \alpha y^2, \quad \alpha > 0 \tag{6.5-18}$$

subject to the constraint described by the set

$$C = \{ (x,y) : x \geq 0, y \geq 0, x+y = 1 \} . \tag{6.5-19}$$

If we run the gradient algorithm first we obtain a minimum at $x = y = 0$. Subsequent projection onto C yields as constrained minimum the point $x = y = \frac{1}{2}$ (no α dependence). However, if we apply the constraint *before* carrying out the gradient descent we obtain as the constrained minimum the point $x = \frac{\alpha}{1+\alpha}, y = \frac{1}{1+\alpha}$ which is the correct solution. Thus, what seems to be logical at the first glance turns out to be incorrect upon reflection.

Step 3 can be done in a similar fashion:

$$h_k^{l+1} = P_h(h_k^l - \beta_k \nabla J_{(f_{(k+1)},h_k^l)}), \tag{6.5-20}$$

where $h_0 = h_{(k-1)}$, and P_h is the projector onto C_h. The iterations in Eqs. (6.5-16) and (6.5-20) converges linearly [6-26]. The convergence becomes very slow when the solutions are close to the true minimum. In order to avoid this kind of behavior, we propose an algorithm of which a block diagram is given in Fig. 6.5-1.

In Fig. 6.5-1, M is a prescribed number, for example 10 or 20, and ϵ is a prescribed constant serving as the convergence criterion. From the proof of Theorem 6.5-1 above, the convergence of the algorithm presented in Fig. 6.5-1 is guaranteed.

C. Numerical Example

To illustrate how well the projection based blind deconvolution algorithm works, we applied it to the following example. The original source signal $f(\mathbf{x})$ is a 64×64 pixel-square image of a triple star modeled by three circles as shown in Fig. 6.5-2a).

The blurring function is modeled by a truncated Gaussian shaped function:

$$h(x_1,x_2) = \begin{cases} \frac{1}{\sqrt{2\pi}\sigma} \exp(-\frac{x_1^2+x_2^2}{2\sigma^2}) & \text{for } x_1^2+x_2^2 \leq 4\sigma^2 \\ 0 & \text{otherwise,} \end{cases} \tag{6.5-21}$$

where σ is chosen to be 7 in this example. The blurring function $h(\mathbf{x})$ is shown in Fig. 6.5-2b), and the blurred image $g(\mathbf{x})$ is shown in Fig. 6.5-2c). To reconstruct $f(\mathbf{x})$ from the data $g(\mathbf{x})$, the following constraints are used:

$$C_f = \{ (u,v) : u(\mathbf{x}) \text{ is non-negative for } \mathbf{x} \in W_f \text{ and } 0 \text{ otherwise} \}, \tag{6.5-22}$$

where W_f is a square region centered at (0, 0) with dimension 33×33, as shown in Fig. 6.5-3a). In other words, C_f implies that the original image is spatially limited.

Similarly, for the blurring function we have

$$C_h = \{ (u, v) : v(\mathbf{x}) = 0, \forall \mathbf{x} \notin W_h \} \qquad (6.5\text{-}23)$$

where W_h is a square region centered at (0,0) with dimension 29×29, as shown in Fig. 6.5-3b).

In this example, an initial guess for f_0 is taken as

$$f_0(\mathbf{x}) = \begin{cases} g(\mathbf{x}) & \text{if } \mathbf{x} \in W_f \\ 0 & \text{otherwise,} \end{cases} \qquad (6.5\text{-}24)$$

and 0 is used for h_0.

In practice, the exact support of the original image and the blurring function is rarely known and therefore it would be unfair to assume perfect knowledge of these in evaluating the algorithm. Nevertheless, when both the signal and the blurring function are positive, the sum of the spatial supports of the signal and blurring function is equal to the spatial support of the observed data. In this example, the proposed algorithm was tested using three different spatial support constraints:

Case 1. W_f: 33×33 and W_h: 29×29, i.e., both W_f and W_h are the smallest squares that contain f and h, respectively.

Case 2. W_f: 37×37 and W_h: 25×25, i.e. W_f is overestimated, while W_h is underestimated.

Case 3. W_f: 29×29 and W_h 33×33, i.e. W_f is underestimated, while W_h is overestimated. The reconstructed images of the source and blur functions are shown, respectively, for cases 1–3 above in Figs. 6.5-4a) and b), c) and d), and e) and f).

The results show the projection based blind deconvolution algorithm can be highly effective in recovering the source and blurring functions in blind deconvolution. Indeed, the deconvolved results furnished show significant improvement over the raw data, even when the prior knowledge contained errors. A more extensive study of this approach including a comparative study with the method of conjugate gradients, described by Lane in [6-24] is furnished in [6-28].

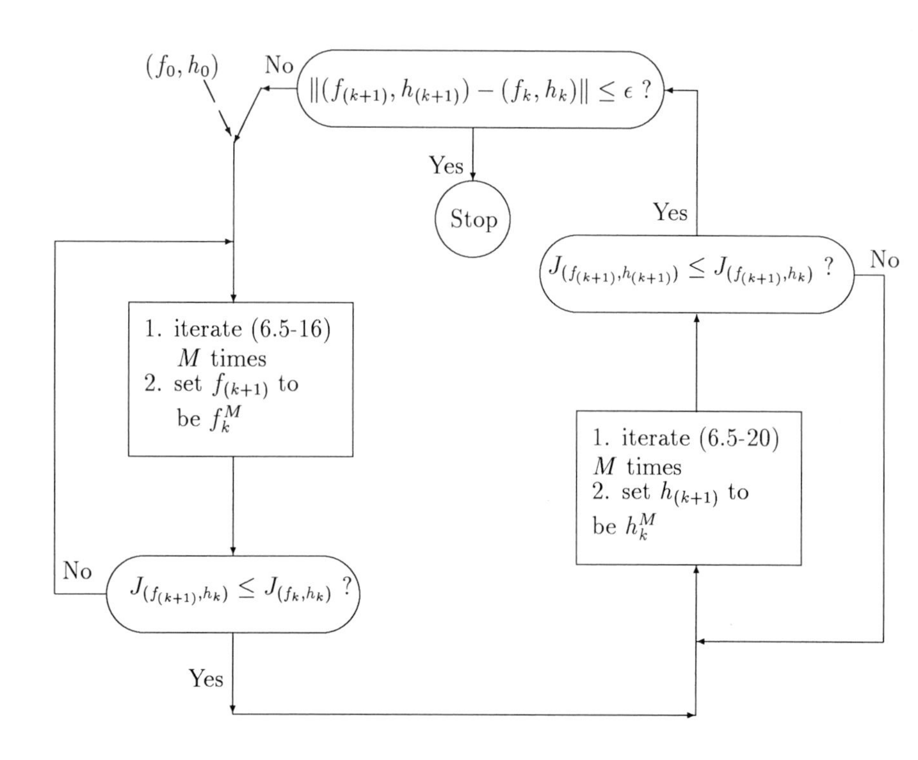

Figure 6.5-1: Block diagram of the projection based blind deconvolution algorithm.

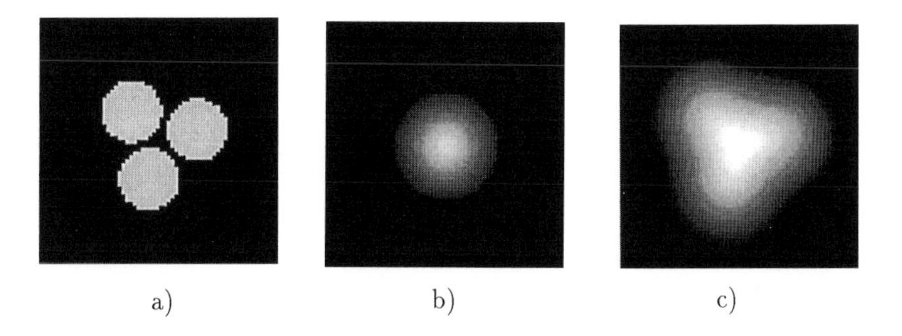

Figure 6.5-2: Illustration of blind deconvolution. a) The original triple-star; b) impulse response of imaging system; c) blurred image obtained at detector.

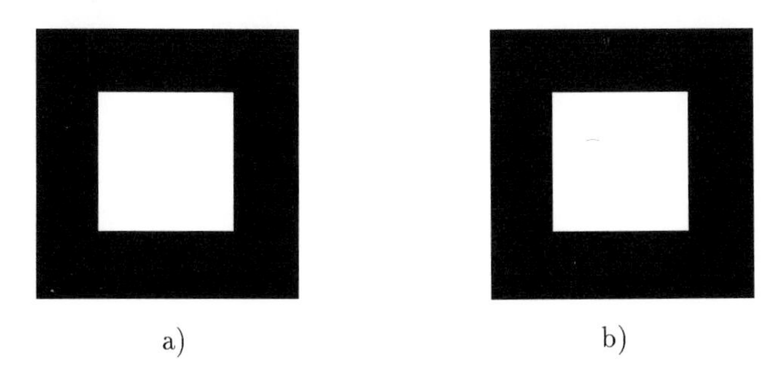

Figure 6.5-3: a) Correct support for spatially bounded image; b) correct support for spatially bounded impulse response (blurring function).

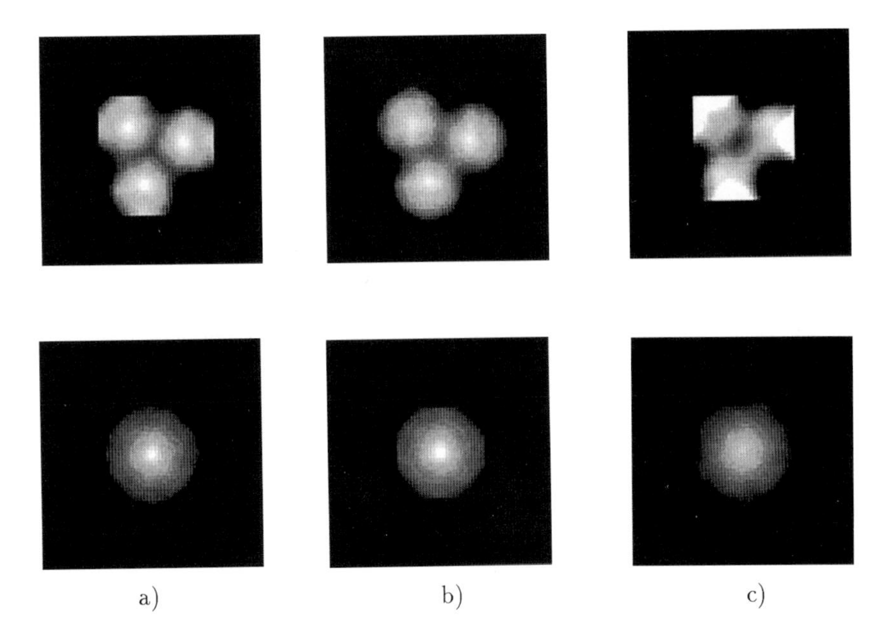

Figure 6.5-4: Reconstructed source and blurring functions, respectively, under different conditions of prior knowledge. a) Exact prior knowledge of W_f and W_h; b) W_f overestimated, W_h underestimated; c) W_f underestimated, W_h overestimated.

Summary

In this chapter we discussed the use of projection methods for solving various problems in imaging science. We began by reviewing the basic ideas behind the method of convex projections and described the alternating projections algorithm that converges to a function satisfying all prior (convex) constraints. Anticipating the some readers might be unfamiliar with these powerful methods, we furnished a simple, hand-computable, example of how convex projections work.

The rest of the chapter illustrated how projection methods can be used to solve practical problems in color, resolution enhancement, and signal restoration in so-called blind deconvolution. In the color matching problem, we showed that the method of convex projections can be used to modify an existing spectrum so that it matches a visual criterion. In the resolution enhancement problem, the method of convex projections was used to recover a high-resolution image from low-resolution detectors. Even when not enough data were furnished to allow for the extraction of a high-resolution image, the use of prior knowledge constraints enabled the recovery of improved images.

The last problem considered, blind deconvolution, involved non-convex constraint sets. Many of the nice properties of convex projections are not retained when non-convex sets are involved. We furnished a brief review of *generalized projections* and tried to apply the latter to blind deconvolutions. However, the numerical difficulties in computing the required, non-convex, projection reminded us that discretion is the better part of valor, and we settled for a mixed gradient-projection algorithm that enabled noise-free deconvolution despite inexact prior knowledge.

In Chapters 7 and 10, the method of convex projections is applied to important problems in image compression and estimation of probability density functions, respectively.

References

[6-1] L.G. Gubin, B.T. Polyak and E.V. Raik, "The method of projections for finding the common point of convex sets," *USSR Comput. Math. Phys.*, vol. 7, 1-24 1967.

[6-2] E. Kreyszig, *Introductory Functional Analysis with Applications*, John Wiley and Sons, N.Y., N.Y. 1978.

[6-3] D.C. Youla, "Mathematical theory of image restoration by the method of convex projections," in *Image Recovery*, H. Stark (ed.), pp. 29-77, Academic Press, Orlando, FL 1987.

[6-4] Z. Opial, "Weak convergence of the sequence of successive approximations for non-expansive mappings," *Bull. Am. Math. Soc.*, vol. 27, pp. 571-575, 1967.

[6-5] R.C. Gonzalez and R.E. Wood, *Digital Image Processing,* New York: Addison-Wesley, 1992.

[6-6] R.C. Buck and E.F. Buck, *Advanced Calculus*, 2nd ed. New York: McGraw-Hill, 1965.

[6-7] E.H. Land and J.J. McCann, "Lightness and retinex theory," *Jour. Opt. Soc. Am.*, vol. 61, pp. 1-11, Jan. 1971.

[6-8] A.K. Jain, *Fundamentals of Digital Image Processing*, Englewood-Cliffs, N.J.: Prentice-Hall, 1969.

[6-9] R.W.G. Hunt, *Measuring Color,* Chichester, UK: Ellis Horwood, 1987.

[6-10] A.K. Katsaggelos (ed.), *Digital Image Restoration*, New York: Springer Verlag, 1991.

[6-11] H.C. Andrews and B.R. Hunt, *Digital Image Restoration*, Englewood Cliffs, NJ: Prentice-Hall, 1971.

[6-12] H. Stark (ed.), *Image Recovery and Application*, Orlando, FL: Academic Press, 1987.

[6-13] J.W. Goodman, *Introduction to Fourier Optics*, New York: McGraw-Hill, 1968.

[6-14] J.A. Richards, *Remote Sensing Digital Image Analysis*, New York: Springer Verlag, 1986.

[6-15] B.R. Frieden and H.H.G. Aumann, "Image reconstruction from multiple I-D scans using filtered localized projections," *Appl. Opt.* vol. 26, pp. 3615-3621, 1987.

[6-16] H. Stark and P. Oskoui, "High-resolution image recovery from image-plane arrays, using convex projections," *J. Opt. Soc. Am., A*, vol. 6, No. 11, pp. 1715-1726, 1989.

[6-17] M.N. Wernick and C.-T. Chen, "Superresolved tomography by convex projections and detector notion," *J. Opt. Soc. Am. A*, vol. 9, pp. 1574-1553, 1991.

[6-18] M.I. Sezan and H. Stark, "Tomographic image reconstruction from incomplete data by convex projections and direct Fourier method," *IEEE Trans. Med. Imag.*, vol. MI-3, pp. 91-98, 1984.

[6-19] D.C. Youla and H. Webb, "Image restoration by the method of convex projections: Part 1-theory," *IEEE Trans. Med. Imag.*, vol. MI-1, pp. 81-94, 1982.

[6-20] L.A. Shepp and B.F. Logan, "The Fourier reconstruction of a head section," *IEEE Trans. Nucl. Sci.*, vol. NS-21, pp. 21-43, 1974.

[6-21] A. Levi "Image restoration by the method of projections with applications to the phase and magnitude retrieval problems," Ph.D. dissertation, Department of Electrical, Computer and System Engineering, Rensselaer Polytechnic Institute, Troy, N.Y., 1983.

[6-22] A. Levi and H. Stark, "Image restoration by the method of generalized projections with applications to restoration from magnitude," *J. Opt. Soc. Am. A*, vol. 1, pp. 932-943, 1984.

[6-23] G. Ayers and J.C. Dainty, "Iterative blind deconvolution method and its applications," *Opt. Lett.*, vol. 13, pp. 547-549, 1988.

[6-24] R.G. Lane, "Blind deconvolution of speckle images," *J. Opt. Soc. Am. A*, vol. 9, No. 9, pp. 1508-1514, Sept. 1992.

[6-25] B.C. McCallum, "Blind deconvolution by simulated annealing," *Optics Communications*, vol. 75, Nov. 2, 1990.

[6-26] A.K. Katsaggelos, R.L. Lagendijk and J. Biemond, "Constrained iterative identification and restoration of images," *Proc. of EURASIP*, Elsevier Publishers, pp. 1585-1588, 1988.

[6-27] J.R. Fienup, "Phase retrieval algorithms: a comparison," *Appl. Opt.*, vol. 21, pp. 2758-2769, 1982.

[6-28] Y. Yang, "Set-theoretic approach to visual communication problems," Ph.D. dissertation, Department of Electrical and Computer Engineering, Illinois Institute of Technology, 1994.

Chapter 7

Projection Based Image Reconstruction from Compressed Data

YONGI YANG and NIKOLAS P. GALATSANOS

Introduction

The recent advances in integrated circuits and display devices have lately given us many new capabilities in visual communications. This has resulted in unprecedented growth in this field. As a result, the demand for communication bandwidth in order to transmit the large amounts of data required by these applications has increased. At this point, bandwidth and channel capacity are the most serious limiting constraints for many new applications in visual communications. However, it is rooted in human nature to try to overcome such limitations in order to facilitate and promote civilization. In this context, image compression is the field where humans try to circumvent the barriers imposed by Shannon's basic law of information in order to develop new visual communication applications [7-1]. Therefore, image compression is a very important problem at the present time because it directly affects many emerging applications in this field.

Over the last 10 years image compression has experienced a flurry of activity and the state-of-the-art in this field has improved drastically. At the present time transform-based image coders are the most popular and commonly used. The basic principle that makes image compression possible is that images are highly correlated. Thus, in transform coding a transform that decorrelates the image is used and the coefficients of this transform are used to represent the image [7-2]. There are a plethora of transforms that have been proposed for image compression.

At the present time the most prominent ones are the *discrete cosine transform* (DCT) and different variants of the *wavelet transform* (WT) [7-2], [7-3]. The *block DCT* (BDCT) is the recommended transform in both the *joint pictures expert group* (JPEG) and *motion pictures expert group* (MPEG) standards [7-4], [7-5]. For this transform the image is divided into small blocks 8×8 or 16×16 and for each one the DCT is taken separately. However, recent advances have widened the performance difference between WT and BDCT based methods in favor of the former, [7-3] and [7-6]. Thus, we expect to soon see WT coders as part of an upcoming industrial standard.

In order to achieve high compression ratios in transform coding, part of the information about the transform coefficients which is not deemed important is *discarded* in the encoder. Quantization is the most simple and widely used approach to discard unimportant information. Other approaches, however, have been also introduced for this purpose, see for example [7-7], [7-3] and [7-6]. As a result the encoded image is only an approximation to the original image. In general, the coding error gets bigger as the compression ratio gets higher. The remaining data is compressed *losslessly* using entropy based coding and then transmitted to the decoder [7-8]. In the decoder, after entropy decompression the compressed image is reconstructed by taking the inverse transform of the transmitted data.

Due to the lossy nature of the encoding process, the reconstructed images display, depending on the transform used, a number of annoying artifacts. The best-known of these are the so-called "blocking artifacts" which result from BDCT based coding, and "ringing artifacts" which are artifacts common to all transform coding techniques. The blocking artifact manifests itself as an artificial discontinuity between adjacent blocks and is a direct result of the independent processing of the blocks which does not take into account the between-block pixel correlations. The ringing artifact manifests itself as a sharp oscillation near edges, and is due to the higher priority that is assigned to low-pass information in all approaches that are used to discard unimportant components in transform coding. In other words, in most transform coding approaches high-frequency information is thrown away, thus, the reconstructed images are less sharp and display ringing near the edges.

Reconstructing the compressed image by just taking the inverse transform of the quantized data is a very simplistic approach. Although the reconstructed images suffer from the previously mentioned artifacts, this reconstruction approach has been extensively used because of decoder complexity constraints. However, due to advances in VLSI technology these constraints are gradually being relaxed. Therefore, more sophisticated approaches can be used to reconstruct the compressed image.

In the past, various post-processing algorithms have been proposed to improve the quality of the compressed images in the decoder without increasing the bit rate. In [7-9] the decoded image is processed directly using space-invariant filters. In [7-10], [7-11], the decoded image is processed directly using space-variant filters.

In [7-12], an image recovery approach is essentially proposed to reconstruct images from the transmitted data. It is only loosely based on the theory of *projections onto convex sets* (POCS) and its convergence cannot be rigorously justified. In the JPEG standard [7-4], a technique for predicting the low-frequency (except the d.c.) BDCT coefficients is recommended as an option at the decoder in order to suppress the between-block discontinuities of the decoded images. For this approach, the image is assumed to be a quadratic surface and the missing low-frequency coefficients are predicted to fit this surface. However, in areas with sharp intensity transitions, a quadratic surface model is no longer valid and the proposed prediction scheme fails. In [7-13], a probabilistic model is assumed and the compressed image is reconstructed using a *maximum a posteriori probability* (MAP) approach.

In this chapter, we present an image recovery approach to reconstruct images in the decoder. Our approach is based on the theory of projections onto convex sets [7-14]. According to this approach, the image in the decoder is recovered using not only the transmitted data, but also other prior knowledge that is available in the decoder. All known properties of the original image are expressed in the form of convex constraint sets. These properties include both the information that is conveyed by the transmitted data and the information that is known *a priori*.

Since the BDCT is presently the industrial standard, in this work we will present our results using a BDCT based coder. However, the principle that we present is applicable to any transform-based coder. The rest of this chapter is organized as follows. In Section 7.1 the underlying principle and its justification are reviewed. In Section 7.2 the convex set that we used, which is based on the transmitted data, is discussed. In Section 7.3 the convex sets which are based on our prior knowledge about the original image in the decoder are discussed. In Section 7.4 issues associated with the implementation of our algorithm are discussed. In Section 7.5 we describe a simplified recovery algorithm. In Section 7.6 we present a computational complexity analysis of the proposed simplified algorithm. In Section 7.7 we present some numerical experiments where we tested this approach.

7.1 Principles of the Proposed Recovery Approach

The information-lossy nature of the encoder implies that the encoding process is not invertible. This fact is reflected in the decoder, since the received data cannot completely characterize the original image. Therefore, image decoding naturally poses itself as an image *recovery* or *inverse* problem. There exist many rather that one solutions to this problem. Specifically, there exist a set of "images," including the original, that will yield the same data as the received image if they are compressed by the same encoding algorithm. Typically in a conventional transform decoder, an image, i.e., a solution to this inverse problem, is found through simple "inverse" operations opposite to the encoding process. However, the judge of the compressed image quality is presumably a human observer in almost every application. Unfortunately, this fact is not taken into consideration in a conventional decoder. As a result, the decoded image suffers from undesired artifacts as previously described.

As mentioned above, in general there exist a set of images satisfying the compressed data. Therefore a legitimate question is: from this set of images can we pick the ones that are more meaningful to a human observer, rather than merely finding one image as a conventional decoder does? The answer to this question is the focal point of this study. First, let's consider again the "blocking artifact" for example in BDCT based image coding. "Blocking artifact" is known to be very annoying to a human observer. But why? The presence of "blocking artifact" in the decoded image contradicts what a human observer expects from a natural image. In other words, the experience, i.e., the *prior knowledge*, of a human observer is that most of the natural images do not have a "blocky" structure. One important conclusion from this observation is that prior knowledge can help to pick out an image from the set of admissible images based on the transmitted data which is favorable to a human observer. Therefore, the characteristics of the human observer should be taken into consideration in image decoding in order to obtain good-quality images. Toward this end, in this chapter we propose a new image decoding approach that incorporates the psychological characteristics of the human visual system.

The proposed algorithm is an image recovery algorithm based on the theory of *projection onto convex sets* (POCS).[1] POCS has been successfully used in a number of image recovery problems [7-14]. According to this approach, all the information available about the unknown image is represented as convex constraint sets in a vector space. Then an element in the intersection of all the available constraint sets satisfies all the available information about the unknown and is legitimately taken as a solution to this problem. One advantage of the POCS approach is that there exist numerical algorithms to find such a solution. Another advantage of this approach is that it provides a flexible framework to incorporate different types of *prior* knowledge about the unknown image in the reconstruction algorithm. A POCS algorithm is typically of iterative nature. A complete description of a POCS algorithm requires the definition of the convex constraint sets that characterize the available information and the computation of their projections.

In this formulation, the unknown image is treated as a vector in a vector space. Constraint sets are defined in this space based on the available knowledge about the image. They fall into two broad categories: *data* sets and *property* sets. Data sets are used to specify that the recovered image should yield the same data as the original if it is run through the encoder. The property sets are defined based on prior knowledge. One simple example of such a set is the set defined on the range of the pixel intensities of an image or the set of smooth between block boundary images

[1]See Chapter 6 for a detailed review of POCS.

[7-15]. A pictorial representation of the proposed approach is given in Fig. 7.1-1.

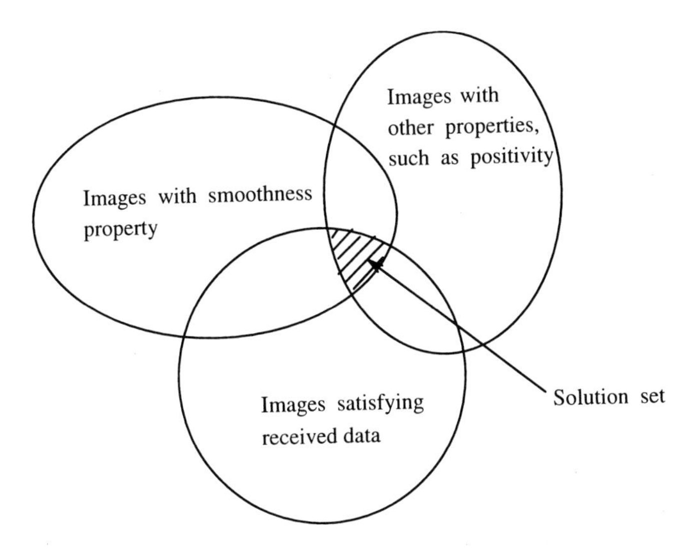

Figure 7.1-1: Illustration of the recovery principle.

7.2 Constraint Sets Based on the Transmitted Data

Before we can quantitatively describe the set of transmitted data we need: (i) to define the encoding algorithm that is used; and (ii) to introduce the mathematical notation that will be used subsequently. In what follows we first give a brief description of JPEG based compression. That description focuses on the aspects of JPEG which are relevant to the context of this problem. More specifically, the entropy encoding part of the algorithm is completely ignored in our description since it is lossless.

A. Basic Principle of Baseline JPEG Image Compression

Since a BDCT/JPEG based coder will be used throughout this chapter, we review the basic principles of this coding approach. A baseline JPEG encoder is illustrated in Fig. 7.2-1, ignoring the entropy encoding process since it is an information lossless process.

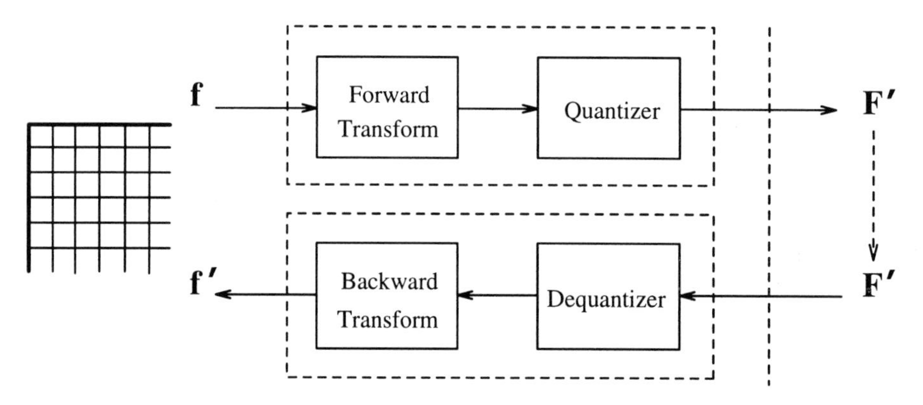

Figure 7.2-1: JPEG baseline coder/decoder.

In the coding process, an image is first partitioned into contiguous 8×8 blocks. Then each block is transformed into the transform domain by the so-called 2-D 8×8 forward DCT transform:

$$F_{(u,v)} = \frac{1}{4} C_u C_v \sum_{x=0}^{7} \sum_{y=0}^{7} f_{(x,y)} \cos\left[\frac{(2x+1)u\pi}{16}\right] \cos\left[\frac{(2y+1)v\pi}{16}\right], \tag{7.2-1}$$

where $(f_{(x,y)}), 0 \le x, y \le 7$, denote the 8×8 array of image pixel intensities; and $(F_{(u,v)}), 0 \le u, v \le 7$, the array of DCT coefficients. Also in Eq. (7.2-1),

$$C_i = \begin{cases} \frac{1}{\sqrt{2}} & \text{for } i = 0 \\ 1 & \text{otherwise.} \end{cases} \tag{7.2-2}$$

Afterwards, the whole array of DCT coefficients for each block is quantized using uniform quantizers whose step sizes are determined by the DCT quantization table: $(Q_{(u,v)}), 0 \le u, v \le 7$. The output of the uniform quantizer at position (u, v) is defined by

$$F_{q(u,v)} = \text{round}\left(\frac{F_{(u,v)}}{Q_{(u,v)}}\right), \tag{7.2-3}$$

where "round" means rounding to the nearest integer. The quantized data are then encoded further by entropy coding which is an invertible process at the decoder.

At the decoder after entropy decoding, the received data is dequantized by

$$F_{(u,v)} = F_{q(u,v)} \cdot Q_{(u,v)}, \tag{7.2-4}$$

for $0 \leq u, v \leq 7$. Then, each block of the image is reconstructed by the inverse 2-D 8×8 DCT transform:

$$f_{(x,y)} = \frac{1}{4} \sum_{u=0}^{7} \sum_{v=0}^{7} C_u C_v F_{(u,v)} \cos\left[\frac{(2x+1)u\pi}{16}\right] \cos\left[\frac{(2y+1)v\pi}{16}\right]. \tag{7.2-5}$$

As we can see, in both the encoding and decoding processes, the image is processed on a block-by-block basis. The existing strong correlation among the neighboring blocks is never taken into consideration in either the encoder or the decoder. Hence, the decoded image exhibits discontinuities at the block boundaries, which is the blocking artifact. The objective of this approach is to take advantage of this fact at the decoder and to enforce correlation among neighboring blocks during the decoding process.

B. Mathematical Notation

For the sake of convenience, we next state some notation that will be used throughout the chapter. Other notation will be introduced as needed.

A digital $N \times N$ image is treated as an $N^2 \times 1$ vector $\mathbf{f}$ in the Euclidean space R^{N^2} by its lexicographic ordering either by rows or columns. Correspondingly, we use $\mathbf{F}$ to represent the vector form of the BDCT of $\mathbf{f}$.

Then the BDCT is viewed as a linear transformation, denoted by matrix $\mathbf{B}$, from R^{N^2} to R^{N^2}, i.e.,

$$\mathbf{F} = \mathbf{B}\,\mathbf{f}. \tag{7.2-6}$$

Due to the unitary property of the 2-D DCT transform for each block, the BDCT matrix is also unitary and the inverse transform can be simply expressed by $\mathbf{B}^T$ where T denotes the transpose of a matrix. Then, the inverse BDCT can be written as

$$\mathbf{f} = \mathbf{B}^T\,\mathbf{F}. \tag{7.2-7}$$

In a BDCT based coder, each of the BDCT coefficients $\mathbf{F}$ is quantized in order to achieve bit rate reduction for transmission. At the receiver the received data is a quantized version of $\mathbf{F}$, say $\mathbf{F}'$. This quantization process can be described mathematically by an operator, say $\mathcal{Q}$, from R^{N^2} to R^{N^2}, then we have

$$\mathbf{F}' = \mathcal{Q}\,\mathbf{F}. \tag{7.2-8}$$

Using this notation, the whole process from original image $\mathbf{f}$ to the received data $\mathbf{F}'$ can be conveniently written as

$$\mathbf{F}' = \mathcal{Q}\,\mathbf{B}\,\mathbf{f}. \tag{7.2-9}$$

In the following, for further simplicity we let $\mathcal{T}$ denote the concatenation of $\mathbf{B}$ and $\mathcal{Q}$, i.e., $\mathcal{T} \triangleq \mathcal{Q}\mathbf{B}$. Then Eq. (7.2-9) can be rewritten as

$$\mathbf{F}' = \mathcal{T}\,\mathbf{f}. \tag{7.2-10}$$

With the above notation, the image decoding problem can be phrased as: given data $\mathbf{F}'$, find $\mathbf{f}$. A seemingly straightforward approach would be to solve for $\mathbf{f}$ from Eq. (7.2-10) by inverting $\mathcal{T}$. Unfortunately, due to the *many-to-one* mapping nature of Q, $\mathcal{T}$ is also a *many-to-one* mapping. Therefore, Eq. (7.2-10) is not invertible. In general there are many $\mathbf{f}$'s that satisfy Eq. (7.2-10).

In a JPEG based decoder the image, say $\mathbf{f}'$, is reconstructed by the inverse BDCT, i.e.,

$$\mathbf{f}' = \mathbf{B}^T\,\mathbf{F}', \tag{7.2-11}$$

where, because $\mathbf{B}^T$ is unitary, $\mathbf{B}^{-1} = \mathbf{B}^T$. As mentioned earlier, the image $\mathbf{f}'$ suffers from blocking artifacts at low bit rate.

C. The Constraint Set Based on the Quantized Transform Coefficients

Let C_T denote the set of all such solutions to Eq. (7.2-10). That is, for a prescribed $\mathbf{F}'$,

$$C_T \triangleq \{\,\mathbf{f} : \mathcal{Q}\,\mathbf{B}\,\mathbf{f} = \mathbf{F}'\,\}. \tag{7.2-12}$$

It follows immediately that the JPEG solution in Eq. (7.2-11) lies in this set.

Let $\mathcal{I}$ denote the index set of all the BDCT coefficients. The set C_T can be rewritten as

$$C_T \triangleq \{\mathbf{f} : (\mathcal{Q}\mathbf{B}\,\mathbf{f})_n = F_n^{'}, \forall n \in \mathcal{I}\}. \tag{7.2-13}$$

In Eq. (7.2-13), knowledge of each $F_n^{'}$ will restrict $(\mathbf{B}\,\mathbf{f})_n$ within some quantization interval. Take the uniform quantizer in Fig. 7.2-2, for example, the knowledge of a 2 in the output implies that the input must be within the interval $(\,1.5q,\ 2.5q\,]$. Therefore, due to the non-closedness of the quantization intervals of the quantizer, the set C_T is also not closed. Nevertheless, we can enlarge this set by adding to it all its interval end points, i.e., all limit points such that we get its closure set which can be written as

$$C_1 \triangleq \overline{C_T} = \{\mathbf{f} : F_n^{min} \le (\mathbf{Bf})_n \le F_n^{max}, \forall n \in \mathcal{I}\}, \tag{7.2-14}$$

where F_n^{min} and F_n^{max} are determined by the quantizer used.

C_1 is a constraint set that captures all known information from the received BDCT coefficients. From Eq. (7.2-10) we know that the original image at the coder (which is not known at the decoder) is an element of the set C_T. It is not difficult to see that the blocky image $\mathbf{f}'$ given in Eq. (7.2-11) is also an element of

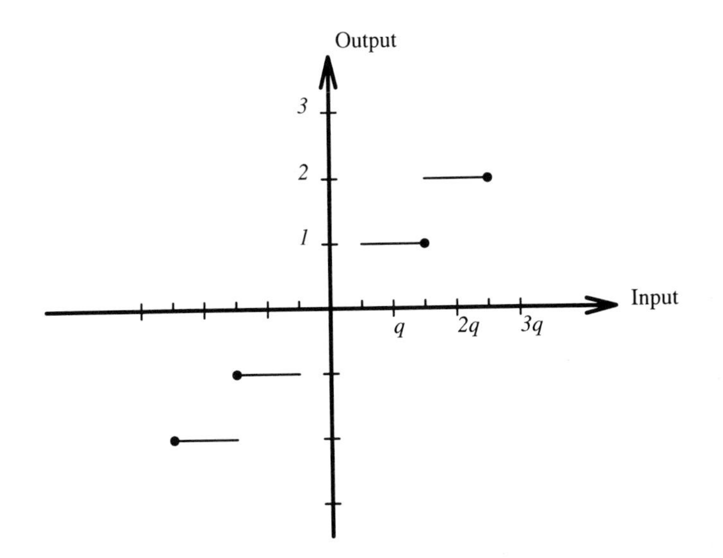

Figure 7.2-2: Input/output relation of a uniform quantizer.

the set $\mathcal{C}_T$, due to the idempotent property of $\mathcal{Q}$, that is, $\mathcal{Q} = \mathcal{Q}^2$. Therefore, the set C_1 alone does not guarantee the quality of the decoded image.

The projection $P_1\mathbf{f}$ of an arbitrary vector $\mathbf{f}$ in R^{N^2} onto C_1 is given by [7-12]

$$P_1\,\mathbf{f} = \mathbf{B}^T \cdot \mathbf{F}, \tag{7.2-15}$$

where $\mathbf{B}^T$ is defined in Eq. (7.2-7) and $\mathbf{F}$ is determined by

$$F_n = \begin{cases} F_n^{min} & \text{if } (\mathbf{B}\,\mathbf{f})_n < F_n^{min} \\ F_n^{max} & \text{if } (\mathbf{B}\,\mathbf{f})_n > F_n^{max} \\ (\mathbf{B}\,\mathbf{f})_n & \text{if } F_n^{min} \leq (\mathbf{B}\,\mathbf{f})_n \leq F_n^{max}. \end{cases} \quad \forall n \in \mathcal{I}. \tag{7.2-16}$$

7.3 Constraint Sets Based on Prior Knowledge

From our previous discussion, it is clear that the set of the transmitted data alone is not sufficient to completely characterize a good-quality image in the decoder. Since blocking is the main artifact that we observe, if the transmitted data are used alone, it is only natural to introduce another constraint set that will *complement* the information about the image conveyed by our data. In other words, a set that describes the between-block smoothness properties of the image would be introducing information which is missing in the transmitted data.

The idea of using, along with the data, information that complements it, is not new in recovery problems. It is the idea on which the principle of *regularization* is based [7-16]. This idea has been extensively used in image recovery problems, [7-17] and [7-18]. In what follows we introduce two types of such sets. The first is a simple space-invariant set and is used to introduce the basic idea. The second is more sophisticated and its properties are discussed in detail.

A. A Spatially-Invariant Smoothness Constraint Set

Let C_s denote the set of images which do not exhibit blocking artifacts. That is,

$$C_s \triangleq \{ \mathbf{f} : \mathbf{f} \text{ is smooth in the block boundaries} \}. \tag{7.3-1}$$

Then the set $C_0 = C_T \bigcap C_s$ contains all the images that satisfy both the received data and also are smooth between blocks. It is therefore clear that an element from C_0 is a better choice than $\mathbf{f}'$ as the recovered image. The theory of POCS provides the mathematical tools for obtaining an element in C_0, [7-14] and [7-19].

It is worth pointing out that if the set C_0 contains more than one point, as it usually does, then the solution is not unique and is influenced by the order of the projections, and the starting point $\mathbf{f}_0$ of the algorithm. The key to applying this theory to image recovery problems lies in expressing every known property of the unknown image by a closed convex set. Then, the POCS theory automatically yields a recovery algorithm. As we can see, this approach is very flexible in incorporating any type of *prior* knowledge into the recovery process which is one of the main advantages of this method.

The definition of the set C_s in Eq. (7.3-1) is only qualitative. A set that can be used in the POCS context requires a rigorous mathematical definition. In [7-15], the smoothness C_s was defined by

$$C_s \triangleq \{\mathbf{f} : \|\mathbf{Q}\,\mathbf{f}\,\| \leq E\}, \tag{7.3-2}$$

where E is a scalar upper bound that defines the size of this set; $\mathbf{Q}$ is a linear operator defined as follows: writing an $N \times N$ image $\mathbf{f}$ in its column vector form

$$\mathbf{f} = \{\, \mathbf{f}_1, \mathbf{f}_2, ..., \mathbf{f}_N \,\}, \tag{7.3-3}$$

where $\mathbf{f}_i$ denotes the ith column of the image; then, $\mathbf{Q}\,\mathbf{f}$ gives the difference between adjacent columns at the block boundaries of $\mathbf{f}$. For example, for the case of $N = 512$ and 8×8 blocks,

$$\mathbf{Q}\,\mathbf{f} = \begin{bmatrix} \mathbf{f}_8 & - & \mathbf{f}_9 \\ \mathbf{f}_{16} & - & \mathbf{f}_{17} \\ & \cdot & \\ & \cdot & \\ & \cdot & \\ \mathbf{f}_{504} & - & \mathbf{f}_{505} \end{bmatrix}. \tag{7.3-4}$$

The norm of $\mathbf{Q}\,\mathbf{f}$

$$\|\mathbf{Q}\,\mathbf{f}\,\| = \left[\sum_{i=1}^{63} \|\,\mathbf{f}_{8\cdot i} - \mathbf{f}_{8\cdot i+1}\,\|^2\right]^{\frac{1}{2}} \tag{7.3-5}$$

is a measure of the total intensity variation between the boundary columns of adjacent blocks.

The projection onto C_s is computed in Appendix 7.A. For an image $\mathbf{f} \notin C_s$ in column form $\{\,\mathbf{f}_1, \mathbf{f}_2, ..., \mathbf{f}_N\,\}$, its projection $\tilde{\mathbf{f}} = P_s\mathbf{f}$ onto set C_s is represented in column form also by $\{\tilde{\mathbf{f}}_1, \tilde{\mathbf{f}}_2, ..., \tilde{\mathbf{f}}_N\}$. For a 512×512 image and 8×8 blocks

$$\tilde{\mathbf{f}}_i = \mathbf{f}_{i+1} + \alpha \cdot (\mathbf{f}_i - \mathbf{f}_{i+1}) \quad \text{and} \quad \tilde{\mathbf{f}}_{i+1} = \mathbf{f}_i - \alpha \cdot (\mathbf{f}_i - \mathbf{f}_{i+1}) \tag{7.3-6}$$
$$\text{for } i = 8 \cdot k \;\text{ and }\; k = 1, 2, ..., 63; \quad \text{otherwise,} \quad \tilde{\mathbf{f}}_i = \mathbf{f}_i,$$

where $\alpha = \frac{1}{2}\left[\frac{E}{\|\mathbf{Q}\cdot\mathbf{f}\|} + 1\right]$.

In a similar fashion, a set $C_s^{'}$ which captures the intensity variations between the rows of the block boundaries can also be defined. When the quantizer $\mathcal{Q}$ is known, the projector P_T onto the data set in Eq. (7.2-12) is easy to find [7-15], [7-20]. Using these convex sets a POCS based recovery algorithm can be defined in the decoder to reconstruct the compressed images. The experimental results show that this approach works well. The recovered images are both visually, and objectively using a mean-square distance measure, better than images produced by traditional decoders. However, according to this approach an entire block boundary column/row vector $\mathbf{f}_i$ is treated uniformly. Thus, the local properties of the image along this column/row are not explicitly used during the recovery process. In what follows we present a *spatially-adaptive* recovery approach that explicitly uses the local image properties to impose smoothness constraints along the columns/rows of the reconstructed image.

B. A Spatially-Adaptive Smoothness Constraint Set

A well-known fact in image processing is that an image is typically a non-stationary random field [7-2]. The statistics of the pixel intensity of an image may vary depending on the region it lies in. For example, in a smooth area, strong correlations exist among the neighboring pixels, while in a textured or edge region the local correlations are weaker. The set defined in Eq. (7.3-4) treats the block variations in the entire image equally, and consequently the processing in Eq. (7.3-6) does not adapt to the local statistics of the image. In addition, it is also well known that the noise visibility in images is not space-invariant and depends on the characteristics of the region under observation. For example, noise in smooth regions is more visible than that in textures or edge regions [7-21]. The blocking artifacts can be viewed as noise with a regular pattern. Therefore, from both points of view, i.e., fidelity to the true statistical nature of real images, and consideration of human perceptual characteristics, a spatially-adaptive smoothness constraint set would be very helpful in recovering high-quality images.

From the definition in Eq. (7.3-4), for a 512×512 image and 8×8 blocks, $\mathbf{Q}\,\mathbf{f}$ is a $(512 \cdot 63) \times 1$ vector. $\mathbf{Q}\,\mathbf{f}$ captures the variations between all the pixels at the neighboring vertical block boundaries. Let $\mathbf{W}$ be a $(512 \cdot 63) \times (512 \cdot 63)$ diagonal matrix of the form

$$\begin{bmatrix} w_1 & 0 & 0 & \cdots & 0 \\ 0 & w_2 & 0 & \cdots & \vdots \\ 0 & 0 & \ddots & 0 & \vdots \\ \vdots & \vdots & 0 & \ddots & 0 \\ 0 & \cdots & \cdots & 0 & w_{512 \cdot 63} \end{bmatrix}, \tag{7.3-7}$$

where the w_i's, $i = 1, 2, \cdots, (512 \cdot 63)$, are weights which capture the local properties of the image. Then, the vector $\mathbf{W}\mathbf{Q}\,\mathbf{f}$ is a weighted version of the vertical between-block variations. Define the set

$$C_w \triangleq \{\mathbf{f} : \|\mathbf{W}\,\mathbf{Q}\,\mathbf{f}\,\| \leq E\}. \tag{7.3-8}$$

It is straightforward to show that the set C_w is both convex and closed. Before describing how the weights w_i are obtained we will examine in detail the projection operation onto the set C_w.

For an image $\mathbf{f} \notin C_w$ represented in column vector form as in Eq. (7.3-3), the projection onto C_w will also be given in column form. Let P_w be the projector onto C_w. Then, we can write

$$\tilde{\mathbf{f}} = P_w\,\mathbf{f} = \{\,\tilde{\mathbf{f}}_1, \tilde{\mathbf{f}}_2, ..., \tilde{\mathbf{f}}_{512}\,\}. \tag{7.3-9}$$

Define

$$\mathbf{x} = \begin{bmatrix} \mathbf{f}_8 \\ \mathbf{f}_{16} \\ \cdot \\ \cdot \\ \cdot \\ \mathbf{f}_{504} \end{bmatrix} \quad \mathbf{y} = \begin{bmatrix} \mathbf{f}_9 \\ \mathbf{f}_{17} \\ \cdot \\ \cdot \\ \cdot \\ \mathbf{f}_{505} \end{bmatrix} \quad \tilde{\mathbf{x}} = \begin{bmatrix} \tilde{\mathbf{f}}_8 \\ \tilde{\mathbf{f}}_{16} \\ \cdot \\ \cdot \\ \cdot \\ \tilde{\mathbf{f}}_{504} \end{bmatrix} \quad \text{and } \tilde{\mathbf{y}} = \begin{bmatrix} \tilde{\mathbf{f}}_9 \\ \tilde{\mathbf{f}}_{17} \\ \cdot \\ \cdot \\ \cdot \\ \tilde{\mathbf{f}}_{505} \end{bmatrix}. \tag{7.3-10}$$

We show in Appendix 7.B that the projection $\tilde{\mathbf{f}}$ is given by

$$\begin{aligned} \tilde{\mathbf{x}} &= \tfrac{1}{2}\,[\,I + (I + 2\lambda\mathbf{W}^T\mathbf{W})^{-1}]\,\mathbf{x}_0 + \tfrac{1}{2}\,[\,I - (I + 2\lambda\mathbf{W}^T\mathbf{W})^{-1}]\,\mathbf{y}_0 \\ \tilde{\mathbf{y}} &= \tfrac{1}{2}\,[\,I - (I + 2\lambda\mathbf{W}^T\mathbf{W})^{-1}]\,\mathbf{x}_0 + \tfrac{1}{2}\,[\,I + (I + 2\lambda\mathbf{W}^T\mathbf{W})^{-1}]\,\mathbf{y}_0, \end{aligned} \tag{7.3-11}$$

and the rest of columns are unchanged, i.e.,

$$\tilde{\mathbf{f}}_i = \mathbf{f}_i \text{ for } i \neq 8 \cdot k \text{ or } 8 \cdot k + 1, k = 1, 2, \cdots, 63.$$

The scalar λ in Eq. (7.3-11) is the positive root of the nonlinear equation

$$\| \mathbf{W}\,\mathbf{Q}\,\tilde{\mathbf{f}} \,\| = E. \tag{7.3-12}$$

In order to better understand the action of the projector P_w, we rewrite $\mathbf{W}$ in a block form as

$$\begin{bmatrix} \mathbf{w}_1 & \mathbf{0} & \mathbf{0} & \cdots & \mathbf{0} \\ \mathbf{0} & \mathbf{w}_2 & \mathbf{0} & \cdots & \vdots \\ \mathbf{0} & \mathbf{0} & \ddots & \mathbf{0} & \vdots \\ \vdots & \vdots & \mathbf{0} & \ddots & \mathbf{0} \\ \mathbf{0} & \cdots & \cdots & \mathbf{0} & \mathbf{w}_{63} \end{bmatrix}, \tag{7.3-13}$$

where $\mathbf{w}_k$ is a 512×512 diagonal matrix corresponding to the kth block column boundary. For $k = 1, 2, \cdots, 63$, $\mathbf{w}_k$ is written in the form

$$\mathbf{w}_k = \begin{bmatrix} w_1^k & 0 & 0 & \cdots & 0 \\ 0 & w_2^k & 0 & \cdots & \vdots \\ 0 & 0 & \ddots & 0 & \vdots \\ \vdots & \vdots & 0 & \ddots & 0 \\ 0 & \cdots & \cdots & 0 & w_{512}^k \end{bmatrix}. \tag{7.3-14}$$

Note that the factors in $\mathbf{w}_k$ are related to those in $\mathbf{W}$ by the relation

$$w_j^k = w_{512\cdot k+j}, \tag{7.3-15}$$

for $j = 1, 2, \cdots, 512$.

Using the notations in Eqs. (7.3-13) and (7.3-14), Eq. (7.3-11) can be rewritten as

$$\begin{aligned} \tilde{\mathbf{f}}_i &= \tfrac{1}{2}\left[\mathbf{I} + (\mathbf{I} + 2\lambda \mathbf{w}_k^T \mathbf{w}_k)^{-1}\right] \mathbf{f}_i + \tfrac{1}{2}\left[\mathbf{I} - (\mathbf{I} + 2\lambda \mathbf{w}_k^T \mathbf{w}_k)^{-1}\right] \mathbf{f}_{i+1} \\ \tilde{\mathbf{f}}_{i+1} &= \tfrac{1}{2}\left[\mathbf{I} - (\mathbf{I} + 2\lambda \mathbf{w}_k^T \mathbf{w}_k)^{-1}\right] \mathbf{f}_i + \tfrac{1}{2}\left[\mathbf{I} + (\mathbf{I} + 2\lambda \mathbf{w}_k^T \mathbf{w}_k)^{-1}\right] \mathbf{f}_{i+1} \\ &\text{for } i = 8 \cdot k \text{ and } k = 1, 2, ..., 63; \text{ otherwise, } \tilde{\mathbf{f}}_i = \mathbf{f}_i. \end{aligned} \tag{7.3-16}$$

The spatially adaptive nature of the projector P_w can be seen by rewriting Eq. (7.3-16) as

$$\begin{aligned} \tilde{\mathbf{f}}_i &= \frac{\mathbf{f}_i + \mathbf{f}_{i+1}}{2} + (\mathbf{I} + 2\lambda \mathbf{w}_k^T \mathbf{w}_k)^{-1} \frac{\mathbf{f}_i - \mathbf{f}_{i+1}}{2} \\ \tilde{\mathbf{f}}_{i+1} &= \frac{\mathbf{f}_i + \mathbf{f}_{i+1}}{2} - (\mathbf{I} + 2\lambda \mathbf{w}_k^T \mathbf{w}_k)^{-1} \frac{\mathbf{f}_i - \mathbf{f}_{i+1}}{2}. \end{aligned} \tag{7.3-17}$$

From Eq. (7.3-14), matrix $(I + 2\lambda \mathbf{w}_k^T \mathbf{w}_k)^{-1}$ can be written as

$$\begin{bmatrix} \frac{1}{1+2\lambda(w_1^k)^2} & 0 & 0 & \cdots & 0 \\ 0 & \frac{1}{1+2\lambda(w_2^k)^2} & 0 & \cdots & \vdots \\ 0 & 0 & \ddots & 0 & \vdots \\ \vdots & \vdots & 0 & \ddots & 0 \\ 0 & \cdots & \cdots & 0 & \frac{1}{1+2\lambda(w_{512}^k)^2} \end{bmatrix}. \tag{7.3-18}$$

From Eqs. (7.3-17) and (7.3-18) it follows that in areas where the weighting factors w_i (i.e. w_j^k) are large, the difference between the neighboring pixels of the projected image is reduced more than in areas where the w_i are small. It is useful to note the following extreme cases: when $w_i = \infty$, the projected pixels are simply the average of the two neighboring pixels; when $w_i = 0$, the projected pixels remain unchanged.

Comparing the projector onto C_w in Eq. (7.3-17) with the projector onto C_s in Eq. (7.3-6), we see that the spatially-adaptive nature of the new constraint set becomes clear. All pixels along a column in the projector P_s in Eq. (7.3-16) are processed using the same weights without taking into account the local properties of the image. However, the penalty for the more sophisticated model used in C_w is that the projection cannot be expressed in closed form. The parameter λ must be found by solving numerically the nonlinear equation in Eq. (7.3-12). Finally, we would like to point out that a similar, in-form, spatially-adaptive smoothness constraint has also been used in regularized image restoration [7-22]–[7-24].

C. On the Properties of the Constant λ in the Projector P_w

The definition of the projector in Eq. (7.3-16) requires the computation of the parameter λ from Eq. (7.3-12). Therefore, in what follows the properties of the roots of Eq. (7.3-12) and the numerical techniques used to find them are examined.

Using the notation in Eq. (7.3-10), we have

$$\mathbf{Qf} = \mathbf{x} - \mathbf{y} \quad \text{and} \quad \mathbf{Q\tilde{f}} = \tilde{\mathbf{x}} - \tilde{\mathbf{y}}. \tag{7.3-19}$$

From Eq. (7.3-11), we have

$$\tilde{\mathbf{x}} - \tilde{\mathbf{y}} = (\mathbf{I} + 2\lambda \mathbf{W}^T\mathbf{W})^{-1} (\mathbf{x} - \mathbf{y}), \tag{7.3-20}$$

that is,

$$\mathbf{Q}\,\tilde{\mathbf{f}} = (\mathbf{I} + 2\lambda \mathbf{W}^T\mathbf{W})^{-1}\,\mathbf{Q}\,\mathbf{f}. \tag{7.3-21}$$

Equation (7.3-12) can be rewritten as

$$\|\,\mathbf{W}\,(\mathbf{I} + 2\lambda \mathbf{W}^T\mathbf{W})^{-1}\,\mathbf{Q}\,\mathbf{f}\,\| = E. \tag{7.3-22}$$

Define

$$\mathbf{d} = \begin{bmatrix} d_1 \\ d_2 \\ \vdots \\ d_{512\times 63} \end{bmatrix} = \mathbf{Q}\,\mathbf{f}. \tag{7.3-23}$$

Then, Eq. (7.3-22) can be rewritten as

$$\|\,\mathbf{W}\,(\mathbf{I} + 2\lambda \mathbf{W}^T\mathbf{W})^{-1}\,\mathbf{d}\,\| = E. \tag{7.3-24}$$

or

$$\sum_{i=1}^{N^2} \frac{w_i^2 d_i^2}{(1+2\lambda w_i^2)^2} = E^2. \tag{7.3-25}$$

Equation (7.3-25) can be expanded further into a polynomial of degree $2N^2$ in λ. However, before this equation can be used, two questions are in order: (i) does it have any solutions? (ii) if it has more than one solution which one corresponds to the projection in Eq. (7.3-16)? These questions are answered by the following theorem.

Theorem 7.3-1 For all w_i's real in Eq. (7.3-25),

1. Eq. (7.3-25) has one and only one positive root in λ.
2. Eq. (7.3-25) has at least one negative root in λ.
3. The negative roots of Eq. (7.3-25) are all greater in magnitude than the positive root.
4. Only the positive root corresponds to the projector in Eq. (7.3-16).

The proof of Theorem 7.3-1 is given in Appendix 7.C. ∎

This theorem establishes the existence of a positive root for the projector in Eq. (7.3-16). In order to find this root, however, numerical methods (for example, Newton's method) have to be used.

Since there is more than one solution to Eq. (7.3-25), a legitimate question is: How can we guarantee that the numerical iterations will converge to the correct root? This question is answered by the following theorem:

Theorem 7.3-2 Let

$$g_{(\lambda)} = \sum_{i=1}^{n} \frac{w_i^2 d_i^2}{(1+2\lambda w_i^2)^2} - E^2. \tag{7.3-26}$$

Then, the iterations generated by Newton's method:

$$\lambda_{k+1} = \lambda_k - \frac{g_{(\lambda_k)}}{g'_{(\lambda_k)}}, \qquad k = 0, 1, 2, \cdots \tag{7.3-27}$$

with $\lambda_0 = 0$ will always converge to the positive root of Eq. (7.3-25). Furthermore, $\lambda_{k+1} > \lambda_k$ and $|\lambda_{k+1} - \lambda^*| < |\lambda_k - \lambda^*|$ where λ^* is the true solution. The proof of Theorem 7.3-2 is given in Appendix 7.D. ∎

A similar equation has been previously studied in [7-25] in the context of image restoration. However, the properties of its roots were not rigorously established.

D. On the Choice of the Weights in W

From the previous discussion it is clear that the weights w_i should be chosen based on the local statistics of the image and the properties of the human visual system. The pixel intensity at location (i, j) can be treated as a random variable with mean $\mu_{i,j}$ and variance $\sigma_{i,j}$. The mean serves as a measure of the local brightness, and the variance is a measure of the variability in a neighborhood around local pixel (i, j).

From the nature of both P_w and the visibility of the blocking artifacts, the weights w_i should be a decreasing function of $\sigma_{i,j}$ [7-21]. An example of such a function is

$$w_{(i,j)} = \frac{1}{1 + \sigma_{i,j}}, \tag{7.3-28}$$

where 1 is added in the denominator to avoid mathematical difficulties when $\sigma_{i,j} = 0$. A range-compressed form of this function is given by

$$w_{(i,j)} = \ln\left(1 + \frac{1}{1 + \sigma_{i,j}}\right). \tag{7.3-29}$$

In our experiments, we noticed that the blocking artifact is more visible in bright rather than in dark areas of the image. A function which captures this property is

$$w_{(i,j)} = \ln\left(1 + \frac{\sqrt{\mu_{i,j}}}{1 + \sigma_{i,j}}\right). \tag{7.3-30}$$

Several forms of the weighting function were previously suggested. Since all of them were based on the human visual system, we shall refer to them as the *visibility functions of the blocking artifact* (VFBA). The above definitions are examples of VFBAs and illustrate how human perceptual characteristics can be incorporated in a spatially-adaptive smoothness constraint set. A study of the exact form of VFBAs is beyond the scope of this paper. The appropriate form of VFBA depends also on the medium through which the images are presented to the users, e.g., hard copy or CRT monitor.

7.4 The Recovery Algorithm

The constraint set defined in Eq. (7.3-8) captures the weighted variations between the columns at the block boundaries. From the projection in Eq. (7.3-16) we see that this constraint set results in the processing of only the immediate columns at the block boundaries. This processing will yield new intensity discontinuities between the columns at the block boundaries and their neighboring columns inside the blocks. To avoid this undesired effect, constraint sets that capture the variations between the columns at block boundaries and their neighboring columns inside the blocks can be introduced in a similar fashion. This type of set can be defined also for the columns inside the blocks.

The set C_w captures only the image smoothness (horizontally) between-block columns. The smoothness constraint set C'_w captures the between block row smoothness. Similarly, constraint sets capturing also off-block boundaries between row smoothness can also be defined.

Besides the sets defined previously, the set C_T in Eq. (7.2-12) is also used. Another valuable set is the set that captures the information about the range of the pixel intensity of an image. This set is defined by

$$C_p \triangleq \{\mathbf{f} : 0 \leq f_{i,j} \leq 255, 1 \leq i, j \leq N; \}. \tag{7.4-1}$$

The projectors of sets C_T and C_p, P_T and P_p, respectively, are well known, see for example [7-20], [7-15].

Using the previous sets the POCS theory yields the following recovery algorithm:

- Set $\mathbf{f}_0 = \mathbf{f}'$.
- For $k = 1, 2, \cdots$, compute $\mathbf{f}_k$ from

$$\mathbf{f}_k = P_w P'_w P_T P_p \, \mathbf{f}_{k-1}, \tag{7.4-2}$$

  where $P_w, P_{w'}, P_T$ and P_p denote the projections onto the sets C_w, C'_w, C_T and C_p, respectively.

- Continue this iteration until $\|\mathbf{f}_k - \mathbf{f}_{k-1}\|$ is less than some prescribed bound.

7.5 A Simplified Algorithm

The algorithm in Eq. (7.4-2) involves the projectors P_w and $P_{w'}$ which requires knowledge of λ which is not available in explicit form and requires the numerical solution of Eq. (7.3-12). From an application point of view, a less computationally intensive algorithm is always preferred. Therefore, a valid question is: can we avoid some of the numerical computations required to find the projection in Eq. (7.3-16) and still maintain the adaptive nature of the recovery algorithm?

The weighting factors w_i in $\mathbf{W}$ have a continuous range of values. A natural simplification is to quantize the continuous range of the weights to a finite number of values. Assuming that $\mathbf{W}$ has only M quantized levels for its weighting factors which are denoted by $q_1, q_2, \cdots, q_M$, then, for each i, $i = 1, 2, \cdots, M$, define a set $\mathcal{I}_i$ using the following rules:

1. Assign all the pixels in the set of block boundary pixels whose corresponding weightings have values equal to q_i to set $\mathcal{I}_i$.

2. Assign all pixels off the block boundaries to the same set to which the closest boundary pixel belongs to.

For this segmentation rule we can write

$$\bigcap_{i=1}^{M} \mathcal{I}_i = \emptyset, \qquad \text{and} \qquad \bigcup_{i=1}^{M} \mathcal{I}_i = \text{the entire image} . \tag{7.5-1}$$

Define $\mathbf{I}_i$ to be the matrix operator which picks out only the pixels in the set $\mathcal{I}_i$ and sets the rest to 0. Then, we have

$$\mathbf{I} = \mathbf{I}_1 + \mathbf{I}_2 + \cdots + \mathbf{I}_M, \tag{7.5-2}$$

where $\mathbf{I}$ is the identity operator. Hence

$$\mathbf{f} = \mathbf{I}_1\mathbf{f} + \mathbf{I}_2\mathbf{f} + \cdots + \mathbf{I}_M\mathbf{f}, \tag{7.5-3}$$

and therefore,

$$\mathbf{WQ}\,\mathbf{f} = \mathbf{WQI}_1\mathbf{f} + \mathbf{WQI}_2\mathbf{f} + \cdots + \mathbf{WQI}_M\mathbf{f}. \tag{7.5-4}$$

By the definition of the $\mathbf{I}_i$'s, we have

$$\mathbf{WQf} = q_1\mathbf{QI}_1\mathbf{f} + q_2\mathbf{QI}_2\mathbf{f} + \cdots + q_M\mathbf{QI}_M\mathbf{f}, \tag{7.5-5}$$

and

$$\| \mathbf{WQf} \|^2 = q_1\| \mathbf{QI}_1\mathbf{f} \|^2 + q_2 \| \mathbf{QI}_2\mathbf{f} \|^2 + \cdots + q_M \| \mathbf{QI}_M\mathbf{f} \|^2. \tag{7.5-6}$$

Note that the set C_w defined in Eq. (7.3-8) puts an upper bound on the total variation $\|\mathbf{WQf}\|$. From Eq. (7.5-6) this constraint can be expressed in a functionally equivalent form by

$$\begin{aligned} \| \mathbf{QI}_1\mathbf{f} \| &\leq E_1 \\ \| \mathbf{QI}_2\mathbf{f} \| &\leq E_2 \\ &\vdots \\ \| \mathbf{QI}_M\mathbf{f} \| &\leq E_M, \end{aligned} \tag{7.5-7}$$

where $E_i, i = 1, 2, \cdots, M$, are constants. Consequently, we have the following constraint sets

$$\begin{aligned} C_{w1} &= \{ \mathbf{f} : \| \mathbf{QI}_1\mathbf{f} \| \leq E_1 \} \\ C_{w2} &= \{ \mathbf{f} : \| \mathbf{QI}_2\mathbf{f} \| \leq E_2 \} \\ &\vdots \\ C_{wM} &= \{ \mathbf{f} : \| \mathbf{QI}_M\mathbf{f} \| \leq E_M \}. \end{aligned} \tag{7.5-8}$$

Note that for each i, $\mathbf{I}_i\mathbf{f}$ is only a segment of the entire image and the set C_{wi} only puts a constraint on this segment of the image. The projection P_i onto the set C_{wi} is given in Eq. (7.3-6) when $\mathbf{I}_i\mathbf{f}$ is the image.

From Eq. (7.5-3), we see that in essence the image is partitioned into different segments based on the value of the VFBA. The image pixels within a segment are treated uniformly since they have nearly the same local statistical properties. Therefore, the spatially-adaptive nature of the algorithm is preserved and the computational complexity in the new approach is almost the same as that of the POCS algorithm that uses the projection in Eq. (7.3-6).

The previously defined sets capture only the smoothness between the columns at block boundaries. It is straightforward to define sets C'_{wi} that capture the smoothness between the rows of the block boundaries and sets that capture column/row off-block-boundary smoothness. Based on these sets a POCS recovery algorithm can be defined in similar fashion to Eq. (7.4-2).

7.6 Computational Complexity Analysis

In this section the computational cost of the simplified algorithm presented in Section 7.5 is compared to the cost of a traditional decoder. The complexity cost is measured in terms of numbers of real additions and multiplications. The computational cost of the segmentation process is not included in the following calculations.

Since the analyzed algorithm is iterative, the cost of each iteration will be examined first. Each iteration of the simplified algorithm in Section 7.5 is a concatenation of projection operators. Thus, the computational cost per iteration is obtained by summing the computational costs for each projector.

1. The projectors $(P_1 \cdots P_M)$ for each of the constraint sets in Eq. (7.5-8) are only defined for one segment of the entire image. Thus, the total cost for $(P_1 \cdots P_M)$ is the same as for the projection in Eq. (7.3-6). Hence, we will count the operations required in Eq. (7.3-6). This equation can be rewritten as

$$\tilde{\mathbf{f}}_i = \mathbf{f}_{i+1} + \alpha \cdot (\mathbf{f}_i - \mathbf{f}_{i+1}) \quad \text{and} \quad \tilde{\mathbf{f}}_{i+1} = \mathbf{f}_i - \alpha \cdot (\mathbf{f}_i - \mathbf{f}_{i+1}). \tag{7.6-1}$$

   For $N \times N$ images and $K \times K$ blocks, Eq. (7.6-1) requires N real multiplications and $3N$ real additions. There are $(\frac{N}{K} - 1)$ such block boundaries, hence the total cost is: $N(\frac{N}{K} - 1)$ real multiplications and $3N(\frac{N}{K} - 1)$ real additions. In addition to this, the computation of α in Eq. (7.6-1) requires approximately $N(\frac{N}{K} - 1)$ multiplications and $2N(\frac{N}{K} - 1)$ additions.

   The above calculations are only for the smoothness constraint applied only at the block boundaries. Assume that a total of $(m - 1)$ off-block-boundary smoothness constraints are also enforced on the columns inside the blocks. Then, the total computational cost for the smoothness constraints is: $2mN(\frac{N}{K} - 1)$ real multiplications and $5mN(\frac{N}{K} - 1)$ real additions. The maximum value for m is K.

2. The projectors $(P_1' \cdots P_M')$ are of the same nature as $(P_1 \cdots P_M)$, and their computational costs are identical.

3. The projector P_p is simply a thresholding operation. Every pixel outside the intensity range is thresholded to the correct value. No additions and multiplications are required for this projection.

4. The projector P_T is also a thresholding operation except that it is performed in the BDCT domain. Therefore, both forward and inverse BDCT are required to convert the data from the spatial domain to the transform domain and vice versa. The computational cost of BDCT is dependent on the specific implementations. In the algorithm reported in [7-26], for each $K \times K$ DCT transform, $K^2 \log_2 K$ real multiplications and $K(3K \log_2 K - K + 1)$ real additions are required. For an $N \times N$ image there are $(\frac{N}{K})^2$ blocks and the total number of real multiplications is $N^2 \log_2 K$ and the total number of real additions is $(3N^2 \log_2 K - N^2 + \frac{N^2}{K})$.

Therefore, the total number of operations in each iteration are: $2[2mN(\frac{N}{K} - 1)] + 2[N^2 \log_2 K]$ real multiplications and $2[5mN(\frac{N}{K} - 1) + 2[3N^2 \log_2 K - N^2 + \frac{N^2}{K}]$ real additions.

For example, when $N = 512$, $K = 8$ and for the worst case $m = K$, the computations required for each iteration are: 10 real multiplications per pixel (6 of which are for the BDCT) and 26 real additions per pixel (16 of which are for the BDCT). As we will see in Section 7.7, it only takes $3 - 5$ iterations for this algorithm to converge. Therefore, the total cost of this recovery algorithm is: $(3 \sim 5) \cdot 10$ real multiplications per pixel and $(3 \sim 5) \cdot 26$ real additions per pixel. For a traditional decoder, an inverse BDCT is required, this takes about 3 real multiplications and 8 real additions. Therefore, the proposed algorithm requires 10–17 times the computations of a conventional decoder without any post-processing.

The computations in this algorithm are dominated by the BDCT transforms. Therefore, a hardware implementation for the BDCT could speed up this algorithm significantly. Furthermore, the multiplication operation in Eq. (7.5-7) is only a scaling operation of a vector which is very amenable to parallel implementation. Finally, the total cost of the proposed recovery process is approximately the same as that of a 2-D 512×512 *fast Fourier transform* (FFT).

7.7 Experiments

In this section experiments are presented in order to test the proposed algorithms and compare them to previous approaches. The 512×512 "Lena" image is used as a test image. This image was compressed using a JPEG based coder-decoder with quantization table shown in Fig. 7.7-1 which yields a bit-rate of 0.24

bpp [7-12]. For presentation purposes the center 256×256 section of this image is shown in Fig. 7.7-2. The same section of the processed images will also be shown in what follows.

50	60	70	70	90	120	255	255
60	60	70	96	130	255	255	255
70	70	80	120	200	255	255	255
70	96	120	145	255	255	255	255
90	130	200	255	255	255	255	255
120	255	255	255	255	255	255	255
255	255	255	255	255	255	255	255
255	255	255	255	255	255	255	255

Figure 7.7-1: The quantization table used for JPEG based compression [7-12].

Figure 7.7-2: Blocky "Lena" from JPEG compression at 0.24 bpp.

As an objective measure of the distance between a reconstructed image $\mathbf{g}$ and its original image $\mathbf{f}$, we used the *peak-signal-to-noise-ratio* (PSNR). For $N \times N$ images with $[0, 255]$ gray-level range, PSNR is defined in dB by

$$\text{PSNR} = 10 \log_{10} \left[\frac{N^2 \times 255^2}{\|\mathbf{g} - \mathbf{f}\|^2} \right]. \tag{7.7-1}$$

The PSNR of the blocky image in Fig. 7.7-2 is 29.579 dB.

First the algorithm using the spatially-invariant smoothness sets C_s and C_s' in Section 7.3 is tested. The upper bounds E and E' used to define sets C_s and C_s' are determined using the received data, i.e., the blocky image $\mathbf{f}'$. Write the blocky image $\mathbf{f}'$ in column vector form as $\mathbf{f}' = \{\mathbf{f}_1', \mathbf{f}_2', ..., \mathbf{f}_{512}'\}$. Then E is estimated by

$$E = \frac{1}{7}\sum_{k=1}^{7} S_k, \tag{7.7-2}$$

where

$$S_k \triangleq \left[\sum_{i=1}^{63} \| \mathbf{f}'_{8\cdot i+k} - \mathbf{f}'_{8\cdot i+k+1} \|^2\right]^{\frac{1}{2}} \quad \text{for } k = 1, 2, \cdots, 7.$$

This E is a measure of the average vertical discontinuity between adjacent columns of the entire image and is found to yield satisfactory results when used as a bound for the set C_s. In a similar fashion the bound E' is obtained. The reconstructed image from the JPEG compressed blocky image is shown in Fig. 7.7-3. The corresponding PSNR is equal to 30.35 dB.

Figure 7.7-3: Reconstructed image using the spatially-invariant algorithm in Section 7.3.

To implement the algorithm using the spatially-adaptive smoothness sets C_w and C'_w in Section 7.3, the weights in Eq. (7.3-7) are computed based on Eq. (7.3-30) using the transmitted data. The weight matrices for the column and row smoothness constraint sets are $\mathbf{W}$ and $\mathbf{W}'$, respectively. Thus, both means μ and μ' corresponding to $\mathbf{W}$ and $\mathbf{W}'$, respectively, and both variances σ and σ' must be estimated. The weights w_{ij} and w'_{ij} used for the implementation of our algorithm were constant within an 8×8 region surrounding a vertical or horizontal block boundary, respectively.

Since the DC coefficient in each block is the average (up to a constant) of the pixel intensity within this block, the mean μ was estimated as follows. Consider the vertical boundary l and let DC_L, DC_R denote the DC coefficients of the left and right blocks, respectively. Then, the estimate of the mean μ_l used for the computation of the weights in $\mathbf{W}$ is given by

$$\hat{\mu}_l = \frac{DC_L + DC_R}{2 \times 8^2}. \tag{7.7-3}$$

A similar equation is used to estimate the mean μ'_l used for the computation of the weights in $\mathbf{W}'$. However, in this case the DC coefficients of the blocks above and below this horizontal boundary l' are used.

The variance σ_l at the vertical block boundary l is estimated by

$$\hat{\sigma}_l = \sqrt{\frac{VAC_L + VAC_R}{2 \times \ 8^2}}, \tag{7.7-4}$$

where VAC_L and VAC_R are the sums of the squared AC coefficients in the first column of the block left and right of the boundary l, respectively. In a similar fashion $\sigma'_{l'}$ is estimated using the sums of the squared AC coefficients in the first row of the blocks above and below the horizontal boundary l'. This choice of estimators for the variances captures the local properties of the image in the direction that smoothing is performed.

After proper scaling, the weight maps for the "Lena" image corresponding to $\mathbf{W}$ and $\mathbf{W}'$ are shown in Figs. 7.7-4a) and b), respectively. In the bright areas of these maps the blocking artifact is more visible than in the dark areas.

For the implementation of the spatially-adaptive smoothness sets, the upper bound E in Eq. (7.3-8) has to be determined. It was estimated from the received data as follows. Write the blocky image $\mathbf{f}'$ in Eq. (7.2-11) in its column vector form

$$\mathbf{f}' = \{ \mathbf{f}'_1, \mathbf{f}'_2, ..., \mathbf{f}'_{512} \}. \tag{7.7-5}$$

For $k = 0, 1, 2, \cdots, 7$, define

$$S_k \triangleq \left[\sum_{i=1}^{63} \| \mathbf{w}_k \left(\mathbf{f}'_{8 \cdot i + k} - \mathbf{f}'_{8 \cdot i + k + 1} \right) \|^2 \right]^{\frac{1}{2}}, \tag{7.7-6}$$

where $\mathbf{w}_k$ is defined in Eq. (7.3-14). Note that $S_0 = \| \mathbf{W\,Q\,f}' \|$. In our experiments, we found that S_0 is about 10 to 100 times larger than the rest of the S_k's. We determined E by

$$E = \frac{1}{7}\sum_{k=1}^{7} S_k. \tag{7.7-7}$$

A simpler approach is to choose E such that $0.01S_0 < E < 0.1S_0$. The bound E' can be determined similarly. The algorithm using the spatially-adaptive smoothness sets is implemented and the center part of the reconstructed image is shown in Fig. 7.7-5. The corresponding PSNR is 30.426 dB.

To implement the algorithm in Section 7.5, the image pixels are classified according to their corresponding weighting factors. In this experiment, we found that $M = 3$ levels yields satisfactory results. The image is segmented into three regions: one with high, one with medium and one with low values of the weights. The segmentation scheme used is given below:

Define

$$\bar{w} = \frac{1}{7}\sum_{i=1}^{63\cdot 512} w_i \quad \text{and} \quad \bar{s} = \frac{1}{63\cdot 512}\left[\sum_{i=1}^{63\cdot 512}(w_i - \bar{w})^2\right]^{\frac{1}{2}}. \tag{7.7-8}$$

Then $w_i, i = 1, 2, \cdots, 63\cdot 512$, are classified according to following rule:

- if $w_i < \bar{w} - \bar{s}/2$, then assign it to class 1;
- if $w_i > \bar{w} + \bar{s}/2$, then assign it to class 3;
- otherwise, assign it to class 2.

The resulting weight maps obtained by quantizing the weight maps in Figs. 7.7-4a) and b) are shown in Figs. 7.7-6a) and b), respectively. As one can see, the maps in Figs. 7.7-6a) and b) are very close to those in Figs. 7.7-4a) and b), respectively. After the image is segmented, the upper bounds E_i in Eq. (7.5-8) have to be determined. This can be done by the scheme used to determine E. Then, the recovery algorithm can be applied and the reconstructed image is shown in Fig. 7.7-7. The corresponding PSNR is 30.385 dB.

For comparison purposes, in Fig. 7.7-8 we show the reconstructed image using the JPEG AC prediction recommendation in Annex-K.8.2. The corresponding PSNR is 29.525 dB.

All the previously shown images are obtained after 5 iterations of the respective algorithms. To illustrate the convergence properties of the proposed algorithms, the average difference per pixel between-iterates given by $\frac{\| \mathbf{f}_k - \mathbf{f}_{k-1} \|}{N^2}$ and the PSNR of the recovered images are plotted versus the number of iterations in Figs. 7.7-10 and 7.7-9, respectively. The proposed algorithms are tested in a number of other experiments using both higher and lower compression ratios. In all cases the proposed approaches yield satisfactory reconstructions.

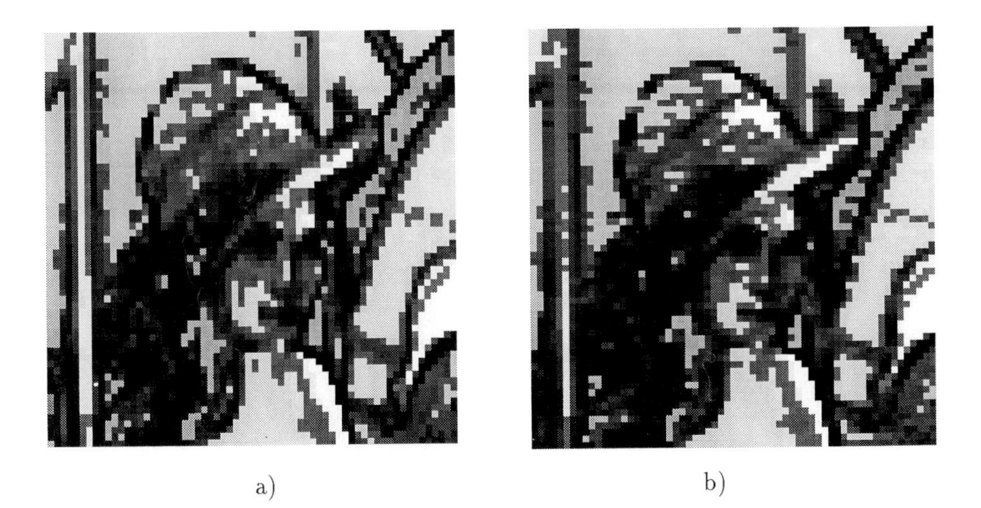

a) b)

Figure 7.7-4: The weight maps of the blocky "Lena": a) horizontal; b) vertical.

Figure 7.7-5: Reconstructed image using the spatially-variant algorithm in Section 7.3.

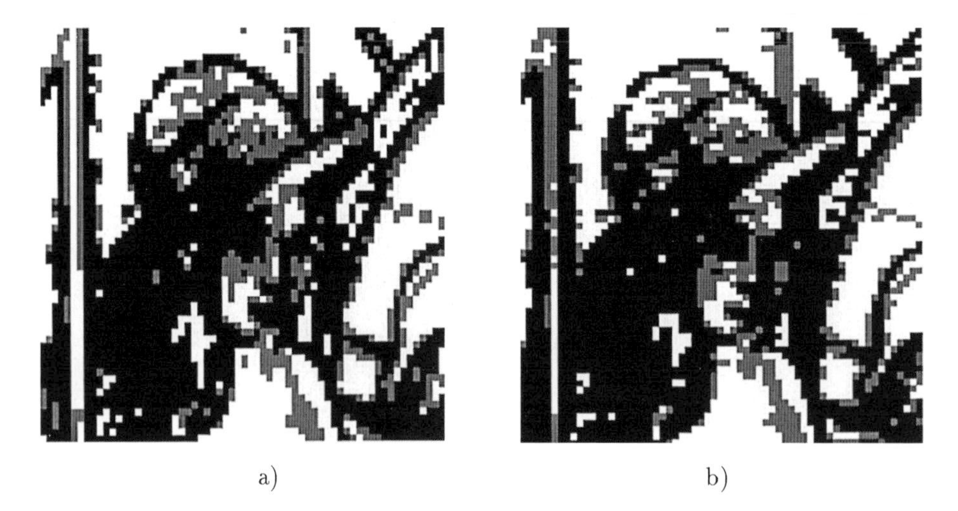

Figure 7.7-6: The segmented map of the blocky "Lena": a) horizontal; b) vertical.

Figure 7.7-7: Reconstructed image using the simplified algorithm in Section 7.5.

Figure 7.7-8: Reconstructed image using JPEG recommended AC prediction.

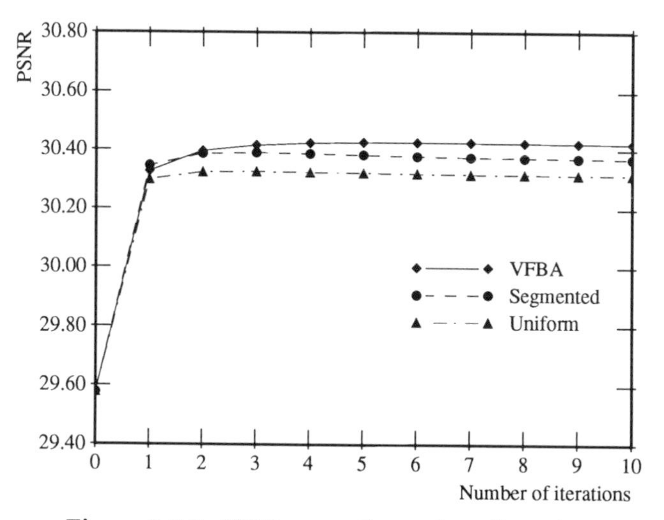

Figure 7.7-9: PSNR versus the number of iterations.

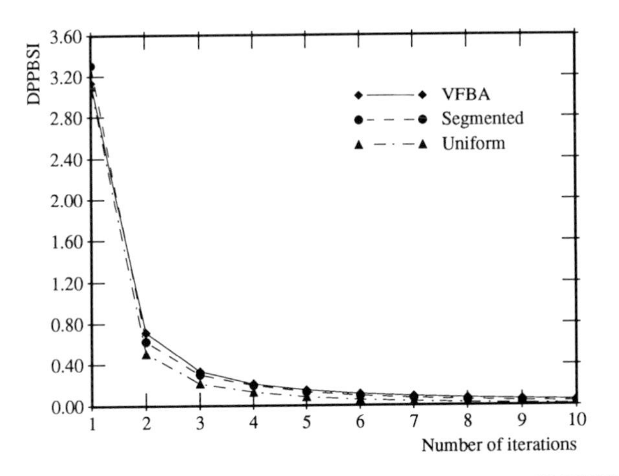

Figure 7.7-10: Difference per pixel between successive iterates (DPPBSI) versus the number of iterations.

Summary

A new approach based on the theory of POCS is proposed for image decompression. According to this approach the image is *recovered* using both the transmitted data and prior knowledge about the original image. This prior knowledge is available in the decoder, does not burden the transmission with overhead, and complements the transmitted data. Thus, at the same bit rates the reconstructed images by this approach are of higher quality than those of traditional decoders which only use the transmitted data.

We also presented a new spatially-adaptive smoothness constraint set that captures the properties of the human visual system and the local statistical properties of the image. Using this spatially adaptive set we were able to recover better images, both visually and closer to the original based on an objective metric, than when the spatially-invariant set in Eq. (7.3-2) is used. The penalty for this improvement in image quality is an increase in the complexity of the recovery algorithm. In an effort to combine the advantages of adaptive weighting and simplicity of recovery, we introduced a simplified but still spatially adaptive recovery algorithm that yielded only slightly inferior results.

This work is one more example of the wide range of problems that can be solved using the theory of POCS. However, the flexibility of this approach is also its main difficulty. This approach is too general and does not provide any guidelines for the selection of the convex sets that have to be used. It is up to the individual researcher to define the sets that best capture the properties to impose on the solution. This has become apparent to the authors during the course of this work. One of the main shortcomings of the recovery algorithm given in this paper is that it does not explicitly incorporate edge information. As is well known, edges are very important visual features in images. Thus, the next step in the "evolution" of this work is to explicitly incorporate edge information in this algorithm.

Appendix 7.A: Derivation of the projection onto C_2

In the vector space $R^n \times R^n$, a vector is expressed as a 2-tuple of vectors in R^n, i.e., $(\mathbf{x}, \mathbf{y})$, where $\mathbf{x}, \mathbf{y} \in R^n$. For $(\mathbf{x}, \mathbf{y}), (\mathbf{u}, \mathbf{v}) \in R^n \times R^n$, their natural inner product is defined by

$$\langle (\mathbf{x}, \mathbf{y}), (\mathbf{u}, \mathbf{v}) \rangle = \langle \mathbf{x}, \mathbf{u} \rangle + \langle \mathbf{y}, \mathbf{v} \rangle, \tag{7.A-1}$$

where the inner product in the right-hand side is the natural inner product defined in R^n. The inner product defined in Eq. (7.A-1) also induces a norm given by

$$\| (\mathbf{x}, \mathbf{y}) \| = \left[\|\mathbf{x}\|^2 + \|\mathbf{y}\|^2 \right]^{\frac{1}{2}}, \tag{7.A-2}$$

where $\| \cdot \|$ in the right-hand side is the l_2 norm in R^n.

Define

$$C = \{ (\mathbf{x}, \mathbf{y}) : \|\mathbf{x} - \mathbf{y}\| \le E, \, \mathbf{x} \in R^n, \, \mathbf{y} \in R^n \}.$$

Clearly, C is a subset of the vector space $R^n \times R^n$. It is straightforward to show that the set C is closed and convex. Then we have the following.

Lemma 7.A-1 For an arbitrary vector $(\mathbf{x}, \mathbf{y}) \in R^n \times R^n$, the projection $P(\mathbf{x}, \mathbf{y})$ onto the set C is given by

$$P(\mathbf{x}, \mathbf{y}) = \begin{cases} (\mathbf{x}, \mathbf{y}) & \text{if} \quad \|\mathbf{x} - \mathbf{y}\| \le E \\ (\alpha \mathbf{x} + (1-\alpha)\mathbf{y}, \, (1-\alpha)\mathbf{x} + \alpha \mathbf{y}) & \text{otherwise} \end{cases}$$

where $\alpha = \left[\frac{E}{\|\mathbf{x}-\mathbf{y}\|} + 1 \right] / 2$.

Proof: By definition, $P(\mathbf{x}, \mathbf{y})$ is the vector in set C satisfying

$$\|(\mathbf{x}, \mathbf{y}) - P(\mathbf{x}, \mathbf{y})\| = \min_{(\mathbf{u},\mathbf{v}) \in R^n \times R^n} \|(\mathbf{x}, \mathbf{y}) - (\mathbf{u}, \mathbf{v})\|,$$

where $(\mathbf{x}, \mathbf{y}) - (\mathbf{u}, \mathbf{v}) = (\mathbf{x} - \mathbf{u}, \mathbf{y} - \mathbf{v})$, and $(\mathbf{u}, \mathbf{v}) \in C$. Clearly, if $(\mathbf{x}, \mathbf{y}) \in C$, i.e., $\|\mathbf{x} - \mathbf{y}\| \le E$, then $P(\mathbf{x}, \mathbf{y}) = (\mathbf{x}, \mathbf{y})$.

Suppose that $(\mathbf{x}, \mathbf{y}) \notin C$; then $P(\mathbf{x}, \mathbf{y})$ can be found by minimizing the Lagrange auxiliary function

$$J_\lambda = \|(\mathbf{x}, \mathbf{y}) - (\mathbf{u}, \mathbf{v})\|^2 + \lambda [\|\mathbf{u} - \mathbf{v}\|^2 - E^2].$$

Taking the gradients of J_λ with respect to vectors $\mathbf{u}$ and $\mathbf{v}$ and setting them to 0, yields

$$\mathbf{u} - \mathbf{x} + \lambda(\mathbf{u} - \mathbf{v}) = 0 \qquad \text{and} \qquad \mathbf{v} - \mathbf{y} + \lambda(\mathbf{v} - \mathbf{u}) = 0.$$

Therefore

$$\mathbf{u} = \alpha\,\mathbf{x} + (1-\alpha)\,\mathbf{y} \qquad \text{and} \qquad \mathbf{v} = (1-\alpha)\,\mathbf{x} + \alpha\,\mathbf{y},$$

where $\alpha = \frac{1+\lambda}{1+2\lambda}$. Furthermore, setting $\|(\mathbf{u}-\mathbf{v})\| = E$, λ can be found to be $\lambda = \left[\frac{\|\mathbf{x}-\mathbf{y}\|}{E} - 1\right]/2$. ∎

Using the above lemma the projector onto C_2 can be found as follows: Writing the $N \times N$ image $\mathbf{f}$ in its column vector form as in Eq. (7.3-3) , the projection of $\mathbf{f}$ onto the set C_2 is the vector $\tilde{\mathbf{f}}$ in C_2 that minimizes the distance function

$$\|\mathbf{f} - \tilde{\mathbf{f}}\|^2 = \sum_{i=1}^{N} \|\mathbf{f}_i - \tilde{\mathbf{f}}_i\|^2, \tag{7.A-3}$$

where $\tilde{\mathbf{f}} = \{\tilde{\mathbf{f}}_1, \tilde{\mathbf{f}}_2, ..., \tilde{\mathbf{f}}_N\} \in C_2$. In the case of $N = 512$ and 8×8 blocks $Q\mathbf{f}$ is given by Eq. (7.3-4). Thus, $\|Q\,\tilde{\mathbf{f}}\| \leq E$ only constrains columns in the block boundaries. Define

$$\mathbf{x} = \begin{bmatrix} \mathbf{f}_8 \\ \mathbf{f}_{16} \\ \cdot \\ \cdot \\ \cdot \\ \mathbf{f}_{504} \end{bmatrix} \qquad \text{and} \qquad \mathbf{y} = \begin{bmatrix} \mathbf{f}_9 \\ \mathbf{f}_{17} \\ \cdot \\ \cdot \\ \cdot \\ \mathbf{f}_{505} \end{bmatrix}.$$

Then $Q\,\tilde{\mathbf{f}} = \mathbf{x} - \mathbf{y}$. It follows immediately from Lemma 7.A-1 that the distance function defined in Eq. (7.A-3) is minimized when

$$\tilde{\mathbf{f}}_i = \alpha \cdot \mathbf{f}_i + (1-\alpha) \cdot \mathbf{f}_{i+1} \qquad \text{and} \qquad \tilde{\mathbf{f}}_{i+1} = (1-\alpha) \cdot \mathbf{f}_i + \alpha \cdot \mathbf{f}_{i+1}$$

for $i = 8 \cdot k$ and $k = 1, 2, ..., 31$, where $\alpha = \frac{1}{2}\left[\frac{E}{\|Q \cdot \mathbf{f}\|} + 1\right]$; otherwise, $\tilde{\mathbf{f}}_i = \mathbf{f}_i$.

Appendix 7.B: Derivation of the projector onto C_w

First, consider the following lemma.

Lemma 7.B-1 Let $\mathbf{x}_0, , \mathbf{y}_0, \mathbf{u}$ and $\mathbf{v}$ be vectors in the space R^n with the Euclidean norm $\|\cdot\|$. Let also $\mathbf{W}$ be an arbitrary $n \times n$ matrix. Then under the constraint that $\|\,\mathbf{W}(\mathbf{u}-\mathbf{v})\,\| \leq E$, the functional

$$\psi(\mathbf{u}, \mathbf{v}) = \|\,\mathbf{u} - \mathbf{x}_0\,\|^2 + \|\,\mathbf{v} - \mathbf{y}_0\,\|^2$$

is minimized when

$$\begin{aligned}\mathbf{u} &= \tfrac{1}{2}\,[\mathbf{I} + (\mathbf{I} + 2\lambda\mathbf{W}^T\mathbf{W})^{-1}]\,\mathbf{x}_0 + [\mathbf{I} - (\mathbf{I} + 2\lambda\mathbf{W}^T\mathbf{W})^{-1}]\,\mathbf{y}_0 \\ \mathbf{v} &= \tfrac{1}{2}\,[\mathbf{I} - (\mathbf{I} + 2\lambda\mathbf{W}^T\mathbf{W})^{-1}]\,\mathbf{x}_0 + [\mathbf{I} + (\mathbf{I} + 2\lambda\mathbf{W}^T\mathbf{W})^{-1}]\,\mathbf{y}_0,\end{aligned}$$

where λ is to be solved by $\| W(\mathbf{u} - \mathbf{v}) \| = E$.

Proof: Form the Lagrange auxiliary function

$$J_\lambda(\mathbf{u}, \mathbf{v}) = \psi(\mathbf{u}, \mathbf{v}) + \lambda\,(\,\| \mathbf{W}(\mathbf{u} - \mathbf{v}) \|^2 - E^2\,).$$

Taking the gradients of $J_\lambda(\mathbf{u}, \mathbf{v})$ with respect to vectors $\mathbf{u}$ and $\mathbf{v}$ and setting them to 0, yields

$$(\mathbf{u} - \mathbf{x}_0) + \lambda\,(\,\mathbf{W}^T\mathbf{W}(\mathbf{u} - \mathbf{v})\,) = 0,$$

and

$$(\mathbf{v} - \mathbf{y}_0) + \lambda\,(\,\mathbf{W}^T\mathbf{W}(\mathbf{v} - \mathbf{u})\,) = 0.$$

Therefore

$$\mathbf{u} + \mathbf{v} = \mathbf{x}_0 + \mathbf{y}_0 \quad \text{and} \quad (\mathbf{I} + 2\lambda\mathbf{W}^T\mathbf{W})(\mathbf{u} - \mathbf{v}) = \mathbf{x}_0 - \mathbf{y}_0,$$

and the result follows immediately. ∎

Using Lemma 7.B-1, the projector P_w can be derived as follows. Writing the image $\mathbf{f}$ in its column vector form, the projection of $\mathbf{f}$ onto the set C_w is the vector $\tilde{\mathbf{f}}$ in C_w that minimizes the distance function

$$\| \mathbf{f} - \tilde{\mathbf{f}} \|^2 = \sum_{i=1}^{N} \| \mathbf{f}_i - \tilde{\mathbf{f}}_i \|^2, \tag{7.B-1}$$

where $\tilde{\mathbf{f}} = \{\tilde{\mathbf{f}}_1, \tilde{\mathbf{f}}_2, ..., \tilde{\mathbf{f}}_N\} \in C_w$.

By the definition of $Q\mathbf{f}$ given in Eq. (7.3-4), the relation $\|WQ\tilde{\mathbf{f}}\| \leq E$ only puts constraint columns at the block boundaries of $\tilde{\mathbf{f}}$. Therefore, for all columns not at block boundaries, $\tilde{\mathbf{f}}_i = \mathbf{f}_i$. Also, with the notations in Eq. (7.3-10), we have $Q\tilde{\mathbf{f}} = \mathbf{u} - \mathbf{v}$. The projection in Eq. (7.3-11) follows from Lemma 7.B-1.

Appendix 7.C: Proof of Theorem 7.3-1

First, consider the following lemma.

Lemma 7.C-2 For any real numbers $x < 0, y > 0$ and $|x| > y$, the following holds,

$$\left| \frac{x}{1 + x} \right| > \frac{y}{1 + y}$$

Proof: Consider the following two cases:

Case 1: $x < -1$. We have $|\frac{x}{1+x}| = |1 - \frac{1}{1+x}| = 1 + |\frac{1}{1+x}| > 1 > \frac{y}{1+y}$.

Case 2: $x > -1$. Then $0 < x + 1 < 1$. We have $|\frac{x}{1+x}| = \frac{|x|}{1+x} > |x| > y > \frac{y}{1+y}$. ∎

Proof of Theorem 7.3-1:

1. Let
$$g_{(\lambda)} = \sum_{i=1}^{n} \frac{w_i^2 d_i^2}{(1+2\lambda w_i^2)^2} - E^2.$$
Then
$$g_{(0)} = \parallel WQ\mathbf{f} \parallel - E \geq 0,$$
since $\mathbf{f} \notin C_w$. Also
$$\lim_{\lambda \to +\infty} g_{(\lambda)} = -E \leq 0.$$
Since $g_{(\lambda)}$ is continuous for all $\lambda \geq 0$, there must be a $\lambda_+ \in (0, \infty)$ such that $g_{(\lambda_+)} = 0$. Note also that $g_{(\lambda)}$ is strictly decreasing for all $\lambda \geq 0$. The uniqueness of λ_+ follows.

2. Let
$$w_{min}^2 = \min\{w_i^2, i = 1, 2, \cdots, n\}$$
and let
$$\lambda_{max} = -\frac{1}{2w_{min}^2}.$$
Then $g_{(\lambda)}$ is continuous for all $\lambda \in (-\infty, \lambda_{max})$. Note that
$$\lim_{\lambda \to -\infty} g_{(\lambda)} = -E \leq 0 \quad \text{and} \quad \lim_{\lambda \to \lambda_{max}} g_{(\lambda)} = +\infty.$$
Therefore, there must be a $\lambda_- \in (-\infty, \lambda_{max})$ such that $g_{(\lambda_-)} = 0$. $g_{(\lambda)}$ may possibly have some other negative roots.

3. For $\lambda \geq 0$, we have
$$(1 - 2\lambda w_i^2)^2 < (1 + 2\lambda w_i^2)^2,$$
for each w_i. Therefore, for $\lambda > 0$, $g_{(-\lambda)} > g_{(\lambda)}$.
Let λ_+ denote the positive root of Eq. (7.3-25). Since $g_{(\lambda)} > 0$ for all $\lambda \in (0, \lambda_+)$, there won't be any root for $g_{(\lambda)}$ in the interval $[-\lambda_+, 0)$. Therefore, all the negative roots have a larger magnitude than the positive root.

4. From Eq. (7.3-11), the projection $\tilde{\mathbf{f}}$ of the image $\mathbf{f}$ is a function of λ. Let
$$D_{(\lambda)} = \parallel \tilde{\mathbf{f}} - \mathbf{f} \parallel^2.$$
Then from Eq. (7.3-11), we have
$$D_{(\lambda)} = \parallel \mathbf{u} - \mathbf{x}_0 \parallel^2 + \parallel \mathbf{v} - \mathbf{y}_0 \parallel^2. \tag{7.C-1}$$
From Eq. (7.3-11), we also have
$$\mathbf{u} + \mathbf{v} = \mathbf{x}_0 + \mathbf{y}_0 \quad \text{and} \quad \mathbf{u} - \mathbf{v} = (\mathbf{I} + 2\lambda \mathbf{W}^T\mathbf{W})^{-1}(\mathbf{x}_0 - \mathbf{y}_0).$$

Therefore,

$$\mathbf{u} - \mathbf{x}_0 = \mathbf{y}_0 - \mathbf{v},$$

and Eq. (7.C-1) can be rewritten as

$$\begin{aligned} D_{(\lambda)} &= 2\| \mathbf{u} - \mathbf{x}_0 \|^2 \\ &= 2\| \frac{1}{2}[(\mathbf{u} - \mathbf{x}_0) + (\mathbf{u} - \mathbf{x}_0)] \|^2 \\ &= \frac{1}{2}\| (\mathbf{u} - \mathbf{x}_0) + (\mathbf{y}_0 - \mathbf{v}) \|^2 \\ &= \frac{1}{2}\| (\mathbf{u} - \mathbf{v}) + (\mathbf{x}_0 - \mathbf{y}_0) \|^2 \\ &= \frac{1}{2}\| [(\mathbf{I} + 2\lambda \mathbf{W}^T\mathbf{W})^{-1} - \mathbf{I}](\mathbf{x}_0 - \mathbf{y}_0) \|^2 \\ &= \frac{1}{2}\| [(\mathbf{I} + 2\lambda \mathbf{W}^T\mathbf{W})^{-1} - \mathbf{I}]Q\mathbf{f} \|^2. \end{aligned}$$

From Eq. (7.3-23), we have

$$\begin{aligned} D_{(\lambda)} &= \frac{1}{2}\| [(\mathbf{I} + 2\lambda \mathbf{W}^T\mathbf{W})^{-1} - \mathbf{I}]\mathbf{d} \|^2 \\ &= \sum_{i=1}^{n} (\frac{2\lambda w_i^2}{1 + 2\lambda w_i^2})^2 d_i^2. \end{aligned}$$

Let λ_- denote a negative root of Eq. (7.3-25) and λ_+ be its positive root. Then we have $|\lambda_-| > \lambda_+$. Hence from Lemma 7.C-2 we have

$$(\frac{2\lambda_- w_i^2}{1 + 2\lambda_- w_i^2})^2 > (\frac{2\lambda_+ w_i^2}{1 + 2\lambda_+ w_i^2})^2,$$

for each i. Therefore, we have

$$D_{(\lambda_-)} > D_{(\lambda_+)}.$$

The theorem follows since λ_- is an arbitrary negative root. ∎

Appendix 7.D: Proof of Theorem 7.3-2

Note that the function $g_{(\lambda)}$ has derivative

$$g'_{(\lambda)} = -4 \sum_{i=1}^{n} \frac{w_i^4 d_i^2}{(1 + 2\lambda w_i^2)^3}$$

which is continuous for $\lambda \geq 0$. Note also that $g'_{(\lambda)}$ is an increasing function of λ and $g'_{(\lambda)} < 0$ for all $\lambda \geq 0$. If $\lambda_k \geq 0$ is such that $g_{(\lambda_k)} > 0$, then from Newton's iteration

$$\lambda_{k+1} = \lambda_k - \frac{g_{(\lambda_k)}}{g'_{(\lambda_k)}},$$

we have $\lambda_{k+1} > \lambda_k$. On the other hand,

$$g_{(\lambda_{k+1})} = \int_{\lambda_k}^{\lambda_{k+1}} g'_{(\lambda)}\, d\lambda + g_{(\lambda_k)}.$$

For $\lambda \in [\lambda_k, \lambda_{k+1}]$, $g'_{(\lambda)} > g'_{(\lambda_k)}$. It follows that

$$g_{(\lambda_{k+1})} > \int_{\lambda_k}^{\lambda_{k+1}} g'_{(\lambda_k)}\, d\lambda + g_{(\lambda_k)} = g'_{(\lambda_k)}(\lambda_{k+1} - \lambda_k) + g_{(\lambda_k)} = 0.$$

Therefore, $g_{(\lambda_{k+1})} > 0$ also. Hence, for $\lambda_0 = 0$, Newton's iteration

$$\lambda_{k+1} = \lambda_k - \frac{g_{(\lambda_k)}}{g'_{(\lambda_k)}},$$

will generate a sequence

$$0 = \lambda_0 < \lambda_1 < \lambda_2 < \cdots$$

which will converge to the positive root λ_+, since $g_{(\lambda_k)} > 0 = g_{(\lambda_+)}$ guarantees that $\lambda_k < \lambda_+$ for all $k = 0, 1, 2, \cdots$. ∎

References

[7-1] C.E. Shannon, "A mathematical theory of communications," *Bell Syst. Tech. J.*, vol. 27, pt.I, pp. 379-423, pt. II, pp. 626-656, 1948.

[7-2] A.K. Jain, *Fundamentals of Digital Image Processing,* Prentice Hall, 1989.

[7-3] K. Ramchandran and M. Vetterli, "Best wavelet packet bases in a rate-distortion sense," *IEEE Trans. Image Processing,* vol. IP-2, No. 2, pp. 160-175, April 1993.

[7-4] Committee Draft ISO/IEC CD 10918-1, *Digital Compression and Coding of Continuous-Tone Still Images, Part 1: Requirements and Guidelines*, Mar. 15, 1991.

[7-5] ISO/IEC DIS 11172, *Coding of Moving Pictures and Associated Audio for Digital Storage Media up to about 1.5 Mbits/s*, 1992.

[7-6] J.M. Shapiro, "Embedded image coding using zero trees of wavelet coefficients," *IEEE Trans. Signal Processing*, vol. SP-41, No. 12, pp. 3445-3462, Dec. 1993.

[7-7] Y. Huang, H.M. Dreizen and N.P. Galatsanos, "Prioritized DCT for compression and progressive transmission of images," *IEEE Trans. Image Processing,* vol. 1, No. 4, Oct. 1992.

[7-8] M. Rabani and P. Jones, *Digital Image Compression*, Bellingham, WA: SPIE, 1991.

[7-9] H.C. Reeves and J.S. Lim, "Reduction of blocking effects in image coding," *Optical Eng.*, vol. 23, No. 1, pp. 34-37, Jan./Feb. 1984.

[7-10] B. Ramamurthi and A. Gersho, "Nonlinear space-variant postprocessing of block coded images," *IEEE Trans. Acoust., Speech and Signal Processing,* vol. ASSP-34, No. 5, pp. 1258-1267, Oct. 1986.

[7-11] K. Sauer, "Enhancement of low bit-rate coded images using edge detection and estimation," *Computer Vision Graphics and Image Processing: Graphical Models and Image Processing*, vol. 53, No.1, pp. 52-62., Jan. 1991.

[7-12] R. Rosenholtz and A. Zakhor, "Iterative procedures for reduction of blocking effects in transform image coding," *IEEE Trans. Circts. and Systs. for Video Tech.*, vol. 2, No. 1, pp. 91-94, Mar. 1992.

[7-13] R.L. Stevenson, "Reduction of coding artifacts in transform image coding," *Proc. of the IEEE ICASSP*, pp. 401-404, 1993.

[7-14] D.C. Youla, "Mathematical theory of image restoration by the method of convex projections," Chapter 2 in *Image Recovery: Theory and Application,* H. Stark, (Ed.), Academic Press, 1987.

[7-15] Y. Yang, N. Galatsanos and A. Katsaggelos, "Regularized reconstruction to reduce blocking artifacts of block discrete cosine transform compressed images," *IEEE Trans. Circts. and Systs. for Video Tech.*, vol. 3, No. 6, pp. 421-432, Dec. 1993.

[7-16] A. Tikhonov and V. Arsenin, *Solution of Ill-Posed Problems*, John Wiley and Sons, 1977.

[7-17] G. Demoment, "Image reconstruction and restoration: overview of common estimation problems," *IEEE Trans. Acoust., Speech and Signal Processing,* vol. ASSP-37, pp. 2024-2036, Dec. 1989.

[7-18] N.P. Galatsanos, and A.K. Katsaggelos, "Methods for choosing the regularization parameter and estimating the noise variance in image restoration and their relation," *IEEE Trans. Image Processing*, vol. IP-1, No. 3, pp. 322-336, July 1992.

[7-19] P.L. Combettes, "The foundations of set theoretic estimation," *Proc. of the IEEE*, vol. 81, No. 2, Feb. 1993.

[7-20] H. Stark, (Ed.), *Image Recovery: Theory and Application,* Academic Press, 1987.

[7-21] G. Anderson and A. Netravali, "Image restoration based on a subjective criterion," *IEEE Trans. Systems, Man, and Cybernetics,* vol. 6, No. 12, pp. 845-853, Dec. 1976.

[7-22] A.K. Katsaggelos, J. Biemond, R.M. Mersereau, and R.W. Schafer, "Nonstationary iterative image restoration ," *Proc. of the IEEE ICASSP*, pp. 696-699, Mar. 1985.

[7-23] R.L. Lagendijk, J. Biemond and D.E. Boekee, "Regularized iterative restoration with ringing reduction," *IEEE Trans. Acoust., Speech and Signal Processing*, vol. ASSP-36, No. 12, pp. 1874-1888, Dec. 1988.

[7-24] A.K. Katsaggelos, J. Biemond, R.M. Mersereau, and R.W. Schafer, "A regularized iterative image restoration algorithm," *IEEE Trans. Signal Processing*, vol. 39, No. 4, pp. 914-929, April 1991.

[7-25] H.J. Trussell and M.R. Civanlar, "The feasible solution in signal restoration," *IEEE Trans. Acoust., Speech and Signal Processing,* vol. ASSP-32, No. 2, pp. 201-212, April 1984.

[7-26] B.G. Lee, "A new algorithm to compute the discrete cosine transform," *IEEE Trans. Acoust., Speech, and Signal Processing*, vol. ASSP-32, No. 6, pp. 1243-1245, Dec. 1984.

Chapter 8

Non-Orthogonal Expansion for Template Matching and Edge Detection†

K. RAGHUNATH RAO and JEZEKIEL BEN-ARIE

Introduction

Template matching is a fundamental operation in any application that involves detection or recognition of patterns in multi-dimensional signals. In template matching, one is given an input image (which could be visual, infrared, radar, sonar or any other kind of sensory data) and is asked to detect and locate a known template in this input image. The word template refers to the sub-image or pattern which is to be recognized in the input image. In vision and image understanding tasks, the first stage of extracting any kind of image description from raw image data involves template matching in some form [8-1]–[8-4]. Even traditional methods for detecting edges, corners, junctions and other features rely upon template matching with appropriate detectors.

Matching can be regarded as the process of finding the similarity between two signals. An optimal approach to template matching is to quantify the similarity between the input image and the template image, and develop a matching method that maximizes this similarity measure for the signal of interest. Conversely, one could minimize a dissimilarity measure (or error) between the two images. The non-linear filters in Chapter 3 can give good results. However, in this chapter, we concern ourselves with linear methods and show that linear filters can also give

†This work was supported by the Advanced Research Projects Agency under ARPA/ONR Grant No. N00014-93-1-1088

dramatic results.

If one considers the absolute difference between two images as a measure of their dissimilarity, one can use image subtraction [8-1], [8-5] to detect known templates in the input image. However, this simple method is very sensitive to variations in gray level (brightness), noise, and deformations/distortions of the template. Due to this lack of robustness, image subtraction is not widely used except in controlled environments (lighting, etc.) and in applications that require very simple processing.

Perhaps the most widely used similarity measure is the *signal-to-noise ratio* (SNR) which is maximized by the matched filter, [8-5], [8-6], sometimes referred to as the prewhitening matched filter. When the additive noise is already white, matched filtering reverts to direct correlation of the image with the template. Assuming that the input image consists of the template with additive noise, the SNR is defined by the ratio of the peak filter response to the template, versus the power of the response to the noise alone. Note that this measure does not consider the off-center response to the template itself. Thus, the widely used matched filter produces broad peaks and spurious off-center responses which hamper good detection and localization. Furthermore, correlation frequently fails to match templates that are partially occluded (as might occur in visual images) or linearly superposed (as might occur in radar or sonar signals). Another problem with correlation and correlation-based techniques is that the result is sensitive to variation in the brightness of the input image. An improved method is the normalized correlation [8-1], which reduces this sensitivity but does not alleviate the problems of occlusion and superposition.

The method of *expansion matching* (EXM) [8-7]–[8-9] overcomes the above problems of the matched filtering approach. EXM optimizes a new similarity measure called the *discriminative signal-to-noise ratio* (DSNR). In contrast to the SNR, the DSNR is defined by the ratio of the response to the template at its center to the power of all off-center responses including the off-center response to the template itself. Thus, the ideal response sought is an impulse at the template-center. Such a response would considerably simplify the detection process and improve localization. The experiments in Fig. 8.5-1 and Fig. 8.5-2 demonstrate these effects.

This chapter is organized as follows. Section 8.1 presents some background on the SNR criterion and matched filtering (the correlation approach). Section 8.2 introduces the DSNR criterion and presents a generalized derivation of the optimal DSNR EXM filter which is not limited to real signals but is expanded to the complex-valued signal domain for matching multiple dissimilar templates with a single filter. Here, given a set of complex-valued templates to be matched, we seek the optimal filter that elicits user-defined responses from each template while maximizing the DSNR criterion. Such a method could be used to match any signal with vectorial elements of two orthogonal components, such as the edge gradient images resulting from edge detection along two orthogonal directions. Another application is the shift-invariant matching of complex spectra in the frequency domain. When considering real signals, the result simplifies to the previously developed EXM method for matching real templates. This formulation is generalized and includes additive noise as a parameter of the input image. Special cases of this optimal DSNR filter [8-9] correspond to previous methods which are limited to real images, such as the *synthetic discriminant function* (SDF) [8-10] and *minimum average correlation energy* (MACE) [8-11].

Section 8.3 addresses the special case of matching a single template and shows that it corresponds exactly to *minimum squared error* (MSE) restoration. This implies that to optimally match a template in an image, one is required to 'restore' the image using the template itself as the given blurring function. It is important to note that the input image is usually sharp and does not require any deblurring as such. Section 8.3 elaborates on this issue and presents a restoration model of the EXM approach. The MSE restoration analogy is extended according to multiple template EXM and a new generalized MSE restoration paradigm is introduced.

Another interesting relationship is presented in Section 8.4, which shows that the underlying principle in the single-template optimal DSNR matching corresponds to a non-orthogonal expansion of the input image using basis functions that are all shifted versions of the template. Intuitively, if one wishes to detect a template in an image, it is reasonable to expand the image in terms of basis functions that are similar to the template itself, i.e., versions of the template translated to all the candidate locations. Then the expansion coefficients of such a representation would indicate the presence of the template at the corresponding locations. Section 8.4 formally shows that the above template-similar non-orthogonal expansion of the image yields exactly the same result as the optimal DSNR filter. Since the shifted versions of the template are almost always mutually non-orthogonal, the expansion is also non-orthogonal.

Section 8.5 presents experimental comparisons of correlation and EXM. It has been shown in [8-7], [8-8] that EXM outperforms correlation in matching templates in conditions of noise, linear superposition and severe occlusion. Correlation yields broad peaks and substantial off-center response and spurious peaks at unrelated locations. On the other hand, the peaks generated by EXM are sharp and easy to detect and there is minimal off-center response. Also, EXM does not respond to other unrelated features in the image. More insight into the effectiveness of EXM can be gained by viewing the actual filters generated by EXM for some simple shapes. It can be seen from these results that the EXM filters are concentrated at the innovations of the shape, i.e., the corners and other high-curvature areas, while being substantially zero elsewhere. This relates to psychophysical evidence [8-13] that high-curvature edges form the major features of the shape to be recognized. Also, the impulse response consists of dipole-like elements at each high-curvature region. This intuitively explains the reason for the effectiveness of EXM in matching occluded templates. As long as sufficient shape innovation regions (high-curvature areas) are unoccluded in the template, the EXM filter responds with a sharp peak since the corresponding dipoles match the visible regions.

Edge detection is a very common preprocessing used in many applications. If one defines an edge model, then edge detection can be considered as a template matching problem, where the template being sought is the edge model itself. Section 8.6 presents such EXM based edge detectors using this paradigm. These optimal DSNR edge detectors are easily defined for any edge model, and examples of step, roof, and ramp edge detectors are shown. The step edge detector (which we call the *step expansion filter* or SEF) is formulated analytically and we have shown [8-12] that the SEF outperforms the widely used Canny edge detector [8-21] in terms of DSNR, accuracy of edge maps, localization and noise resistance.

8.1 The Correlation Approach: A Brief Review

To formally define the SNR, we first formulate the problem of template matching as follows. Given an $M \times M$ discrete image $s(x, y)$ that contains the two-dimensional template $\psi(x, y)$ at a certain location (x_0, y_0) and additive noise $\lambda(x, y)$, we wish to detect and locate the template $\psi(x, y)$. In template matching based on convolution, a filter $h(x, y)$ is convolved with the signal to yield the filter response $z(x, y)$:

$$\begin{aligned} z(x,y) &= s(x,y) * h(x,y) \\ &= c_\psi \psi(x - x_0, y - y_0) * h(x,y) + \lambda(x,y) * h(x,y) \\ &= z_\psi(x,y) + z_\lambda(x,y) , \end{aligned} \tag{8.1-1}$$

where $*$ denotes discrete convolution and c_ψ is an amplitude scale factor.

The SNR is defined [8-1], [8-5], [8-6] by the ratio between the energy of the response at the template center (x_0, y_0) and the energy of the overall noisy response

$$\mathrm{SNR} = 10 \log \frac{[z_\psi(x_0, y_0)]^2}{\frac{1}{M^2} \sum_M \sum_M [z_\lambda(i,j)]^2} . \tag{8.1-2}$$

Optimization of the SNR yields the matched filter $h(x, y)$, which has a Fourier transform given by [8-1], [8-5], [8-6]

$$H(u,v) = c_h \frac{\overline{\Psi}(u,v)}{S_{\lambda\lambda}(u,v)} , \tag{8.1-3}$$

where $\Psi(u, v)$ is the Fourier transform of $\psi(x, y)$, $S_{\lambda\lambda}(u, v)$ is the spectral density of the noise $\lambda(x, y)$, c_h is a constant and the bar symbol denotes complex conjugation. If the noise $\lambda(x, y)$ is wide-sense stationary and white, $S_{\lambda\lambda}(u, v)$ is a constant, and the optimal filter $h(x, y)$ is simply the mirror image of the template $\psi(x, y)$, and thus matched filtering is equivalent to correlating the signal with the template.

The matched filtering approach is probably the most widely used method for template recognition. However, there is a major drawback in its formulation. The response $z_\psi(x,y)$ at locations other than (x_0, y_0) is completely overlooked in the definition of the SNR. For the purpose of template matching, these responses are also unwanted and should also be considered as 'noise'. Experimental results in Section 8.5 confirm that correlation can generate broad peaks and that z_ψ can be quite substantial in the neighborhood of (x_0, y_0) and should not be overlooked. In addition, correlation generates many spurious peaks that do not correspond to the correct feature, and interfere with the detection process. Moreover, the correct peaks – which ideally correspond to the peaks of the autocorrelation function of the template – are not sharp enough and their localization is often inaccurate. The peak response of correlation also depends on the magnitude of the template in the image (expressed here as c_ψ). In order to eliminate this effect and to somewhat enhance the sharpness of the peaks, it has been suggested [8-1] to employ, as the output matching result, the normalized correlation coefficient $\rho(x,y)$ given by

$$\rho(x,y) = \frac{s(x,y) * h(x,y)}{\left[\sum_{M_x}\sum_{M_y} h^2(i,j) \cdot \sum_{M_x}\sum_{M_y} s^2(x+i, y+j)\right]^{1/2}}, \tag{8.1-4}$$

where M_x and M_y define the region of support of the template. The correlation coefficient is bounded by $-1 \leq \rho \leq 1$. Moreover, $\|\rho\| = 1 \iff h(x,y) = c_\psi s(x,y)$ within the region of the template. As is demonstrated in Section 8.5, even normalized correlation does not produce ideal results and is inferior to recognition by the expansion approach. Furthermore, normalized correlation requires a much larger amount of computation compared to unnormalized correlation.

8.2 The Discriminative Signal-to-Noise Ratio and Expansion Matching

In contrast to the widely-used SNR, the new similarity measure discussed here, *discriminative signal-to-noise ratio* (DSNR), penalizes even off-center responses due to the template, in the matching result. Thus, optimization of this measure would lead to a sharp and well-localized matching response at the template center and minimum off-center response everywhere else. For a filter $\theta(x,y)$, the DSNR of the matching result (filter output) $c(x,y)$ is defined at the template-center (x_0, y_0) as

$$\text{DSNR} = 10 \log \frac{[c(x_0, y_0)]^2}{\frac{1}{M^2-1} \sum \sum_{(i,j) \neq (x_0,y_0)} c^2(i,j)}, \tag{8.2-1}$$

where the input to the filter is the given signal $s(x,y)$. The objective is to find the optimal filter $\theta(x,y)$ that maximizes this DSNR.

In contrast to the traditional SNR of Eq. (8.1-2), the DSNR penalizes *all* off-center responses in the output $c(x, y)$, irrespective of whether the response is due to noise or the template itself. Thus, matching results with high DSNR imply a sharp peak with minimal main and side lobes, i.e., minimal off-center responses. The ideal matching result (infinite DSNR) is when the output $c(x, y)$ has only an impulse function at the center of the template.

To generalize our approach to include complex images we present the optimization of the DSNR for a generalized multiple-template matching problem in the complex image domain. The problem is formulated in one-dimension using matrix-vector notation. All the variables are considered complex-valued unless otherwise stated. The results can easily be generalized into multiple dimensions by using lexicographically ordered vectors and block-circulant matrices [8-1], [8-5].

Given a known set of M-point complex-valued templates $\boldsymbol{\psi}_i$; $i = 1, \cdots, N$, the input is assumed to be one of the set of noisy templates $\boldsymbol{s}_i = \boldsymbol{\psi}_i + \boldsymbol{\lambda}$, where $\boldsymbol{\lambda}$ is a complex-valued random noise vector. The desired filter $\boldsymbol{\theta}$ is required to yield a user-defined response u_i as its expected peak response to each template $\boldsymbol{s}_i$:

$$E[\boldsymbol{\theta}^H \boldsymbol{s}_i] = u_i \; ; \quad i = 1, \cdots, N, \tag{8.2-2}$$

where $\boldsymbol{\theta}^H \equiv \overline{\boldsymbol{\theta}}^T$, i.e., the transpose conjugate.

The correlation[1] of the desired filter $\boldsymbol{\theta}$ with the ith template results in the vector $\boldsymbol{c}_i$:

$$\boldsymbol{c}_i = \boldsymbol{S}_i \boldsymbol{\theta} \quad ; \; i = 1, \cdots, N, \tag{8.2-3}$$

where $\boldsymbol{S}_i$ denotes the circulant correlation matrix obtained by circulating the noisy input template's conjugate $\overline{\boldsymbol{s}}_i$.

The desired filter $\boldsymbol{\theta}$ is required to optimize the DSNR criterion in its correlation response to each template, in addition to satisfying the constraints of Eq. (8.2-2). The DSNR to be maximized for the ith template is defined as

$$\mathrm{DSNR}_i = 10 \log \frac{E[(\boldsymbol{\theta}^H \boldsymbol{s}_i)^2]}{E[\boldsymbol{c}_i^H \boldsymbol{c}_i] - E[(\boldsymbol{\theta}^H \boldsymbol{s}_i)^2]} \; \rightarrow \; \max . \tag{8.2-4}$$

Thus, high DSNR implies a large response at the template-center and relatively small response off the center of the template. Ideally $\boldsymbol{c}_i = [0 \cdots 0, u_i, 0 \cdots 0]^T$ where the location of u_i corresponds to the ith pattern's center. Since the center-response power is constrained by Eq. (8.2-2), to maximize DSNR_i we need only minimize the first term in the denominator:

$$E[\boldsymbol{c}_i^H \boldsymbol{c}_i] \; \rightarrow \; \min . \tag{8.2-5}$$

[1] Without loss of generality, we use correlation to represent linear filtering rather than convolution.

Since a global minimization of all the DSNR criteria for each input template is not possible [8-11], we seek to minimize a convex weighted sum:

$$E[\sum_{i=1}^{N} \alpha_i \boldsymbol{c}_i^H \boldsymbol{c}_i] \rightarrow \min \; ; \quad \sum_{i=1}^{N} \alpha_i = 1, \tag{8.2-6}$$

where α_i are weighting factors, typically chosen to be $1/N$. Intuitively, α_i imparts a relative importance to the DSNR maximization for matching the ith template. The weights α_i signify the relative energy in the matching peak for the ith template.

To perform the actual minimization in Eq. (8.2-6), we need to compute the derivative of complex matrices and vectors with respect to vectors. For this purpose, we invoke the identities derived by Haykin [8-14]:

$$\frac{\partial \boldsymbol{p}^H \boldsymbol{q}}{\partial \overline{\boldsymbol{q}}} = 0, \tag{8.2-7}$$

where $\boldsymbol{p}$ and $\boldsymbol{q}$ are complex-valued vectors. Also,

$$\frac{\partial \boldsymbol{q}^H \boldsymbol{p}}{\partial \overline{\boldsymbol{q}}} = \boldsymbol{p}, \tag{8.2-8}$$

and for a quadratic form,

$$\frac{\partial \boldsymbol{q}^H \boldsymbol{R} \boldsymbol{q}}{\partial \overline{\boldsymbol{q}}} = \boldsymbol{R} \boldsymbol{q}, \tag{8.2-9}$$

where $\boldsymbol{R} = \boldsymbol{R}^H$ is a Hermitian matrix.

The noise $\boldsymbol{\lambda}$ is assumed to be zero-mean and stationary, and thus the first-order noise terms in the expectations drop out, yielding

$$E[\boldsymbol{s}_i] = E[\boldsymbol{\psi}_i + \boldsymbol{\lambda}] = \boldsymbol{\psi}_i. \tag{8.2-10}$$

Also, the correlation matrix can be split as $\boldsymbol{S}_i = \boldsymbol{\Psi}_i + \boldsymbol{\Lambda}$, where $\boldsymbol{\Psi}_i$ and $\boldsymbol{\Lambda}$ represent the circulant correlation matrices for the ith template $\boldsymbol{\psi}_i$ and the noise vector $\boldsymbol{\lambda}$, respectively. Thus, we can write

$$\begin{aligned} E[\boldsymbol{c}_i^H \boldsymbol{c}_i] &= \boldsymbol{\theta}^H E[\boldsymbol{S}_i^H \boldsymbol{S}_i] \boldsymbol{\theta} \\ &= \boldsymbol{\theta}^H [\boldsymbol{R}_{\psi\psi i} + \boldsymbol{R}_{\lambda\lambda}] \boldsymbol{\theta}, \end{aligned} \tag{8.2-11}$$

where $\boldsymbol{R}_{\psi\psi i} \equiv \boldsymbol{\Psi}_i^H \boldsymbol{\Psi}_i$ and $\boldsymbol{R}_{\lambda\lambda} \equiv E[\boldsymbol{\Lambda}^H \boldsymbol{\Lambda}]$ are defined as the autocorrelation matrices of the ith template and the noise, respectively.

Combining the requirements of Eq. (8.2-2) and Eq. (8.2-6), and using a modified method of Lagrange multipliers [8-14], a single objective function to be minimized can be defined as

$$\begin{aligned} J(\boldsymbol{\theta}) &= \sum_{j=1}^{N} \left[\boldsymbol{\theta}^H \left(\alpha_j \boldsymbol{R}_{\psi\psi j} + \boldsymbol{R}_{\lambda\lambda} \right) \boldsymbol{\theta} + \Re \left(\overline{\xi}_j (\boldsymbol{\theta}^H \boldsymbol{s}_j - u_j) \right) \right] \rightarrow \min \\ &= \boldsymbol{\theta}^H \boldsymbol{R} \boldsymbol{\theta} + \sum_{j=1}^{N} \Re \left(\overline{\xi}_j (\boldsymbol{\theta}^H \boldsymbol{\psi}_j - u_j) \right) \rightarrow \min, \end{aligned} \tag{8.2-12}$$

where $\boldsymbol{R} = \sum_{j=1}^{N} \alpha_j \boldsymbol{R}_{\psi\psi j} + \boldsymbol{R}_{\lambda\lambda}$, and $\Re(\cdot)$ denotes the real part of a complex quantity.

Using the identities of Eq. (8.2-7) – (8.2.9) and setting the first derivative of Eq. (8.2-12) to zero, we obtain

$$\frac{\partial J(\boldsymbol{\theta})}{\partial \overline{\boldsymbol{\theta}}} = \boldsymbol{R}\boldsymbol{\theta} + \sum_{j=1}^{N} \overline{\xi}_j \boldsymbol{\psi}_j = 0, \tag{8.2-13}$$

which yields the solution for $\boldsymbol{\theta}$ as

$$\boldsymbol{\theta} = -\boldsymbol{R}^{-1} \sum_{j=1}^{N} \overline{\xi}_j \boldsymbol{\psi}_j. \tag{8.2-14}$$

To find $\boldsymbol{\xi} = [\xi_1 \cdots \xi_N]^T$ we enforce the constraints of Eq. (8.2-2)

$$\begin{aligned} u_i &= \boldsymbol{\theta}^H \boldsymbol{\psi}_i \\ &= \left(-\boldsymbol{R}^{-1} \sum_{j=1}^{N} \overline{\xi}_j \boldsymbol{\psi}_j \right)^H \boldsymbol{\psi}_i \\ &= -\sum_{j=1}^{N} \xi_j \boldsymbol{\psi}_j^H \boldsymbol{R}^{-1} \boldsymbol{\psi}_i \; ; \; i = 1, \cdots, N, \end{aligned} \tag{8.2-15}$$

or in matrix form, the solution to this set of equations is written as

$$\boldsymbol{\xi} = \boldsymbol{A}^{-1} \boldsymbol{u} \; ; \quad [\boldsymbol{A}]_{ij} = -\boldsymbol{\psi}_i^H \boldsymbol{R}^{-1} \boldsymbol{\psi}_j \, , \tag{8.2-16}$$

where $\boldsymbol{A}$ is a Hermitian $N \times N$ matrix which is non-singular for distinct and non-zero $\boldsymbol{\psi}_i$, and $\boldsymbol{u} = [u_1 \cdots u_N]^T$ defines the constraints.

The circulant matrices involved in Eq. (8.2-14) can be diagonalized [8-15] using the $M \times M$ unitary *discrete Fourier transform* (DFT) matrix

$$[\boldsymbol{W}]_{mn} = M^{-0.5} \exp(j\pi mn/M)$$

and an efficient relationship can be found [8-9] in the frequency domain for the optimal DSNR or EXM filter. Writing the DFT of a vector $\boldsymbol{\eta}$ as $\boldsymbol{\eta}^f \equiv \mathcal{F}[\boldsymbol{\eta}] = \boldsymbol{W}^{-1}\boldsymbol{\eta}$, the EXM filter in the frequency domain is

$$\boldsymbol{\theta}^f(k) = \frac{-\sum_{j=1}^{N} \overline{\xi}_j \boldsymbol{\psi}_j^f(k)}{\sum_{i=1}^{N} \alpha_i \boldsymbol{S}_{\psi\psi i}(k) + \boldsymbol{S}_{\lambda\lambda}(k)} \; ; \quad k = 1, \cdots, M, \tag{8.2-17}$$

where $\boldsymbol{\psi}_i^f$ is the DFT of the ith template $\boldsymbol{\psi}_i$, $\boldsymbol{\psi}_i^f(k)$ refers to the kth component of the vector $\boldsymbol{\psi}_i^f$, $\boldsymbol{S}_{\psi\psi i}(k) = \boldsymbol{\psi}_i^f(k)\overline{\boldsymbol{\psi}}_i^f(k)$, and $\boldsymbol{S}_{\lambda\lambda}(k)$ is the power spectrum of the noise. Equation (8.2-17) offers an efficient computational method for the design of the expansion filter $\boldsymbol{\theta}$. With $N = 1$, this is easily recognized as the Wiener filter result.

Similarly, in solving the linear system of equations of (8.2-16) the following efficient frequency domain result can be implemented:

$$[\boldsymbol{A}]_{ij} = \sum_{k=1}^{M} \left(\frac{-\overline{\boldsymbol{\psi}}_i^f(k)\boldsymbol{\psi}_j^f(k)}{\sum_{l=1}^{N} \alpha_k \boldsymbol{S}_{\psi\psi l}(k) + \boldsymbol{S}_{\lambda\lambda}(k)} \right). \tag{8.2-18}$$

Implementing the design in the frequency domain offers considerable savings in computation since the DFTs of the templates need be computed only once. The matrix inversion in Eq. (8.2-16) involves a dimension equal to the number of templates N which is typically very small compared to M, the number of points in the signal. Thus, the inversion of the matrix $\boldsymbol{A}$ in Eq. (8.2-16), is not a major computational concern.

The above multiple-template EXM formulation yields a number of previously known solutions as special cases. When the general complex result is reduced to the real signal domain, we have shown [8-9] that in the noiseless case the solution is identical to the *minimum average correlation energy* (MACE) filter [8-11]. At the other extreme, if the noise power tends to infinity the EXM filter approaches the *synthetic discriminant function* (SDF) [8-10], or *minimum variance* SDF approach [8-16] depending on the noise color. For the case of a single template, we shall see in the following section that the EXM filter is exactly the Wiener filter for restoration of blurred images. Also, when the noise power in the formulation tends to infinity, EXM for matching a single template tends to matched filtering [8-8].

8.3 EXM Related to Minimum Squared Error Restoration

Consider the case of matching only a single template $\psi(x)$ in the continuous domain with Fourier transform $\Psi(\omega)$. The result of Eq. (8.2-17) is easily extended to continuous variables and simplifies with $N = 1$ to

$$\Theta(\omega) = \frac{\Psi(\omega)}{S_{\psi\psi}(\omega) + S_{\lambda\lambda}(\omega)} \tag{8.3-1}$$

within an amplitude scaling constant. This can be recognized to be exactly the Wiener filter [8-6] for *minimum squared error* (MSE) restoration, when the blurring function is assumed to be the template itself. Thus, single-template EXM can be regarded as MSE restoration of the input image using the template as the blurring

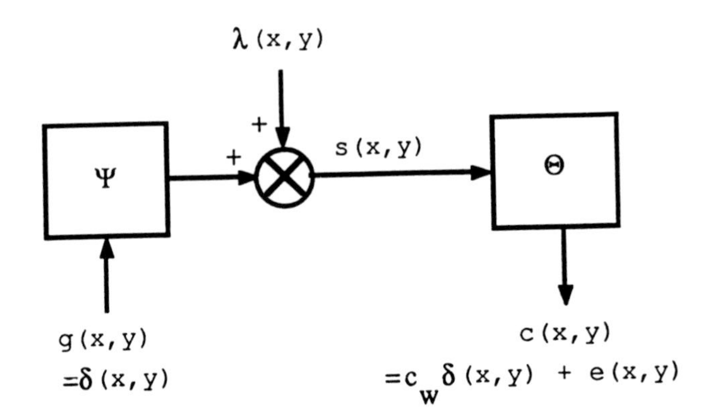

Figure 8.3-1: Restoration model of single-template EXM. Ψ denotes the template used as a blurring function. The input image is $s(x, y)$, from which the delta function is restored with minimal MSE by the EXM filter Θ.

function, where the ideal restored signal is a delta function at the center of the template. In contrast to its conventional use, restoration is employed here as an efficient implementation of EXM. Note that the input image is usually sharp and does not require any restoration.

It is easy to see this relationship if the input image is assumed to consist only of the template shifted to some location with no additive noise. In this case, the result of the MSE restoration (or deconvolution) assuming the template as the blurring function will simply be a delta function (impulse) at that location, and zero elsewhere. In practice, there is always additive noise or background clutter in the image, and the Wiener filter – which is a regularized, stable solution – yields the solution that is closest (in terms of MSE) to the ideal delta function.

The above relationship leads to a different perspective of the matching problem as illustrated in Fig. 8.3-1. Figure 8.3-1 shows a block-diagram model of the input image where the input $g(x, y)$ is an impulse function $\delta(x, y)$, which is smoothed by a blurring function which equals the template itself, and noise is added to yield the input image $s(x, y)$. Noise generally refers to the background and other residual parts of the image not corresponding to the template itself. In the previous section, the noise was assumed to be stationary and uncorrelated with the template. This assumption is not unreasonable, since, in practice, the background in the image has very little to do with the template itself, and to a large part the background has roughly a low-pass power spectrum. In fact, even assuming that the noise is white, very good results are obtained in Section 8.5 (also see [8-7]–[8-9]).

Thus, the matching problem has been reformulated as one of image restoration, i.e., recovering the input impulse function from the noisy, template-blurred version, and the classical solution is that of the Wiener MSE restoration filter. The output $c(x, y)$ will consist of the input delta function (amplitude scaled by a user-defined constant c_W) and an error $e(x, y)$ whose energy has been minimized. Note that DSNR maximization minimizes the energy of the off-center response which is exactly equivalent to minimizing the squared error in the restoration. If there are more than one, say L, instances of the template at different locations, the input to the system of Fig. 8.3-1 would be

$$g(x,y) = \sum_{j=1}^{L} \beta_j \delta(x - x_j, y - y_j), \tag{8.3-2}$$

where (x_j, y_j) is the location of the jth instance of the template in the image, and β_j is its corresponding amplitude. The output of the optimal-DSNR filter $\boldsymbol{\theta}$ would then be

$$c(x,y) = c_W \sum_{j=1}^{L} \beta_j \delta(x - x_j, y - y_j) + e(x,y). \tag{8.3-3}$$

Thus, the output $c(x, y)$ consists of the input combination of impulse functions (with any desired amplitude scaling) at the centers of the different instances of the template, and an error term $e(x, y)$ whose energy has been minimized.

The above restoration analogy can be extended further, and the multiple-template EXM of Section 8.2 can be related to the block-diagram of Fig. 8.3-2. In this model, there are N blurring blocks, corresponding to each of the N given templates $\boldsymbol{\psi}_l$. In general, the lth template ($l = 1, \cdots, N$) has $L^{(l)}$ instances in the input image, the jth instance being located at $(x_j^{(l)}, y_j^{(l)})$ with an amplitude of $\beta_j^{(l)}$, where $(j = 1, \cdots, L^{(l)})$. Thus, the input to the lth blurring block is

$$g^{(l)}(x,y) = \sum_{j=1}^{L^{(l)}} \beta_j^{(l)} \delta(x - x_j^{(l)}, y - y_j^{(l)}). \tag{8.3-4}$$

The sum output of all the N blurring blocks will simply be the linear superposition of all the instances of all the templates in the image which are to be detected. The additive noise $\lambda(x, y)$ represents everything else in the signal $s(x, y)$ (including background imagery, distortion, and occlusion, if any). The multiple template EXM filter $\boldsymbol{\theta}$ of Section 8.2 will perform the optimal task (in the sense of squared error) in restoring the impulse functions $\beta_j^{(l)} \delta(x - x_j^{(l)}, y - y_j^{(l)})$ scaled by the user

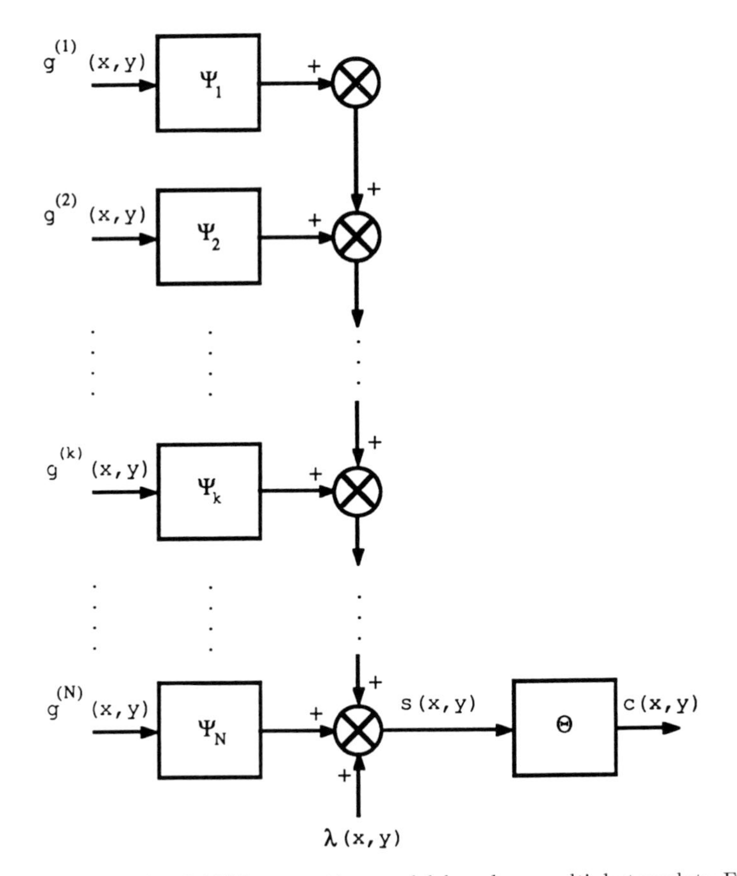

Figure 8.3-2: Generalized MSE restoration model based on multiple-template EXM. $\boldsymbol{\Psi}_i$ denote the multiple templates used as blurring functions. The input image is $s(x, y)$, from which the different delta functions are restored with minimal MSE by the multiple-template EXM filter $\boldsymbol{\Theta}$.

defined constants u_l, i.e.

$$\begin{aligned} c(x,y) &= \sum_{l=1}^{N} u_l \left(\sum_{j=1}^{L} \beta_j^{(l)} \delta(x - x_j, y - y_j) \right) + e(x,y) \\ &= \sum_{l=1}^{N} u_l g^{(l)}(x,y) + e(x,y). \end{aligned} \tag{8.3-5}$$

Thus, the output $c(x,y)$ consists of the input delta functions, linearly scaled (by user-defined constants) and superposed, with an additional error term $e(x,y)$ whose energy has been minimized.

The above interpretation of multiple-template EXM is actually a generalized formulation of MSE restoration, where the special case of $N = 1$ corresponds to the conventional Wiener MSE filter. Thus, using this new formulation it is possible to choose u_l so as to restore only selected $g^{(l)}(x,y)$ while rejecting the others. Of course, this approach has its limitations. For example, if all the templates $\{\psi_l\}$ are highly correlated, the effectiveness of this approach is reduced.

8.4 EXM Related to Non-Orthogonal Expansion

Signal expansions are extensively used for data compression purposes [8-5]. Non-orthogonal image expansions [8-17], [8-18] implemented by neural lattice architectures [8-19], [8-20] can yield compression ratios superior to state-of-the-art compression techniques, especially for low bit rates. The basic principle behind this non-orthogonal expansion scheme is that if the *basis functions* (BFs) can be made to resemble the image features, then one can obtain efficient compaction of the signal energy and therefore better image compression. Orthogonal expansions place severe restrictions on the shape of the BFs due to the mutual orthogonality constraint. If this constraint is removed, the BFs are free to assume the shape of features in the image as long as they are all mutually independent [8-6]. Non-orthogonal expansions using Gaussian and Gabor BFs [8-17]–[8-20] have been shown to yield good compression results.

A similar argument can be made in favor of using non-orthogonal expansions for matching as well. If one wants to detect a template in an image, it might be possible to create a set of BFs that correspond to every candidate location of the template in the image. This set of BFs would typically be non-orthogonal (except for very uncommon and special template functions). If the input image were expanded by this set of BFs, then in the coefficients of the expansion, the BF corresponding to the location of the template in the image would have a high value, while others would have lower values. In other words, the values of the expansion coefficients would directly signify the presence of the template at the corresponding locations.

In [8-7] it is shown that such a set of template-similar BFs are indeed independent subject to very minor conditions. A two-dimensional image of size $M \times M$ has dimensionality $M^2 < \infty$, and therefore any M^2 independent functions can serve as a complete basis. In what follows, it is formally shown that the underlying principle in single-template EXM is actually a non-orthogonal expansion of the input image with template-similar BFs.

Non-orthogonal expansion can be formulated as follows. Suppose one wishes to represent a discrete, finite-support real-valued image $s(x,y)$ by a set of BFs $\{\psi_i(x,y)\}$ as

$$\hat{s}(x,y) = \sum_{i=1}^{m} c_i \psi_i(x,y) \; ; \quad x,y = 1,\cdots,M, \tag{8.4-1}$$

where $\{c_i\}$ are the coefficients of the expansion. Using vector notation as in Section 8.2, $\hat{s}(x,y)$ is translated into an M^2-dimensional vector $\hat{\boldsymbol{s}}$, and similarly $\psi_i(x,y) \equiv \boldsymbol{\psi}_i$ and $s(x,y) \equiv \boldsymbol{s}$. We want to minimize the squared error e in the representation as

$$e = \|\boldsymbol{s} - \hat{\boldsymbol{s}}\|^2 = \|\boldsymbol{s} - \sum_{i=1}^{m} c_i \boldsymbol{\psi}_i\|^2 \;\rightarrow\; \min \tag{8.4-2}$$

by choosing the appropriate c_i. Using the orthogonality principle [8-6] (or setting the partial derivative of Eq. (8.4-2) with respect to c_i to zero), the representation error $\boldsymbol{s} - \hat{\boldsymbol{s}}$ has to be orthogonal to all the BFs, i.e.

$$\langle (\boldsymbol{s} - \hat{\boldsymbol{s}}), \boldsymbol{\psi}_i \rangle = 0 \; ; \quad i = 1,\cdots,m. \tag{8.4-3}$$

This leads to a set of equations represented in matrix-vector form as

$$\boldsymbol{R}_{\psi\psi} \boldsymbol{c} = \boldsymbol{\Psi} \boldsymbol{s}, \tag{8.4-4}$$

where $[\boldsymbol{R}_{\psi\psi}]_{ij} = \langle \boldsymbol{\psi}_i, \boldsymbol{\psi}_j \rangle$ is the autocorrelation matrix of the BFs, $\boldsymbol{c} = [c_1 \; c_2 \; \cdots \; c_m]$ represents the expansion coefficients, and $\boldsymbol{\Psi} = [\boldsymbol{\psi}_1 \; \boldsymbol{\psi}_2 \; \cdots \; \boldsymbol{\psi}_m]^T$. If the set of BFs $\{\boldsymbol{\psi}_i\}$ is linearly independent, then the matrix $\boldsymbol{R}_{\psi\psi}$ is positive definite (since it is also Hermitian) and therefore invertible and yields the solution for $\boldsymbol{c}$ as

$$\boldsymbol{c} = \boldsymbol{R}_{\psi\psi}^{-1} \boldsymbol{\Psi} \boldsymbol{s}. \tag{8.4-5}$$

Suppose we choose the BFs $\{\psi_i(x,y)\}$ to be different shifted versions of the template $\psi(x,y)$ as

$$\psi_i(x,y) = \psi(x - x_i, y - y_i) \; ; \; i = 1 \cdots M^2 , \tag{8.4-6}$$

where (x_i, y_i) spans the region of support $(1 \cdots M, 1 \cdots M)$ for $i = 1,\cdots,M^2$. This could correspond to a lexicographic ordering of a discrete image into a vector. Note that in this case the number of BFs m is equal to M^2, the number of possible shifted locations of the template. It has been shown that this BF set is complete and independent subject to very modest conditions [8-7]. For the case of circularly shifted BFs, the condition for completeness is that the template should not have any zeros in its DFT. If there are any zeros, a small regularizing constant can be added to the DFT. In practice, almost all templates can be used to form such a complete and independent template-similar basis. If we assume that the signals are circulant, the matrix $\boldsymbol{\Psi}$ is a circulant matrix, which exactly represents the correlation operation using the template $\boldsymbol{\psi}$ as the associated kernel. Furthermore, the autocorrelation matrix $\boldsymbol{R}_{\psi\psi}$ is exactly equivalent to the discrete autocorrelation matrix of the template $\boldsymbol{\psi}$, and Eq. (8.4-5) becomes

$$\boldsymbol{c} = \left[\boldsymbol{\Psi}^H \boldsymbol{\Psi}\right]^{-1} \boldsymbol{\Psi} \boldsymbol{s} \, . \tag{8.4-7}$$

If we assume zero noise and $N = 1$ in the result of Eq. (8.2-14), then $\boldsymbol{\theta} = \xi \boldsymbol{R}_{\psi\psi}^{-1} \boldsymbol{\psi}$, and we can write out the output $\boldsymbol{c}$ of the filter $\boldsymbol{\theta}$ as

$$\boldsymbol{c} = \boldsymbol{\Theta} \boldsymbol{s}, \tag{8.4-8}$$

where the circulant correlation matrix $\boldsymbol{\Theta}$ represents filtering by $\boldsymbol{\theta}$. This result has been shown to be exactly the same (ignoring the user-defined amplitude scaling ξ) as the non-orthogonal expansion coefficients of Eq. (8.4-7) [8-7]. Thus, the underlying principle in single-template EXM is actually that the image undergoes decomposition into BFs that are all shifted versions of the template itself. An in-depth analysis of the stability of the expansion is not necessary, since the above relationship is only an analogy that helps to clarify the understanding of EXM. In practice, the assumed noise is always some non-zero quantity and thus the expansion output $\boldsymbol{c}$ is always stable and regularized.

The superior performance of EXM in matching linearly superimposed templates (as might occur in sonar or radar signals) is easily explained using the non-orthogonal expansion interpretation of EXM. Since EXM is a linear decomposition technique, superimposed templates can be ideally detected and located.

A simple two-vector illustration in Fig. 8.4-1 explains the superior performance of EXM over correlation. Here, the BFs $\boldsymbol{\psi}_1$ and $\boldsymbol{\psi}_2$ represent two shifted, candidate versions of the template, and the input signal $\boldsymbol{s}$ is assumed to be closer to $\boldsymbol{\psi}_1$. In correlation or matched filtering, the signal $\boldsymbol{s}$ is projected on to the BFs to yield the correlation coefficients $\boldsymbol{z}(1) = \langle \boldsymbol{\psi}_1, \boldsymbol{s} \rangle$ and $\boldsymbol{z}(2) = \langle \boldsymbol{\psi}_2, \boldsymbol{s} \rangle$ as in Eq. (8.1-1). On the other hand, in EXM, the two BFs *compete for the energy* of the signal, and $\boldsymbol{s}$ is decomposed into its components $\boldsymbol{c}(1)$ and $\boldsymbol{c}(2)$ (see Eq. (8.4-5)) along the corresponding two BFs. The closer BF $\boldsymbol{\psi}_1$ extracts almost all the signal energy and the small residual is shared among the remaining BFs (in our example, only $\boldsymbol{\psi}_2$). In fact, it has been formally shown [8-8] that $[\boldsymbol{c}(1)/\boldsymbol{c}(2)]^2 \geq [\boldsymbol{z}(1)/\boldsymbol{z}(2)]^2$. Equality exists only when all the shifted BFs $\{\boldsymbol{\psi}_i\}$ are mutually orthogonal and thus projection (or correlation) itself performs the required expansion. Evidently, such

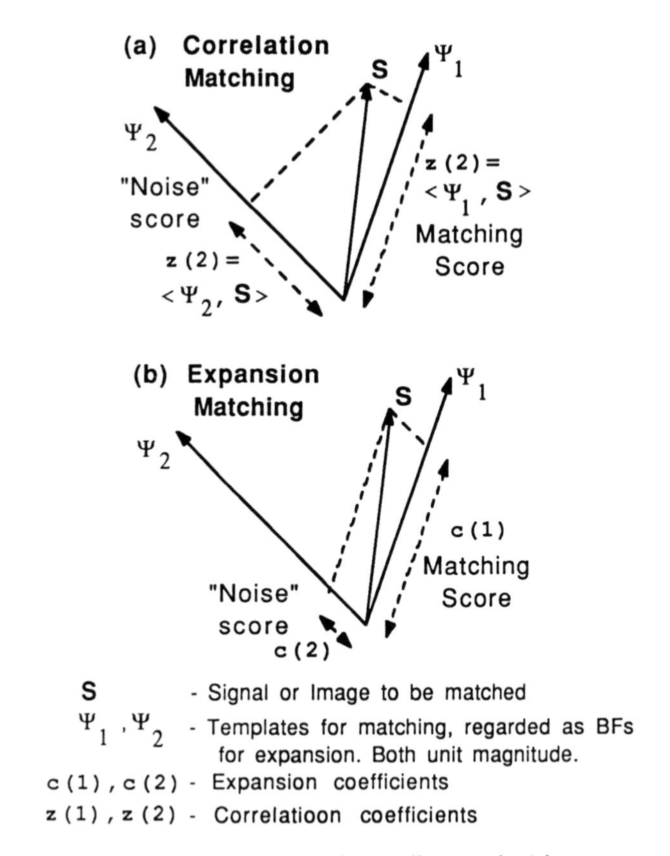

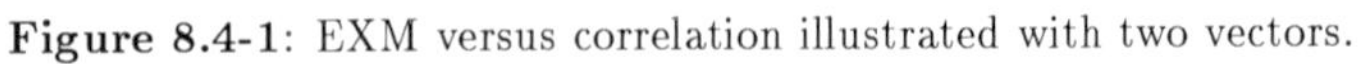

Figure 8.4-1: EXM versus correlation illustrated with two vectors.

a case is impractical, since practical templates in general do not form a complete, orthogonal basis when shifted to all locations in the image.

8.5 Experimental Results

This section presents examples of template matching comparing the traditional correlation or matched filtering method with the EXM approach. In all the experimental results presented, the images used are 128×128 with 256 gray levels. The filtering operations are performed by computing the *fast Fourier transform* (FFT) of the input image and the kernel, multiplying these FFTs and obtaining the inverse FFT as the result. We assume that the template is small enough to neglect the effects of wrap-around. For correlation, the kernel is simply the template itself (flipped around the origin). For EXM, the noise is assumed to be white and the frequency domain design technique is used with the FFT of the template to generate the EXM filter $\boldsymbol{\Theta}$ in the frequency domain. This is directly multiplied with the FFT of the input image and the inverse FFT yields the EXM coefficients. For experiments involving normalized correlation, the correlation result is normalized by the local power of the template and signal as per Eq. (8.1-4).

Figure 8.5-1a) shows an image of three cars against a natural background. Two of these cars are occluded up to 60%. Using the template of the unoccluded car, the correlation approach yields the result in Fig. 8.5-1b). Note the broad peaks and large spurious response marked by '?'. Figure 8.5-1c) shows the corresponding single-template EXM result and displays three sharp peaks (marked by arrows) and minimal off-center response. The example in Fig. 8.5-2 shows that the multiple-template EXM is also successful in recognizing occluded objects. Figure 8.5-2a) shows pliers and scissors that are substantially occluded by other objects in the scene. Using the multiple template approach, a single filter is designed to elicit unit response from unoccluded templates of the scissors and the pliers. The matching result in Fig. 8.5-2b) uses a zero-noise parameter which equals the result of the MACE approach. The peak marked by the arrow corresponds to the center of the scissors template in the image. The pliers template has only a weak response. Figure 8.5-2c) shows the optimal EXM result which uses a noise parameter of 18 dB and can be seen to have two sharp peaks and the off-center response is less than that in the MACE result of Fig. 8.5-2b). The noise parameter of 18 dB was chosen empirically. It is seen that the matching result is not sensitive to this parameter, and even rough estimates of the noise power yield comparable results. Figure 8.5-2d) corresponds to the SDF result with very large noise parameter (tending to infinity). The peaks are broad and there is substantial off-center response and false peaks (marked by '?').

As explained earlier, EXM is ideally suited for matching linearly superposed templates since it is a linear decomposition process. Linearly superposed templates occur in speech or radar signals. Figure 8.5-3b) is composed of two shifted tem-

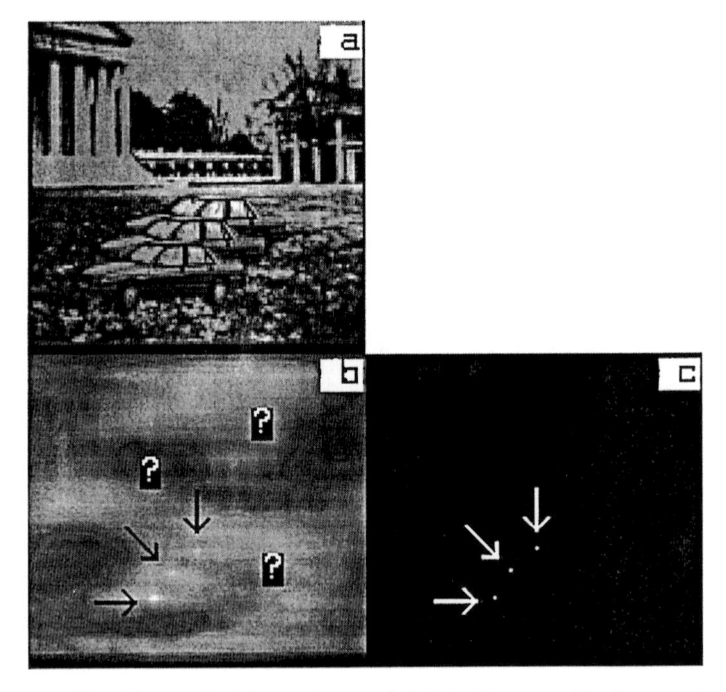

Figure 8.5-1: Matching occluded templates. a) Original image with three occluded cars (up to 60% occlusion). b) Correlation result with one broad peak (marked by an arrow) and substantial spurious response (marked by '?'). c) EXM result with three sharp peaks and negligible off-center response.

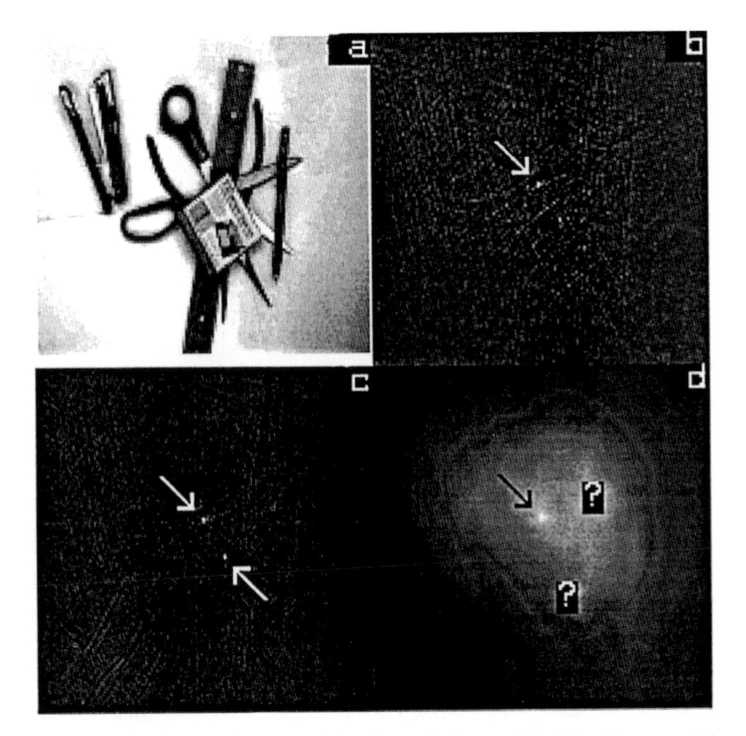

Figure 8.5-2: Multiple template matching in occlusion. a) Original image with occluded scissors and pliers. b) MACE result with one peak (marked by an arrow) and noisy off-center response. c) Optimal DSNR EXM result with two sharp peaks and minimal off-center response. d) SDF result with one broad peak and substantial off-center response (marked by '?').

plates of Fig. 8.5-3a) and additive white noise of SNR = 19 dB. If this test signal is correlated with the template itself, the result of Fig. 8.5-3c) (dashed line) is obtained (normalized correlation yields the solid line result with DSNR=9.8 dB). It can be seen that correlation yields a broad plateau instead of two peaks and cannot discriminate between the two templates. EXM yields the near-perfect result (DSNR=30.1 dB) in Fig. 8.5-3d) since the only two BFs that receive strong coefficients are those that compose the signal itself.

In other experiments with different geometric shapes, the EXM filters for a given shape are seen to enhance the innovations of the shape. In general, the EXM filters have an impulse response roughly proportional to the curvature of the shape. For example, the EXM filter for a square in Fig. 8.5-4a) and b) consists of four quadpole structures at the corners and substantially nothing elsewhere. The quadpoles at each corner of the square are two-dimensional (2-D) versions of dipoles and each consist of four delta functions of alternating sign. Similarly, the EXM filter for the triangle in Fig. 8.5-4c) and d) has strong quadpoles only at the vertices and is zero elsewhere, and for the shape of the digit '2' in Fig. 8.5-4e), the EXM filter impulse response in Fig. 8.5-4f) has approximately constant value along the circular section of the shape, zero along the straight lines, and strong bipolar response at the corners.

Observing Fig. 8.5-4, the effectiveness of EXM in suppressing off-center response can be explained by the quadpole structures at the corners of the square/triangle in the EXM filter. When such a filter is applied to an image, the response of a quadpole is zero for any region that is uniform or not matching the corresponding corner. Thus, the output of the EXM filter will have a high positive value only at locations where all the quadpoles of the filter match corresponding corners of the square/triangle, and the output will be negligible elsewhere. If a higher noise parameter is used, the quadpoles become dilated (with an exponential decay), the off-center response increases, the peak becomes less sharp, and thus DSNR is traded off for noise resistance.

The above explanation also reveals how EXM is effective under conditions of severe occlusion. Since the quadpoles are effectively detecting only the corners (high-curvature regions) of the template, as long as sufficient number of these corners are visible in the scene, EXM is able to yield a sharp matching peak (albeit somewhat weaker) and recognize the template, while the quadpoles which correspond to occluded corners have negligible contribution.

Thus, the EXM approach presents a systematic method to extract a given template's shape innovations. EXM filters capture the intrinsic features of the template which are its high-curvature edges. This observation is supported by the results of psychophysical experiments [8-13]. The optimization of the novel DSNR criterion yields the EXM filter, that mathematically confirms the importance of high-curvature regions of a shape in recognition – a well-known psychophysical fact [8-13].

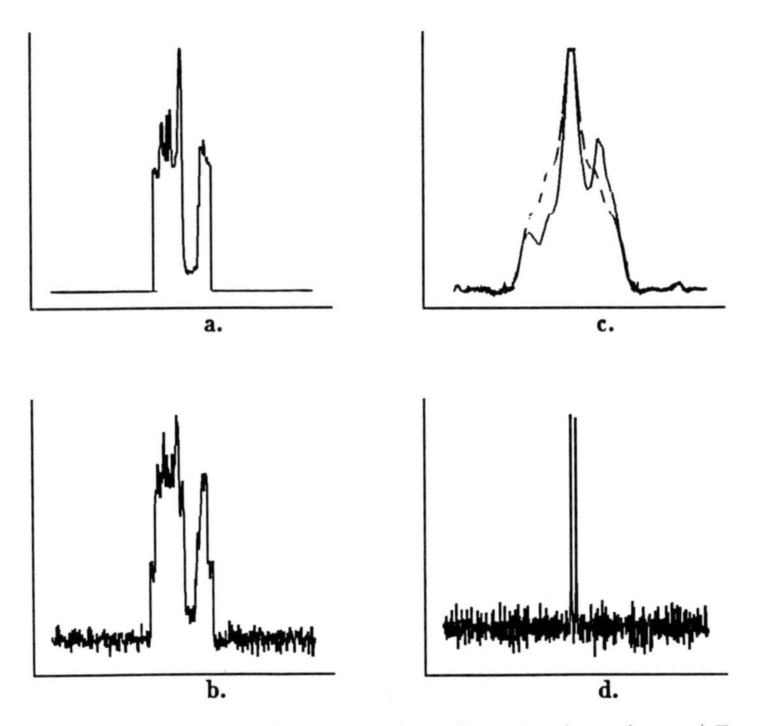

Figure 8.5-3: Matching of linearly superposed one-dimensional templates. a) Feature to be matched. b) Superposition of two signal features with additive noise (SNR=18 dB). c) Normalized (solid) and unnormalized correlation (dashed) fail to detect the two features (DSNR=9.8 dB). d) EXM coefficients display two distinct and sharp peaks (DSNR=30.1 dB).

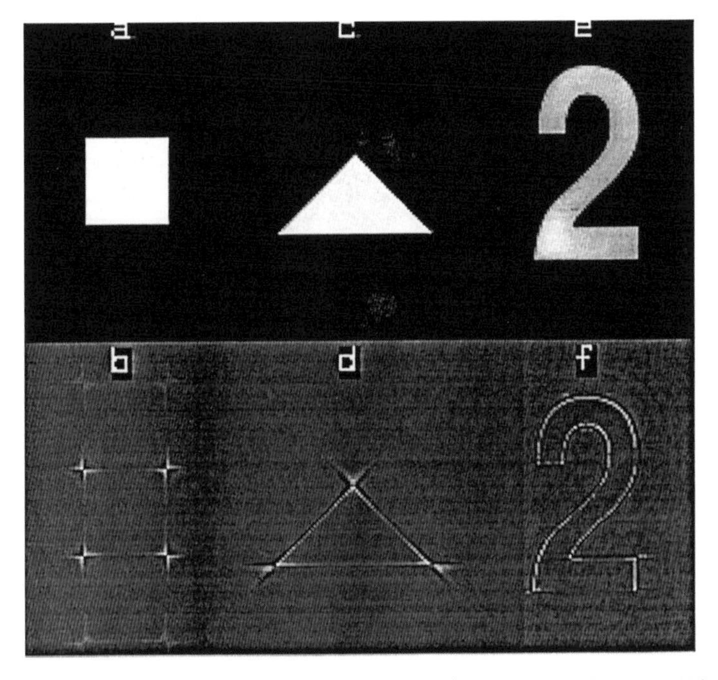

Figure 8.5-4: EXM filters and shape innovations. a) The square shape and b) its EXM filter. c) The triangle shape and d) its EXM filter. e) The digit '2' and f) its EXM filter.

8.6 EXM Based Optimal DSNR Edge Detection

Edges capture most of the relevant image information required for recognition tasks, and thus edge detection is a fundamental preprocessing task in vision systems. Among the numerous methods proposed for edge detection [8-1], [8-5], most popular are those that are based on correlating the given signal with an edge detector. Many works have been performed on the design of an 'optimal' detector and many optimality criteria have been formulated as well. A more detailed discussion and literature survey can be found in [8-12].

Edge detection can also be cast as a template matching problem, where the template being sought is the edge model. Using this paradigm, EXM can be used to generate optimal DSNR edge detectors for any given edge model. The discrete-domain EXM results of the previous sections quite naturally extend to the continuous domain. The continuous version of single-template EXM presented in Section 8.3 is used to derive the results in this section.

A. Edge Detection in One Dimension

To obtain the optimal DSNR step edge detector, it is first necessary to set up a model for the edge. Suppose we want to detect step edges. A bipolar step (signum) function can be used itself as the edge model. In one dimension this is defined as

$$\mathrm{EDGE}_{1-D}(x) = \mathrm{sgn}(x) = \begin{cases} -1 & : & x < 0 \\ 0 & : & x = 0 \\ 1 & : & x > 0. \end{cases} \tag{8.6-1}$$

This model is more convenient for analysis than the unit step function and has a Fourier transform given by

$$\mathcal{F}[\mathrm{sgn}(x)] = \frac{2}{j\omega}. \tag{8.6-2}$$

Using this edge model for the template $\Psi(\omega)$ in Eq. (8.3-1), the *step expansion filter* (SEF) for this template can be analytically derived. Assuming $\lambda(x)$ to be white noise with variance σ_λ^2, the SEF in the frequency domain is given by [8-12]

$$\Theta_s(\omega) = \frac{2j\omega\sigma_\lambda^{-2}}{4\sigma_\lambda^{-2} + \omega^2}. \tag{8.6-3}$$

The inverse Fourier transform of Eq. (8.6-3) yields the impulse response $\theta_s(x)$ of the SEF as

$$\theta_s(x) = \alpha_0 \exp(- \mid \frac{2x}{\sigma_\lambda} \mid)\mathrm{sgn}(x), \tag{8.6-4}$$

where $\alpha_0 = -\sigma_\lambda^{-2}$ affects only the output amplitude. The above SEF is optimal in terms of DSNR for detecting step edges for the specified noise variance σ_λ^2.

Canny [8-21] formulated an edge detector which optimized a combination of three criteria: detection or SNR, localization, and multiple response suppression. Although this edge detection formulation includes a parameter that yields various detector shapes, he suggests a close approximation by the first derivative of the Gaussian as the most practical choice. This is most widely used, and thus we present as the *Canny edge detector* (CED) $\phi(x)$, the first derivative of a Gaussian

$$\phi(x) = \gamma x \exp(\frac{-x^2}{2\sigma_c^2}), \tag{8.6-5}$$

where γ is an amplitude scale factor and σ_c^2 is the variance (width) of the detector.

The SEF impulse response (solid line) is illustrated in Fig. 8.6-1b) along with the CED (dashed line). It is quite evident that the two edge detectors are very different. The width of the exponential side lobes of the SEF increases with the parameter σ_λ and in the limiting case of infinite additive noise, the SEF leads to the step edge model itself. This corresponds to the fact that the matched filter is the limiting case of the single-template EXM filter with infinite noise [8-8]. Larger widths of the CED – or any other detector for that matter – yield smoother outputs and thus less detail in the edge maps. For a given noise variance σ_λ^2, it has been analytically determined [8-12] that the optimal CED width (in terms of DSNR) is

$$\sigma_c^2 = \frac{\sigma_\lambda^2}{8}. \tag{8.6-6}$$

The response of the SEF edge detector to a comparable CED (one that is DSNR-optimal for the same noise level) can be expected to be sharper, due to the discontinuity at the origin. Consequently, one would also expect the CED to be more noise resistant. However, analytical and experimental results [8-12] show that the SEF performs significantly better than the CED (in terms of DSNR) under extreme noise conditions as well.

Figure 8.6-1a) shows a noisy step edge with an SNR of −9.2 dB which is considered as quite a high noise level. Figure 8.6-1b) shows the SEF for this noise level (solid line) and also the corresponding CED (dashed line). The CED is designed (optimized in terms of DSNR) according to Eq. (8.6-6). Figure 8.6-1c) shows the response of the SEF to this noisy step and Fig. 8.6-1d) shows the corresponding CED response. It can be seen that the response of the SEF edge detector is sharper and the peak is accurately localized even for such a high noise level. Furthermore, the off-center SEF response is small compared to the CED off-center response. A more detailed analysis and comparison of the detectors can be found in [8-12], and reveals that the SEF consistently yields 5 dB higher DSNR than the CED over a wide range of noise levels.

Using different edge models, it is possible to create a family of optimal DSNR edge detectors based on EXM. Figure 8.6-2 shows a ramp edge model (dashed line) and the corresponding optimal DSNR edge detector. It can be seen that the edge

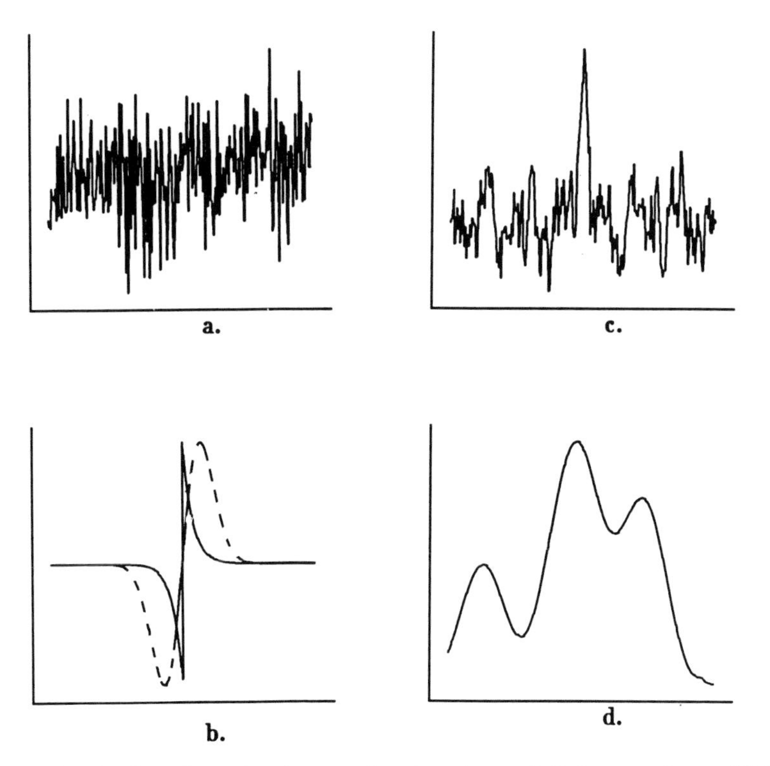

Figure 8.6-1: One-dimensional step edge detection. a) Noisy step input (SNR=-9.2 dB). b) Optimal DSNR step expansion filter (solid) and Canny edge detector (dashed). c) Step expansion filter response. d) Canny edge detector response.

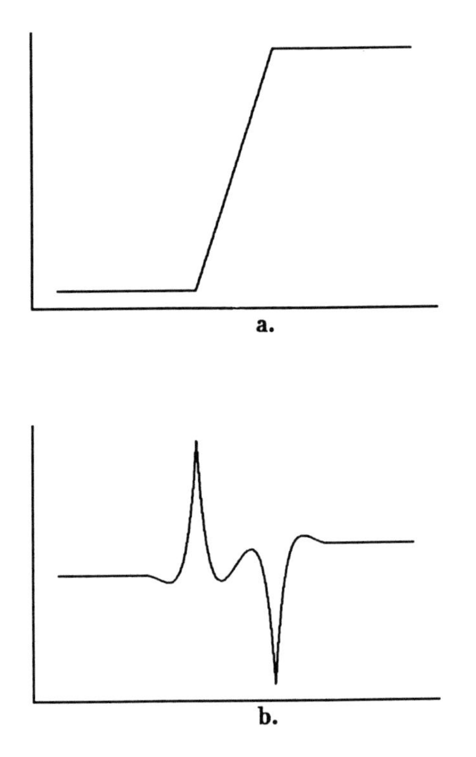

Figure 8.6-2: a) Ramp edge model. b) Optimal DSNR ramp edge EXM filter.

detector consists of two major sections, each corresponding to one discontinuity in the slope of the ramp. The widths of these sections vary with the amount of assumed input noise.

The roof edge model in Fig. 8.6-3 (dashed line) has the optimal DSNR detector shown as a solid line. Here also, one can see that there are three predominant sections to the detector, corresponding to the three discontinuities in the derivative of the roof edge model.

B. Edge Detection in Two Dimensions

One possible way of extending the one-dimensional (1-D) results of the previous section is to design a set of two-dimensional (2-D) detectors, one for every rotated step edge. However, it is sufficient to design two edge detectors for step edges in orthogonal directions, and combine these responses to get the gradient magnitude and direction of the step edge response. In detecting 2-D edges, one question that arises is how wide is the edge in the direction orthogonal to the step edge? No specific answer is possible for this question and thus we need to localize the edge detector to respond only over a certain width of the edge. This localization is done by applying a separable window in the direction parallel to the edge. The window is necessarily low-pass since we need to average the noise along the edge. A gracefully decaying and smooth window is preferable and we therefore use the Gaussian window. A more detailed discussion of this point can be found in [8-12].

Since the slope of a surface in any direction can be determined from the slope in two orthogonal directions, we apply the 2-D separable CED defined in the X direction as

$$\phi_X(x,y) = \beta x \exp(-\frac{x^2}{2\sigma_c^2}) \exp(-\frac{y^2}{2\sigma_W^2}), \tag{8.6-7}$$

where σ_W^2 is the variance of the Gaussian window in the Y direction. The gradient magnitude (square root of the sum of the squares of the individual X and Y detector responses) is employed. One could threshold and detect local peaks in this output to create a binary edge map.

Invoking the same principles, the two-dimensional version of the SEF consists of the 1-D SEF in the direction of edge detection and the same Gaussian window in the orthogonal direction to localize the edge detector. The 2-D SEF for a step in the X direction is given by

$$\theta_{X_s}(x,y) = \alpha_0 \mathrm{sgn}(x) \exp(-\mid \frac{2x}{\sigma_\lambda} \mid) \exp(-\frac{y^2}{2\sigma_W^2}). \tag{8.6-8}$$

Thus, both the 2-D CED and the 2-D SEF have identical windows along the direction of the edge, facilitating a common ground for comparison. The variance σ_W^2 is chosen to be the same as the CED variance σ_c^2 designed for the given noise level as in Eq. (8.6-6). This is because σ_c^2 is also the variance of the Gaussian corresponding

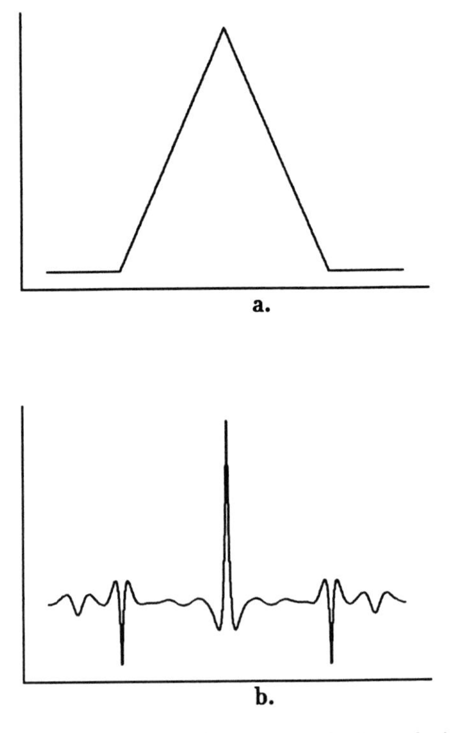

Figure 8.6-3: a) Roof edge model. b) Optimal DSNR roof edge EXM filter.

to the integral of the 1-D CED. Intuitively, σ_c^2 is the optimal width (in terms of DSNR) of the CED for the given amount of input noise σ_λ^2. This is equivalent to the optimal amount of Gaussian smoothing required for that noise level.

An example of the 2-D SEF impulse response for vertical edge detection is shown in Fig. 8.6-4. Here also, one can see Fig. 8.6-4a) shows the test image of a 2-D pulse embedded in additive white noise with SNR of -6.8 dB. Fig. 8.6-4b) shows the 2-D SEF used in this experiment. The 2-D SEF response (36.7 dB DSNR) in Fig. 8.6-4d) is clearly seen to be sharper than the corresponding 2-D CED response (31.8 dB DSNR) in Fig. 8.6-4c). The horizontal profile of the CED response in Fig. 8.6-4c) is taken at the image center and shown in Fig. 8.6-4e). The corresponding SEF response profile taken from Fig. 8.6-4d) is shown in Fig. 8.6-4f). It can be seen that the SEF response is much sharper than the CED response and that the spurious peaks in the SEF response are significantly smaller in amplitude though they are more in number.

A more detailed and objective comparison of the performance of the 2-D SEF and CED can be found in [8-12]. By generating actual edge maps from synthetic test images over a wide range of noise, it has been shown that the SEF edge maps are better localized and more accurate, especially for high noise levels. The CED edge maps get distorted at large noise levels (detector widths) but are better connected (less broken elements). The CED's disadvantage is that its excessive smoothing results in rounded features, while the SEF yields sharper features while still maintaining low spurious responses. It is also observed that the SEF is less sensitive to the design parameter, i.e., a single width of the SEF caters to a larger range of noise levels. Edge maps of real images furnished by the SEF are found to be more accurate and have less spurious edge elements than the corresponding CED edge maps.

Summary

In this chapter, we discussed the problems of conventional template matching (broad peaks and large spurious response) and the SNR criterion. Optimizing a new *discriminative signal-to-noise ratio* (DSNR) results in the *expansion matching* (EXM) method for template matching. The DSNR criterion penalizes all response off the center of the template and thus EXM yields sharp matching peaks and minimal off-center response. Using the generalized multiple-template EXM method, it is possible to design a single filter to match complex-valued multiple templates by eliciting user-defined responses for each template, while optimizing the DSNR criterion. Special cases of the multiple-template EXM filter correspond to previous approaches such as MACE, SDF and MVSDF [8-9], which are all restricted to real signals.

For the special case of matching a single, real, template the EXM filter reduces to the Wiener filter for *minimum squared error* (MSE) restoration. This

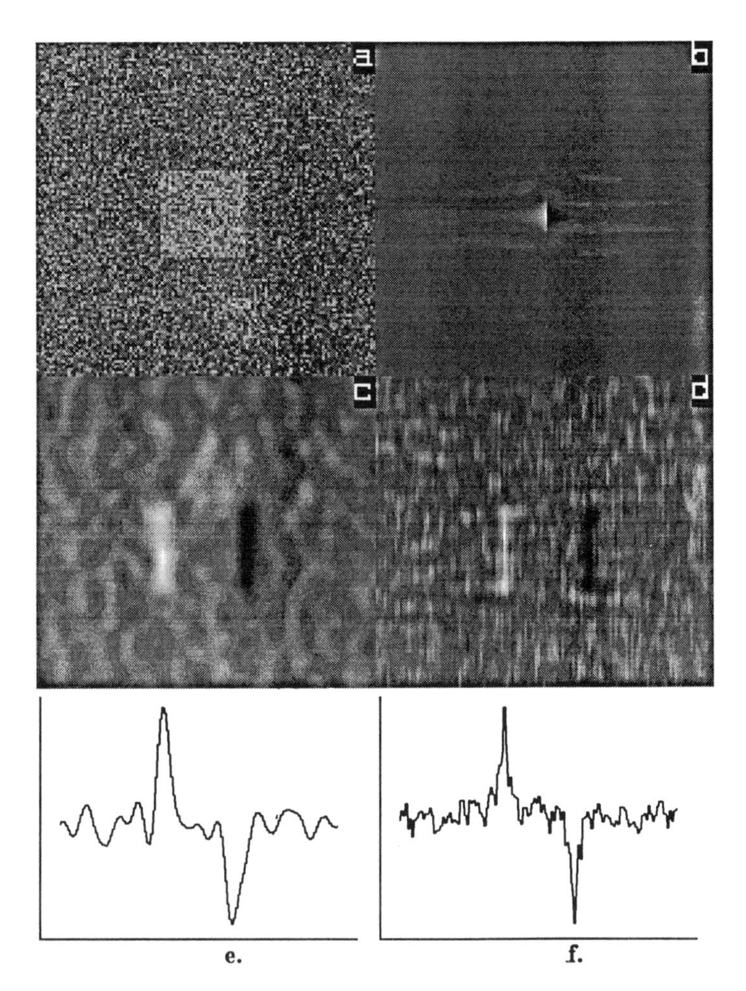

Figure 8.6-4: Two dimensional step edge detection. a) 2-D step edge in white noise (SNR=−6.8 dB). b) 2-D step expansion filter. c) Canny edge detector response (optimized for the highest possible DSNR). d) Step expansion filter response. e) Horizontal profile of CED response in c). f) Horizontal profile of SEF response in d).

means that to obtain the DSNR-optimal response of a delta function, one has to deblur the input image using the template as the blurring function. This leads to a blurring/restoration model of template matching which is extended to multiple templates as well. When the noise level becomes very large, the single-template EXM filter tends to the matched filter itself [8-8].

The underlying principle for the effectiveness of single-template EXM is a non-orthogonal expansion of the input image with *basis functions* (BFs) that are all shifted versions of the template. The BF that matches the template in the input image receives a high coefficient and all the others are very small. Such an expansion is generally non-orthogonal since the shifted set of templates form a non-orthogonal basis. Thus, the difference between correlation and EXM is, principally, that in correlation, the image is projected onto the BFs (shifted templates), while in EXM the image is decomposed into its components along each BF. Thus, EXM is ideally suited for matching linearly superposed templates (as in speech or radar signals), and this is demonstrated with a simple example.

Experimental results comparing EXM to the conventional matched filtering or correlation show that EXM yields sharp matching peaks and minimal off-center response, while correlation yields broad peaks and has substantial spurious responses to irrelevant features in the image. The EXM filter yields superior results under conditions of noise, superposition and severe occlusion of the templates. The effectiveness of EXM can be better understood by examining the EXM filters for simple geometric shapes. It is seen that the EXM filters concentrate on the shape innovations or high-curvature regions of the shape, and are substantially zero elsewhere. Thus, EXM is able to recognize occluded shapes as long as the innovations of the shape are still visible. These observations conform to well-known results of psychophysical experiments [8-13].

Edge detection can be considered as a template matching problem, where the template to be recognized is the edge model. Using different edge models, one can very easily create a family of optimal DSNR edge detectors using EXM. The *step expansion filter* (SEF) for detecting step edges is analytically derived and compared to the widely used Canny edge detector. Other EXM-based detectors for ramp and roof edges are also presented.

References

[8-1] A. Rosenfeld and A. Kak, *Digital Picture Processing*, Chapters 2, 7, 9 and 10, New York: Academic Press, 1982.

[8-2] R. Haralick and L. Shapiro, *Computer and Robotic Vision*, Chapters 16 and 18, New York: Addison Wesley, 1992.

[8-3] H. Wechsler, *Computational Vision*, Chapter 3, New York: Academic Press, 1990.

[8-4] R.J. Schalkoff, *Digital Image Processing and Computer Vision*, Chapters 3 and 4, New York: John Wiley and Sons Inc., 1989.

[8-5] A. Jain, *Fundamentals of Digital Image Processing*, Chapters 5, 8 and 9, Englewood Cliffs, NJ: Prentice Hall, 1989.

[8-6] A. Papoulis, *Probability, Random Variables, and Stochastic Processes*, Chapter 14, New York: McGraw-Hill, 1989.

[8-7] J. Ben-Arie and K.R. Rao, "A novel approach for template matching by non-orthogonal image expansion," *IEEE Trans. Circts. and Systs. for Video Tech.*, vol. 3, no. 1, pp. 71-84, Feb. 1993.

[8-8] J. Ben-Arie and K.R. Rao, "Optimal template matching by non-orthogonal image expansion using restoration," *Int. J. Machine Vision and Apps.*, vol. 7, pp. 69-81, Mar. 1994.

[8-9] K.R. Rao and J. Ben-Arie, "Multiple template matching using expansion matching," *IEEE Trans. Circts. and Systs. for Video Tech.*, vol. 4, no. 6, pp. tbd, Dec. 1994.

[8-10] D. Casasent and W.T. Chang, "Correlation synthetic discriminant functions," *Applied Optics*, vol. 25, pp. 2343-2350, 1986.

[8-11] A. Mahalanobis, B.V.K.V. Kumar and D. Casasent, "Minimum average correlation energy filters," *Applied Optics* vol. 26, no. 17, pp. 3633-3640, Sept. 1987.

[8-12] K.R. Rao and J. Ben-Arie, "Optimal edge detection using expansion matching and restoration," *IEEE Trans. Pattern Analysis and Machine Intelligence*, vol. PAMI-16, no. 12, pp. tbd, Dec. 1994.

[8-13] I. Biederman, "Matching image edges to object memory," *Proc. IEEE Conf. on Computer Vision and Pattern Recognition*, pp. 384-392, 1989.

[8-14] S. Haykin, *Adaptive Filter Theory*, Appendices B and C, (2nd edn.), Englewood Cliffs: Prentice Hall, 1991.

[8-15] B. Noble, *Applied Linear Algebra*, Chapter 5, Englewood Cliffs, NJ: Prentice-Hall, 1969.

[8-16] B.V.K. Vijaya Kumar, "Minimum variance synthetic discriminant functions," *J. Opt. Soc. Am.* A 3, pp. 1579-1586, 1986.

[8-17] J. Ben-Arie and K.R. Rao, "Signal representation by generalized non-orthogonal Gaussian wavelet groups using lattice networks," *Proc. IEEE Intl. Jt. Conf. on Neural Networks*, Singapore, pp. 968-973, Nov. 1991.

[8-18] J. Ben-Arie and K.R. Rao, "Parallel generation of Fourier and Gabor transforms and other shape descriptors by Gaussian wavelet groups using a set of multi-dimensional lattices," *Proc., IEEE 7th Workshop on Multidimensional Signal Processing* Session 10, Lake Placid, NY Sept. 1991.

[8-19] K.R. Rao and J. Ben-Arie, "Lattice architectures for multiple-scale Gaussian convolution, image processing, sinusoid-based transforms and Gabor filtering," *Analog Integrated Circuits and Signal Processing*, vol. 4, no. 2, pp. 141-160, Aug. 1993.

[8-20] J. Ben-Arie, "Multi-dimensional linear lattice for Fourier and Gabor transforms, multiple-scale Gaussian filtering, and edge detection," in *Neural Networks for Human and Machine Perception*, H. Wechsler (ed.), New York: Academic Press, pp. 231-252, 1992.

[8-21] J. Canny, "A computational approach to edge detection," *IEEE Trans. on Pattern Analysis and Machine Intelligence*, Vol. PAMI-8, No. 6, pp. 679-697, Nov. 1986.

Chapter 9

Locally Optimum Detection and its Application to Communications and Signal Processing[†]

WILLIAM E. JACKLIN and DONALD R. UCCI

Introduction

In many communications and radar applications the interference is of a non-Gaussian nature that may be the result of a naturally occurring phenomenon, or a man-made source. In either case, traditional *linear* detection schemes (usually matched filters or correlator receivers) typically exhibit a performance degradation making the output of the receiver unreliable. Thus, it is necessary to develop detectors that are able to perform reliably in these scenarios. In general, these detectors are nonlinear in nature and may be difficult to implement. However, if the information signal is small relative to the interference (known as the "small-signal assumption") then a *locally optimum* (LO) version of the detector may be implemented, and is usually simpler in design.

This chapter is concerned primarily with the problem of binary detection of discrete signals. Basic concepts of statistical detection and probability theory are used to develop the optimum detectors from which their LO versions are derived. All detector derivations are based on the likelihood ratio. The two main types of LO detectors discussed are those without memory and those with memory. The

[†]This work was supported in part by the United States Department of the Air Force, Rome Laboratory, NY.

type of LO detector employed in a given system is chosen primarily in response to the assumptions made regarding the interference.

Once the various LO detector algorithms are identified, the actual implementation of the detector is addressed. There are two possible implementation strategies. If the *probability density function* (pdf) of the received signal is already known, then the desired LO detector may be implemented directly. However, if the interference pdf is not completely known, or is possibly changing, then other techniques must be investigated. In particular, *robust* LO detection techniques are employed in which an estimate of the interference pdf is used to implement the detector. Thus, a variety of pdf estimation techniques, as well as other robust implementation methods, are addressed.

Finally, the application of LO detection to direct-sequence spread-spectrum communication systems is examined. First, a brief tutorial of spread-spectrum systems is provided. Next, the application of LO detection to a spread-spectrum system and the various resulting receiver structures are discussed. Finally, sample results showing the performance improvement gained through the use of LO detection are presented.

9.1 Derivation of the Memoryless Locally Optimum Detector

In binary-signal detection in additive noise, the goal is to decide which of two possible information signals is present at the detector. For example, in a *binary phased shift keyed* (BPSK) communications system, the receiver must decide whether a value of $+1$ or -1 was sent by the transmitter. Stated more formally, the goal of the detector is to correctly choose one of the following two hypotheses:

$$H_1 : \text{Signal } s_1 \text{ present}$$
$$H_0 : \text{Signal } s_0 \text{ present.}$$

For example, if the received signal is given by the scalar equation

$$r = s_m + n \tag{9.1-1}$$

where n is the noise and $m = 0$ or 1, and the observed value of the received signal is $r = \rho$, then the optimum detector has the form ([9-1], Chapters 3 and 4)

$$\lambda(\rho) \underset{\text{choose } H_0}{\overset{\text{choose } H_1}{\gtrless}} \gamma, \tag{9.1-2}$$

where $\lambda(\rho)$ is the likelihood ratio. The likelihood ratio in this case is the ratio of two pdfs:

$$\lambda(\rho) = \frac{f_r(\rho|H_1)}{f_r(\rho|H_0)}, \tag{9.1-3}$$

where $f_r(\rho|H_m)$ is the pdf of r given that H_m is true, $m = 0$ or 1. The constant γ is chosen depending on the type of hypothesis test being used, for example, *maximum likelihood* (ML), *maximum a posteriori* (MAP), *Neyman–Pearson* (NP), or general Bayesian ([9-1], Chapter 3). If the information signals are independent of the noise then $f_r(\rho|H_m) = f_n(\rho - s_m)$, where $f_n(\eta)$ is the noise pdf,[1] and Eq. (9.1-3) becomes

$$\lambda(\rho) = \frac{f_n(\rho - s_1)}{f_n(\rho - s_0)}. \tag{9.1-4}$$

Equivalently, one may examine the log likelihood ratio, which in this case is

$$\ln[\lambda(\rho)] = \ln\left[\frac{f_n(\rho - s_1)}{f_n(\rho - s_0)}\right] = \ln[f_n(\rho - s_1)] - \ln[f_n(\rho - s_0)], \tag{9.1-5}$$

where $\ln[\cdot]$ denotes the natural logarithm.

One method of deriving the LO detector is to approximate each of the two terms on the right side of Eq. (9.1-5) with its Taylor series expansion around the received point, ρ. Using the first two terms in the expansion yields the approximation $l(\rho) \approx \ln(\rho)$, where

$$\begin{aligned} \ln(\rho) \approx l(\rho) &= \left[\ln[f_n(\rho)] - s_1 \frac{d}{d\rho}\ln[f_n(\rho)]\right] - \left[\ln[f_n(\rho)] - s_0 \frac{d}{d\rho}\ln[f_n(\rho)]\right] \\ &= -(s_1 - s_0)\frac{d}{d\rho}\ln[f_n(\rho)] \\ &= (s_1 - s_0)\, g(\rho), \end{aligned} \tag{9.1-6}$$

where

$$g(\rho) \triangleq -\frac{d}{d\rho}\ln[f_n(\rho)] = -\frac{\frac{d}{d\rho}f_n(\rho)}{f_n(\rho)}. \tag{9.1-7}$$

The function $g(\rho)$ is called the LO nonlinearity and in general it is a nonlinear function of ρ. Using Eq. (9.1-6) and Eq. (9.1-7) in Eq. (9.1-2) yields the LO detector for the binary detection of a discrete scalar signal in additive noise:

$$l(\rho) = (s_1 - s_0)\, g(\rho) \quad \underset{\text{choose } H_0}{\overset{\text{choose } H_1}{\gtrless}} \quad \tilde{\gamma}, \tag{9.1-8}$$

where $l(\rho)$ is the LO test statistic and $\tilde{\gamma} = \ln(\gamma)$.

A discussion of the LO nonlinearity is necessary at this point. In manipulating the expression for $\ln[\lambda(\rho)]$ from Eq. (9.1-5) to Eq. (9.1-6) an important assumption was made, namely the *small signal assumption.* Since only the first-order Taylor

[1]This result can be seen by noting that $P[r \le \rho|H_m] = P[n \le \rho - s_m|H_m] = P[n \le \rho - s_m]$ since n is assumed to be independent of s_m.

series was used to approximate the two terms in Eq. (9.1-5), it is necessary that the higher-order terms of each Taylor series decay to zero. These higher-order terms are of the form $(-1)^k (s_m)^k \frac{d^k}{d\rho^k} \ln[f_n(\rho)]$. If $|s_m| \ll 1$ (or, with a slight modification to the problem, if $|n| \gg |s_m|$), then $(s_m)^k \to 0$, and these higher-order terms become negligible. Due to the small signal requirement on s_m, another commonly used term for LO detection is "threshold detection."

Example 9.1-1

Assume that the noise has a Cauchy pdf, given by

$$f_n(\eta) = \frac{1}{\pi(1+\eta^2)} \quad \text{for} \quad -\infty < \eta < \infty.$$

Let $s_1 = -s_0 = s$ so that under H_1, $r = s + n$ and under H_0, $r = -s + n$. The likelihood ratio is given by

$$\lambda(\rho) = \frac{1+(\rho+s)^2}{1+(\rho-s)^2}.$$

The LO nonlinearity is given by the simpler expression

$$g(\rho) = -\frac{f_n{}'(\rho)}{f_n(\rho)} = -\frac{-2\rho/[\pi(1+\rho^2)^2]}{1/[\pi(1+\rho^2)]} = \frac{2\rho}{(1+\rho^2)}.$$

Thus, the LO detector for this test is

$$l(\rho) = 2s\frac{2\rho}{1+\rho^2} \underset{\text{choose } H_0}{\overset{\text{choose } H_1}{\gtrless}} \tilde{\gamma}.$$

A graph of $g(\rho)$ is provided in Fig. 9.1-1. If $\tilde{\gamma} = 0$ (corresponding to an ML test) and s_1 and s_0 are assumed to be equally likely, the probability of error, P_e, corresponding to this test is

$$P_e = \frac{1}{2}\left[P(l(\rho) > 0 \,|\, H_0) + P(l(\rho) < 0 \,|\, H_1)\right].$$

If it is assumed that $s > 0$, then from Fig. 9.1-1 it can be seen that

$$\begin{aligned} P(l(\rho) < 0 \,|\, H_1) &= P(\rho < 0 \,|\, H_1) = \int_{-\infty}^{0} \frac{1}{\pi(1+(\rho-s)^2)}\, d\rho \\ &= \frac{1}{2} - \frac{1}{\pi}\tan^{-1} s. \end{aligned}$$

Similarly

$$\begin{aligned} P(l(\rho) > 0 \,|\, H_0) &= P(\rho > 0 \,|\, H_0) = \int_{0}^{\infty} \frac{1}{\pi(1+(\rho+s)^2)}\, d\rho \\ &= \frac{1}{2} - \frac{1}{\pi}\tan^{-1} s. \end{aligned}$$

Thus

$$P_e = \frac{1}{2} - \frac{1}{\pi} \tan^{-1} s$$

which is shown in Fig. 9.1-2.[2]

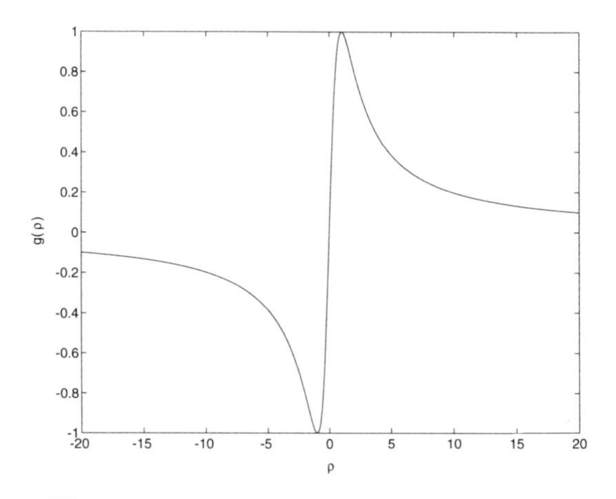

Figure 9.1-1: LO nonlinearity for Example 9.1-1.

The more general binary signal detection problem is to choose between the two hypotheses:

$$H_1 : \text{Signal } \mathbf{s_1} \text{ present}$$
$$H_0 : \text{Signal } \mathbf{s_0} \text{ present,}$$

where $\mathbf{s_0}$ and $\mathbf{s_1}$ are vectors with elements s_{0i} and s_{1i}, respectively, and the notation $\mathbf{x} = [x_1 \cdots x_N]^T$ denotes a vector of length N. The received signal under the two hypotheses is

$$\begin{aligned} \text{under } H_1 : \quad & \mathbf{r} = \mathbf{s_1} + \mathbf{n} \\ \text{under } H_0 : \quad & \mathbf{r} = \mathbf{s_0} + \mathbf{n}\,. \end{aligned} \tag{9.1-9}$$

[2]For the case when $\tilde{\gamma} = 0$, the LO detector and the optimum likelihood ratio detector have the same P_e and can both be reduced to a linear detector of the form

$$s\rho \quad \mathop{\gtrless}_{\text{choose } H_0}^{\text{choose } H_1} \quad 0\,.$$

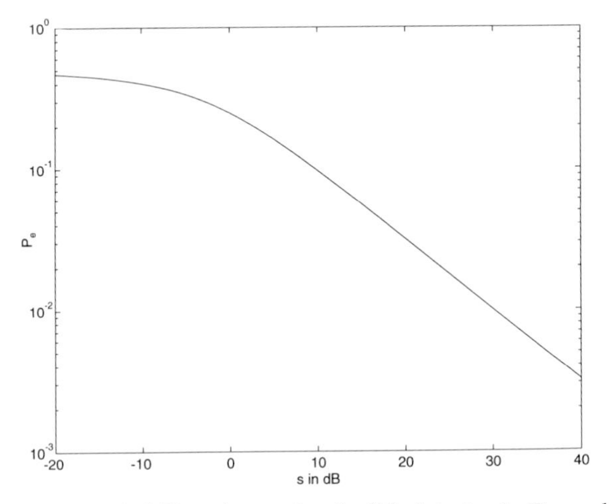

Figure 9.1-2: Probability of error for the LO detector in Example 9.1-1.

If the noise samples are assumed to be *independent and identically distributed* (iid), and $\boldsymbol{\rho}$ is the observed value of $\mathbf{r}$, then the likelihood ratio is

$$\begin{aligned}\lambda(\boldsymbol{\rho}) &= \frac{f_{\mathbf{r}}(\boldsymbol{\rho} \mid H_1)}{f_{\mathbf{r}}(\boldsymbol{\rho} \mid H_0)} = \frac{\prod_{i=1}^{N} f_r(\rho_i \mid H_1)}{\prod_{i=1}^{N} f_r(\rho_i \mid H_0)} \\ &= \prod_{i=1}^{N} \frac{f_r(\rho_i \mid H_1)}{f_r(\rho_i \mid H_0)} .\end{aligned} \tag{9.1-10}$$

Furthermore, if $\mathbf{s_m}$, $m = 0$ or 1, is independent of $\mathbf{n}$, then $f_r(\rho_i \mid H_m) = f_n(\rho_i - s_m)$ and the log likelihood ratio becomes

$$\ln[\lambda(\boldsymbol{\rho})] = \sum_{i=1}^{N} \{ \ln[f_n(\rho_i - s_1)] - \ln[f_n(\rho_i - s_0)] \} . \tag{9.1-11}$$

Using a development similar to that of Eq. (9.1-6) yields the general *memoryless* LO detector

$$l(\boldsymbol{\rho}) = \sum_{i=1}^{N} (s_{1i} - s_{0i}) g(\rho_i) \underset{\text{choose } H_0}{\overset{\text{choose } H_1}{\gtrless}} \tilde{\gamma}, \tag{9.1-12}$$

where the memoryless LO nonlinearity is given by

$$g(\rho_i) \triangleq -\frac{d}{d\rho_i} \ln[f_n(\rho_i)] = -\frac{\frac{d}{d\rho_i} f_n(\rho_i)}{f_n(\rho_i)} . \tag{9.1-13}$$

The LO nonlinearity in Eq. (9.1-13) is called memoryless because the output of the nonlinearity function, $g(\rho_i)$, depends upon only the current sample ρ_i, and is not affected by past or future samples.

Example 9.1-2

Let the received signal be given by ($m = 0$ or 1)

$$r_i = s_{mi} + \sin(u_i) \qquad i = 1, ..., N$$

where the u_i are iid samples distributed uniformly over $[-\pi, \pi]$. Then $n_i = \sin(u_i)$ has the following pdf ([9-2], Chapter 2)

$$f_{n_i}(\eta) = f_n(\eta) = \begin{cases} \frac{1}{\pi\sqrt{1-\eta^2}} & \text{for } |\eta| < 1 \\ 0 & \text{otherwise.} \end{cases}$$

The corresponding memoryless LO nonlinearity is

$$g(\rho_i) = -\frac{\rho_i \,/\, [\pi(1-\rho_i^2)^{3/2}]}{1 \,/\, [\pi(1-\rho_i^2)^{1/2}]} I_{(-1,1)}(\rho_i) = -\frac{\rho_i}{1-\rho_i^2} I_{(-1,1)}(\rho_i)$$

and is shown in Fig. 9.1-3. The function $I_A(x)$ is the set indicator function, i.e., $I_A(x) = 1$ for $x \in A$ and 0 otherwise. Furthermore, let $s_{0i} = 0$ for all i (indicative of a radar problem where H_1 is the hypothesis that a target is present and H_0 is the hypothesis that a target is absent). Then the resulting memoryless LO detector is given by

$$l(\boldsymbol{\rho}) = \sum_{i=1}^{N} s_{1i} \left[-\frac{\rho_i}{1-\rho_i^2} \right] I_{(-1,1)}(\rho_i) \quad \underset{\text{choose } H_0}{\overset{\text{choose } H_1}{\gtrless}} \quad \tilde{\gamma}\,.$$

Example 9.1-3

Let the received signal be given by ($m = 0$ or 1)

$$r_i = s_{mi} + n_i$$

where the noise samples are iid and have a Cauchy pdf (see Example 9.1-1). Thus, the memoryless LO nonlinearity is given by (from Example 9.1-1)

$$g(\rho_i) = \frac{2\,\rho_i}{1+\rho_i^2}\,.$$

Furthermore, assume that $\mathbf{s_1} = \mathbf{s_0} = \mathbf{s}$ and that each signal is equally likely. In this case, the resulting memoryless LO detector is

$$l(\boldsymbol{\rho}) = 2\,s \sum_{i=1}^{N} \frac{2\rho_i}{1+\rho_i^2} \quad \underset{\text{choose } H_0}{\overset{\text{choose } H_1}{\gtrless}} \quad \tilde{\gamma}\,.$$

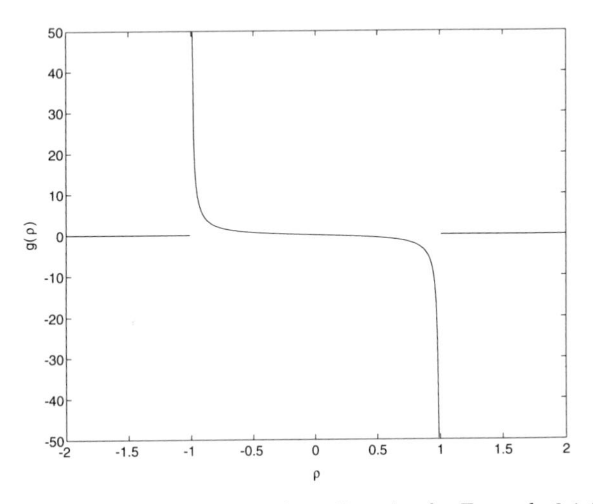

Figure 9.1-3: Memoryless LO nonlinearity for Example 9.1-2.

To compute the P_e corresponding to this problem, first examine the first and second-order conditional expectations of $g(\rho_i)$, given by[3]

$$
\begin{aligned}
E[g(r_i)\,|\,H_0] &= \int_{-\infty}^{\infty} \frac{2\,\rho}{1+\rho^2}\,\frac{1}{\pi(1+(\rho+s)^2)}\,d\rho \;=\; -\frac{2\,s}{s^2\,+\,4} \\
E[g^2(r_i)\,|\,H_0] &= \int_{-\infty}^{\infty} \left[\frac{2\,\rho}{1+\rho^2}\right]^2 \frac{1}{\pi(1+(\rho+s)^2)}\,d\rho \;=\; \frac{6\,s^2\,+\,8}{(s^2\,+\,4)^2} \\
E[g(r_i)\,|\,H_1] &= \int_{-\infty}^{\infty} \frac{2\,\rho}{1+\rho^2}\,\frac{1}{\pi(1+(\rho-s)^2)}\,d\rho \;=\; \frac{2\,s}{s^2\,+\,4} \\
E[g^2(r_i)\ H_1] &= \int_{-\infty}^{\infty} \left[\frac{2\,\rho}{1+\rho^2}\right]^2 \frac{1}{\pi(1+(\rho-s)^2)}\,d\rho \;=\; \frac{6\,s^2\,+\,8}{(s^2\,+\,4)^2}
\end{aligned}
$$

and

$$
\mathrm{Var}[g(r_i)\,|\,H_0] \;=\; \mathrm{Var}[g(r_i)\,|\,H_1] \;=\; \frac{2}{s^2\,+\,4}\,.
$$

Thus, the conditional means and variances of $l(\boldsymbol{\rho})$ are

$$
\begin{aligned}
E[l(\boldsymbol{\rho})\,|\,H_0] \;=\; 2\,N\,s\,E[g(r_i)\,|\,H_0] &\;=\; -\frac{4\,N\,s^2}{s^2\,+\,4} \\
E[l(\boldsymbol{\rho})\,|\,H_1] \;=\; 2\,N\,s\,E[g(r_i)\,|\,H_1] &\;=\; \frac{4\,N\,s^2}{s^2\,+\,4} \\
\mathrm{Var}[l(\boldsymbol{\rho})\,|\,H_0] \;=\; 4\,N\,s^2\,\mathrm{Var}[g(r_i)\,|\,H_0] &\;=\; \frac{8\,N\,s^2}{s^2\,+\,4}
\end{aligned}
$$

[3]Computed symbolically using Mathcad 4.0 by MathSoft.

$$\mathrm{Var}[l(\boldsymbol{\rho}) \mid H_1] = 4\,N\,s^2\,\mathrm{Var}[g(r_i) \mid H_1] = \frac{8\,N\,s^2}{s^2 + 4}.$$

Because $\mathbf{s_1}$ and $\mathbf{s_0}$ are equally likely, P_e for this detector and noise scenario can be computed as follows:

$$P_e = \frac{1}{2}[P(l(\boldsymbol{\rho}) > \tilde{\gamma} \mid H_0) + P(l(\boldsymbol{\rho}) < \tilde{\gamma} \mid H_1)]$$

with

$$P(l(\boldsymbol{\rho}) > \tilde{\gamma} \mid H_0) = P\left(\frac{l(\boldsymbol{\rho}) + (4Ns^2)/(s^2+4)}{\sqrt{(8Ns^2)/(s^2+4)}} > \frac{\tilde{\gamma} + (4Ns^2)/(s^2+4)}{\sqrt{(8Ns^2)/(s^2+4)}} \mid H_0\right)$$

and

$$P(l(\boldsymbol{\rho}) < \tilde{\gamma} \mid H_1) = P\left(\frac{l(\boldsymbol{\rho}) - (4Ns^2)/(s^2+4)}{\sqrt{(8Ns^2)/(s^2+4)}} < \frac{\tilde{\gamma} - (4Ns^2)/(s^2+4)}{\sqrt{(8Ns^2)/(s^2+4)}} \mid H_1\right).$$

Assuming that $\tilde{\gamma} = 0$ (corresponding to an ML test), and defining

$$z_0 \triangleq \frac{l(\boldsymbol{\rho}) + (4Ns^2)/(s^2+4)}{\sqrt{(8Ns^2)/(s^2+4)}}$$

and

$$z_1 \triangleq \frac{l(\boldsymbol{\rho}) - (4Ns^2)/(s^2+4)}{\sqrt{(8Ns^2)/(s^2+4)}},$$

then

$$\begin{aligned} P(l(\boldsymbol{\rho}) > \tilde{\gamma} \mid H_0) &= P\left(z_0 > \sqrt{\tfrac{2Ns^2}{s^2+4}} | H_0\right) \\ &\approx \tfrac{1}{2} - \tfrac{1}{2}\,\mathrm{erf}\left(\sqrt{\tfrac{N\,s^2}{s^2+4}}\right) \end{aligned}$$

and

$$\begin{aligned} P(l(\boldsymbol{\rho}) < \tilde{\gamma} \mid H_1) &= P\left(z_1 < -\sqrt{\tfrac{2Ns^2}{s^2+4}} | H_1\right) \\ &\approx \tfrac{1}{2} - \tfrac{1}{2}\,\mathrm{erf}\left(\sqrt{\tfrac{N\,s^2}{s^2+4}}\right). \end{aligned}$$

Here, the central limit theorem ([9-3], Chapter 5) has been employed and hence, for sufficiently large N, z_0 and z_1 are approximately Gaussian with zero mean and unit variance when H_0 and H_1 are true, respectively. The erf function in the previous equations is defined as

$$\mathrm{erf}(x) \triangleq \frac{2}{\sqrt{\pi}} \int_0^x e^{-t^2}\,dt.$$

Plots of P_e for various values of N, including the case when $N = 1$ discussed in Example 9.1-1, are shown in Fig. 9.1-4.[4]

[4]It should be remembered that the expression for P_e derived in this example is only an approximation for large N. The effects of the approximation become more evident as s increases. For the case when $N = 1$ and $\tilde{\gamma} = 0$ discussed in Example 9.1-1, $P_e \to 0$ as $s \to \infty$. However, when the approximation for large N is used, $P_e \to \frac{1}{2} - \frac{1}{2}\mathrm{erf}\sqrt{N}$ as $s \to \infty$, and $P_e \to 0$ only as $N \to \infty$.

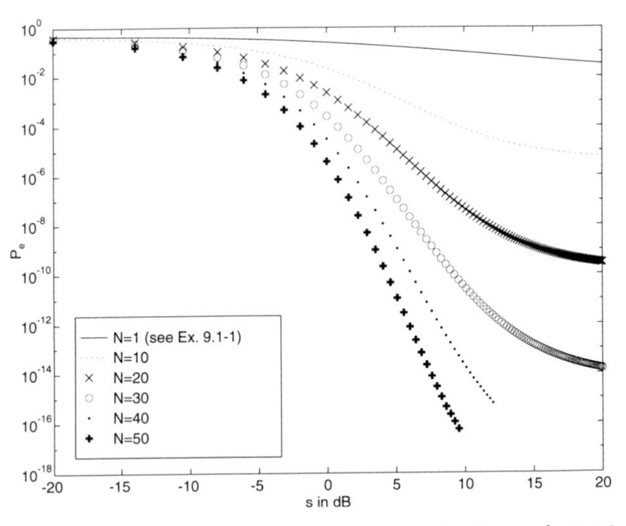

Figure 9.1-4: Probability of error curves for Example 9.1-3.

Note that the procedure described above for calculating P_e cannot be used for the problem described in Example 9.1-2 for the following reason. The value of $E[g^2(r_i) \mid H_0]$ for the sinusoidal-type noise is given by

$$\begin{aligned} E[g^2(r_i) \mid H_0] &= \int_{-1}^{1} \left[\frac{\rho}{(1-\rho^2)}\right]^2 \frac{1}{\pi\sqrt{1-\rho^2}}\, d\rho = \frac{1}{\pi}\int_{-1}^{1} \frac{\rho^2}{(1-\rho^2)^{5/2}}\, d\rho \\ &\Rightarrow \infty . \end{aligned}$$

Since $E[g^2(r_i) \mid H_0]$ does not converge, the central limit theorem cannot be applied to the calculation of P_e.

The procedure for computing P_e discussed in Example 9.1-3 can be extremely useful for computing the performance of the general memoryless LO detector when N, the number of available samples, is sufficiently large. In this case, and when $s_{0i} = s_{0j} = s_0$ and $s_{1i} = s_{1j} = s_1$ ($s_1 > s_0$), P_e can be computed as follows. Compute the first and second-order conditional expectations of the memoryless LO nonlinearity via the equations

$$\mu_{gm} = E[g(r_i) \mid H_m] = \int_{-\infty}^{\infty} g(\rho)\, f_n(\rho - s_m)\, d\rho \tag{9.1-14}$$

$$E[g^2(r_i) \mid H_m] = \int_{-\infty}^{\infty} g^2(\rho)\, f_n(\rho - s_m)\, d\rho \tag{9.1-15}$$

$$\sigma_{gm}^2 = E[g^2(r_i) \mid H_m] - \mu_{gm}^2 \tag{9.1-16}$$

for $m = 0$ and 1. Then, from Eq. (9.1-12), P_e for the memoryless LO detector is given by

$$P_e = P(\mathbf{s_m} = \mathbf{s_0})\, P(l(\boldsymbol{\rho}) > \tilde{\gamma} \mid H_0) + P(\mathbf{s_m} = \mathbf{s_1})\, P(l(\boldsymbol{\rho}) < \tilde{\gamma} \mid H_1) \tag{9.1-17}$$

where, by using the central limit theorem,

$$\begin{aligned} P(l(\boldsymbol{\rho}) > \tilde{\gamma} \,|\, H_0) &= P\left(\frac{l(\boldsymbol{\rho}) - n(s_1 - s_0)\mu_{g0}}{\sqrt{n(s_1 - s_0)^2\sigma_{g0}^2}} > \frac{\tilde{\gamma} - n(s_1 - s_0)\mu_{g0}}{\sqrt{n(s_1 - s_0)^2\sigma_{g0}^2}} \,|\, H_0\right) \\ &\approx \tfrac{1}{2} - \tfrac{1}{2}\,\mathrm{erf}\left(\frac{\tilde{\gamma} - n\,(s_1 - s_0)\mu_{g0}}{\sqrt{2\,n}(s_1 - s_0)\sigma_{g0}}\right) \end{aligned} \tag{9.1-18}$$

and

$$\begin{aligned} P(l(\boldsymbol{\rho}) < \tilde{\gamma} \,|\, H_1) &= P\left(\frac{l(\boldsymbol{\rho}) - n(s_1 - s_0)\mu_{g1}}{\sqrt{n(s_1 - s_0)^2\sigma_{g1}^2}} < \frac{\tilde{\gamma} - n(s_1 - s_0)\mu_{g1}}{\sqrt{n(s_1 - s_0)^2\sigma_{g1}^2}} \,|\, H_1\right) \\ &\approx \tfrac{1}{2} + \tfrac{1}{2}\,\mathrm{erf}\left(\frac{\tilde{\gamma} - n\,(s_1 - s_0)\mu_{g1}}{\sqrt{2\,n}(s_1 - s_0)\sigma_{g1}}\right). \end{aligned} \tag{9.1-19}$$

9.2 Derivation of the Locally Optimum Detector with Memory

There are many applications in which the noise samples are not iid. For example, in a communications system subjected to a narrowband jammer the noise samples are highly correlated. Another example of highly correlated samples is found in many signal processing problems where the noise may be modeled as the output of an *autoregressive* (AR) process. In situations like these, the memoryless LO detector is unable to take advantage of the information available in the correlation structure of the noise. Thus, a detector with memory, i.e., one which uses other data samples in addition to the current sample, is required.

The LO detector with memory can be developed in a manner similar to the memoryless detector. Consider the known signal binary detection problem:

$$\begin{aligned} H_1 &: \text{Signal } \mathbf{s_1} \text{ present} \\ H_0 &: \text{Signal } \mathbf{s_0} \text{ present.} \end{aligned}$$

Let the received signal vector of length N be given by

$$\mathbf{r} = \mathbf{s_m} + \mathbf{n}\,, \tag{9.2-1}$$

where $m = 0$ or 1 and $\mathbf{n}$ is a random noise vector. Given the observation $\mathbf{r} = \boldsymbol{\rho}$, the optimum detector is of the form

$$\ln[\lambda(\boldsymbol{\rho})] \quad \underset{\text{choose } H_0}{\overset{\text{choose } H_1}{\gtrless}} \quad \tilde{\gamma}\,, \tag{9.2-2}$$

where the log likelihood ratio, as before, is given by

$$\ln[\lambda(\boldsymbol{\rho})] = \ln\left[\frac{f_{\mathbf{r}}(\boldsymbol{\rho}\,|\,H_1)}{f_{\mathbf{r}}(\boldsymbol{\rho}\,|\,H_0)}\right] = \ln\left[\frac{f_{\mathbf{n}}(\boldsymbol{\rho} - \mathbf{s}_1)}{f_{\mathbf{n}}(\boldsymbol{\rho} - \mathbf{s}_0)}\right] \tag{9.2-3}$$

assuming that $\mathbf{s_m}$ and $\mathbf{n}$ are independent. Approximating Eq. (9.2-3) using first-order Taylor series expansions yields the following result:

$$\begin{aligned} \ln[\lambda(\boldsymbol{\rho})] \approx l(\boldsymbol{\rho}) &= \ln[f_{\mathbf{n}}(\boldsymbol{\rho} - \mathbf{s}_1)] - \ln[f_{\mathbf{n}}(\boldsymbol{\rho} - \mathbf{s}_0)] \qquad (9.2\text{-}4) \\ &= \ln[f_{\mathbf{n}}(\boldsymbol{\rho})] - \sum_{i=1}^{N} s_{1i}\frac{\partial}{\partial\rho_i}\ln[f_{\mathbf{n}}(\boldsymbol{\rho})] \\ &\quad - \left\{\ln[f_{\mathbf{n}}(\boldsymbol{\rho})] - \sum_{i=1}^{N} s_{0i}\frac{\partial}{\partial\rho_i}\ln[f_{\mathbf{n}}(\boldsymbol{\rho})]\right\} \\ &= -\sum_{i=1}^{N}(s_{1i} - s_{0i})\,\frac{\partial}{\partial\rho_i}\ln[f_{\mathbf{n}}(\boldsymbol{\rho})]\,, \end{aligned}$$

and the resulting decision statistic becomes

$$l(\boldsymbol{\rho}) \triangleq \sum_{i=1}^{N}(s_{1i} - s_{0i})\,g_i(\boldsymbol{\rho})\,, \tag{9.2-5}$$

where

$$g_i(\boldsymbol{\rho}) \triangleq -\frac{\partial}{\partial\rho_i}\ln[f_{\mathbf{n}}(\boldsymbol{\rho})] = -\frac{\frac{\partial}{\partial\rho_i}f_{\mathbf{n}}(\boldsymbol{\rho})}{f_{\mathbf{n}}(\boldsymbol{\rho})} \tag{9.2-6}$$

is called the LO nonlinearity *with memory.* Thus, the LO detector with memory for the binary detection of known signals in additive noise is (compare with Eq. (9.1-12))

$$l(\boldsymbol{\rho}) = \sum_{i=1}^{N}(s_{1i} - s_{0i})\,g_i(\boldsymbol{\rho}) \quad \underset{\text{choose } H_0}{\overset{\text{choose } H_1}{\gtrless}} \quad \tilde{\gamma}\,. \tag{9.2-7}$$

Equation (9.2-4) can be thought of as the sum of N scalar tests, one for each ρ_i. Thus, the statistic $(s_{1i} - s_{0i})\,g_i(\boldsymbol{\rho})$ for the ith component involves $g_i(\rho_1, \rho_2, ..., \rho_N)$, i.e., *all* observations of the data. In this sense the detector of Eq. (9.2-7) has memory. The reason for the dependence on the $(i-1)$ previous and $(N-i)$ future samples of the received signal is that no assumptions concerning independence of the noise vector, $\mathbf{n}$, are made. Instead, the entire N-variate joint pdf of $\mathbf{n}$ is used in deriving the detector of Eq. (9.2-7).

If certain assumptions concerning the dependence structure of $\mathbf{n}$ can be made, the complexity of Eq. (9.2-6) can be reduced. For example, if $\mathbf{n}$ is a Markov random sequence ([9-2], Chapter 7), then

$$f_{n_i}(\eta_i\,|\,\eta_{i-1}, \eta_{i-2}, ..., \eta_1) = f_{n_i}(\eta_i\,|\,\eta_{i-1})\,. \tag{9.2-8}$$

Using Eq. (9.2-8) and recalling that

$$\begin{aligned} f_{\mathbf{n}}(\boldsymbol{\eta}) &= f_{\mathbf{n}}(\eta_1, \eta_2, ..., \eta_N) \\ &= f_{n_N}(\eta_N \,|\, \eta_{N-1}, ..., \eta_1)\, f_{n_{N-1}}(\eta_{N-1} \,|\, \eta_{N-2}, ..., \eta_1) \\ &\qquad \cdots f_{n_2}(\eta_2 \,|\, \eta_1)\, f_{n_1}(\eta_1) \end{aligned} \tag{9.2-9}$$

the joint pdf of $\mathbf{n}$ can be reduced to

$$f_{\mathbf{n}}(\boldsymbol{\eta}) = \prod_{i=1}^{N} f_{n_i}(\eta_i \,|\, \eta_{i-1}), \tag{9.2-10}$$

where $f_{n_1}(\eta_1 \,|\, \eta_0) \triangleq f_{n_1}(\eta_1)$. The LO nonlinearity requires $f_{\mathbf{n}}(\boldsymbol{\rho})$, the noise pdf evaluated at the value of the received signal. From Eq. (9.2-10) this is seen to be

$$f_{\mathbf{n}}(\boldsymbol{\rho}) = \prod_{i=1}^{N} f_{n_i}(\rho_i \,|\, \rho_{i-1}). \tag{9.2-11}$$

From Eq. (9.2-11) the LO nonlinearity with memory, $g_i(\boldsymbol{\rho}) = -\frac{\partial}{\partial \rho_i} \ln[f_{\mathbf{n}}(\boldsymbol{\rho})]$, is

$$g_i(\boldsymbol{\rho}) = \begin{cases} -\frac{\partial}{\partial \rho_i}\Big(\ln[f_{n_i}(\rho_i \,|\, \rho_{i-1})] + \ln[f_{n_{i+1}}(\rho_{i+1} \,|\, \rho_i)]\Big) & ; \text{for } i = 1, ..., N-1 \\ -\frac{\partial}{\partial \rho_N} \ln[f_{n_N}(\rho_N \,|\, \rho_{N-1})] & ; \text{for } i = N\,. \end{cases} \tag{9.2-12}$$

Thus, the original N-variate function $g_i(\boldsymbol{\rho})$ reduces to a function of at most three variables: ρ_{i-1}, ρ_i, and ρ_{i+1}.

Equation (9.2-12) can be written in a simpler form by noting that

$$\begin{aligned} \frac{\partial}{\partial \rho_i} \ln[f_{n_i}(\rho_i \,|\, \rho_{i-1})] &= \frac{\partial}{\partial \rho_i} \ln\left[\frac{f_{n_i\, n_{i-1}}(\rho_i, \rho_{i-1})}{f_{n_{i-1}}(\rho_{i-1})}\right] \\ &= \frac{\partial}{\partial \rho_i} \ln[f_{n_{i-1}\, n_i}(\rho_{i-1}, \rho_i)] \end{aligned} \tag{9.2-13}$$

and

$$\begin{aligned} \frac{\partial}{\partial \rho_i} \ln[f_{n_{i+1}}(\rho_{i+1} \,|\, \rho_i)] &= \frac{\partial}{\partial \rho_i} \ln\left[\frac{f_{n_i\, n_{i+1}}(\rho_i, \rho_{i+1})}{f_{n_i}(\rho_i)}\right] \\ &= \frac{\partial}{\partial \rho_i} \ln[f_{n_i\, n_{i+1}}(\rho_i, \rho_{i+1})] - \frac{\partial}{\partial \rho_i} \ln[f_{n_i}(\rho_i)]. \end{aligned} \tag{9.2-14}$$

Define the following functions as

$$\begin{aligned} g_i(\rho_{i-1}, \rho_i) &\triangleq -\tfrac{\partial}{\partial \rho_i} \ln[f_{n_{i-1}\, n_i}(\rho_{i-1}, \rho_i)], \\ g_i(\rho_i, \rho_{i+1}) &\triangleq -\tfrac{\partial}{\partial \rho_i} \ln[f_{n_i\, n_{i+1}}(\rho_i, \rho_{i+1})], \\ g_i(\rho_i) &\triangleq -\tfrac{\partial}{\partial \rho_i} \ln[f_{n_i}(\rho_i)]. \end{aligned} \tag{9.2-15}$$

Then, the LO nonlinearity of Eq. (9.2-12) becomes

$$g_i(\boldsymbol{\rho}) = \begin{cases} g_i(\rho_{i-1}, \rho_i, \rho_{i+1}) = g_i(\rho_{i-1}, \rho_i) + g_i(\rho_i, \rho_{i+1}) - g_i(\rho_i) & ; i = 1, \ldots, N-1 \\ g_N(\rho_{N-1}, \rho_N) & ; i = N \,. \end{cases} \tag{9.2-16}$$

As can be seen by Eq. (9.2-12) or Eq. (9.2-15) and Eq. (9.2-16), if $\mathbf{n}$ can be modeled as a Markov random sequence, the general N-variate LO nonlinearity, $g_i(\boldsymbol{\rho})$, of Eq. (9.2-6) reduces to the simpler three-variable function $g_i(\rho_{i-1}, \rho_i, \rho_{i+1})$.[5]

Example 9.2-1

Let the received signal be given by $\mathbf{r} = [r_1, r_2, ..., r_N]^T$, with

$$r_i = s_{mi} + n_i \,,$$

where $n_i = a\,n_{i-1} + w_i$, a is a known constant, and w_i is iid noise with marginal pdf $f_{w_i}(\alpha)$. The resulting noise conditional pdfs are

$$\begin{aligned} f_{n_i}(\eta_i \,|\, \eta_{i-1}) &= f_{w_i}(\eta_i - a\eta_{i-1}) \\ f_{n_{i+1}}(\eta_{i+1} \,|\, \eta_i) &= f_{w_{i+1}}(\eta_{i+1} - a\eta_i) \end{aligned}$$

and the conditional noise pdfs evaluated at the received signal are

$$\begin{aligned} f_{n_i}(\rho_i \,|\, \rho_{i-1}) &= f_{w_i}(\rho_i - a\,\rho_{i-1}) \\ f_{n_{i+1}}(\rho_{i+1} \,|\, \rho_i) &= f_{w_{i+1}}(\rho_{i+1} - a\,\rho_i). \end{aligned}$$

Defining $\rho_0 \triangleq 0$, $g_i(\rho_{i-1}, \rho_i, \rho_{i+1})$ for $i = 1, \ldots, N-1$ is given by

$$\begin{aligned} g_i(\rho_{i-1}, \rho_i, \rho_{i+1}) &= -\frac{\partial}{\partial \rho_i} \ln[f_{w_i}(\rho_i - a\rho_{i-1})] - \frac{\partial}{\partial \rho_i} \ln[f_{w_{i+1}}(\rho_{i+1} - a\rho_i)] \\ &= -\frac{f'_{w_i}(\rho_i - a\rho_{i-1})}{f_{w_i}(\rho_i - a\rho_{i-1})} + a\frac{f'_{w_{i+1}}(\rho_{i+1} - a\rho_i)}{f_{w_{i+1}}(\rho_{i+1} - a\rho_i)}. \end{aligned}$$

and

$$g_N(\rho_{N-1}, \rho_N) = -\frac{f'_{w_N}(\rho_N - a\,\rho_{N-1})}{f_{w_N}(\rho_N - a\,\rho_{N-1})} \,,$$

where the notation $f'(x)$ denotes differentiation of f with respect to x.

There are many applications in communications, radar, and signal processing in which a more general form of the LO detector with memory is required. The detector described by Eq. (9.2-6) and Eq. (9.2-7) corresponds to the case when $\mathbf{s}_1$ and $\mathbf{s}_0$ are known signals. However, there are times when these signals are random

[5] See [9-4] and [9-5] for further discussion concerning LO detection in Markov noise.

or have random components. For example, $\mathbf{s_1}$ and $\mathbf{s_0}$ may have the same mean but possess different pdfs. Or, as in the case of a radar problem, $\mathbf{s_0}$ may be the all zero vector and $\mathbf{s_1}$ may be a signal with zero mean stochastic components obeying some multivariate probability distribution. In scenarios similar to these two examples, it will be shown that the first-order term in the Taylor series expansion yielding the LO detector will become zero. Thus, a second-order Taylor series expansion is required, and results in the general form of the LO detector with memory.

To begin the derivation consider the binary detection problem:

$$\begin{aligned} H_1 &: \text{Signal } \mathbf{s_1} \text{ present} \\ H_0 &: \text{Signal } \mathbf{s_0} \text{ present,} \end{aligned}$$

where $\mathbf{s_1}$ and $\mathbf{s_0}$ are *random* vectors with observed values $\boldsymbol{\alpha}_1$ and $\boldsymbol{\alpha}_0$ and corresponding pdfs, $f_{\mathbf{s_1}}(\boldsymbol{\alpha}_1)$ and $f_{\mathbf{s_0}}(\boldsymbol{\alpha}_0)$. Let the received signal vector of length N be given by Eq. (9.2-1) with an observed value of $\boldsymbol{\rho}$. The optimum detector is of the form given by Eq. (9.2-2) which in this case is

$$\ln[\lambda(\boldsymbol{\rho})] = \ln\left[\frac{f_{\mathbf{r}}(\boldsymbol{\rho}\,|\,H_1)}{f_{\mathbf{r}}(\boldsymbol{\rho}\,|\,H_0)}\right] = \ln\left[\frac{\int_{-\infty}^{\infty} f_{\mathbf{n}}(\boldsymbol{\rho}-\boldsymbol{\alpha}_1)\, f_{\mathbf{s_1}}(\boldsymbol{\alpha}_1)\, d\boldsymbol{\alpha}_1}{\int_{-\infty}^{\infty} f_{\mathbf{n}}(\boldsymbol{\rho}-\boldsymbol{\alpha}_0)\, f_{\mathbf{s_0}}(\boldsymbol{\alpha}_0)\, d\boldsymbol{\alpha}_0}\right]. \qquad (9.2\text{-}17)$$

As before, Eq. (9.2-17) will be approximated using a truncated Taylor series. However, this time a second-order Taylor series expansion of $f_{\mathbf{n}}(\boldsymbol{\rho} - \boldsymbol{\alpha}_m)$, $m = 0$ or 1, is needed and is given by

$$f_{\mathbf{n}}(\boldsymbol{\rho}-\boldsymbol{\alpha}_m) \approx f_{\mathbf{n}}(\boldsymbol{\rho}) - \sum_{i=1}^{N} \alpha_{mi}\frac{\partial}{\partial\rho_i} f_{\mathbf{n}}(\boldsymbol{\rho}) + \frac{1}{2}\sum_{i=1}^{N}\sum_{j=1}^{N} \alpha_{mi}\alpha_{mj}\frac{\partial^2}{\partial\rho_i\partial\rho_j} f_{\mathbf{n}}(\boldsymbol{\rho}). \qquad (9.2\text{-}18)$$

Thus,

$$\begin{aligned} \int_{-\infty}^{\infty} f_{\mathbf{n}}(\boldsymbol{\rho}-\boldsymbol{\alpha}_m)\, f_{\mathbf{s_m}}(\boldsymbol{\alpha}_m)\, d\boldsymbol{\alpha}_m &\approx f_{\mathbf{n}}(\boldsymbol{\rho}) - \sum_{i=1}^{N} E[s_{mi}]\frac{\partial}{\partial\rho_i} f_{\mathbf{n}}(\boldsymbol{\rho}) \\ &\quad + \frac{1}{2}\sum_{i=1}^{N}\sum_{j=1}^{N} E[s_{mi}s_{mj}]\frac{\partial^2}{\partial\rho_i\partial\rho_j} f_{\mathbf{n}}(\boldsymbol{\rho}), \end{aligned} \qquad (9.2\text{-}19)$$

where $E[\cdot]$ denotes expectation. Rewriting Eq. (9.2-17) as

$$\begin{aligned} \ln[\lambda(\boldsymbol{\rho})] &= \ln\left[\frac{\int_{-\infty}^{\infty} f_{\mathbf{n}}(\boldsymbol{\rho}-\boldsymbol{\alpha}_1)\, f_{\mathbf{s_1}}(\boldsymbol{\alpha}_1)\, d\boldsymbol{\alpha}_1}{f_{\mathbf{n}}(\boldsymbol{\rho})}\right] \\ &\quad - \ln\left[\frac{\int_{-\infty}^{\infty} f_{\mathbf{n}}(\boldsymbol{\rho}-\boldsymbol{\alpha}_0)\, f_{\mathbf{s_0}}(\boldsymbol{\alpha}_0)\, d\boldsymbol{\alpha}_0}{f_{\mathbf{n}}(\boldsymbol{\rho})}\right], \end{aligned} \qquad (9.2\text{-}20)$$

then substituting Eq. (9.2-19) into Eq. (9.2-20), and noting that $\ln(1-x) \approx -x$ for $x \approx 0$, Eq. (9.2-17) may be approximated as (for sufficiently small $\mathbf{s_1}$ and $\mathbf{s_0}$)

$$\ln[\lambda(\boldsymbol{\rho})] \approx \sum_{i=1}^{N} (E[s_{1i}] - E[s_{0i}]) \left[-\frac{\frac{\partial}{\partial \rho_i} f_{\mathbf{n}}(\boldsymbol{\rho})}{f_{\mathbf{n}}(\boldsymbol{\rho})} \right] + \frac{1}{2} \sum_{i=1}^{N} \sum_{j=1}^{N} (E[s_{1i}\, s_{1j}] - E[s_{0i}\, s_{0j}]) \left[\frac{\frac{\partial^2}{\partial \rho_i \partial \rho_j} f_{\mathbf{n}}(\boldsymbol{\rho})}{f_{\mathbf{n}}(\boldsymbol{\rho})} \right]. \quad (9.2\text{-}21)$$

To facilitate the derivation, $\ln[\lambda(\boldsymbol{\rho})]$ will be expressed in matrix notation. Define the following vectors as

$$\begin{aligned} \mathbf{g}(\boldsymbol{\rho}) &\triangleq [g_1(\boldsymbol{\rho}), \ldots, g_N(\boldsymbol{\rho})]^T \triangleq \left[-\frac{\frac{\partial}{\partial \rho_1} f_{\mathbf{n}}(\boldsymbol{\rho})}{f_{\mathbf{n}}(\boldsymbol{\rho})}, \ldots, -\frac{\frac{\partial}{\partial \rho_N} f_{\mathbf{n}}(\boldsymbol{\rho})}{f_{\mathbf{n}}(\boldsymbol{\rho})} \right]^T \\ \boldsymbol{\mu}_1 &\triangleq E[\mathbf{s_1}] \\ \boldsymbol{\mu}_0 &\triangleq E[\mathbf{s_0}], \end{aligned} \quad (9.2\text{-}22)$$

where $g_i(\boldsymbol{\rho})$ is the LO nonlinearity with memory given by Eq. (9.2-6) and $\boldsymbol{\mu}_m$ is the mean vector of $\mathbf{s_m}$. Next, define the covariance matrix of the signal $\mathbf{s_m}$ to be $\mathbf{K_m}$, given by

$$\mathbf{K_m} \triangleq E\left\{ [\mathbf{s_m} - \boldsymbol{\mu}_m]\, [\mathbf{s_m} - \boldsymbol{\mu}_m]^T \right\}. \quad (9.2\text{-}23)$$

Also, let $\mathbf{H}_f(\mathbf{x})$ denote the Hessian matrix [9-6] of a function f evaluated at the vector $\mathbf{x}$. The (i, j)th element of $\mathbf{H}_f(\mathbf{x})$ is defined to be

$$[\mathbf{H}_f(\mathbf{x})]_{ij} \triangleq \frac{\partial^2}{\partial x_i \partial x_j} f(\mathbf{x}). \quad (9.2\text{-}24)$$

Using this notation, Eq. (9.2-21) can be written as

$$\ln[\lambda(\boldsymbol{\rho})] = l(\boldsymbol{\rho}) \approx (\boldsymbol{\mu}_1 - \boldsymbol{\mu}_0)^T \mathbf{g}(\boldsymbol{\rho}) + \frac{1}{2 f_{\mathbf{n}}(\boldsymbol{\rho})} \left\{ E[\mathbf{s_1}^T\, \mathbf{H}_{f_n}(\boldsymbol{\rho})\, \mathbf{s_1}] - E[\mathbf{s_0}^T\, \mathbf{H}_{f_n}(\boldsymbol{\rho})\, \mathbf{s_0}] \right\}. \quad (9.2\text{-}25)$$

To facilitate implementation, each term inside the braces of Eq. (9.2-25) and its factor can be rewritten after some algebra as

$$\frac{1}{2 f_{\mathbf{n}}(\boldsymbol{\rho})} E[\mathbf{s_m}^T\, \mathbf{H}_{f_n}(\boldsymbol{\rho})\, \mathbf{s_m}] = [\mathbf{g}(\boldsymbol{\rho})]^T\, \mathbf{K_m}\, \mathbf{g}(\boldsymbol{\rho}) + \operatorname{tr}\left\{ [\mathbf{K_m} + \boldsymbol{\mu}_m \boldsymbol{\mu}_m^T]\, \mathbf{H}_{\ln f_n}(\boldsymbol{\rho}) \right\}, \quad (9.2\text{-}26)$$

where $\text{tr}\{\cdot\}$ indicates the trace of a matrix, and $m = 0$ or 1. Thus, substituting Eq. (9.2-26) into Eq. (9.2-25) yields the general form of the LO detector with memory:

$$l(\boldsymbol{\rho}) \underset{\text{choose } H_0}{\overset{\text{choose } H_1}{\gtrless}} \tilde{\gamma}, \tag{9.2-27}$$

where[6]

$$\begin{aligned} l(\boldsymbol{\rho}) &= (\boldsymbol{\mu}_1 - \boldsymbol{\mu}_0)^T \mathbf{g}(\boldsymbol{\rho}) + \frac{1}{2}\left[[\mathbf{g}(\boldsymbol{\rho})]^T (\mathbf{K}_1 - \mathbf{K}_0)\, \mathbf{g}(\boldsymbol{\rho})\right] \\ &\quad + \frac{1}{2}\,\text{tr}\left\{[(\mathbf{K}_1 - \mathbf{K}_0) + \boldsymbol{\mu}_1\boldsymbol{\mu}_1{}^T - \boldsymbol{\mu}_0\boldsymbol{\mu}_0{}^T]\, \mathbf{H}_{\ln f_n}(\boldsymbol{\rho})\right\}. \end{aligned} \tag{9.2-28}$$

The form of the LO detector given in Eq. (9.2-28) is usually preferred over that of Eq. (9.2-25) for the following reasons. If the information signal statistics, i.e., $\boldsymbol{\mu}_1$, $\boldsymbol{\mu}_0$, $\mathbf{K}_1$, and $\mathbf{K}_0$, are known in advance, they can be computed once at the beginning of the detection process, as is indicated by Eq. (9.2-28). Then, only $\mathbf{g}(\boldsymbol{\rho})$ and $\mathbf{H}_{\ln f_n}(\boldsymbol{\rho})$ need be computed for each value of $\boldsymbol{\rho}$. The detector of Eq. (9.2-25), on the other hand, requires that the computationally intensive expectation operation, $E\left\{\mathbf{s_m}^T\, \mathbf{H}_{f_n}(\boldsymbol{\rho})\, \mathbf{s_m}\right\}$, be performed for each value of $\boldsymbol{\rho}$. Thus, the detector described by Eq. (9.2-28) is a more efficient implementation.

As discussed at the beginning of this derivation, a first-order Taylor series implementation is insufficient for problems where the random signals $\mathbf{s_1}$ and $\mathbf{s_0}$ have the same mean, or when $\mathbf{s_0} = \mathbf{0}$ and $\mathbf{s_1}$ has zero mean. From Eq. (9.2-28) it is evident that for these two cases the first-order term, $(\boldsymbol{\mu}_1 - \boldsymbol{\mu}_0)\, \mathbf{g}(\boldsymbol{\rho})$, of $l(\boldsymbol{\rho})$ is zero and provides no useful information in deciding between the two hypotheses, H_1 and H_0. However, even if $\boldsymbol{\mu}_1 \neq \boldsymbol{\mu}_0$, or if $\mathbf{s_1}$ and $\mathbf{s_0}$ are known, one may still wish to use the form of the LO detector given in Eq. (9.2-28) since it is a closer approximation (in the small-signal case[7]) to $\ln[\lambda(\boldsymbol{\rho})]$ than is the detector of Eq. (9.2-6) and Eq. (9.2-7).

Finally it should be noted that the memoryless LO detector of Eq. (9.1-12) and Eq. (9.1-13) and the detector with memory in Eq. (9.2-6) and Eq. (9.2-7) are subclasses of the general LO detector described by Eq. (9.2-27) and Eq. (9.2-28). Beginning with Eq. (9.2-28), if one ignores the second-order terms, and $\mathbf{s_1}$ and $\mathbf{s_0}$ are known, the result is Eq. (9.2-6). Furthermore, if the noise samples are iid, the resulting form of the LO nonlinearity is given by Eq. (9.1-13). Thus, the designer of an LO detector can begin with Eq. (9.2-28) and then, by appropriate choice of assumptions, arrive at the LO nonlinearity most suitable for the given application.

---

[6]This expression is equivalent to Eq. (18) of [9-6] and Eq. (19) of [9-7].

[7]The convergence properties of the Taylor series guarantee that the approximation of a function improves as more terms of the series are utilized. However, the small-signal assumption is required so that the approximation $\ln(1 - x) \approx -x$ used in Eq. (9.2-21) does not offset the improvements gained by including the second-order term in the expansion.

9.3 Detector Implementation Methods

The preceding two sections presented the derivations of a number of LO detectors. These detectors have either a scalar or a vector structure, and thus are considered to be memoryless or to have memory, respectively. The type of LO detector chosen for a particular application is primarily a consequence of the assumptions made concerning the noise samples. If the noise samples are iid, then a memoryless LO detector, such as that of Eq. (9.1-12) and Eq. (9.1-13) may be used. For noise samples that are correlated, on the other hand, LO detectors such as those described by Eq. (9.2-6) and Eq. (9.2-7), Eq. (9.2-12), or Eq. (9.2-27) and Eq. (9.2-28) may be employed depending on the relationship between these samples. In any case, after a good model of the noise is obtained, or at least some initial assumptions are determined, the next step is to implement the corresponding LO detector.

If an accurate model of the noise pdf is obtainable *a priori* the LO detector may be implemented directly, as shown in Examples 9.1-1, 9.1-2, and 9.1-3. The detector implementation then becomes a matter of solving the corresponding LO detector equations in terms of the parameters of the noise pdf, e.g., mean and variance.

However, if a noise model is not available then other methods of implementation are required. These methods primarily concentrate on estimating the noise pdf from the available noise samples, and then using the result to approximate the LO detector. If the uncorrupted noise samples are unavailable (as is the usual case in a communications system) and the small-signal assumption is valid, the received signal samples may be used in many cases to give a crude approximation to the noise pdf. Thus, various univariate and multivariate pdf estimation techniques and their use in implementing the LO detector are discussed in the remainder of this section.[8]

A. Univariate pdf Estimation via the Histogram

Univariate pdf estimates are required to implement the memoryless LO nonlinearity of Eq. (9.1-13). One of the simplest techniques for estimating a pdf is the histogram ([9-8], Chapter 3). The histogram algorithm, in essence, assigns the data samples to one of K intervals and computes the sample probability associated with each interval. Stated somewhat more rigorously, let $\{x_i\}$ be the set of M observed samples of a random variable X, and let $x_{\max}$ and $x_{\min}$ be the maximum and minimum values of $\{x_i\}$, respectively. Furthermore, divide the range of $\{x_i\}$ into K intervals, or "bins," B_k where

$$B_k = \{\, x_i \colon b_k \le x_i < b_{k+1}, \quad k = 0, \ldots, K-1\} \,. \tag{9.3-1}$$

[8]Estimation of pdfs is also the subject of Chapter 10.

The values $\{b_k\}$, with $b_0 = x_{\min}$ and $b_K = x_{\max}$, are called the "breakpoints" of the histogram and are determined heuristically. For example, the breakpoints may be chosen so that the bins have equal width, or so that each bin contains the same number of observed samples. Next, the probability of each bin is approximated via its relative frequency:

$$P[B_k] = P[b_k \le X < b_{k+1}] \approx \frac{\sum_{i=1}^{M} I_{B_k}(x_i)}{M} = \hat{P}[B_k], \tag{9.3-2}$$

where again $I_A(y)$ is the set indicator function. Equation (9.3-2) simply means that $\hat{P}[B_k]$ is the ratio of the number of samples falling in the kth bin to the total number of samples. Finally, the histogram estimate of the pdf of X, denoted as $\hat{f}_X(x)$, is constructed as:

$$\hat{f}_X(x) \triangleq \sum_{k=0}^{K-1} \frac{\hat{P}[B_k]}{b_{k+1} - b_k} I_{B_k}(x). \tag{9.3-3}$$

The scale factor $1/(b_{k+1} - b_k)$ is included so that the area under $\hat{f}_X(x)$ equals one.[9] An example of a histogram is shown in Fig. 9.3-1 and is characterized by its bar graph appearance.

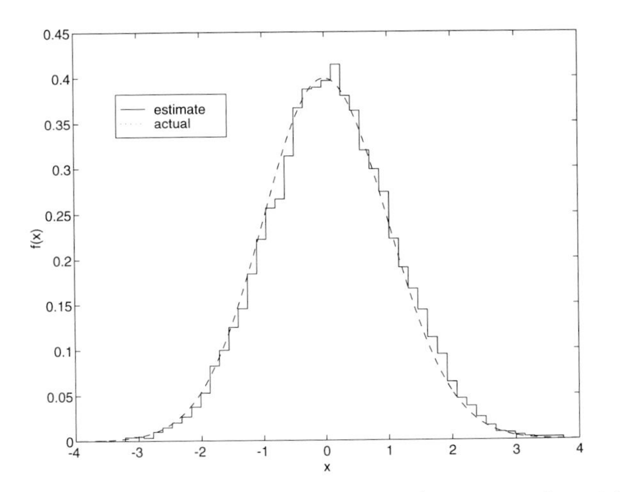

Figure 9.3-1: Histogram estimate of a pdf (10,000 samples, 50 bins).

Recall that the memoryless LO nonlinearity is defined by

$$g(\rho_i) \triangleq -\frac{d}{d\rho_i} \ln[f_n(\rho_i)]. \tag{9.3-4}$$

[9]Note that for $\hat{f}_X(x)$ to be well defined, $b_k \neq b_{k+1}$.

To approximate $g(\rho_i)$ using the histogram algorithm, the first step is to approximate $f_n(\rho_i)$ using Eq. (9.3-3). Let $\{\eta_i\}$ be the set of M observed noise samples. Then:

$$\begin{aligned} B_k &= \{\eta_i\colon \quad b_k \leq \eta_i < b_{k+1}, \quad k = 0, \ldots, K-1\} \\ \hat{P}[B_k] &= \frac{1}{M}\sum_{i=1}^{M} I_{B_k}(\eta_i) \end{aligned} \tag{9.3-5}$$

and

$$\hat{f}_n(\rho_i) = \sum_{k=0}^{K-1} \frac{\hat{P}\{B_k\}}{b_{k+1} - b_k} I_{B_k}(\rho_i)\,. \tag{9.3-6}$$

Next, the derivative of $\ln[\hat{f}_n(\rho_i)]$ must be computed. If it is assumed that the actual pdf $f_n(\rho_i)$ is continuous, then the impulses that arise from differentiating Eq. (9.3-6) do not accurately model $\frac{d}{d\rho_i}\ln[f_n(\rho_i)]$. To remedy this dilemma another way of viewing the histogram is utilized. One can think of the process of assigning samples to a bin as a form of quantization. In other words, all ρ_i in the range $(b_k \leq \rho_i < b_{k+1})$ are quantized to b_k. Then a numerical approximation of the derivative of $\ln[f_n(\rho_i)]$ evaluated at b_k is used, i.e.,

$$\begin{aligned} \frac{d}{d\rho_i}\ln[f_n(\rho_i)]|_{\rho_i=b_k} &\approx \frac{\ln[\hat{f}_n(b_{k+1})] - \ln[\hat{f}_n(b_{k-1})]}{b_{k+1} - b_{k-1}} \\ &\approx \frac{\ln[\hat{P}(B_{k+1})] - \ln[\hat{P}(B_{k-1})]}{b_{k+1} - b_{k-1}} \\ &\quad - \frac{\ln[b_{k+2} - b_{k+1}] - \ln[b_k - b_{k-1}]}{b_{k+1} - b_{k-1}}\,. \end{aligned} \tag{9.3-7}$$

Note, if the bins have equal width then the right side of Eq. (9.3-7) reduces to $\{\ln[\hat{P}\{B_{k+1}\}] - \ln[\hat{P}\{B_{k-1}\}]\}/2\Delta$ where Δ is the width of a bin. Finally, the value of the approximation is extended from b_k to all ρ_i in the range $(b_k \leq \rho_i < b_{k+1})$, yielding the histogram implementation of the memoryless LO nonlinearity:

$$\begin{aligned} \hat{g}(\rho_i) &= -\sum_{k=0}^{K-1}\left\{\frac{\ln[\hat{P}\{B_{k+1}\}] - \ln[\hat{P}\{B_{k-1}\}]}{b_{k+1} - b_{k-1}}\right. \\ &\quad \left. - \frac{\ln[b_{k+2} - b_{k+1}] - \ln[b_k - b_{k-1}]}{b_{k+1} - b_{k-1}}\right\} I_{B_k}(\rho_i)\,. \end{aligned} \tag{9.3-8}$$

A plot of $\hat{g}(\rho_i)$ corresponding to the histogram pdf of Fig. 9.3-1 is shown in Fig. 9.3-2. The diagram illustrates the major drawback associated with the histogram method. Since data samples associated with tail regions of a pdf occur rather infrequently (in the usual case when the tail regions have low probability), the error of the histogram method in these regions may be large. Furthermore, since tail probabilities are typically small, and $g(\rho) \propto 1/f_n(\rho)$, the tail estimate errors

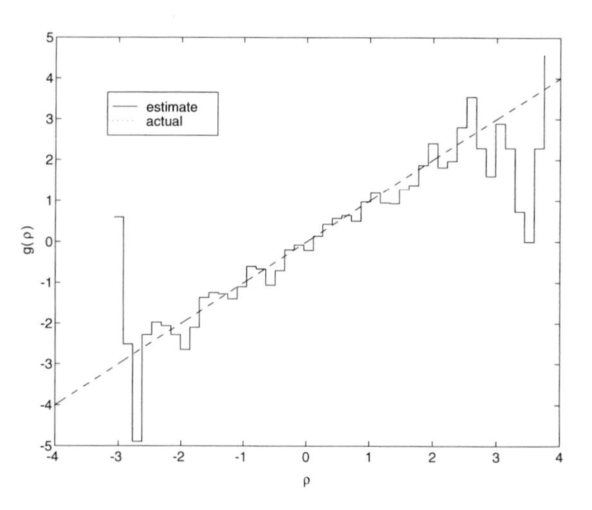

Figure 9.3-2: Histogram estimate of the LO nonlinearity corresponding to Fig. 9.3-1.

will be amplified in the approximation of the LO nonlinearity. The possibility of this type of error should be taken into account when implementing $\hat{g}(\rho)$ using a histogram.

Histogram estimation of pdfs usually requires a large number of samples ([9-8], Chapter 3). Thus M, the total number of observed samples used to construct the histogram, may be much larger than N, the number of samples in each signal vector. To obtain the total number of required observed samples, the value of M should be chosen such that $M = QN$, where Q is an integer. The Q vectors can be stored and all the available samples used to compute the histogram. Detection can then be performed on each data vector using the resulting histogram to compute the approximate nonlinearity of Eq. (9.3-8).

A number of more sophisticated algorithms for estimating pdfs make use of the histogram. Three of these are the *continuous polynomial approximation* (CPA) [9-9], cubic splines ([9-10], Chapter 18), and the method of convex projections (discussed in the next chapter). In the first two of these techniques a unique polynomial function is fitted to each bin of a histogram. The overall estimate of the pdf is formed by concatenating each of these component functions. In both methods the polynomials are chosen so as to meet given continuity requirements at the breakpoints.

The major difference between the CPA and cubic spline techniques is that, given the bin values and the derivatives at b_0 and b_K, the cubic splines algorithm computes the third-order polynomial coefficients *and* the first derivatives at $\{b_0, \ldots, b_{K-1}\}$ so that the resulting spline approximation has a continuous first *and* second derivative throughout its domain. In contrast, in CPA all the required derivatives are first computed numerically from the bin breakpoint values and the approximate bin probabilities, $\hat{P}[B_k]$, and then the polynomial coefficients are computed so as to satisfy the continuity constraints. However, a CPA of third order has only a continuous first derivative throughout its domain; a fifth-order CPA is required for continuous first and second derivatives.

A fourth type of polynomial approximation also utilizes the histogram. Let $\hat{f}_X(x) = \sum_{k=0}^{K-1} y_k(x)\, I_{B_k}(x)$, where $y_k(x)$ is a polynomial of order p. The coefficients of each polynomial are determined by minimizing the *integrated squared error* (ISE), or L_2 norm:

$$\text{ISE} = \int [\hat{f}_X(x) - f_X(x)]^2\, dx\,. \tag{9.3-9}$$

For each component polynomial this translates to minimizing

$$\text{ISE}_k = \int_{b_k}^{b_{k+1}} [y_k(x) - f_X(x)]^2\, dx\,. \tag{9.3-10}$$

Define the random variable Z to be $Z = X$ if $b_k \leq X < b_{k+1}$, and 0 otherwise. It can be shown that the coefficients of $y_k(x)$ are a function of the first p moments of Z. The advantage of this pdf estimation method over the CPA or cubic splines algorithm is that this method actually minimizes an error metric (the ISE) between the estimated pdf and the actual pdf. The disadvantage of this method is that it is not guaranteed to be continuous (since there are no continuity requirements at the breakpoints), or to have continuous derivatives. Thus, even though this method minimizes the ISE between $\hat{f}_X(x)$ and $f_X(x)$, there may be a much larger error associated with the approximation of the LO nonlinearity, particularly at the breakpoints.

B. Density Estimation using the Fourier Series

A major drawback associated with the histogram algorithm for estimating pdfs, as with any quantization algorithm, is the loss of information incurred through the binning process. While the global shape of the histogram approximates the actual pdf, locally the result might be poor. This section presents an algorithm for estimating pdfs that is based on the Fourier series ([9-10], Chapter 8). Its major features are that: (1) the approximation is continuous throughout its domain; (2) it possesses an arbitrary number of continuous derivatives; and (3) no quantization process is required, i.e., each data sample contributes uniquely to the overall estimate.

Given the set of M observed samples $\{x_i\}$ with minimum value $x_{\min}$ and maximum value $x_{\max}$, the p-term *Fourier series approximation* (FSA) of the pdf of X, denoted $\hat{f}_X(x)$, is given by the following expression:

$$\hat{f}_X(x) = \begin{cases} \frac{a_0}{2} + \sum_{k=1}^{p} a_k \cos(k\omega_0 x) + \sum_{k=1}^{p} b_k \sin(k\omega_0 x) & ; \quad \text{for } x_{\min} \leq x \leq x_{\max} \\ 0 & ; \quad \text{otherwise.} \end{cases} \tag{9.3-11}$$

The natural frequency of the FSA is given by $\omega_0 = 2\pi / T$ where $T = x_{\max} - x_{\min}$ is the "period." The coefficients $\{a_k\}$ and $\{b_k\}$ are determined by minimizing the ISE:

$$\begin{aligned} \text{ISE} &= \int_{x_{\min}}^{x_{\max}} [\hat{f}_X(x) - f_X(x)]^2 dx \\ &= \int_{x_{\min}}^{x_{\max}} \left[f_X(x) - \frac{a_0}{2} - \sum_{k=1}^{p} a_k \cos(k\omega_0 x) - \sum_{k=1}^{p} b_k \sin(k\omega_0 x) \right]^2 dx \,. \end{aligned} \tag{9.3-12}$$

Differentiating Eq. (9.3-12) with respect to a_0 and equating to zero yields the following expression:

$$\int_{x_{\min}}^{x_{\max}} \left[\frac{a_0}{2} + \sum_{k=1}^{p} a_k \cos(k\omega_0 x) + \sum_{k=1}^{p} b_k \sin(k\omega_0 x) \right] dx = \int_{x_{\min}}^{x_{\max}} f_X(x)\, dx \,. \tag{9.3-13}$$

By noting that the integration of the sine or cosine functions over an integer multiple of its period is equal to zero and that $\int_{x_{\min}}^{x_{\max}} f_X(x)\, dx = 1$,[10] Eq. (9.3-13) becomes

$$a_0 = \frac{2}{x_{\max} - x_{\min}} \,. \tag{9.3-14}$$

To find the coefficients $\{a_k\}$, differentiation of Eq. (9.3-12) is performed with respect to a_m and the result is equated to zero. This gives the following equation:

$$\int_{x_{\min}}^{x_{\max}} \cos(m\omega_0 x) \left[f_X(x) - \frac{a_0}{2} - \sum_{k=1}^{p} a_k \cos(k\omega_0 x) - \sum_{k=1}^{p} b_k \sin(k\omega_0 x) \right] dx = 0. \tag{9.3-15}$$

After some manipulation, the expression for a_m is found to be

$$a_m = \frac{2}{x_{\max} - x_{\min}} \int_{x_{\min}}^{x_{\max}} \cos(m\omega_0 x) f_X(x) dx \; ; \quad m = 1, ..., p. \tag{9.3-16}$$

Similarly, the coefficients $\{b_k\}$ are found by differentiating Eq. (9.3-12) with respect to b_m and equating to zero, resulting in:

$$\int_{x_{\min}}^{x_{\max}} \sin(m\omega_0 x) \left[f_X(x) - \frac{a_0}{2} - \sum_{k=1}^{p} a_k \cos(k\omega_0 x) - \sum_{k=1}^{p} b_k \sin(k\omega_0 x) \right] dx = 0 \,. \tag{9.3-17}$$

[10] It is assumed that the support length of the actual pdf $f_X(x)$ is $x_{\max} - x_{\min}$. However, this is not a necessary constraint, since truncation can be used to approximate a pdf with infinite support.

Again, after some manipulation, the expression for b_m is found to be

$$b_m = \frac{2}{x_{\max} - x_{\min}} \int_{x_{\min}}^{x_{\max}} \sin(m\omega_0 x) f_X(x) dx \; ; \qquad m = 1, ..., p. \tag{9.3-18}$$

Recalling that the expectation of a function $h(X)$ of the random variable X is given by $E\{h(X)\} = \int_x h(x)\, f_X(x)\, dx$, Eq. (9.3-16) and Eq. (9.3-18) can be written as

$$\begin{aligned} a_k &= \frac{2}{x_{\max} - x_{\min}} E\{\cos(\omega_0\, k\, X)\} \; ; \quad k = 1, \ldots, p \\ b_k &= \frac{2}{x_{\max} - x_{\min}} E\{\sin(\omega_0\, k\, X)\} \; ; \quad k = 1, \ldots, p. \end{aligned} \tag{9.3-19}$$

Since the actual pdf of X is unavailable, Eq. (B.) can be approximated using the sample averages, resulting in the estimated FSA coefficients:

$$\begin{aligned} \hat{a}_0 &= a_0 = \frac{2}{x_{\max} - x_{\min}} \\ \hat{a}_k &= \frac{2}{x_{\max} - x_{\min}} \frac{1}{M} \sum_{i=1}^{M} \cos(k\, \omega_0\, x_i), \; ; \quad k = 1, \ldots, p \\ \hat{b}_k &= \frac{2}{x_{\max} - x_{\min}} \frac{1}{M} \sum_{i=1}^{M} \sin(k\, \omega_0\, x_i), \; ; \quad k = 1, \ldots, p. \end{aligned} \tag{9.3-20}$$

Thus, the coefficients given by Eq. (9.3-20) are used in Eq. (9.3-11) to yield the FSA estimate of $f_X(x)$.

To implement the memoryless LO nonlinearity, Eq. (9.3-11) and Eq. (9.3-20) are used directly in the expression for $g(\rho_i)$ given in Eq. (9.1-13). Given the set of noise samples $\{\eta_i\}$, with maximum value $\eta_{\max}$ and minimum value $\eta_{\min}$, the result is the following expression:

$$\hat{g}(\rho_i) = \frac{\sum_{k=1}^{p} \hat{a}_k k\omega_0 \sin(k\omega_0\rho_i) - \sum_{k=1}^{p} \hat{b}_k k\omega_0 \cos(k\omega_0\rho_i)}{\frac{a_0}{2} + \sum_{k=1}^{p} \hat{a}_k \cos(k\omega_0\rho_i) + \sum_{k=1}^{p} \hat{b}_k \sin(k\omega_0\rho_i)} I_{[\eta_{\min}, \eta_{\max}]}(\rho_i), \tag{9.3-21}$$

where

$$\begin{aligned} \hat{a}_0 &= a_0 = \frac{2}{\eta_{\max} - \eta_{\min}} \\ \hat{a}_k &= \frac{2}{\eta_{\max} - \eta_{\min}} \frac{1}{M} \sum_{i=1}^{M} \cos(k\, \omega_0\, \eta_i), \quad ; \; k = 1, \ldots, p \\ \hat{b}_k &= \frac{2}{\eta_{\max} - \eta_{\min}} \frac{1}{M} \sum_{i=1}^{M} \sin(k\, \omega_0\, \eta_i), \quad ; \; k = 1, \ldots, p. \end{aligned} \tag{9.3-22}$$

The major features of the FSA algorithm can be seen from Eq. (9.3-11) and Eq. (9.3-20). First, since the FSA estimate of a pdf is a sum of continuous functions the overall result is a continuous function. Second, since the sine and cosine functions are infinitely differentiable, the resulting pdf estimate is also. Finally, each data sample contributes uniquely (no quantization) to the FSA estimate through Eq. (9.3-20).

Example 9.3-1

Let $\{x_i\}$ be a set of M iid samples of a random variable X having the following pdf (related to the beta pdf family):

$$f_X(x) = \frac{3}{4}(x+1)(1-x)\, I_{(-1,1)}(x) = \frac{3}{4}(1 - x^2) I_{(-1,1)}(x) .$$

Furthermore, assume $x_{\min} = -1$ and $x_{\max} = 1$, so that $\omega_0 = \frac{2\pi}{T} = \frac{2\pi}{1-(-1)} = \pi$. Since $f_X(x)$ is an even function and $\sin(k\omega_0 x)$ is an odd function, $b_k = 0$ for $k = 1, \ldots, p$. The coefficient a_0 is given by $a_0 = \frac{2}{x_{\max} - x_{\min}} = 1$. The coefficients $\{a_k\}$ are given by

$$a_k = \int_{-1}^{1} \frac{3}{4}(1 - x^2)\cos(\pi\, k\, x)\, dx = \frac{3\,(-1)^{k+1}}{\pi^2\, k^2} .$$

Thus, the FSA estimate of $f_X(x)$ is

$$\hat{f}_X(x) = \frac{1}{2} + \sum_{k=1}^{p} \frac{3\,(-1)^{k+1}}{\pi^2\, k^2} \cos(k\,\pi\, x) ,$$

for $x_{\min} \le x \le x_{\max}$ and 0 otherwise. Plots of $\hat{f}_X(x)$ for various values of p are shown in Figs. 9.3-3 and 9.3-4. In each case the dashed line is the plot of the actual pdf and the solid line is the plot of the FSA estimate.

The LO nonlinearity, $g(x) = -f'(x)/f(x)$, is given by[11]

$$g(x) = \frac{2\,x^2}{1 - x^2} I_{(-1,1)}(x) .$$

Figures 9.3-5 and 9.3-6 illustrate the LO nonlinearities corresponding to the FSA pdf estimates of Figs. 9.3-3 and 9.3-4.

Given a few assumptions, a Fourier series may be used to directly estimate the memoryless LO nonlinearity. Recall that the memoryless LO nonlinearity has the

[11] Note that this nonlinearity differs by only a constant multiple (-2) from the LO nonlinearity in Example 9.1-2. However, the FSA estimate of the LO nonlinearity resulting from the pdf in Example 9.2-1 converges much more slowly to the actual result than does the FSA estimate in this example.

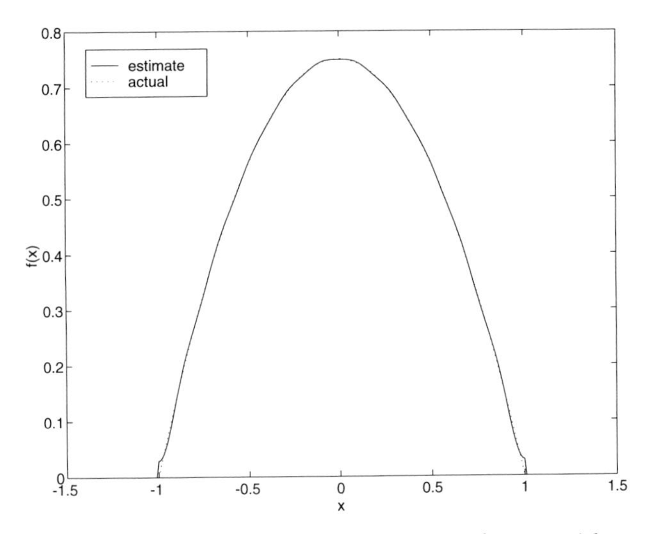

Figure 9.3-3: FSA estimate of the pdf in Example 9.3-1 with $p = 10$.

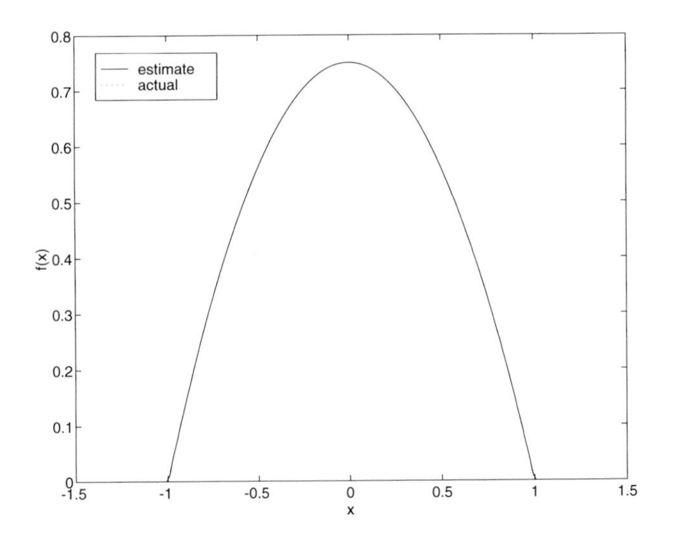

Figure 9.3-4: FSA estimate of the pdf in Example 9.3-1 with $p = 50$.

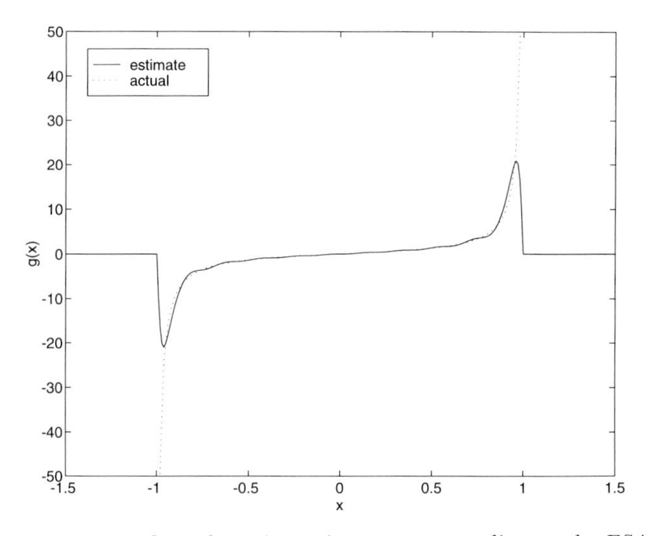

Figure 9.3-5: Resulting LO nonlinearity estimate corresponding to the FSA estimate of the pdf in Example 9.3-1 with $p = 10$.

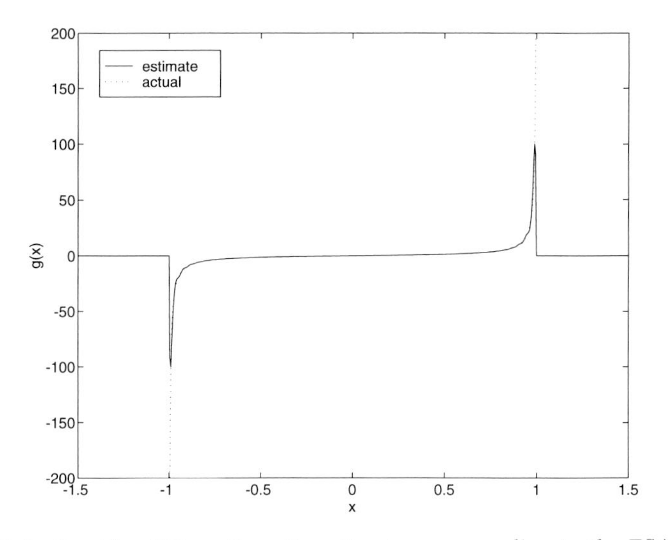

Figure 9.3-6: Resulting LO nonlinearity estimate corresponding to the FSA estimate of the pdf in Example 9.3-1 with $p = 50$.

form $g(\rho) = -f_n'(\rho) / f_n(\rho)$. To derive the FSA estimate of $g(\rho)$ the following assumptions are made concerning the noise pdf, $f_n(\eta)$: $f_n(\eta)$ is even symmetric about the origin (and thus has zero mean), and is bounded by $\eta_{\max}$ and $\eta_{\min}$, where $\eta_{\max} = -\eta_{\min}$.[12] Begin the derivation by approximating $g(\rho)$ as

$$\hat{g}(\rho) = \begin{cases} \frac{a_0}{2} + \sum_{k=1}^{p} a_k \cos(k\omega_0\rho) + \sum_{k=1}^{p} b_k \sin(k\omega_0\rho) & ; \quad \text{for } \eta_{\min} \le \rho \le \eta_{\max} \\ 0 & ; \quad \text{otherwise} \end{cases} \tag{9.3-23}$$

where $T = \eta_{\max} - \eta_{\min}$ and $\omega_0 = (2\pi) / T$. The coefficients $\{a_k\}$ and $\{b_k\}$ (which are different from those in Eq. (9.3-11)) are chosen so as to minimize the *mean squared error* (MSE) when the observation consists of noise only, i.e., $r = n$. The MSE (denoted by the function J) is given by:

$$\begin{aligned} J &= E\left\{[g(r) - \hat{g}(r)]^2\right\}\Big|_{r=n} = \int_{-\infty}^{\infty} [g(\rho) - \hat{g}(\rho)]^2 f_n(\rho)\, d\rho \\ &= \int_{\eta_{\min}}^{\eta_{\max}} \left[g(\rho) - \frac{a_o}{2} - \sum_{k=1}^{p} a_k \cos(k\omega_0\rho) - \sum_{k=1}^{p} b_k \sin(k\omega_0\rho)\right]^2 f_n(\rho)\, d\rho\,. \end{aligned} \tag{9.3-24}$$

To find a_0 which minimizes J, the expression $\partial J / \partial a_0$ is computed and equated to zero. By symmetry,

$$\int_{\eta_{\min}}^{\eta_{\max}} \sin(k\omega_0\rho)\, f_n(\rho)\, d\rho = 0$$

and

$$f_n(\eta_{\min}) = f_n(\eta_{\max}),$$

resulting in the following equation:

$$\frac{a_0}{2} + \sum_{k=1}^{p} a_k\, E[\cos(k\omega_0 n)] = f_n(\eta_{\min}) - f_n(\eta_{\max}) = 0\,. \tag{9.3-25}$$

Next, the expression $\partial J / \partial a_m$ is computed for $m \neq 0$ to find the coefficients $\{a_k\}$ which minimize J. Noting that

$$\int_{\eta_{\min}}^{\eta_{\max}} f_n'(\rho) \cos(m\omega_0\rho) d\rho = 0$$

and

$$\int_{\eta_{\min}}^{\eta_{\max}} \sin(k\omega_0\rho) \cos(m\omega_0\rho) f_n(\rho) d\rho = 0,$$

[12] These assumptions are valid as long as $f_n(\eta)$ is bounded and is symmetric about its mean since a simple linear transformation may be used to shift the pdf to the origin.

the resulting system of equations is:

$$a_0\, E[\cos(m\omega_o n)] + \sum_{k=1}^{p} a_k \,\{E[\cos((k-m)\omega_o n)] + E[\cos((k+m)\omega_0 n)]\} = 0 \;; \qquad m = 1,\ldots,p. \tag{9.3-26}$$

Finally, J is minimized with respect to $\{b_k\}$ by computing $\partial J/\partial b_k$ and equating it to zero. The result is the following system of equations:

$$\begin{aligned}\frac{1}{2}\sum_{k=1}^{p} b_k\{E[\cos((k-m)\omega_0 n)] - E[\cos((k+m)\omega_0 n)]\} \\ = f_n(\eta_{\min})\sin(m\omega_0\eta_{\min}) - f_n(\eta_{\max})\sin(m\omega_0\eta_{\max}) + m\omega_0 E[\cos(m\omega_0 n)] \\ = m\omega_0 E[\cos(m\omega_0 n)] \;; \qquad m = 1,2,\ldots,p\end{aligned} \tag{9.3-27}$$

since $f_n(\eta_{\min}) = f_n(\eta_{\max})$ and $\sin(m\omega_0\eta_{\min}) = \sin(m\omega_0\eta_{\max}) = 0$.[13] To simplify notation, define the vectors

$$\begin{aligned}\mathbf{a} &\triangleq [a_0 \quad a_1 \quad \ldots \quad a_p]^T \\ \mathbf{b} &\triangleq [b_1 \quad b_2 \quad \ldots \quad b_p]^T \\ \mathbf{c} &\triangleq \begin{bmatrix} E\{\cos(\omega_0 n)\} \\ 2\,E\{\cos(2\omega_0 n)\} \\ \vdots \\ p\,E\{\cos(p\omega_0 n)\}\end{bmatrix},\end{aligned} \tag{9.3-28}$$

and the matrix

$$\mathbf{T} \triangleq \begin{bmatrix} 1 - E\{\cos(2\omega_0 n)\} & \begin{matrix}E\{\cos(\omega_0 n)\} \\ -E\{\cos(3\omega_0 n)\}\end{matrix} & \cdots & \begin{matrix}E\{\cos[(P-1)\omega_0 n]\} \\ -E\{\cos[(P+1)\omega_0 n]\}\end{matrix} \\ \begin{matrix}E\{\cos(\omega_0 n)\} \\ -E\{\cos(3\omega_0 n)\}\end{matrix} & 1 - E\{\cos(4\omega_0 n)\} & \cdots & \begin{matrix}E\{\cos[(P-2)\omega_0 n]\} \\ -E\{\cos[(P+2)\omega_0 n]\}\end{matrix} \\ \vdots & & \ddots & \\ \begin{matrix}E\{\cos[(P-1)\omega_0 n]\} \\ -E\{\cos[(P+1)\omega_0 n]\}\end{matrix} & \cdots & \cdots & 1 - E\{\cos(2P\omega_0 n)\}\end{bmatrix}. \tag{9.3-29}$$

Then, assuming $\mathbf{T}$ is invertible and that the set of equations described by Eq. (9.3-25) and Eq. (9.3-26) are linearly independent, the coefficients $\{a_k\}$ and $\{b_k\}$ which minimize J are found via the matrix equations

$$\begin{aligned}\mathbf{a} &= \mathbf{0} \\ \mathbf{b} &= 2\,\omega_0\,\mathbf{T}^{-1}\,\mathbf{c}.\end{aligned} \tag{9.3-30}$$

---

[13]To see this, recall that $\omega_0 = 2\pi/(\eta_{\max} - \eta_{\min}) = \pi/\eta_{\max} = -\pi/\eta_{\min}$.

Thus the resulting FSA estimate of $g(\rho)$ is given by

$$\hat{g}(\rho) = \sum_{k=1}^{p} b_k \sin(k\omega_0\rho) \qquad (9.3\text{-}31)$$

for $\eta_{\min} \leq \rho \leq \eta_{\max}$ and 0 otherwise, where the set of coefficients $\{b_k\}$ is given by Eq. (9.3-30).

Example 9.3-2

Using the pdf given in Example 9.3-1 as the noise pdf, $f_n(\eta)$, the goal is to compute the FSA estimate of $g(\rho)$. The vector $\mathbf{c}$ in Eq. (9.3-28) is given by $\mathbf{c} = \frac{3}{\pi^2}\left[1 \quad -\frac{1}{2} \;\; \frac{1}{3} \cdots \frac{(-1)^{p+1}}{p}\right]^T$ and the matrix $\mathbf{T}$ is

$$\mathbf{T} = \frac{3}{\pi^2}\begin{bmatrix} \pi^2/3 - (-1/4) & 1 - 1/9 & \cdots & \begin{array}{c}(-1)^p/(p-1)^2 \\ -(-1)^{p+2}/(p+1)^2\end{array} \\ 1 - 1/9 & \pi^2/3 - (-1/16) & \cdots & \begin{array}{c}(-1)^{p-1}/(p-2)^2 \\ -(-1)^{p+3}/(p+2)^2\end{array} \\ \vdots & & & \vdots \\ \begin{array}{c}(-1)^p/(p-1)^2 \\ -(-1)^{p+2}/(p+1)^2\end{array} & \cdots & \cdots & \pi^2/3 - (-1)^{2p+1}/(2p)^2 \end{bmatrix}.$$

The required FSA coefficients are given by the equation $\mathbf{b} = 2\pi\mathbf{T}^{-1}\mathbf{c}$. Figure 9.3-7 shows a plot of the resulting memoryless LO nonlinearity estimate when $p = 100$.

The previous derivations for the FSA estimate of a pdf and of the memoryless LO nonlinearity used the assumption that the noise pdf, $f_n(\eta)$, was bounded. However, there are many cases of interest in which $f_n(\eta)$ is unbounded, e.g., the Gaussian pdf ([9-2], Chapter 2). In these instances the FSA algorithm may still be used with the following modifications. The observed received signal in most applications will have a maximum and minimum value. If $\eta_{\max}$ is chosen as $\max(|\rho_{\max}|, |\rho_{\min}|)$, then the FSA estimate with $\omega_0 = \pi/\eta_{\max}$ will encompass all values of the received signal. However, the actual range of the pdf, $(-\infty, \infty)$, can be used to compute the coefficients via Eq. (B.) or Eq. (9.3-30). While the result will not be a true estimate of the actual unbounded noise pdf, in most cases it will provide a suitable approximation given the fact that the range of observed received signal values is bounded.

The FSA algorithm can also be used for the case when the underlying pdf is unknown and knowledge of its domain (bounded or unbounded) is unavailable. Provided that it is valid to assume that $f_n(\eta)$ is even-symmetric one can choose $\eta_{\max}$ as $\max(|\rho_{\max}|, |\rho_{\min}|)$. The required coefficients can then be computed using Eq. (9.3-22) for the pdf estimate, or Eq. (9.3-30) (substituting a sample average for

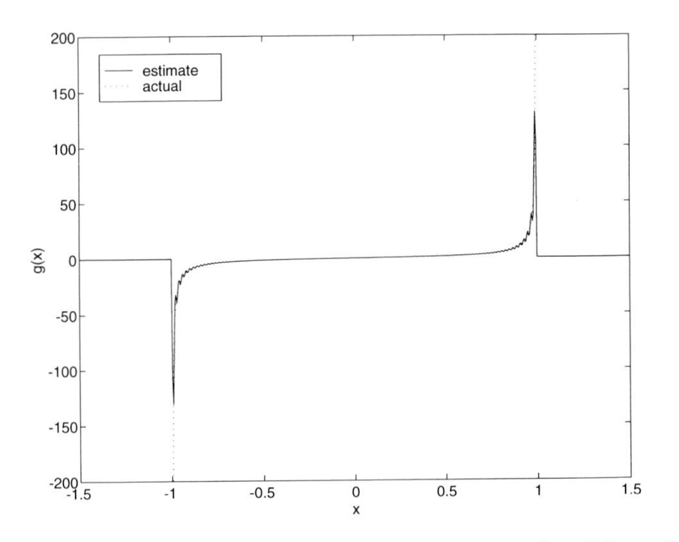

Figure 9.3-7: Plot of the direct FSA estimate of the memoryless LO nonlinearity of Example 9.3-2.

the expectation operation) for an estimate of the LO nonlinearity.

Example 9.3-3

Let $\{x_i\}$ be a set of iid samples from a random variable X having a Cauchy pdf

$$f_X(x) = \frac{1}{\pi\,(1+x^2)}\;; \qquad -\infty < x < \infty.$$

To approximate $f_X(x)$ over the range $-T/2 \leq x \leq T/2$, first compute the coefficients given by Eq. (9.3-14) and Eq. (B.), where $T = x_{\max} - x_{\min}$. Thus, $a_0 = \frac{2}{T}$, $a_k = \frac{2}{T}e^{-2\pi k/T}$ for $k = 1, \ldots, p$, and $b_k = 0$ for all k. Thus, the FSA estimate is

$$\hat{f}_X(x) = \frac{1}{T} + \sum_{k=1}^{p} \frac{2}{T} e^{-\frac{2\pi k}{T}} \cos(\frac{2\pi}{T} k\, x)$$

for $-T/2 \leq x \leq T/2$ and 0 otherwise. Plots of $\hat{f}_X(x)$ and the resulting LO nonlinearity estimate $\hat{g}(x)$ for the case when $p = 20$ and $p = 50$ (with $T = 20$) are shown in Figs. 9.3-8 through 9.3-11.

The LO nonlinearity can also be estimated directly. A plot of $\hat{g}(x)$ estimated directly using the FSA algorithm is shown in Fig. 9.3-12 for the case when $T = 20$ and p is only 20. In all figures, the solid line represents the FSA estimate and the dashed line represents the actual function.

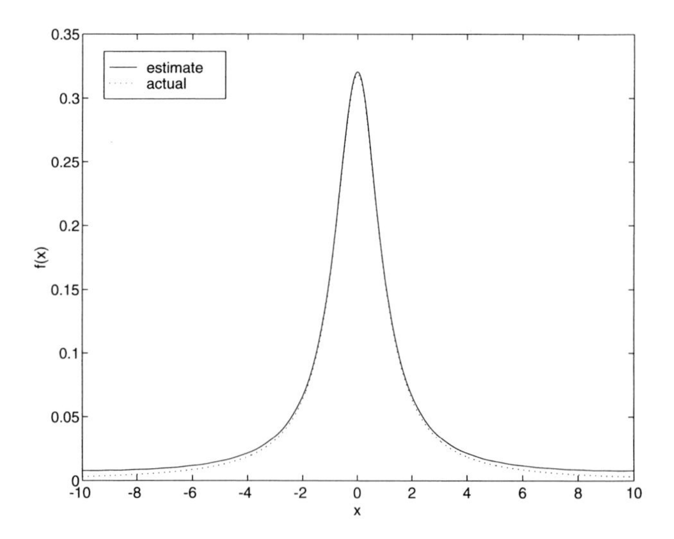

Figure 9.3-8: FSA estimate of a Cauchy pdf with $T = 20$ and $p = 20$.

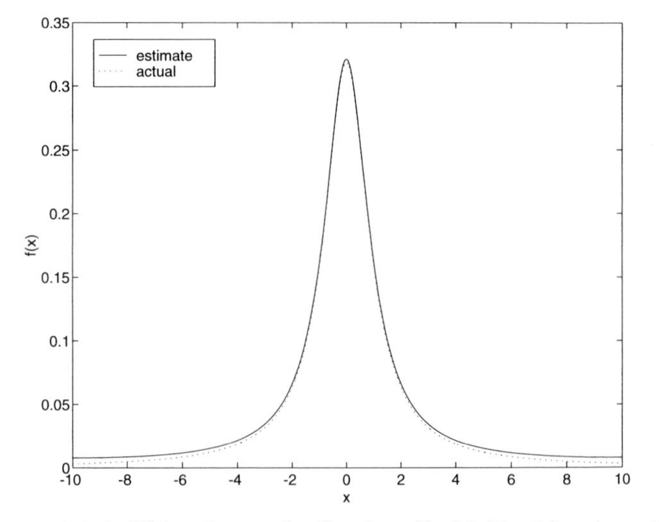

Figure 9.3-9: FSA estimate of a Cauchy pdf with $T = 20$ and $p = 50$.

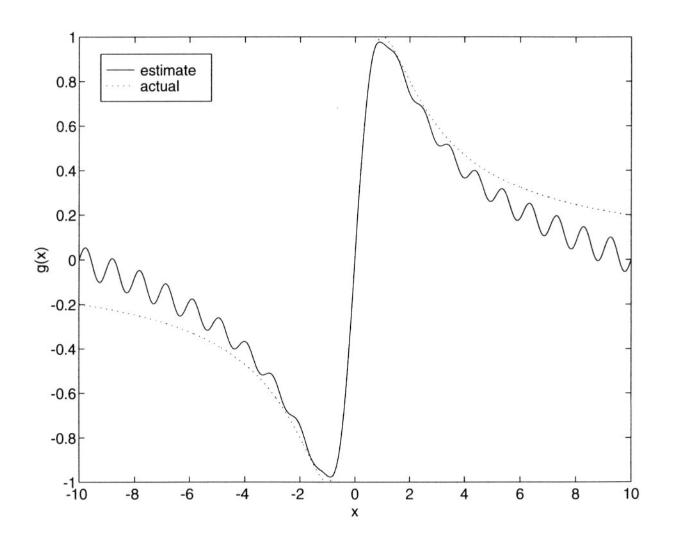

Figure 9.3-10: Indirect FSA estimate of the LO nonlinearity in Example 9.3-3 with $T = 20$ and $p = 20$.

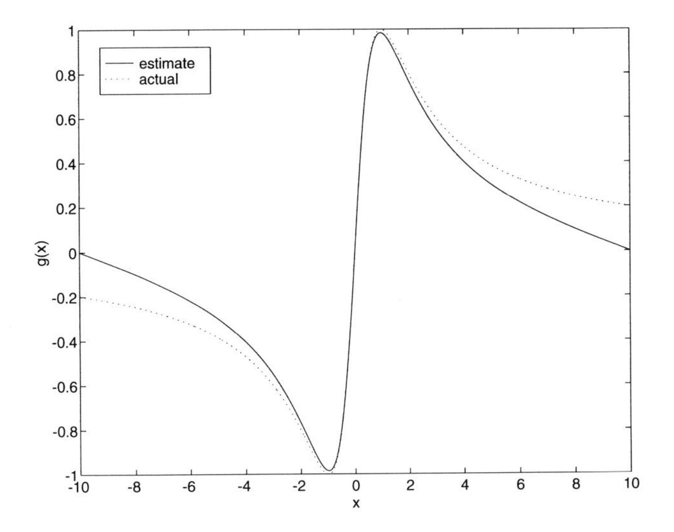

Figure 9.3-11: Indirect FSA estimate of the LO nonlinearity in Example 9.3-3 with $T = 20$ and $p = 50$.

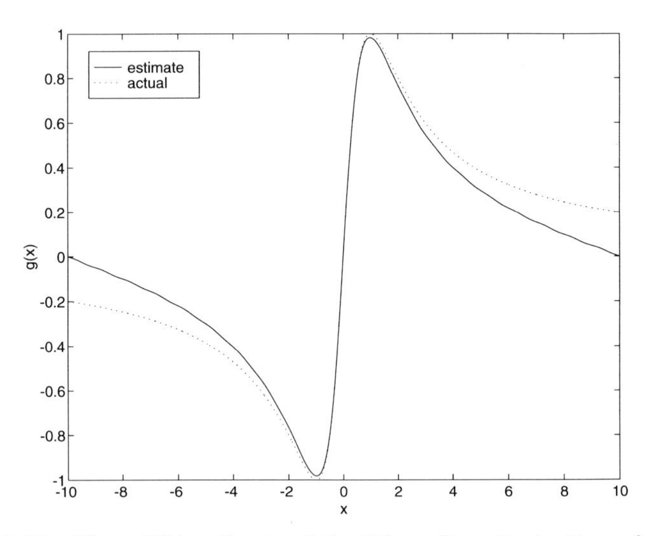

Figure 9.3-12: Direct FSA estimate of the LO nonlinearity in Example 9.3-3 with $T = 20$ and $p = 20$.

C. Kernel Methods for pdf Estimation

Kernel pdf estimators have received considerable interest in the past few decades ([9-8], Chapter 6 and references). Like the histogram, the kernel estimator is a nonparametric method, i.e., it does not assume that the pdf being estimated has a particular form. Instead, the resulting shape of the pdf estimate is solely a function of the observed data samples. The kernel estimator also has a number of advantages over the histogram: (1) it is continuous, (2) it may possess continuous derivatives, (3) it requires, on average, less samples to form an estimate than does the histogram, and (4) it is readily applicable to the estimation of multivariate pdfs and their gradients. Its major disadvantage is that in many cases the kernel estimator requires considerably more computation than does the histogram. However, with the availability of specialized processors, this increased computational burden is probably not an issue.

Since the subject of kernel estimation is such a rich area, the remainder of this section provides a brief overview of the topic. In particular, the kernel estimators for univariate and multivariate pdfs will be presented, with emphasis placed on their application to LO detection.

1. Univariate pdf Estimation

Let $\{x_i\}$ be the set of M observed samples of a random variable X. The kernel estimator of $f_X(x)$, the pdf of X, may be written as

$$\hat{f}_X(x) = \frac{1}{M\,h} \sum_{i=1}^{M} K\left(\frac{x - x_i}{h}\right). \tag{9.3-32}$$

The kernel window $K(t)$ determines the weighting effect that the samples $\{x_i\}$ have on the pdf estimate at the value x. The variable h is known as the "smoothing parameter" and determines the range of samples $\{x_i\}$ that will affect the pdf estimate at the value x.

Recall that the memoryless LO nonlinearity requires the first derivative of $f_X(x)$. Using the kernel estimator, this can be approximated as

$$\frac{d}{dx}\,\hat{f}_X(x) = \frac{d}{dx}\left[\frac{1}{Mh}\sum_{i=1}^{M} K\left(\frac{x - x_i}{h}\right)\right] = \frac{1}{Mh^2}\sum_{i=1}^{M} K'\left(\frac{x - x_i}{h}\right). \tag{9.3-33}$$

As can be seen, the kernel estimator can provide a direct approximation of ${f_X}'(x)$, which relieves the need to use numerical differentiation (as in the histogram method). From Eq. (9.3-33) it can be shown that the kth derivative of $\hat{f}_X(x)$ is given by

$$\frac{d^k}{dx^k}\,\hat{f}_X(x) = \frac{1}{M\,h^{k+1}}\sum_{i=1}^{M} K^{(k)}\left(\frac{x - x_i}{h}\right), \tag{9.3-34}$$

where $K^{(k)}(t) = \frac{d^k}{dt^k}K(t)$.

The characteristics of the kernel pdf estimator of Eq. (9.3-32) are primarily determined by the kernel and the smoothing parameter. Kernels usually satisfy the following constraints:

$$\begin{aligned} K(t) &\geq 0 \\ \int K(t)\,dt &= 1 \\ \int t\,K(t)\,dt &= 0. \end{aligned} \tag{9.3-35}$$

Two kernels which are widely used are (1) the Epanechnikov kernel, given by

$$K(t) = \frac{3}{4}(1 - t^2)\,I_{(-1,1)}(t), \tag{9.3-36}$$

and (2) the Gaussian kernel, given by

$$K(t) = \frac{1}{\sqrt{2\,\pi}}\,e^{-t^2/2}. \tag{9.3-37}$$

The Epanechnikov kernel is particularly useful since it minimizes the *asymptotic mean ISE* (AMISE) between $f_X(x)$ and $\hat{f}_X(x)$. Recalling the definition of ISE in Eq. (9.3-9), the *mean ISE* (MISE) is given by

$$\text{MISE}\,[\hat{f}_X(x)] \triangleq E\left\{\text{ISE}\,[\hat{f}_X(x)]\right\}, \tag{9.3-38}$$

and the AMISE is defined to be the main terms of the Taylor series approximation of the MISE ([9-8], Chapter 2). The Gaussian kernel, on the other hand, is particularly useful in situations when higher-order derivatives of $\hat{f}_X(x)$ are required.[14]

The smoothing parameter h also affects the AMISE of $\hat{f}_X(x)$. A useful result shown in [9-8], Chapter 6 is that for Gaussian data $\{x_i\}$ and a Gaussian kernel, the value of h which minimizes the AMISE is

$$h = \left(\frac{4}{3}\right)^{1/5} \sigma M^{-1/5}, \tag{9.3-39}$$

where σ^2 is the variance of X. Equation (9.3-39) can be a useful starting point for determining a value of h suitable for the given application. In most cases, the value of h will need to be decreased as the data becomes "less Gaussian."

The simplest way to apply the univariate kernel estimator to LO detection is by direct substitution into the memoryless LO nonlinearity, $g(\rho)$, of Eq. (9.1-13). Thus, letting $\{\eta_i\}$ be the set of M observed noise samples, the kernel estimate of $g(\rho)$ is

$$\hat{g}(\rho) = -\frac{\hat{f}_n'(\rho)}{\hat{f}_n(\rho)} = -\frac{1}{h}\frac{\sum_{i=1}^{M} K'\left(\frac{\rho-\eta_i}{h}\right)}{\sum_{i=1}^{M} K\left(\frac{\rho-\eta_i}{h}\right)}. \tag{9.3-40}$$

More sophisticated forms of $\hat{g}(\rho)$ can be derived if characteristics concerning the noise pdf, $f_n(\eta)$, are known prior to detector implementation.

2. Multivariate pdf Estimation

The general form of the multivariate kernel estimator is developed as follows. Let $\{\mathbf{x_i}\} = \{[x_{1i}\ \ x_{2i} \ldots x_{Ni}]^T\}, i = 1, \ldots, L$, be the set of L samples of a random vector $\mathbf{X}$ having length N. Thus, the total number of available samples is $M = LN$. The general multivariate kernel estimator is then defined as ([9-8], Chapter 6):

$$\hat{f}_{\mathbf{X}}(\mathbf{x}) = \frac{1}{L\,|\mathbf{H}|}\sum_{i=1}^{L} K(\mathbf{H}^{-1}(\mathbf{x}-\mathbf{x_i})). \tag{9.3-41}$$

The window $K(\mathbf{t})$ is now a multivariate kernel and $\mathbf{H}$ is a linear transform matrix.

At this point a special form of the general multivariate kernel estimator will be presented: the product kernel estimator. The product kernel estimate of $f_{\mathbf{X}}(\mathbf{x})$ is ([9-8], Chapter 6):

$$\hat{f}_{\mathbf{X}}(\mathbf{x}) = \hat{f}_{\mathbf{X}}(x_1, \ldots, x_j, \ldots, x_N) = \frac{1}{L\,h_1 \cdots h_N}\sum_{i=1}^{L}\left[\prod_{j=1}^{N} K\left(\frac{x_j - x_{ij}}{h_j}\right)\right], \tag{9.3-42}$$

[14] For an extensive list of kernel functions see reference [9-8], Chapter 6.

where x_j is the jth argument of $f_{\mathbf{X}}(\mathbf{x})$ and x_{ij} is the jth sample of the ith observed data vector. The parameters $\{h_j\}$ are the smoothing parameters corresponding to each dimension of the kernel estimate. The kernel function $K(t)$ is the same univariate kernel discussed in the previous section.

Recall that the LO detector with memory requires the partial derivatives of the multivariate noise pdf. The product kernel estimate of $\frac{\partial}{\partial x_q} f_{\mathbf{X}}(\mathbf{x})$ is found by differentiating $\hat{f}_{\mathbf{X}}(\mathbf{x})$ in Eq. (9.3-42). The result is

$$\begin{aligned} \frac{\partial}{\partial x_q} \hat{f}_{\mathbf{X}}(\mathbf{x}) &= \frac{\partial}{\partial x_q} \left\{ \frac{1}{L\, h_1 \cdots h_N} \sum_{i=1}^{L} \left[\prod_{j=1}^{N} K\left(\frac{x_j - x_{ij}}{h_j} \right) \right] \right\} \\ &= \frac{1}{L\, h_q\, h_1 \cdots h_N} \sum_{i=1}^{L} \left[K'\left(\frac{x_q - x_{iq}}{h_q} \right) \prod_{\substack{j=1 \\ j \neq q}}^{N} K\left(\frac{x_j - x_{ij}}{h_j} \right) \right]. \end{aligned} \tag{9.3-43}$$

Note that the expression in Eq. (9.3-43) can be extended to partial derivatives of arbitrary order.

As in the case of the univariate kernel estimator, the product kernel estimator has a useful expression yielding the set $\{h_i\}$ which minimizes the AMISE. If a Gaussian kernel is used and $\mathbf{X}$ is a multivariate Gaussian vector with independent components, then the value of h_j which minimizes the AMISE of $\hat{f}_{\mathbf{X}}(\mathbf{x})$ is ([9-8], Chapter 6)

$$h_j = \left(\frac{4}{N+2} \right)^{1/(N+4)} \sigma_j\, L^{-1/(N+4)}, \tag{9.3-44}$$

where σ_j^2 is the variance of the jth component of $\mathbf{X}$. Equation (9.3-44) provides a starting point for determining $\{h_j\}$ that is most suitable for the given application.

The product kernel estimate can be used to approximate the LO nonlinearity with memory by substituting the kernel estimates of $f_{\mathbf{n}}(\boldsymbol{\eta})$ and $\frac{\partial}{\partial \eta_q} f_{\mathbf{n}}(\boldsymbol{\eta})$ ($q = 1, \ldots, N$) into Eq. (9.2-6). Letting $\{\boldsymbol{\eta}_i\}$ be L samples of the random vector $\mathbf{n}$ having length N, the resulting product kernel estimate of $g_q(\boldsymbol{\rho})$ is:

$$\hat{g}_q(\boldsymbol{\rho}) = -\frac{1}{h_q} \frac{\sum_{i=1}^{L} \left[K'\left(\frac{\rho_q - \eta_{iq}}{h_q} \right) \prod_{\substack{j=1 \\ j \neq q}}^{N} K\left(\frac{\rho_j - \eta_{ij}}{h_j} \right) \right]}{\sum_{i=1}^{L} \left[\prod_{j=1}^{N} K\left(\frac{\rho_j - \eta_{ij}}{h_j} \right) \right]}. \tag{9.3-45}$$

The techniques employed in arriving at Eq. (9.3-45) can be used to develop more sophisticated forms of the LO detector with memory.

D. Other Methods

There are, of course, numerous other methods for estimating the pdf. The method of convex projections, mentioned earlier, uses prior knowledge in addition to the sample realizations to estimate the pdf. This method is discussed in some detail in the last chapter of this text. The method of kernel windows is based on the method introduced by Parzen [9-11]. Also, the penalized maximum likelihood method [9-12] is readily available through the IMSL library [9-13]. A more recent pdf estimation method is the bootstrap method introduced by Efron [9-14]. The use of these methods to estimate the LO nonlinearity function, $g(\rho)$, is a subject for further research.

E. Summary of Techniques

Each of the pdf estimation techniques presented in this section has advantages and disadvantages when applied to LO nonlinearity implementation. The strength of the histogram method is in its simplicity and minimal computational requirements. However, it is prone to error in regions of low probability (e.g., the tails of the pdf) and inaccuracies due to the binning process. In addition, the discrete nature of the histogram and the required differentiation in the implementation of the LO nonlinearity may not accurately model the continuous nature of the actual pdf.

The FSA algorithm, on the other hand, provides a pdf estimate which is continuous throughout its entire domain and which possesses continuous derivatives. Also, each available data sample uniquely contributes to the overall estimate; no quantization is required. However, the result is a moderate increase in computational complexity, particularly if a large value of p is required for convergence to a good approximation. Also, there is no restriction placed on the sign of the pdf estimate (negative regions are possible), and the area under the approximation is not necessarily unity.

The kernel method does not suffer from many of the problems that afflict the histogram and FSA methods. The kernel pdf estimate is continuous and may possess continuous derivatives, with the degree of continuity determined by the kernel function. The kernel estimate also is non-negative and has unit area. In addition, each data sample contributes uniquely to the overall pdf approximation. And finally, kernel pdf estimation is a rich area of study and a wealth of additional information is available. The major drawback of this method is its high computational requirements.

As a final note, LO nonlinearities implemented using pdf estimates are particularly sensitive in regions of low probability since $g(\rho) \propto 1/f_n(\rho)$. Thus, modifications to the original nonlinearity may be required, e.g., discounting values of ρ which fall in regions where $\hat{f}_n(\rho) < C$, a predetermined constant threshold. This sensitivity can be reduced if $g(\rho)$ is estimated directly, for example via the FSA algorithm. Furthermore, direct estimates of $g(\rho)$ are not exposed to the inherently noisy operation of differentiation. However, in the case of the FSA algorithm, the direct estimate of $g(\rho)$ is subject to many of the same limitations encountered when estimating a pdf. In the end, the method of implementing $g(\rho)$ will be a compromise between accuracy and computational complexity.

9.4 LO Detection Applied to a Spread-Spectrum System

The usage of LO detection techniques in a *direct sequence* (DS) spread-spectrum communications system is the focus of this section. Spread-spectrum systems have two primary characteristics: (1) the bandwidth of the transmitted signal is much larger than the information signal bit rate, and (2) the transmitted signal is modified so as to appear as pseudo-random noise, with the pseudo-random feature known, ideally, only by the transmitter and receiver. There are two fundamental types of spread-spectrum systems: DS and *frequency hopped* (FH). In DS spread-spectrum systems the time-domain characteristics, e.g., phase and/or amplitude, of the transmitted signal are modified in a pseudo-random fashion. In FH spread-spectrum systems, the frequency is the signal characteristic which is randomized.

Spread-spectrum systems have a wide variety of applications. For example, their ability to overcome jamming interference and the noise-like appearance of the transmitted signal have made spread-spectrum systems useful in military environments where interference rejection and *low probability of intercept/detection* (LPI/D) are essential [9-15], [9-16]. An example of a commercial application of these systems is multiple-access communications, e.g., mobile communication systems, since many different spread-spectrum signals may be superimposed and still be correctly demodulated by their respective receivers [9-17]. Also, the pseudo-random nature of the transmitted signal is useful in the suppression of multipath interference [9-18].

The remainder of this section is devoted to improving the performance of a DS spread-spectrum system through the use of LO detection. A brief review of DS spread-spectrum communications is provided, primarily as a means of introducing the notation necessary to develop the required LO detector.

A. Review of Direct-sequence Spread-spectrum

A baseband model of a *binary phase shift keyed* (BPSK) DS spread-spectrum system, similar to the one discussed in [9-19], Chapter 8, is shown in Fig. 9.4-1. The message bit sequence, represented by $\mathbf{m} = \{m(l)\}$, is to be transmitted at a rate $1/T_b$. The sequence $\mathbf{m}$ is fed into an $(L_c, 1)$ encoder and a baseband modulator.

The result is a coded information signal having rate $1/T_c$, given by

$$d(t) = \sum_k d(k)\,\Pi(t - k\,T_c), \tag{9.4-1}$$

where $d(k) = \{\pm 1\}$, the coded bit sequence, is dependent on the value of $m(l)$, and $\Pi(t)$ is a pulse of unit amplitude that extends from $t = 0$ to T_c seconds.[15] The information signal $d(t)$ is multiplied by a *pseudo-noise* (PN) "chipping" signal $c(t)$ given by

$$c(t) = \sqrt{\frac{2\,E_c}{T_c}} \sum_k c(k)\,\Pi(t - k\,T_c), \tag{9.4-2}$$

where $c(k) = \{\pm 1\}$ is a PN sequence, commonly termed the "chip" sequence, and E_c, the energy in each chip, is defined as

$$E_c \triangleq \tfrac{1}{2} \int_0^{T_c} |c(t)|^2\,dt. \tag{9.4-3}$$

The value of T_c is chosen such that $T_b/T_c = L_c$, an integer representing the number of chips per information bit. Note from Eq. (9.4-1) and Eq. (9.4-2) that the chip rate is the same as the rate of the coded information signal. These two rates are chosen to be the same to facilitate later computations. Finally, the resulting transmitted signal is

$$s(t) = d(t)\,c(t) = \sqrt{\frac{2\,E_c}{T_c}} \sum_k d(k)\,c(k)\,\Pi(t - kT_c). \tag{9.4-4}$$

The overall channel noise, $n(t)$, is assumed to contain two components: an interference signal, $z(t)$, and (assuming no bandwidth limitations) a white Gaussian background noise signal, $w(t)$, having *power spectral density* (PSD) N_0. Thus, the received signal is given by

$$r(t) = s(t) + n(t) = d(t) \cdot c(t) + z(t) + w(t)\,. \tag{9.4-5}$$

For the case when $n(t) = w(t)$, or $z(t) = 0$, it is well known ([9-19], Chapter 8) that the optimum receiver contains either a matched filter or a correlator. Since the baseband waveform of the PN signal is a pulse of amplitude $\sqrt{2\,E_c\,/\,T_c}$ and $c(k)$ is constant over the interval $[kT_c, (k+1)T_c]$, the correlator can be implemented by: (1) scaling the received signal by $\sqrt{2\,E_c\,/\,T_c}$ and integrating from kT_c to $(k+1)T_c$ seconds, (2) sampling, and (3) multiplying the resulting sequence with a synchronized version of $c(k)$, as shown in Fig. 9.4-1. The output of the correlator is sent to a decoder which provides the estimate of the original message bit stream, $\hat{m}(l)$.

---

[15]This representation may also be used for an uncoded system. In this case, $d(k) = 1$ or -1 in the range $lL_c \le k < (l+1)L_c$ corresponding to $m(l) = 1$ or 0, respectively. Thus, $d(t)$ is held constant over the interval $kT_c \le t < (k + L_c)T_c$, where $L_c = T_b/T_c$

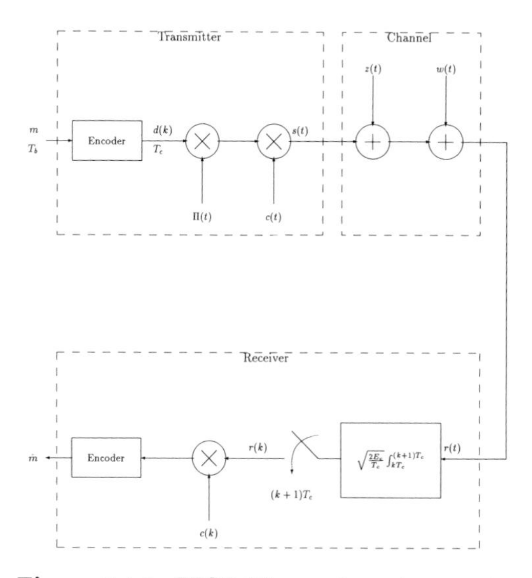

Figure 9.4-1: BPSK DS spread-spectrum system.

The probability of bit error, P_b, for an uncoded BPSK DS spread-spectrum system in additive Gaussian noise is obtained in the following manner. First, the sampled output of the integrator, denoted as $r(k)$, is computed. This is given by[16]

$$\begin{aligned} r(k) &= \sqrt{\frac{2\,E_c}{T_c}} \int_{k\,T_c}^{(k+1)\,T_c} r(t)\,dt = \sqrt{\frac{2\,E_c}{T_c}} \int_{k\,T_c}^{(k+1)\,T_c} [\,d(t)\,c(t) + w(t)]\,dt \\ &= 2\,E_c\,d(k)\,c(k) + w(k)\,, \end{aligned} \tag{9.4-6}$$

where $w(k)$ is a Gaussian random variable with zero mean and variance $2E_cN_0$. Multiplying $r(k)$ by the synchronized PN sequence $c(k)$ yields[17]

$$\hat{d}(k) = r(k)\,c(k) = 2\,E_c\,d(k) + n(k)\,, \tag{9.4-7}$$

where $n(k) = w(k)c(k)$ is an iid Gaussian random sequence whose components have zero mean and variance $2E_cN_0$. Since no coding was used, $d(k) = d = \pm 1$ is constant over the range $l\,L_c \leq k < (l+1)\,L_c$ corresponding to the information bit $m(l)$. Thus, the decoder simply computes the decision variable

$$U = \sum_{k=0}^{L_c-1} \hat{d}(k + l\,L_c) = 2\,L_c\,E_c\,d + \sum_{k=0}^{L_c-1} n(k + l\,L_c) \tag{9.4-8}$$

[16]The discrete time-scale k at the receiver is delayed by one time unit (T_c seconds) from that at the transmitter.

[17]Note that this receiver implementation requires that $c(k)$ at the receiver be delayed by one time unit (T_c seconds) with respect to $c(k)$ at the transmitter.

and compares U to a threshold determined by the type of detection being used. If $P[m(l) = 1] = P[m(l) = 0] = 1/2$, then the minimum P_b occurs when the threshold is zero ([9-1], Chapter 3). Since the samples $\{n(k)\}$ are iid, the noise term in Eq. (9.4-8) is a Gaussian random variable with zero mean and variance $2L_cE_cN_0$. The resulting P_b for this system is given by ([9-19], Chapter 8)

$$P_b = \frac{1}{2} - \frac{1}{2}\,\mathrm{erf}\left(\sqrt{\frac{L_c\,E_c}{N_0}}\right) = \frac{1}{2} - \frac{1}{2}\,\mathrm{erf}\left(\sqrt{\frac{E_b}{N_0}}\right), \tag{9.4-9}$$

where $E_b = L_cE_c$ is the energy per information bit. Thus, P_b for a BPSK DS spread-spectrum system in additive white Gaussian noise is the same as for a standard BPSK communications system ([9-19], Chapter 4).

Since the standard BPSK communications system is designed to provide the minimum P_b attainable in additive white Gaussian noise when no coding is used, the result in Eq. (9.4-9) shows that the spread-spectrum system is also able to provide the minimum P_b, as well as the additional benefits of interference rejection and LPI/LPD. The interference rejection capabilities of the BPSK DS spread-spectrum system, and particularly their enhancement through the use of LO detection, are discussed in the following section.

B. Applying the LO Detector

An LO detector for the uncoded version of the spread-spectrum system shown in Fig. 9.4-1 can be derived using the sampled output of the correlator, denoted as $r(k)$. For the general case when $r(t) = s(t) + z(t) + w(t)$, $r(k)$ is given by

$$r(k) = s(k) + z(k) + w(k), \tag{9.4-10}$$

where from Eq. (9.4-6)

$$s(k) = 2\,E_c\,d(k)\,c(k) \tag{9.4-11}$$

$$w(k) = \sqrt{\frac{2\,E_c}{T_c}} \int_{k\,T_c}^{(k+1)\,T_c} w(t)\,dt$$

and

$$z(k) = \sqrt{\frac{2\,E_c}{T_c}} \int_{k\,T_c}^{(k+1)\,T_c} z(t)\,dt\,. \tag{9.4-12}$$

Recalling that $w(t)$ is a white Gaussian noise process with PSD equal to N_0, $\{w(k)\}$ is therefore an iid random sequence whose elements are zero mean Gaussian random variables with variance $2E_cN_0$.

Define the two detection hypotheses for the uncoded DS spread-spectrum system as:

H_1 : Message bit s_1 was transmitted, and thus $d(k) = d_1 = 1, k = 0, \ldots, L_c - 1$
H_0 : Message bit s_0 was transmitted, and thus $d(k) = d_0 = -1, k = 0, \ldots, L_c - 1$.

To simplify notation the discrete time index k in the definitions of H_1 and H_0 is normalized to the given bit interval under consideration. Let $n(k) = z(k) + w(k)$ and the observed value of $r(k)$ be ρ_k, $k = 0, \ldots, L_c - 1$. Define the following vectors: $\mathbf{r} = \{r(k)\}$, $\boldsymbol{\rho} = \{\rho_k\}$, $\mathbf{c} = \{c(k)\}$, $\mathbf{n} = \{n(k)\}$, $\mathbf{z} = \{z(k)\}$, and $\mathbf{w} = \{w(k)\}$. Since $\mathbf{c}$ is a known sequence at the receiver, $f_{\mathbf{r}}(\boldsymbol{\rho} \,|\, H_m) = f_{\mathbf{n}}(\boldsymbol{\rho} - \mathbf{s_m}) = f_{\mathbf{n}}(\boldsymbol{\rho} - 2\,E_c\,d_m\,\mathbf{c})$,[18] and the LO detector of Eq. (9.2-6) and Eq. (9.2-7) may be used. Thus, the LO detector for an uncoded BPSK DS spread-spectrum system is

$$l(\boldsymbol{\rho}) = \sum_{k=0}^{L_c-1} c(k)\, g_k(\boldsymbol{\rho}) \quad \underset{\text{choose } H_0}{\overset{\text{choose } H_1}{\gtrless}} \quad \frac{\tilde{\gamma}}{4\,E_c}, \tag{9.4-13}$$

where

$$g_k(\boldsymbol{\rho}) = -\frac{\partial}{\partial \rho_k} \ln[f_{\mathbf{n}}(\boldsymbol{\rho})] = -\frac{\partial}{\partial \rho_k} \ln[f_{\mathbf{z}+\mathbf{w}}(\boldsymbol{\rho})]\,. \tag{9.4-14}$$

A block diagram of this system is shown in Fig. 9.4-2.

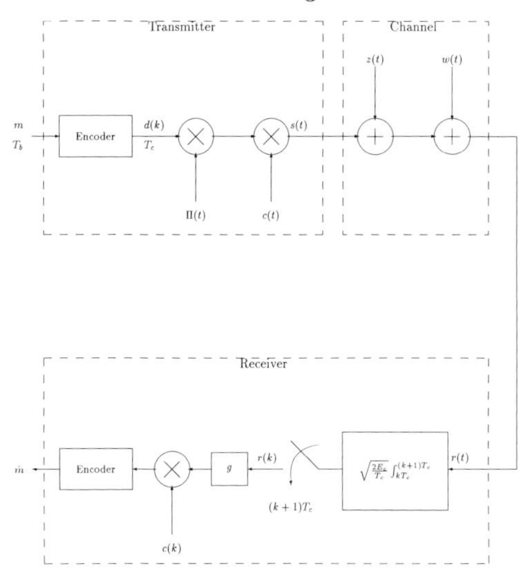

Figure 9.4-2: BPSK DS spread-spectrum system with LO detection.

Since $\mathbf{z}$ and $\mathbf{w}$ are assumed to be independent, the total noise pdf, $f_{\mathbf{n}}(\boldsymbol{\eta})$, is the L_c-dimensional convolution of the pdfs of $\mathbf{z}$ and $\mathbf{w}$, written formally as $f_{\mathbf{n}}(\boldsymbol{\eta}) = f_{\mathbf{z}}(\boldsymbol{\eta}) * f_{\mathbf{w}}(\boldsymbol{\eta})$. Using Leibniz's rule for differentiation of integrals ([9-20], Chapter 16), [19] and assuming that $f_{\mathbf{z}}(\pm\infty) = f_{\mathbf{w}}(\pm\infty) = 0$, $g_k(\boldsymbol{\rho})$ is given by

$$g_k(\boldsymbol{\rho}) = -\frac{\frac{\partial}{\partial \rho_k} \int_{-\infty}^{\infty} f_{\mathbf{z}}(\boldsymbol{\alpha})\, f_{\mathbf{w}}(\boldsymbol{\rho} - \boldsymbol{\alpha})\, d\boldsymbol{\alpha}}{\int_{-\infty}^{\infty} f_{\mathbf{z}}(\boldsymbol{\alpha})\, f_{\mathbf{w}}(\boldsymbol{\rho} - \boldsymbol{\alpha})\, d\boldsymbol{\alpha}}$$

[18] The subscripted index, m, denotes hypothesis H_0 or H_1, not the message bit sequence $\{m(l)\}$.
[19] Assuming that the constraints specified in [9-20], Chapter 16, are satisfied.

$$= -\frac{\int_{-\infty}^{\infty} f_{\mathbf{z}}(\boldsymbol{\alpha}) \frac{\partial}{\partial \rho_k} f_{\mathbf{w}}(\boldsymbol{\rho} - \boldsymbol{\alpha})\, d\boldsymbol{\alpha}}{\int_{-\infty}^{\infty} f_{\mathbf{z}}(\boldsymbol{\alpha})\, f_{\mathbf{w}}(\boldsymbol{\rho} - \boldsymbol{\alpha})\, d\boldsymbol{\alpha}} \tag{9.4-15}$$

$$= \frac{\rho_k}{2\, E_c\, N_0} - \frac{1}{2\, E_c\, N_0} \frac{\int_{-\infty}^{\infty} \alpha_k f_{\mathbf{z}}(\boldsymbol{\alpha})\, f_{\mathbf{w}}(\boldsymbol{\rho} - \boldsymbol{\alpha})\, d\boldsymbol{\alpha}}{\int_{-\infty}^{\infty} f_{\mathbf{z}}(\boldsymbol{\alpha})\, f_{\mathbf{w}}(\boldsymbol{\rho} - \boldsymbol{\alpha})\, d\boldsymbol{\alpha}},$$

where

$$f_{\mathbf{w}}(\boldsymbol{\beta}) = \frac{1}{(4\pi E_c N_0)^{L_c/2}} \exp\left(-\frac{\boldsymbol{\beta}^T \boldsymbol{\beta}}{4 E_c N_0}\right). \tag{9.4-16}$$

The expression in Eq. (9.4-15) illustrates an interesting result. For the case when the total noise is the sum of white Gaussian noise and non-Gaussian interference, the LO nonlinearity, $g_k(\boldsymbol{\rho})$, contains two components: a linear term corresponding to the Gaussian noise, and a nonlinear term corresponding to the interaction between the Gaussian noise and the non-Gaussian interference.

While the expression for $g_k(\boldsymbol{\rho})$ in Eq. (9.4-15) appears formidable, it can be greatly simplified if certain characteristics of $\mathbf{z}$ are known. For example, if the elements of $\mathbf{z}$ are iid then the L_c-dimensional integrals of Eq. (9.4-15) are replaced by one-dimensional integrals, resulting in the expression:

$$g_k(\boldsymbol{\rho}) = g_k(\rho_k) = \frac{\rho_k}{2E_c N_0} - \frac{1}{2E_c N_0} \frac{\int_{-\infty}^{\infty} \alpha_k\, f_z(\alpha_k)\, f_w(\rho_k - \alpha_k)\, d\alpha_k}{\int_{-\infty}^{\infty} f_z(\alpha_k)\, f_w(\rho_k - \alpha_k)\, d\alpha_k}. \tag{9.4-17}$$

Also, if $f_{\mathbf{z}}(\boldsymbol{\alpha})$ has a Markov structure, Eq. (9.4-15) can be simplified in a manner similar to that presented in Section 9.2. Finally, it should be noted that if the pdf of $\mathbf{n} = \mathbf{z} + \mathbf{w}$ is known explicitly, $g_k(\boldsymbol{\rho})$ can be implemented directly using Eq. (9.4-14) with appropriate modifications depending on the characteristics of $\mathbf{n}$, e.g., iid samples or Markov structure.

Example 9.4-1

Let the sampled received signal be given by

$$r(k) = 2\, E_c\, d_m\, c(k) + n(k).$$

The sequence $\{n(k)\}$ is modeled as the output of an AR(1) process and is written as

$$n(k) = a\, n(k-1) + v(k), \qquad |a| < 1,$$

where a is a parameter and $\{v(k)\}$ is an iid sequence of Gaussian random variables having zero mean and variance $\sigma_v^2 = (2E_c\sigma^2)$.[20] Recall that the index k has been normalized to the current (lth) bit interval, i.e., $k = 0, \ldots, L_c - 1$

[20]The term $2E_c$ is included as a reminder that the noise sequence $\{n(k)\}$ is the sampled output of the correlation of $c(t)$ and $n(t)$.

corresponds to $k = lL_c, \ldots, lL_c + L_c - 1$. Assume that l is large (steady state). Then ([9-21], Chapter 2)

$$f_{n(0)}(\eta_0) \approx \frac{1}{\sqrt{2\pi\sigma_n^2}} e^{-\frac{\eta_0^2}{2\sigma_n^2}}$$

and

$$f_{n(k)|n(k-1)}(\eta_k \,|\, \eta_{k-1}) = \frac{1}{\sqrt{2\pi\sigma_v^2}} e^{-\frac{(\eta_k - a\eta_{k-1})^2}{2\sigma_v^2}},$$

where $\sigma_n^2 = \sigma_v^2 \,/\, (1 - a^2)$. Furthermore, for l large the autocorrelation of $n(k)$, $r_n(i) = E\{n(k)\, n(k+i)\}$, approaches ([9-21], Chapter 2)

$$r_n(i) \Rightarrow \frac{a^{|i|}}{1 - a^2}\sigma_v^2 .$$

Since the joint L_c-dimensional pdf of $\mathbf{n}$ can be written as

$$f_{\mathbf{n}}(\boldsymbol{\eta}) = f_{n(L_c-1)\,|\,n(L_c-2)}(\eta_{L_c-1} \,|\, \eta_{L_c-2}) \cdots f_{n(1)\,|\,n(0)}(\eta_1 \,|\, \eta_0)\, f_{n(0)}(\eta_0),$$

and recalling the results in Example 9.2-1, the LO nonlinearity $g_k(\rho_{k-1}, \rho_k, \rho_{k+1})$ for this system is given by

$$g_k(\rho_{k-1}, \rho_k, \rho_{k+1}) = \begin{cases} \frac{\rho_0}{\sigma_n^2} - a\frac{\rho_1 - a\rho_0}{\sigma_v^2} = \frac{\rho_0 - a\rho_1}{\sigma_v^2}\,; & k = 0 \\ \frac{1}{\sigma_v^2}\left[(1 + a^2)\rho_k - a\rho_{k-1} - a\rho_{k+1}\right]\,; & k = 1, \ldots, L_c - 2 \\ \frac{\rho_{L_c-1} - a\rho_{L_c-2}}{\sigma_v^2}\,; & k = L_c - 1 . \end{cases}$$

Thus, using Eq. (9.4-13), the LO detector is given by

$$l(\boldsymbol{\rho}) = (c(0) - ac(1))\rho_0 + \sum_{k=1}^{L_c-2}\left[c(k)(1 + a^2) - (c(k-1) + c(k+1))a\right]\rho_i + (c(L_c - 1) - ac(L_c - 2))\rho_{L_c-1} \;\underset{\text{choose } H_0}{\overset{\text{choose } H_1}{\gtrless}}\; \tilde{\gamma}\frac{\sigma^2}{2}.$$

One method for determining the performance of the LO detector is to compute its output *signal-to-noise ratio* (SNR), defined as ([9-22], Chapter 2)

$$\mathrm{SNR}_{LO} = \frac{\left[E\{l(\mathbf{r}) \,|\, d_1\} - E\{l(\mathbf{r}) \,|\, d_0\}\right]^2}{\mathrm{Var}\{l(\mathbf{r}) \,|\, d_0\}},$$

where in most cases $\mathrm{Var}\{l(\mathbf{r}) \,|\, d_1\} \approx \mathrm{Var}\{l(\mathbf{r}) \,|\, d_0\}$. Since the signal and noise are assumed to be independent with $E\{n(i)\} = 0$, and using the approximation that $E\{c(i)\, c(j)\} = \delta_K(i - j)$, where δ_K is the Kronecker delta function, then

$$E\{l(\mathbf{r}) \,|\, d_m\} = 2\, E_c\, d_m \left[L_c + a^2(L_c - 2)\right],$$

and

$$[E\{l(\mathbf{r})\,|\,d_1\} \;-\; E\{l(\mathbf{r})\,|\,d_0\}]^2 \;=\; 16\,E_c^2\,[L_c \,+\, a^2(L_c \,-\, 2)]^2 .$$

The expression for $\mathrm{Var}\{l(\mathbf{r})\,|\,d_m\}$ can be written as

$$\mathrm{Var}\{l(\mathbf{r})\,|\,d_m\} \;=\; E\left\{l^2(\mathbf{r})\,|\,d_m\right\} \;-\; [E\{l(\mathbf{r})\,|\,d_m\}]^2 ,$$

where

$$\begin{aligned}
E\left\{l^2(\mathbf{r})\,|\,d_m\right\} \;=\; & \sum_{i=1}^{L_c-2}\sum_{j=1}^{L_c-2} E\Big\{[c(i)\,(1+a^2) \,-\, (c(i-1)+c(i+1))\,a]\cdot \\
& \qquad [c(j)\,(1+a^2) \,-\, (c(j-1)+(c(j+1))\,a]\,r_i\,r_j\Big\} \\
& +2\sum_{i=1}^{L_c-2} E\Big\{(c(0)-ac(1))[c(i)(1+a^2)-(c(i-1)+c(i+1))\,a]r_0 r_i\Big\} \\
& +2\sum_{i=1}^{L_c-2} E\,\{(c(L_c-1)-ac(L_c-2))\cdot \\
& \qquad [c(i)(1+a^2)-(c(i-1)+c(i+1))a]r_{L_c}r_i\Big\} \\
& +E\left\{(c(0) \,-\, a\,c(1))^2\,r_0^2\right\} \,+\, E\left\{(c(L_c-1) \,-\, a\,c(L_c-2))^2\,r_{L_c-2}^2\right\} \\
& +2E\,\{(c(0)-ac(1))\,(c(L_c-1)-ac(L_c-2))\,r_0\,r_{L_c-1}\}\;.
\end{aligned}$$

Using the assumptions that (1) the signal and noise are independent, and (2) $E\{c(i)\,c(j)\} \;=\; \delta_K(i-j)$, along with the results that (3) $c^2(i) = 1$, (4) $E\{n(i)\} = 0$, and (5) $E\{n(i)\,n(j)\} \;=\; r_n(i-j)$, yields the result

$$\begin{aligned}
E\{l^2(\mathbf{r})\,|\,d_m\} \;=\; & 4\,E_c^2\,[L_c^2 \,+\, (2L_c^2 \,-\, 2L_c \,-\, 2)\,a^2 \,+\, (L_c^2 \,-4L_c \,+\, 4)\,a^4] \\
& +\sigma_n^2\,[L_c\,(1 \,-\, a^4) \,+\, 6a^2 \,+\, 2a^4] ,
\end{aligned}$$

and hence

$$\mathrm{Var}\{l(\mathbf{r})\,|\,d_m\} \;=\; 8\,E_c^2\,(L_c \,-\, 1)\,a^2 \,+\, \sigma_n^2\,[L_c\,(1 \,-\, a^4) \,+\, 6a^2 \,+\, 2a^4]$$

independent of m. Therefore, the resulting output SNR is given by

$$\mathrm{SNR}_{LO} \;=\; \frac{[L_c \,+\, (L_c \,-\, 2)\,a^2]^2}{\frac{(L_c-1)}{2}\,a^2 \,+\, \frac{1}{E_c/\sigma^2}\,\frac{1}{8\,(1-a^2)}\,[L_c\,(1 \,-\, a^4) \,+\, 6a^2 \,+\, 2a^4]} .$$

The SNR of the LO detector can be compared to that of a linear receiver. For a standard linear receiver the detector is of the form

$$\sum_{k=0}^{L_c-1} c(k)\,\rho_k \quad \underset{\text{choose } H_0}{\overset{\text{choose } H_1}{\gtrless}} \quad \hat{\gamma} .$$

Using a similar development as for the LO detector, the output SNR for the linear receiver can be shown to be

$$\mathrm{SNR}_{LR} = \frac{16\, L_c\, E_c^2}{\sigma_n^2} = \frac{8\,(1 - a^2)\, L_c\, E_c}{\sigma^2}.$$

Expressing SNR_{LO} in terms of SNR_{LR} gives the following result:

$$\mathrm{SNR}_{LO} = \frac{[L_c + (L_c - 2)\, a^2]^2}{\frac{(L_c - 1)}{2}\, a^2 + \frac{1}{\mathrm{SNR}_{LR}}\, L_c\,[L_c\,(1 - a^4) + 6a^2 + 2a^4]}.$$

A plot of SNR_{LO} versus SNR_{LR} for $a = 0.99$ and $L_c = 100$ is shown in Fig. 9.4-3. A more useful measurement of the performance of the LO detector is the ratio of SNR_{LO} to SNR_{LR}, defined as G_{LO}. From the preceding expression, it is seen that G_{LO} is given by

$$G_{LO} = \frac{[L_c + (L_c - 2)\, a^2]^2}{\left[\frac{(L_c - 1)}{2}\, a^2\right] \mathrm{SNR}_{LR} + L_c\,[L_c\,(1 - a^4) + 6a^2 + 2a^4]}.$$

A plot of G_{LO} versus SNR_{LR} for $a = 0.99$ and $L_c = 100$ is shown in Fig. 9.4-4. As can be seen G_{LO} is largest when SNR_{LR} is small, and decreases as SNR_{LR} increases.

This example illustrates a number of important concepts. First, Figs. 9.4-3 and 9.4-4 show that the LO detector provides a large SNR improvement in regions where the SNR of the linear receiver is poor. Thus, in regions where the performance of a linear receiver is unsatisfactory, the LO detector provides an output SNR sufficient to make reliable detection possible. Second, examining the case when the noise is modeled as an AR process is particularly useful since a common method of estimating *power spectral densities* (PSDs) is via AR processes ([9-21], Chapter 7). The PSD of the AR process in this example with $a = 0.99$ is shown in Fig. 9.4-5. Figure 9.4-6 illustrates the PSD for various values of the parameter, a. Note that as a increases, the PSD becomes less "white" and thus the noise is more correlated (narrowband).

Finally, this example shows that an LO detector is not always more complicated than a linear detector. In fact, the LO detector in this example is simply a linear combination of the received signal samples, with the multiplying coefficients a function of the PN sequence and the AR model parameter a. However, while the complexity increase is negligible, the performance gain of the LO detector is large, particularly in regions where the performance of the linear receiver is poor.

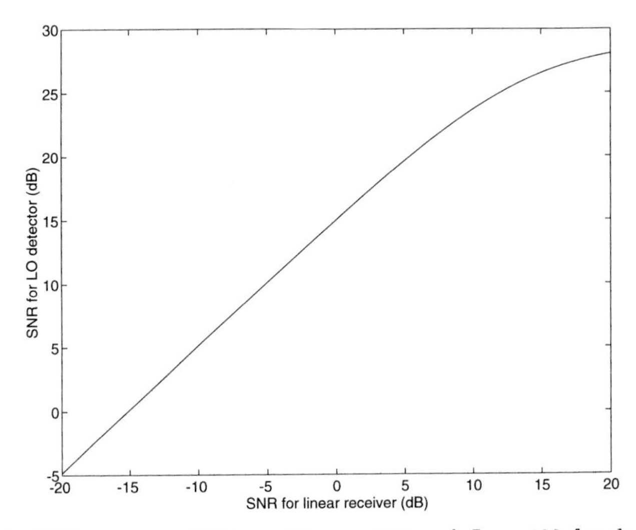

Figure 9.4-3: SNR_{LO} versus SNR_{LR} with $a = 0.99$ and $L_c = 100$ for the system in Example 9.4-1.

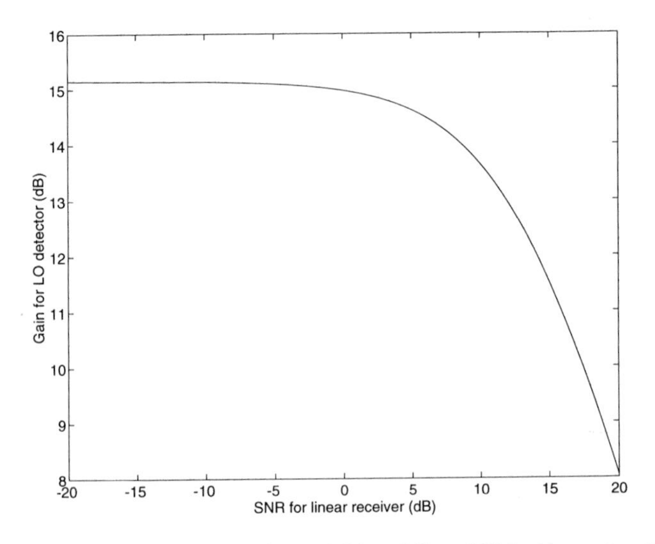

Figure 9.4-4: G_{LO} versus SNR_{LR} with $a = 0.99$ and $L_c = 100$ for the system in Example 9.4-1.

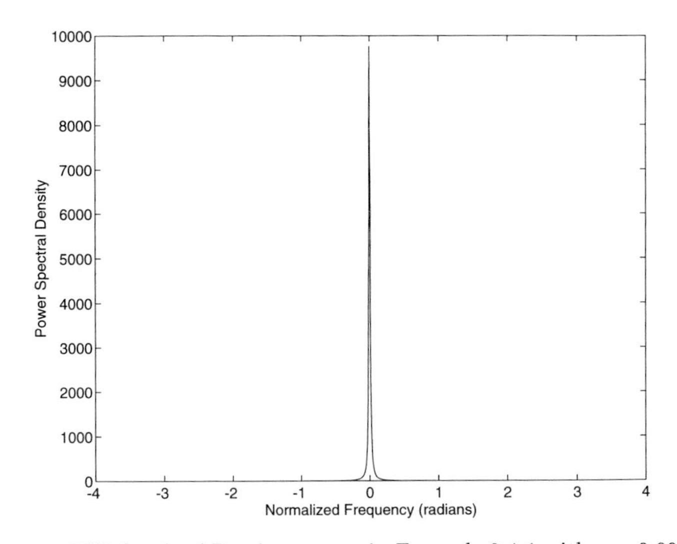

Figure 9.4-5: PSD for the AR noise process in Example 9.4-1 with $a = 0.99$ (assuming $\sigma_v^2 = 1$).

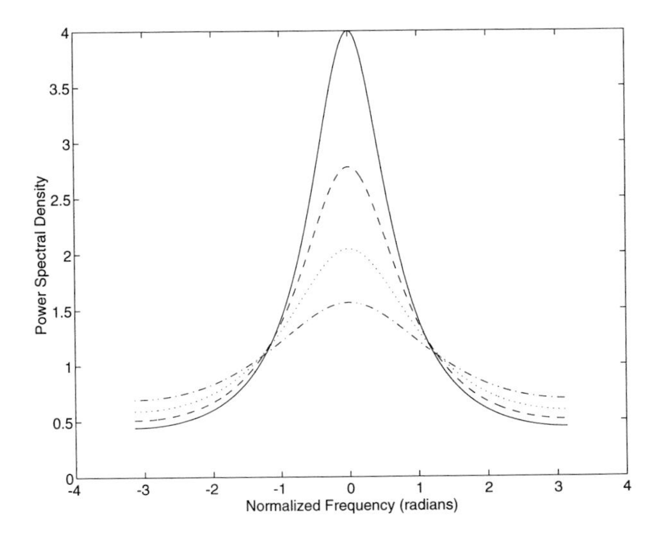

Figure 9.4-6: PSD for the AR noise process in Example 9.4-1 ($-$ $a = 0.5$, - - - $a = 0.4$, . . . $a = 0.3$, - . - $a = 0.2$).

Summary

The focus of this chapter has been *locally optimum* (LO) detection and its application to communication/radar systems and signal processing applications. LO detectors can be divided into two broad classes: those with memory and those without memory. The memoryless detector was derived in Section 9.1 and was shown to depend only on the current sample under consideration. A technique based on the central limit theorem was presented for calculating the probability of error, P_e, for this detector.

The LO detector with memory was the subject of Section 9.2. A number of different detectors were derived, depending on the nature of the noise (or interference) and the type of information signal, i.e., deterministic or random. In its most general form the LO detector with memory can be rather complicated, but is readily simplified if *a priori* information concerning the interference characteristics is available.

Once a type of LO detector is chosen for a given application, the next phase is its implementation. If the *probability density function* (pdf) of the interference is known in advance, the detector can be implemented directly using its describing equations. However, in many cases the pdf is unavailable and must be estimated. A number of these estimation procedures and their use in implementing the required LO nonlinearities were discussed in Section 9.3.

Finally, LO detection in a spread-spectrum communications system was presented in Section 9.4. In particular, a form of the LO detector was derived for the case when the channel noise consists of both interference (possibly non-Gaussian) and background white Gaussian noise. The example provided in this section further clarified the performance gains that can be attained using these methods.

A number of metrics can be used to determine the effectiveness of an LO detector. Two such metrics are P_e and *signal-to-noise ratio* (SNR). When the LO detector is memoryless, and its first and second-order conditional expectations exist and are finite, a central limit theorem technique can be employed to compute P_e (see Section 9.1). When the LO detector has memory, or when computation of the detector statistics is not feasible, the SNR gain may be computed (see Example 9.4-1). A third performance metric that has been widely documented is the *asymptotic relative efficiency* (ARE) ([9-22], Chapter 2 and [9-23]). Simply stated, for a given false alarm probability and as the hypothesis H_1 approaches H_0, the ARE of detector D_1 with respect to detector D_2 is given by

$$\text{ARE}(D_1, D_2) = \lim_{\substack{n_1 \to \infty \\ n_2 \to \infty}} \left\{ \frac{n_2}{n_1} \right\},$$

where n_2 and n_1 are the smallest number of samples required by detectors D_2 and D_1, respectively, to achieve a given probability of false dismissal.[21] Thus, the ef-

[21]The probability of false alarm is the probability of choosing H_1 when H_0 is true; conversely,

fectiveness of the LO detector can be quantified by computing the ARE of the LO detector with respect to the optimum Bayesian detector. It has been shown in [9-23] and [9-7] that for the small-signal case, and under certain conditions, the ARE of the LO detector with respect to the optimum Bayesian detector approaches unity. While a further discussion of the ARE of the LO detector is beyond the scope of this chapter, a number of references are available which present the subject in more detail, including [9-7], [9-22], and [9-23].

Disclaimer

Some of the work presented in this chapter was motivated by research supported under contracts from the United States Department of the Air Force at Rome Laboratory, NY. The views and conclusions contained in this chapter are those of the authors and should not be interpreted as necessarily representing the official policies either expressed or implied of Rome Laboratory or the United States Government.

References

[9-1] J.L. Melsa and D.L. Cohn, *Decision and Estimation Theory*, New York: McGraw-Hill, 1978.

[9-2] H. Stark and J. Woods, *Probability, Random Processes, and Estimation Theory for Engineers*, 2nd edn., Englewood Cliffs, NJ: Prentice-Hall, 1994.

[9-3] R. Hogg and A. Craig, *Introduction to Mathematical Statistics*, 4th edn., New York: Macmillan Publishing Co., 1978.

[9-4] D.M. Hummels and J. Ying, "Locally optimum detection of unknown signals in non-Gaussian Markov noise," *IEEE Midwest Symposium on Circts. and Systs.*, pp. 1098-1101, 1991.

[9-5] A.M. Maras, "Locally optimum Bayes detection in ergodic Markov noise," *IEEE Trans. Information Theory*, vol. IT-40, No. 2, pp. 41-55, Jan. 1994.

[9-6] W. Gardner, "Structural characterization of locally optimum detectors in terms of locally optimum estimators and correlators," *IEEE Trans. Information Theory*, vol. IT-28, No. 6, pp. 924-932, Nov. 1982.

[9-7] D. Middleton, "Canonically optimum threshold detection," *IEEE Trans. Information Theory*, vol. IT -12, No. 2, pp. 230-243, Apr. 1966.

[9-8] D.W. Scott, *Multivariate Density Estimation: Theory, Practice, and Visualization*, New York: John Wiley and Sons, 1992.

[9-9] J.H. Grimm, *et al.*, "Continuous polynomial approximation," *Proc. IEEE MILCOM*, pp. 283-287, Oct. 1993.

the probability of false dismissal is the probability of choosing H_0 when H_1 is true.

[9-10] E. Kreyszig, *Advanced Engineering Mathematics*, 6th edn., New York: John Wiley and Sons, 1988.

[9-11] E. Parzen, "On the estimation of a probability density function and the mode," *Ann. Math. Stat.*, vol. 33, pp. 1065-1076, 1962.

[9-12] R.A. Tabia and J.R. Thompson, *Nonparametric Probability Density Estimation*, Baltimore: Johns Hopkins University Press, 1978.

[9-13] *STAT/LIBRARY: FORTRAN Subroutines for Statistical Analysis*, IMSL, April 1987.

[9-14] B. Efron, *The Jackknife, the Bootstrap, and Other Resampling Plans*, Philadelphia: The Society for Industrial and Applied Mathematics, 1982.

[9-15] D.J. Torrieri, *Principles of Secure Communications*, 2nd edn., Boston: Artech House, 1992.

[9-16] R.L. Pickholtz, *et al.*, "Theory of spread-spectrum communications – a tutorial," *IEEE Trans. Communications*, vol. COM-30, pp. 855-884, May 1982.

[9-17] D.L. Schilling, *et al.*, "Spread-spectrum for commercial communications," *IEEE Communications Magazine*, pp. 66-79, Apr. 1991.

[9-18] G.L. Turin, "Introduction to spread-spectrum antimultipath techniques and their application to urban digital radio," *Proc. IEEE*, pp. 328-353, Mar. 1980.

[9-19] J.G. Proakis, *Digital Communications*, 2nd edn., New York: McGraw-Hill, 1989.

[9-20] A.E. Taylor, *Advanced Calculus*, New York: Ginn and Company, 1955.

[9-21] P.M. Clarkson, *Optimal and Adaptive Signal Processing*, Boca Raton, FL: CRC Press, Inc., 1993.

[9-22] J.D. Gibson and J.L. Melsa, *Introduction to Nonparametric Detection with Applications*, New York: Academic Press, 1975.

[9-23] J. Capon, "On the asymptotic efficiency of locally optimum detectors," *IRE Trans. Information Theory*, pp. 67-71, Apr. 1961.

Additional References Concerning Locally Optimum Detection

[9-24] R.S. Blum and S.A. Kassam, "Optimum distributed detection of weak signals in dependent sensors," *IEEE Trans. Information Theory*, vol. IT-38, No. 3, pp. 1066-1079, May 1992.

[9-25] J.H. Higbie, "Adaptive nonlinear suppression of interference," *Proc. IEEE MILCOM*, pp. 23.3.1-9, 1988.

[9-26] W.E. Jacklin, J.H. Grimm and D.R. Ucci, "The simulation of a two-dimensional spread-spectrum system with locally optimal processing," *Proc. IEEE MILCOM*, pp. 288-292, 1993.

[9-27] S.A. Kassam, *Signal Detection in Non-Gaussian Noise*, New York: Springer-Verlag, 1988.

[9-28] J.F. Kuehls and E. Geraniotis, "Memoryless locally optimum detection of rate lines," *Proc. IEEE MILCOM*, pp. 812-816, 1990.

[9-29] A.B. Martinez and J.B. Thomas, "Detector design using a density fit to non-Gaussian noise," *IEEE Trans. Information Theory*, vol. IT-34, No. 3, pp. 544-550, May/June 1988.

[9-30] A.B. Martinez, *et al.*, "Locally optimal detection in multivariate non-Gaussian noise," *IEEE Trans. Information Theory*, vol. IT-30, No. 6, pp. 815-822, Nov. 1984.

[9-31] H.V. Poor, *An Introduction to Signal Detection and Estimation*, New York: Springer-Verlag, 1988.

[9-32] I. Song and S.A. Kassam, "Locally optimum detection of signals in a generalized observation model: the known signal case," *IEEE Trans. Information Theory*, vol. IT-36, No. 3, pp. 502-515, May 1990.

[9-33] I. Song and S.A. Kassam, "Locally optimum detection of signals in a generalized observation model: the random signal case," *IEEE Trans. Information Theory*, vol. IT-36, No. 3, pp. 516-530, May 1990.

[9-34] I. Song and S.A. Kassam, "Locally optimum rank detection of correlated random signals in additive noise," *IEEE Trans. Information Theory*, vol. IT-38, No. 4., pp. 1311-1322, July 1992.

[9-35] A.D. Spaulding, "Locally optimum and suboptimum detector performance in a non-Gaussian interference environment," *IEEE Trans. Communications*, vol. COM-33, No. 6, pp. 509-517, July 1985.

[9-36] A.D. Spaulding and D. Middleton, "Optimum reception in an impulsive interference environment – part I: coherent detection," *IEEE Trans. Communications*, vol. COM-25, No. 9, pp. 910-923, Sept. 1977.

[9-37] D.R. Ucci, W.E. Jacklin and J.G. Grimm, "Investigation and simulation of nonlinear processors for spread-spectrum receivers," Final Technical Report for Rome Laboratory, USAF, Report No. RL-TR-93-258, 1993.

Chapter 10

Estimation of Probability Density Functions Using Projections Onto Convex Sets

YONGYI YANG and HENRY STARK

Introduction

Knowledge of probability density functions (pdfs) is essential in communication systems engineering and signal processing. For example, the pdf is required in pattern recognition for optimum discrimination of classes [10-1]; it is required to design optimum codes in communication systems [10-2]; it must be known in order to implement optimum detectors (see Chapter 9 in this book); it is essential in designing efficient quantizers [10-3]. Indeed, virtually any optimum design in communications and signal processing requires knowledge of the pdf.

The pdf, however, is rarely known *a priori*. If the pdf is not known, one has several options: (i) take the coward's way out and assume that everything is Gaussian, invoking the central-limit theorem argument; (ii) settle for suboptimum designs using mean-square error type criteria and second order statistics; (iii) settle for suboptimum designs based on heuristic arguments or defaults (e.g., using a uniform quantizer because the pdf of the signal is not known); and (iv) estimate the pdf from data.

In this chapter we propose a method for estimating pdfs based on the method of projections onto convex sets (POCS) described in Chapter 6. While there are other methods of estimating pdfs, POCS has certain unique advantages that other methods do not share. First and foremost, POCS always yields an estimate consistent with the data and prior knowledge. Indeed it is the ease with which POCS can incorporate prior knowledge together with available data that gives it a superior standing among methods for estimating the pdf.

The coarsest and in some ways the least desirable estimate of the pdf is the *histogram.* The histogram is a graphical representation of the relative frequency distribution of data, consisting of vertical rectangles whose width corresponds to a range of outcomes and whose height corresponds to the number of outcomes in that range. In addition to being discontinuous, the histogram's character depends greatly on the size of the data intervals or "bins" as they are sometimes called. A refinement on the histogram idea is the method of *Parzen windows.* In the method of Parzen windows [10-1], the pdf is constructed as the sum of *window kernels* centered at the location of the data. The user supplies both the shape of the window kernel, i.e., the *window function* and the *window width*, the process sometimes being called "window carpentry". Typically an attempt is made to use smooth window functions, e.g., polynomials and try several different widths to see which gives the "best" results. In addition to being highly dependent on a somewhat arbitrary choice of window carpentry, the method of Parzen windows does not easily allow for the incorporation of prior knowledge.

There are many other techniques for estimating pdfs from data. The method of *penalized maximum likelihood* [10-4] attempts to maximize a weighted likelihood function. It is discussed in greater detail in Section 10.3 of this chapter. Other methods use polynomials and splines to smooth the histogram. A method based on Fourier series approximations is described in Chapter 9; it yields a pdf with many continuous derivatives but is not guaranteed to be non-negative. A method for obtaining pdfs which ignores the data altogether (or at least uses the available data in a very indirect way) is the method of *maximum entropy* (ME). The maximum entropy principle states that if we don't know the pdf $f(x)$ of a random variable X, a very good estimate is the function $p(x)$ that maximizes the entropy $H(X) \triangleq -\int_{-\infty}^{\infty} p(x) \ln p(x)dx$ subject to $p(x) \geq 0, \int_{-\infty}^{\infty} p(x)dx = 1, \int_{-\infty}^{\infty} xp(x)dx = \mu$, etc. This method suffers from the obvious drawback that regardless of the distribution of the data, all pdfs constrained by the same number of moments will look the same [10-5]. All-in-all, to the best of the authors' knowledge, only the POCS estimate readily combines prior knowledge and data to yield smooth estimates at some prior determined level of confidence.

The main drawback to using POCS is that it is computationally quite intensive. Nevertheless with the ready availability of powerful number crunchers, the added computational complexity is not a serious problem. Moreover, since the proposed algorithm is iterative, the user can stop the computation when a desired degree of recovery is achieved. Indeed, other than the computational effort associated with the proposed method, the main caveat to its use is the need to apply certain bounds and parameters which must be chosen *a priori.* Fortunately, as we shall demonstrate in subsequent sections and the appendices, these bounds can often be determined by starting out with reasonable assumptions.

We point out that when using POCS to solve a problem, the key to the application being successful is to define the appropriate constraint sets. Thus while POCS has previously been used to recover non-negative pdf-type functions from *moment information*, it turns out that the recovered functions are very sensitive to the accuracy of the moment estimates and the number of moments used [10-6]. Our approach here is entirely different: the constraints on the pdf are directly related to the probability of observing the outcomes in the data pattern.

The chapter is organized as follows. In Section 10.1 we define the constraints for this problem and derive the corresponding sets and projections. Section 10.2 discusses how we might choose the parameters of the constraint sets introduced in Section 10.1. In Section 10.3 we review some competing algorithms for estimating the pdf from available data. Numerical comparisons between these competing algorithms and POCS are given in Section 10.4. A number of appendices dealing with the computations of projectors and other technical details are included at the end.

10.1 Constraint Sets for Probability Density Function Estimation

We begin by defining some basic notation used throughout this chapter. Additional notation is introduced as needed.

a) $f(x)$, the underlying unknown pdf to be recovered;

b) $x_1, x_2, \cdots, x_n$, the sample data obtained as realizations of n independent identically distributed (iid) random variables $X_1, X_2, \cdots, X_n$ respectively, each with pdf $f(x)$;

c) $\bar{x} = \frac{1}{n}\sum_{i=1}^{n} x_i$, the sample mean;

d) $s^2 = \frac{1}{n-1}\sum_{i=1}^{n}(x_i - \bar{x})^2$, the sample variance;

e) μ, the population mean, i.e., $\mu = E[X_i], i = 1, 2, \cdots, n$;

f) σ^2, the population variance, i.e., $\sigma^2 = E[(X_i - \mu)^2], i = 1, 2, \cdots, n$;

g) $z_{\frac{\alpha}{2}}$, the point such that

$$P[Z \geq z_{\frac{\alpha}{2}}] = \frac{\alpha}{2}, \tag{10.1-1}$$

where Z is an N(0,1) random variable and $P[A]$ is the probability of the event A. Equivalently

$$P\left[-z_{\frac{\alpha}{2}} \leq Z \leq z_{\frac{\alpha}{2}}\right] = 1 - \alpha; \tag{10.1-2}$$

h) $y(x)$, an arbitrary element of a set;

i) $g(x)$, an arbitrary element outside a set;

j) $g^*(x)$, the projection of $g(x)$ onto a set C.

As readers of Chapter 6 know, POCS finds solutions at the intersection of constraint sets. Below we describe the constraint sets for this problem.

A. Constraint Based on the Mean

The first constraint set C_1 uses the fact that we can determine a $100(1-\alpha)\%$ confidence interval on μ from the sample data. With α chosen small enough it is reasonable to require that $f(x)$ has membership in the set

$$C_1 \triangleq \left\{ y(x) : \int_\Gamma x\, y(x)\, dx \in [\, \mu_L, \mu_H \,] \right\}, \tag{10.1-3}$$

where μ_L and μ_H are the bounds on an approximate $100(1-\alpha)\%$ confidence interval for μ, i.e.,

$$\mu_L \triangleq \bar{x} - z_{\frac{\alpha}{2}} \frac{s}{\sqrt{n}} \tag{10.1-4}$$

$$\mu_H \triangleq \bar{x} + z_{\frac{\alpha}{2}} \frac{s}{\sqrt{n}} \tag{10.1-5}$$

and $[\, \mu_L, \mu_H \,]$ is the closed interval consisting of all points μ' such that $\mu_L \leq \mu' \leq \mu_H$. The computation of μ_L and μ_H is derived in Appendix 10.B The set Γ in Eq. (10.1-3) is the compact support for $f(x)$ which has to be estimated from the data. We show later how this is done. For the present we assume that $\Gamma = [\, x_m, x_M \,]$ i.e., the set of all x such that $x_m \leq x \leq x_M$ with $x_m < x_M$. For mathematical convenience we shall assume that all functions of interest are real and of compact support Γ. While this assumption is not necessary, it will significantly reduce the amount of mathematical details in the development to follow.

It is quite straightforward to demonstrate that C_1 in Eq. (10.1-3) is convex and closed, and in the interest of brevity, we omit these derivations. They are given in [10-7].

The projection $g^*(x)$ of an arbitrary function (point) $g(x)$ with support Γ onto C_1 is easily computed using standard methods of the calculus of variations. The goal is to find $g^*(x)$ where

$$g^* = P_1 g = \arg[\min_{y \in C_1} \|g - y\|^2\,] \tag{10.1-6}$$

and P_1 is the projector onto C_1. Then the problem is to find

$$\min_y \int_{x_m}^{x_M} [\, g(x) - y(x)\,]^2\, dx \tag{10.1-7}$$

subject to

$$\mu_L \leq \int_{x_m}^{x_M} xy(x)\,dx \leq \mu_H. \tag{10.1-8}$$

Assume first that $\langle x, g\rangle \triangleq \int_{x_m}^{x_M} xg(x)\,dx > \mu_H$. Using the method of Lagrange multipliers we find the minimum of the Lagrange functional

$$J_H(y,\alpha) = \int_{x_m}^{x_M} [\,g(x) - y(x)\,]^2\,dx + \lambda\left[\int_{x_m}^{x_M} xy(x)\,dx - (\mu_H + \delta^2)\right], \tag{10.1-9}$$

where δ is an auxiliary variable and λ is a Lagrange multiplier. As is well known, for problems involving convex functions, the minimum occurs at $\delta = 0$. With this in mind and with $J_H(y) \triangleq J_H(y,0)$ we find that

$$\frac{\partial J_H}{\partial y} = 0 \tag{10.1-10}$$

implies that

$$y(x) = g(x) - \frac{\lambda}{2}x, \tag{10.1-11}$$

subject to

$$\langle x, y\rangle = \langle x, g\rangle - \frac{\lambda}{2}\|x\|_\Gamma^2 = \mu_H, \tag{10.1-12}$$

where

$$\|x\|_\Gamma^2 \triangleq \int_{x_m}^{x_M} x^2\,dx = \frac{1}{3}[x_M^3 - x_m^3]. \tag{10.1-13}$$

From Eq. (10.1-11) and Eq. (10.1-12) we obtain

$$g^*(x) = g(x) - \frac{\langle x, g\rangle - \mu_H}{\|x\|_\Gamma^2}x, \qquad \text{for } \langle x, y\rangle > \mu_H. \tag{10.1-14}$$

For $\mu_L \leq \langle x, g\rangle \leq \mu_H$, g is already in C_1 and the minimum in Eq. (10.1-6) occurs when $y = g^* = g$. Finally, for $\langle x, g\rangle < \mu_L$, we seek a minimum of

$$J_L(y,\beta) = \int_{x_m}^{x_M} [\,g(x) - y(x)\,]^2\,dx + \lambda\left[\int_{x_m}^{x_M} xy(x)\,dx - (\mu_L - \beta^2)\right], \tag{10.1-15}$$

where β is an auxiliary variable. At the minimum, $\beta = 0$. Then repeating the calculation leading up to Eq. (10.1-14) we obtain

$$g^*(x) = g(x) + \frac{\mu_L - \langle x, g\rangle}{\|x\|_\Gamma^2}x, \qquad \text{for } \langle x, y\rangle < \mu_L. \tag{10.1-16}$$

Putting all the results together we obtain the projection onto C_1:

$$g^*(x) = P_1 g = \begin{cases} g(x) - \frac{\langle x,g\rangle - \mu_H}{\|x\|_\Gamma^2}x, & \text{if } \langle x, g\rangle > \mu_H \\ g(x) & \text{if } \mu_L \leq \langle x, g\rangle \leq \mu_H \\ g(x) + \frac{\mu_L - \langle x,g\rangle}{\|x\|_\Gamma^2}x, & \text{if } \langle x, g\rangle < \mu_L \end{cases} \tag{10.1-17}$$

where

$$\langle x, g \rangle = \int_{x_m}^{x_M} x g(x)\, dx. \qquad (10.1\text{-}18)$$

It is not difficult to show that, if we had not made the assumption that all functions had finite support, the resulting projection $\tilde{g}^*$ would have been

$$\tilde{g}^*(x) = P_1 g = \begin{cases} g^*(x) \text{ in } (10.1\text{-}17) & \text{for } x \in \Gamma \\ g(x) & \text{for } x \notin \Gamma. \end{cases} \qquad (10.1\text{-}19)$$

B. Constraints Based on Interval Probabilities

The most important sets that we shall use in the subsequent POCS algorithm are

$$C_{2i} \triangleq \left\{ y(x) : \int_{a_i}^{b_i} y(x)\, dx \in [\, p_i^L, p_i^H \,] \right\}, \quad i = 1, 2, \cdots, K \qquad (10.1\text{-}20)$$

where p_i^L, p_i^H are the bounds determined from the data on a $100(1-\alpha)\%$ confidence interval on

$$p_i = \int_{a_i}^{b_i} f(x)\, dx, \qquad (10.1\text{-}21)$$

where $[a_i, b_i]$ is some interval in the range of the random variable. Note that C_{2i} is a constraint on the unknown pdf to furnish (within limits) the probability of a certain outcome. The determinations of p_i^L and p_i^H are given in Appendix 10.C. The proofs that C_{2i} is convex and closed follows closely the corresponding proofs on the convexity and closedness of C_1 and are omitted. The projector P_{2i} is easily computed using Lagrange multipliers. We seek to find g^* in

$$g^* = P_{2i} g = \arg\{\min_{y \in C_{2i}} \|g - y\|^2\}, \qquad (10.1\text{-}22)$$

equivalently, to find

$$\min_y \|g - y\|^2 \qquad (10.1\text{-}23)$$

subject to

$$p_i^L \le \int_{a_i}^{b_i} y(x)\, dx \le p_i^H. \qquad (10.1\text{-}24)$$

Assume first that $q \triangleq \int_{a_i}^{b_i} g(x)\, dx > p_i^H$. Then let

$$J(y) = \int_{x_m}^{x_M} [\, g(x) - y(x) \,]^2\, dx + \lambda \left[\int_{a_i}^{b_i} y(x)\, dx - p_i^H \right]. \qquad (10.1\text{-}25)$$

In Eq. (10.1-25) we leave out the auxiliary variable since we already know that at the minimum of $J(y)$ in Eq. (10.1-25) the auxiliary variable is zero. Clearly, there being no constraint on $y(x)$ for $x \notin [a_i, b_i] \triangleq \Gamma_i$, we obtain $g^*(x) = g(x)$ for $x \notin [a_i, b_i]$. From

$$\frac{\partial J}{\partial y} = 0 \tag{10.1-26}$$

we immediately obtain

$$y(x) = g(x) - \frac{\lambda}{2} \tag{10.1-27}$$

and from the constraint that

$$\int_{a_i}^{b_i} y(x)\, dx = q - \frac{\lambda}{2}(b_i - a_i) = p_i^H \tag{10.1-28}$$

we obtain, for $q > p_i^H$,

$$g^*(x) = g(x) - \frac{1}{b_i - a_i}(q - p_i^H). \tag{10.1-29}$$

Repeating the computation for $q < p_i^L$, we obtain

$$g^*(x) = g(x) + \frac{1}{b_i - a_i}(p_i^L - q). \tag{10.1-30}$$

If $p_i^L \le q \le p_i^H$, then $g(x)$ is already in the set and its projection is itself. Summarizing all these results, we obtain

$$g^*(x) = P_{2i}g = \begin{cases} g(x) & \text{for } x \notin \Gamma_i \\ g(x) - \frac{1}{b_i - a_i}(q - p_i^H) & \text{for } x \in \Gamma_i, q > p_i^H \\ g(x) & \text{for } x \in \Gamma_i, p_i^L \le q \le p_i^H \\ g(x) + \frac{1}{b_i - a_i}(p_i^L - q) & \text{for } x \in \Gamma_i, q < p_i^L, \end{cases} \tag{10.1-31}$$

where $\Gamma_i \triangleq [a_i, b_i]$. Equation (10.1-31) defines the operator P_{2i} in

$$g^* = P_{2i}g \tag{10.1-32}$$

that projects an arbitrary point onto C_{2i}.

In Eq. (10.1-20) the set C_{2i} is defined on the probability over a single interval. It can, however, be extended to more general cases. One such possibility is on two disjoint intervals. More specifically, let $\Lambda \triangleq [a, b] \cup [c, d]$ where $[a, b]$ and $[c, d]$ are disjoint. On Λ we define a set of the form

$$C_2' \triangleq \left\{ y(x) : \int_\Lambda y(x)\, dx \le p_0 \right\} \tag{10.1-33}$$

for some positive number p_0 within the range between 0 and 1. Applying the same technique as above, we can derive the projection of an arbitrary function $g(x)$ onto C_2' as

$$g^*(x) = P_2'g = \begin{cases} g(x) - \frac{\int_\Lambda g(x)\, dx - p_0}{(b-a)+(d-c)} & \text{if } x \in \Lambda \text{ and } \int_\Lambda g(x)\, dx > p_0 \\ g(x) & \text{otherwise.} \end{cases} \tag{10.1-34}$$

C. Constraints Based on Percentile Points

A special class of interval probabilities is defined on the percentile points. For $0 < \xi < 1$, the ξ percentile point x_ξ is defined as

$$\xi \triangleq P(X \leq x_\xi) = \int_{-\infty}^{x_\xi} f(x)\, dx. \tag{10.1-35}$$

For example, when $\xi = 0.5$, x_ξ is the median of the population. It is shown in Appendix 10.D that an approximate $100(1-\alpha)\%$ confidence interval for x_ξ is

$$\left[x^{\left\lceil n\xi - z_{\frac{\alpha}{2}}\sqrt{n\xi(1-\xi)}\right\rceil}, x^{\left\lfloor n\xi + z_{\frac{\alpha}{2}}\sqrt{n\xi(1-\xi)}\right\rfloor} \right], \tag{10.1-36}$$

where $\lceil x \rceil$ is the smallest integer not smaller than x, $\lfloor x \rfloor$ is the largest integer not larger than x, and the $\{x_i\}$ are ordered so that $x^1 \leq x^2 \leq \cdots \leq x^n$. Given, therefore, any percentile level ξ there is an associated $100(1-\alpha)\%$ confidence interval, say $[\ x_\xi^L, x_\xi^H\ ]$. For a series of levels, say, $\xi_j, j = 1, 2, \cdots, N$, there are correspondingly series of $100(1-\alpha)\%$ confidence intervals $[\ x_{\xi_j}^L, x_{\xi_j}^H\ ]$ for the percentile point x_{ξ_j}, respectively. To convert this observation into a valid constraint set, observe the following: The percentile point x_ξ, i.e.,

$$\int_{-\infty}^{x_\xi} f(x)\, dx = \xi \tag{10.1-37}$$

implies that for $x_\xi^L \leq x_\xi$,

$$\int_{-\infty}^{x_\xi^L} f(x)\, dx \leq \xi. \tag{10.1-38}$$

Likewise

$$\int_{x_\xi}^{\infty} f(x)\, dx = 1 - \xi \tag{10.1-39}$$

implies that for $x_\xi^H \geq x_\xi$,

$$\int_{x_\xi^H}^{\infty} f(x)\, dx \leq 1 - \xi. \tag{10.1-40}$$

If $f(x)$ has compact support $[x_m, x_M]$ then the constraints assume the form

$$\int_{x_m}^{x_\xi^L} f(x)\, dx \leq \xi, \tag{10.1-41}$$

and

$$\int_{x_\xi^H}^{x_M} f(x)\, dx \leq 1 - \xi. \tag{10.1-42}$$

With these observations we can define the appropriate constraint sets based on percentile points x_{ξ_j} as

$$C_{3j} \triangleq \left\{ y(x) : \int_{x_m}^{x_{\xi_j}^L} y(x)\, dx \leq \xi_j \text{ and } \int_{x_{\xi_j}^H}^{x_M} y(x)\, dx \leq 1 - \xi_j \right\}. \tag{10.1-43}$$

The set C_{3j} is closed and convex. Both of these facts can be demonstrated without much difficulty and, in the interest of brevity, we omit these demonstrations. To find the projection onto C_{3j} is very straightforward. We seek to find g^* in

$$g^* = P_{3j}g = \arg\{\min_{y \in C_{3j}} \|g - y\|^2\}. \tag{10.1-44}$$

To this end we construct the Lagrange functional (auxiliary variables have been set to zero *a priori*).

$$\begin{aligned} J(y) &= \int_{x_m}^{x_M} [\, g(x) - y(x)\,]^2 \, dx + \lambda_1 \left[\int_{x_m}^{x^L_{\xi_j}} y(x)\, dx - \xi_j \right] + \lambda_2 \left[\int_{x^H_{\xi_j}}^{x_M} y(x)\, dx - (1 - \xi_j) \right] \\ &= \int_{x_m}^{x^L_{\xi_j}} [\, g(x) - y(x)\,]^2 \, dx + \lambda_1 \left[\int_{x_m}^{x^L_{\xi_j}} y(x)\, dx - \xi_j \right] + \int_{x^L_{\xi_j}}^{x^H_{\xi_j}} [\, g(x) - y(x)\,]^2 \, dx \\ &+ \int_{x^H_{\xi_j}}^{x_M} [\, g(x) - y(x)\,]^2 \, dx + \lambda_2 \left[\int_{x^H_{\xi_j}}^{x_M} y(x)\, dx - (1 - \xi_j) \right]. \end{aligned} \tag{10.1-45}$$

As written in Eq. (10.1-45), the optimization problem has been reduced to extremizing $J(y)$ over three adjusted intervals on the real line, i.e., $[x_m, x^L_{\xi_j}]$, $[x^L_{\xi_j}, x^H_{\xi_j}]$, and $[x^H_{\xi_j}, x_M]$. As such, the projection can be found easily by treating each interval separately from the others. For $[x_m, x^L_{\xi_j}]$, set

$$\frac{\partial J}{\partial y} = 0. \tag{10.1-46}$$

Solving for λ_1 from the constraint in Eq. (10.1-45), we obtain

$$g* = \begin{cases} g + \frac{1}{x^L_{\xi_j} - x_m}[\xi_j - p_g^{L_j}] & \text{if } p_g^{L_j} > \xi_j \\ g & \text{otherwise,} \end{cases} \tag{10.1-47}$$

where

$$p_g^{L_j} \triangleq \int_{x_m}^{x^L_{\xi_j}} g(x)\, dx. \tag{10.1-48}$$

For $[x^L_{\xi_j}, x^H_{\xi_j}]$, since there is no constraint on this interval

$$g^*(x) = g(x). \tag{10.1-49}$$

For $[x^H_{\xi_j}, x_M]$, set

$$\frac{\partial J}{\partial y} = 0 \tag{10.1-50}$$

and solve for λ_2 from the constraint to get

$$g* = \begin{cases} g + \frac{1}{x_M - x^L_{\xi_j}}[1 - \xi_j - p_g^{H_j}] & \text{if } p_g^{H_j} > 1 - \xi_j \\ g & \text{otherwise,} \end{cases} \tag{10.1-51}$$

where

$$p_g^{H_j} \triangleq \int_{x_{\xi_j}^H}^{x_M} g(x)\,dx. \tag{10.1-52}$$

Equations (10.1-47)–(10.1-51) define the projector P_{3j} that projects an arbitrary point g onto C_{3j}.

D. Constraints Based on Non-Negativity, Unit Area, and Finite Support

We force our reconstructed pdf to have finite support $\Gamma = [x_m, x_M]$. The determination of x_m and x_M is discussed later. The non-negativity constraint ensures that our final estimate of the pdf will not display any negative excursion. The appropriate constraint is

$$C_4 \triangleq \{y(x) : y(x) \geq 0 \text{ if } x \in \Gamma \text{ and } y(x) = 0 \text{ otherwise.}\} \tag{10.1-53}$$

This set is easily shown to be convex and closed. The projection onto C_4 of an arbitrary point $g(x)$ is merely the *rectification* of $g(x)$, i.e.,

$$g*(x) = P_4 g = \begin{cases} \frac{g(x)+|g(x)|}{2} & \text{if } x \in \Gamma \\ 0 & \text{otherwise.} \end{cases} \tag{10.1-54}$$

Equation (10.1-54) can be obtained by inspection and the definition of a projection. The constraint set C_5 guarantees that our estimate will have unit area:

$$C_5 \triangleq \left\{ y(x) : \int_\Gamma y(x)\,dx = 1 \right\}. \tag{10.1-55}$$

This set is convex and closed. The projection onto C_5 is computed in a manner quite similar to the projection onto C_{2i}. The result is

$$g* = P_5 g = g + \frac{1}{x_M - x_m}\left[1 - \int_{x_m}^{x_M} g(x)\,dx\right]. \tag{10.1-56}$$

E. Constraints to Enforce Smoothness

1. Smoothness Resulting from First-Order Derivative Constraints

Our underlying assumption on $f(x)$ is that it is continuous in (x_m, x_M), i.e., for every $x \in (x_m, x_M)$:

$$|f(x + \Delta x) - f(x)| \to 0 \text{ as } \Delta x \to 0. \tag{10.1-57}$$

However, in addition to continuity we require that $f(x)$ is reasonably smooth. This requires some constraint on how quickly $f(x)$ and its derivatives change with x. For example the smoothness constraint can be of the form

$$\|Df\|^2 \leq E^2, \tag{10.1-58}$$

where D is a differential operator and E^2 is the energy (norm-squared) of the function Df. A concrete example of a smoothing constraint is

$$\int |f'(x)|^2 \, dx < E^2 \tag{10.1-59}$$

which is of the form of Eq. (10.1-58) with $D = \frac{d}{dx}$ being the derivative operator. Sets of the form

$$C(E) \triangleq \{y(x) : \|Dy\|^2 \leq E^2\} \tag{10.1-60}$$

and their properties and projectors have been derived in the literature [10-8].

Since POCS is realized on a computer, in deriving an algorithm to enforce a smoothness constraint it is necessary to consider the numerical aspects of the problem. This requires that we consider $f(x)$ and $y(x)$ at discrete, equally spaced points, $x_i = x_m + i\Delta x$ and regard $f(x)$ or $y(x)$ as components of vectors, e.g.,

$$\mathbf{f} = (f(x_1), f(x_2), \cdots, f(x_M))^T . \tag{10.1-61}$$

It is convenient to write $f_i \triangleq f(x_i), y_i \triangleq y(x_i)$, etc. Then the analog of Eq. (10.1-60) for the discrete case is

$$C_6 \triangleq \left\{ \mathbf{y} : \sum_i (y_{i+1} - y_i)^2 \leq E^2 \right\}, \tag{10.1-62}$$

where

$$\mathbf{y} \triangleq (y_1, y_2, \cdots, y_M)^T, \tag{10.1-63}$$

and where the derivative operator has been replaced by the first difference operator. Obtaining a closed form solution for projection onto the set C_6 in Eq. (10.1-62) is not an easy task. Hence, without significant loss of generality, we proceed by a somewhat different approach. Let

$$C_{6e} \triangleq \left\{ \mathbf{y} : \sum_{i \ even} (y_{i+1} - y_i)^2 \leq E_{even}^2 \right\} \tag{10.1-64}$$

$$C_{6o} \triangleq \left\{ \mathbf{y} : \sum_{i \ odd} (y_{i+1} - y_i)^2 \leq E_{odd}^2 \right\}. \tag{10.1-65}$$

We would expect the sums

$$S_{even} \triangleq \sum_{i \ even} (y_{i+1} - y_i)^2 \tag{10.1-66}$$

and

$$S_{odd} \triangleq \sum_{i \ odd} (y_{i+1} - y_i)^2 \tag{10.1-67}$$

to be relatively small and have approximately equal value. The sets C_{6e} and C_{6o} are easily shown to be convex and closed. To find the projections onto these sets we use Lagrange multipliers as in the continuous case. Thus, for example, for C_{6e} we write the Lagrange functional

$$J = \|\mathbf{g} - \mathbf{y}\|^2 + \lambda \left(S_{even} - E_{even}^2 \right) \tag{10.1-68}$$

and compute

$$\frac{\partial J}{\partial y_i} = \frac{\partial J}{\partial y_{i+1}} = 0 \quad \text{for } i \text{ even.} \tag{10.1-69}$$

The details are given in Appendix 10.E. The results are, for i even,

$$\mathbf{g}_i^* \triangleq \begin{pmatrix} g_i^* \\ g_{i+1}^* \end{pmatrix} = \begin{pmatrix} 1-\alpha_e & \alpha_e \\ \alpha_e & 1-\alpha_e \end{pmatrix} \begin{pmatrix} g_i \\ g_{i+1} \end{pmatrix} \triangleq P_{6e}\mathbf{g}_i, \tag{10.1-70}$$

where $\mathbf{g}_i \triangleq (g_i, g_{i+1})^T$ and

$$\alpha_e \triangleq \frac{1}{2}\left(1 - \frac{E_{even}}{\sqrt{S_{even}}}\right). \tag{10.1-71}$$

In Eq. (10.1-70) we show the projection of only two components of $\mathbf{g}$. The projection of the entire vector $\mathbf{g}$ onto C_{6e} is

$$\begin{pmatrix} g_1^* \\ g_2^* \\ g_3^* \\ g_4^* \\ g_5^* \\ \cdot \\ \cdot \\ \cdot \\ g_M^* \end{pmatrix} = \begin{pmatrix} 1 & 0 & 0 & 0 & 0 & 0 & \cdots & 0 \\ 0 & 1-\alpha_e & \alpha_e & 0 & 0 & 0 & \cdots & 0 \\ 0 & \alpha_e & 1-\alpha_e & 0 & 0 & 0 & \cdots & 0 \\ 0 & 0 & 0 & 1-\alpha_e & \alpha_e & 0 & \cdots & 0 \\ 0 & 0 & 0 & \alpha_e & 1-\alpha_e & 0 & \cdots & 0 \\ \cdot & \cdot & \cdot & \cdot & \cdot & \cdot & \cdot & \cdot \\ \cdot & \cdot & \cdot & \cdot & \cdot & \cdot & \cdot & \cdot \\ \cdot & \cdot & \cdot & \cdot & \cdot & \cdot & \cdot & \cdot \\ \cdot & \cdot & \cdot & \cdot & \cdot & \cdot & \cdot & \cdot \end{pmatrix} \begin{pmatrix} g_1 \\ g_2 \\ g_3 \\ g_4 \\ g_5 \\ \cdot \\ \cdot \\ \cdot \\ g_M \end{pmatrix}. \tag{10.1-72}$$

A similar calculation for C_{6o} yields

$$\mathbf{g}_i^* \triangleq \begin{pmatrix} g_i^* \\ g_{i+1}^* \end{pmatrix} = \begin{pmatrix} 1-\alpha_o & \alpha_o \\ \alpha_o & 1-\alpha_o \end{pmatrix} \begin{pmatrix} g_i \\ g_{i+1} \end{pmatrix} \triangleq P_{6o}\mathbf{g}_i, \tag{10.1-73}$$

where

$$\alpha_o \triangleq \frac{1}{2}\left(1 - \frac{E_{odd}}{\sqrt{S_{odd}}}\right). \tag{10.1-74}$$

The projection can be written as

$$\begin{pmatrix} g_1^* \\ g_2^* \\ g_3^* \\ g_4^* \\ \cdot \\ \cdot \\ \cdot \\ g_M^* \end{pmatrix} = \begin{pmatrix} 1-\alpha_o & \alpha_o & 0 & 0 & 0 & \cdots & 0 \\ \alpha_o & 1-\alpha_o & 0 & 0 & 0 & \cdots & 0 \\ 0 & 0 & 1-\alpha_o & \alpha_o & 0 & \cdots & 0 \\ 0 & 0 & \alpha_o & 1-\alpha_o & 0 & \cdots & 0 \\ \cdot & \cdot & \cdot & \cdot & \cdot & \cdot & \cdot \\ \cdot & \cdot & \cdot & \cdot & \cdot & \cdot & \cdot \\ \cdot & \cdot & \cdot & \cdot & \cdot & \cdot & \cdot \\ \cdot & \cdot & \cdot & \cdot & \cdot & \cdot & \cdot \end{pmatrix} \begin{pmatrix} g_1 \\ g_2 \\ g_3 \\ g_4 \\ \cdot \\ \cdot \\ \cdot \\ g_M \end{pmatrix}. \quad (10.1\text{-}75)$$

2. Smoothness Resulting from Second Derivative Constraint

With D in Eq. (10.1-58) being the second derivative operator, i.e.,

$$Df \triangleq \frac{d^2 f}{dx^2} = f''(x) \quad (10.1\text{-}76)$$

we arrive at another, quite powerful, constraint set

$$C(\varepsilon) \triangleq \left\{ y(x) : \|Dy\|^2 \leq \varepsilon^2 \right\}. \quad (10.1\text{-}77)$$

When the constraint is imposed on the sampled values of $y(x)$, i.e., $y_1 = y(x_1), y_2 = y(x_2)$, etc., the constraint set $C(\varepsilon)$ takes the form

$$C_{SD}(\varepsilon) \triangleq \left\{ \mathbf{y} : \sum_{i=2}^{M-1} (2y_i - y_{i+1} - y_{i-1})^2 \leq \varepsilon^2 \right\}, \quad (10.1\text{-}78)$$

where the second derivative operator is replaced by the second central difference

$$Dy = (y_{i+1} - y_i) - (y_i - y_{i-1}) = (2y_i - y_{i+1} - y_{i-1}) \quad (10.1\text{-}79)$$

and

$$\|Dy\|^2 = \sum_{i=2}^{M-1} (2y_i - y_{i+1} - y_{i-1})^2. \quad (10.1\text{-}80)$$

As in the set C_6 in Eq. (10.1-62), computing the projection onto $C_{SD}(\varepsilon)$ is not easy and we proceed in a different way. Note that

$$\begin{aligned} \sum_{i=2}^{M-1} (2y_i - y_{i+1} - y_{i-1})^2 &= \sum_{j=1}^{\lfloor M/3 \rfloor} (2y_{3j-1} - y_{3j-2} - y_{3j})^2 \text{ (term 1)} \\ &+ \sum_{j=1}^{\lfloor (M-1)/3 \rfloor} (2y_{3j} - y_{3j-1} - y_{3j-2})^2 \text{ (term 2)} \\ &+ \sum_{j=1}^{\lfloor (M-2)/3 \rfloor} (2y_{3j+1} - y_{3j} - y_{3j+2})^2 \text{ (term 3)}, \end{aligned} \quad (10.1\text{-}81)$$

where $\lfloor x \rfloor$ is the largest integer not larger than x. For example with $M = 10$:

$$\begin{aligned}\sum_{i=2}^{M-1}(2y_i - y_{i+1} - y_{i-1})^2 &= \sum_{i=2,5,8}(2y_i - y_{i+1} - y_{i-1})^2 \\ &\quad + \sum_{i=3,6,9}(2y_i - y_{i+1} - y_{i-1})^2 \\ &\quad + \sum_{i=4,7}(2y_i - y_{i+1} - y_{i-1})^2 \\ &= \sum_{j=1}^{3}(2y_{3j-1} - y_{3j-2} - y_{3j})^2 \\ &\quad + \sum_{j=1}^{3}(2y_{3j} - y_{3j-1} - y_{3j-2})^2 \\ &\quad + \sum_{j=1}^{2}(2y_{3j+1} - y_{3j} - y_{3j+2})^2. \end{aligned} \tag{10.1-82}$$

The alternative procedure involves constraining each of the subsums, i.e., term $1 \leq \varepsilon_1^2$, term $2 \leq \varepsilon_2^2$, term $3 \leq \varepsilon_3^2$. Typically ε_i^2 is set at $\varepsilon^2/3$ for $i = 1, 2, 3$. Thus we have created three new sets

$$C_{7,1}(\varepsilon_1) \triangleq \left\{ y(x) : \sum_{j=1}^{\lfloor M/3 \rfloor}(2y_{3j-1} - y_{3j-2} - y_{3j})^2 \leq \varepsilon_1^2 \right\} \tag{10.1-83}$$

$$C_{7,2}(\varepsilon_2) \triangleq \left\{ y(x) : \sum_{j=1}^{\lfloor (M-1)/3 \rfloor}(2y_{3j} - y_{3j-1} - y_{3j-2})^2 \leq \varepsilon_2^2 \right\} \tag{10.1-84}$$

$$C_{7,3}(\varepsilon_3) \triangleq \left\{ y(x) : \sum_{j=1}^{\lfloor (M-2)/3 \rfloor}(2y_{3j+1} - y_{3j} - y_{3j+2})^2 \leq \varepsilon_3^2 \right\}. \tag{10.1-85}$$

It is not difficult to show that these sets are convex and closed. The projections for these sets are calculated in Appendix 10.F The form of the projections is the same for all three sets $C_{7,i}(\varepsilon_i), i = 1, 2, 3$. We give only the projection onto $C_{7,1}(\varepsilon_1)$ for brevity:

$$\begin{pmatrix} g_1^* \\ g_2^* \\ g_3^* \\ g_4^* \\ g_5^* \\ g_6^* \\ \cdot \\ \cdot \\ \cdot \\ g_M^* \end{pmatrix} = \begin{pmatrix} 1-\theta & 2\theta & -\theta & 0 & 0 & 0 & 0 & \cdots & 0 \\ 2\theta & 1-4\theta & 2\theta & 0 & 0 & 0 & 0 & \cdots & 0 \\ -\theta & 2\theta & 1-\theta & 0 & 0 & 0 & 0 & \cdots & 0 \\ 0 & 0 & 0 & 1-\theta & 2\theta & -\theta & 0 & \cdots & 0 \\ 0 & 0 & 0 & 2\theta & 1-4\theta & 2\theta & 0 & \cdots & 0 \\ 0 & 0 & 0 & -\theta & 2\theta & 1-\theta & 0 & \cdots & 0 \\ \cdot & \cdot & \cdot & \cdot & \cdot & \cdot & \cdot & \cdot & \cdot \\ \cdot & \cdot & \cdot & \cdot & \cdot & \cdot & \cdot & \cdot & \cdot \\ \cdot & \cdot & \cdot & \cdot & \cdot & \cdot & \cdot & \cdot & \cdot \\ \cdot & \cdot & \cdot & \cdot & \cdot & \cdot & \cdot & \cdot & \cdot \end{pmatrix} \begin{pmatrix} g_1 \\ g_2 \\ g_3 \\ g_4 \\ g_5 \\ g_6 \\ \cdot \\ \cdot \\ \cdot \\ g_M \end{pmatrix} \tag{10.1-86}$$

where

$$\theta \triangleq \frac{1}{6}\left[1 - \frac{\varepsilon_1}{\sqrt{\sum_{j=1}^{\lfloor M/3 \rfloor}(y_{3j-2} - 2y_{3j-1} + y_{3j})^2}}\right].$$

We conclude this section by summarizing the complete POCS algorithm for pdf estimation below:

(a) Starting with data $x_1, x_2, \cdots, x_n$, generate the histogram and let this be the initial vector estimate of the unknown pdf vector $\mathbf{f}$. Call this initial estimate $\mathbf{g_0}$.

(b) Apply the following composition of projection operators to $\mathbf{g_0}$ to generate $\mathbf{g_1}$:

$$\mathbf{g_1} = \mathbf{P}\mathbf{g_0}, \quad \mathbf{g}_0 \text{ arbitrary}, \tag{10.1-87}$$

where

$$\mathbf{P} \triangleq P_1 P_{2,1} \cdots P_{2,K} P_{3,1} \cdots P_{3,N} P_4 P_5 P_{6e} P_{6o} \tag{10.1-88}$$

if first derivative smoothness is enforced or

$$\mathbf{P} \triangleq P_1 P_{2,1} \cdots P_{2,K} P_{3,1} \cdots P_{3,N} P_4 P_5 P_{7,1} P_{7,2} P_{7,3} \tag{10.1-89}$$

if second derivative smoothness is enforced.

(c) Apply the operator $\mathbf{P}$ in Eq. (10.1-88) or (10.1-89) in

$$\mathbf{g_{k+1}} = \mathbf{P}\mathbf{g_k} \tag{10.1-90}$$

until convergence is achieved. While other initial vectors can be used, in practice a reasonable choice for $\mathbf{g}_0$ is the histogram of the data. The order of the projectors is somewhat immaterial even though different orders will most likely give different results. This is because all the results will have one thing in common: they will satisfy all the constraints imposed upon the solution.

10.2 Determining the Parameters in Constraint Sets

A. Prescribing the Support

In the development of the POCS algorithm in Section 10.1 we assumed that the underlying pdf $f(x)$ has finite support $[x_m, x_M]$. This does not, however, limit the generality of the proposed POCS algorithm since any pdf function with infinite support can be approximated to an arbitrary degree of accuracy by a function with finite support. In practice, the support information of the pdf under estimation often comes from the specific application. For example, in the study of the distribution of the human life span, it would be safe enough to assume that it is well within the range of 0 and 200 in years (if not, replace 200 by any large number one can imagine). When the support information, i.e., $[x_m, x_M]$ is not available, it has to be estimated from the data. There are a number of ways of doing this, some more successful than others. Let $x^1, x^2, \cdots, x^n$ represent the observed samples, ranked in size such that $x^1 \leq x^2 \leq \cdots \leq x^n$. One way is to let $x_m = x^1$ and $x_M = x^n$. But this is risky as it assumes that future observed data will never be less than x^1 or greater than x^n. Another seemingly reasonable way is to find a $100(1-\alpha)\%$ confidence interval for $[\,x_m,\ x_M\,]$. However, this requires knowing $f(x)$ which is assumed unknown. Since the actual choices of x_m and x_M are not critical, we use a heuristic approach:

$$x_m = x^1 - \frac{x^n - x^1}{2} \tag{10.2-1}$$

$$x_M = x^n + \frac{x^n - x^1}{2} \tag{10.2-2}$$

and hence

$$x_M - x_m = 2(x^n - x^1). \tag{10.2-3}$$

Here x_m and x_M are chosen in such a way that the estimated support is twice as large as the sample support.

B. Choosing the Constraint Intervals $[\,a_i,\ b_i\,]$

We stated earlier that the most important constraint sets in pdf estimation by POCS were of the form

$$C_{2i} \triangleq \left\{ y(x) : \int_{a_i}^{b_i} y(x)\, dx \in [\, p_i^L, p_i^H \,] \right\} \tag{10.2-4}$$

for $i = 1, 2, \cdots, K$. We now discuss how to reasonably choose the intervals $[\,a_i,\ b_i\,]$. First observe that if the interval $[\,a_i,\ b_i\,]$ becomes larger the effectiveness of the constraint is reduced; indeed as $[\,a_i,\ b_i\,] \longrightarrow (-\infty, \infty)$, the constraint serves merely to normalize $y(x)$ and is satisfied by every pdf. At the other extreme, we might want to favor small intervals but when they are too small it may lead to many intervals not containing any samples at all. These considerations favor an adaptive approach which guarantees that a certain number of samples will appear in each interval $[\,a_i,\ b_i\,]$. Such an approach is described below.

Let $m \simeq \lfloor \sqrt{n} \rfloor$, and $K_0 = \lceil \frac{n}{m} \rceil$ where $\lfloor x \rfloor$ is the largest integer no larger than x, and where $\lceil x \rceil$ is the least integer no less than x. Then the intervals $[\,a_i,\ b_i\,]$ are determined in the following way:

(i) For $i = 1$,

$$a_1 = x^1 - \frac{x^2 - x^1}{2} \quad \text{and} \quad b_1 = x^m + \frac{x^{m+1} - x^m}{2}; \tag{10.2-5}$$

(ii) For $i = 2, 3, \cdots, K_0 - 1$,

$$a_i = b_{i-1} \quad \text{and} \quad b_i = x^{im} + \frac{x^{im+1} - x^{im}}{2}; \tag{10.2-6}$$

(iii) For $i = K_0$,

$$a_{K_0} = b_{K_0-1} \quad \text{and} \quad b_{K_0} = x^n + \frac{x^n - x^{n-1}}{2}. \tag{10.2-7}$$

In addition to the intervals defined in Eqs. (10.2-5)–(10.2-7), there are two tail intervals left within the support $[\,x_m,\ x_M\,]$, namely $[\,x_m,\ a_1\,]$ and $[\,b_{K_0},\ x_M\,]$ where no samples are observed. These two intervals can be treated as one in the form given by Eq. (10.1-33). Note that the absence of samples in these two intervals does not necessarily imply a zero probability of their occurrence. Therefore we use a $100(1-\alpha)\%$ interval estimation as discussed in Appendix 10.C.

In this fashion we obtain intervals that contain m points, although the last interval may have fewer samples. The advantage of this approach is clear: it simplifies the task of determining confidence intervals for interval probabilities. To illustrate the construction of the intervals we use the example in Fig. 10.2-1 a). Here, $n = 11$, then $m = 3$ and $K_0 = 4$. The intervals are shown in Fig. 10.2-1 b).

It is necessary to point out that the construction suggested here is essentially a guideline and is not meant to exclude other reasonable approaches. In this approach, the intervals are determined based on the observed data. In non-adaptive methods, the intervals might be determined *prior* to the observations. For example, we can use equally divided intervals. Also, the intervals need not necessarily be mutually exclusive. Therefore it is possible to use overlapping intervals such that the same data points can be used in several intervals rather than one. Nevertheless, to reduce the generation of excessive data and keep our focus on the POCS algorithm, we will use the simple scheme described above for the numerical examples in Section 10.4.

C. Choosing Bounds on the First-Derivative Smoothing Constraint

Equation (10.1-62) describes a smoothing constraint on the first-difference variation which is a numerical approximation for the first-derivative variation. We rewrite Eq. (10.1-62) as

$$C \triangleq \left\{ \mathbf{f} : \sum_i (f_{i+1} - f_i)^2 \leq E^2 \right\}, \tag{10.2-8}$$

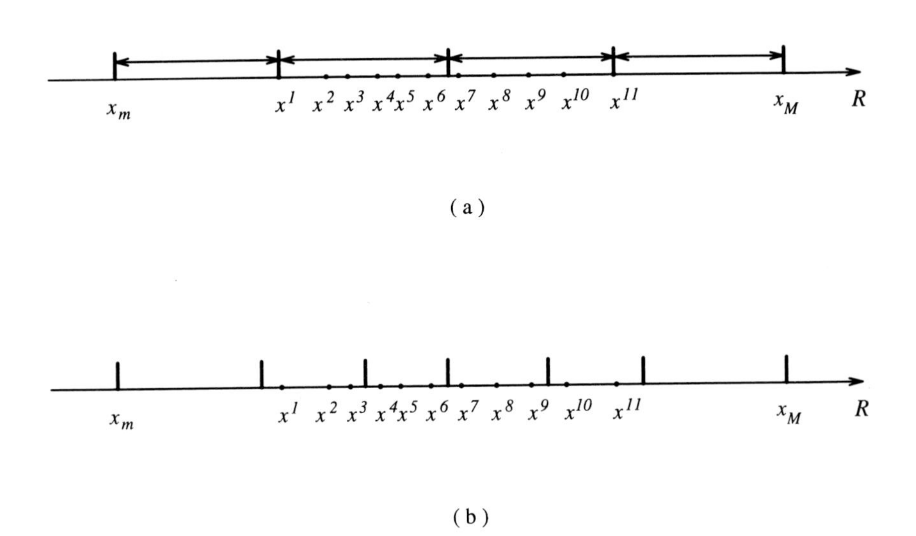

Figure 10.2-1: a) Estimating the population support $[x_m, x_M]$ from the sample support $[x^1, x^{11}]$; b) division of the support $[x_m, x_M]$ into probability constraint intervals.

where the variable **f** has replaced the variable **y** merely to remind us that it is the smoothing of a pdf that we are concerned with, and the usual symbol for a pdf is **f**. What is an appropriate choice for E^2? To solve this problem consider Fig. 10.2-2 which shows a typical histogram $H(x)$ of the data, with the original pdf $f(x)$ superimposed.

We see that $f(x)$ *runs through* $H(x)$ in a *smooth* fashion. It seems meaningful to estimate the total variation of $f(x)$ based on $H(x)$. Toward this end, consider Fig. 10.2-3 in which a small portion of a histogram $H(x)$ is superimposed with a small portion of the desired smooth pdf $f(x)$. We note that $H(x)$ goes from $H(a)$ to $H(b)$ in a "rough", i.e., stepwise, fashion while $f(x)$ goes from $f(a)$ to $f(b)$ in a "smooth" fashion. Assume that $H(x)$ and $f(x)$ increase by the same amount in the interval $[a, b]$, i.e, $H(a) = f(a)$ and $H(b) = f(b)$. Assume further that the "smooth" pdf $f(x)$ in the interval $[a, b]$ is expressed by a vector, say, $(f_1, f_2, \cdots, f_l)^T$, where f_j are values of the pdf on the grid. Then the first-difference variation is given by

$$T = \sum_{i=1}^{l-1}(f_{i+1} - f_i)^2 = \sum_{i=1}^{l-1}(\Delta f_i)^2, \tag{10.2-9}$$

where $\Delta f_i \triangleq f_{i+1} - f_i$. What choice of Δf_i minimizes T? First note that we are constrained by

$$\sum_{i=1}^{l-1} \Delta f_i = f_l - f_1 = f(b) - f(a). \tag{10.2-10}$$

Then writing the functional

$$J = \sum_{i=1}^{l-1}(\Delta f_i)^2 + \lambda\left\{\sum_{i=1}^{l-1} \Delta f_i - [f(b) - f(a)]\right\}, \tag{10.2-11}$$

where λ is a Lagrange multiplier, and solving

$$\frac{\partial J}{\partial \Delta f_i} = 2\Delta f_i + \lambda = 0 \tag{10.2-12}$$

for a minimum, yields

$$\Delta f_i = -\frac{\lambda}{2}, \quad \text{independent of } i. \tag{10.2-13}$$

Substituting into Eq. (10.2-9) shows that the optimum solution is obtained by choosing $\Delta f_i = \Delta f$, i.e., uniform increments. This yields the smallest T as

$$T_{min} = \sum_{i=1}^{l-1}(\Delta f)^2 = \left[\frac{f(b) - f(a)}{l-1}\right]^2 \cdot (l-1) = \frac{[f(b) - f(a)]^2}{l-1}. \tag{10.2-14}$$

But $f(b) - f(a) = H(b) - H(a)$, which furnishes

$$T_{min} = \frac{[H(b) - H(a)]^2}{l-1}. \tag{10.2-15}$$

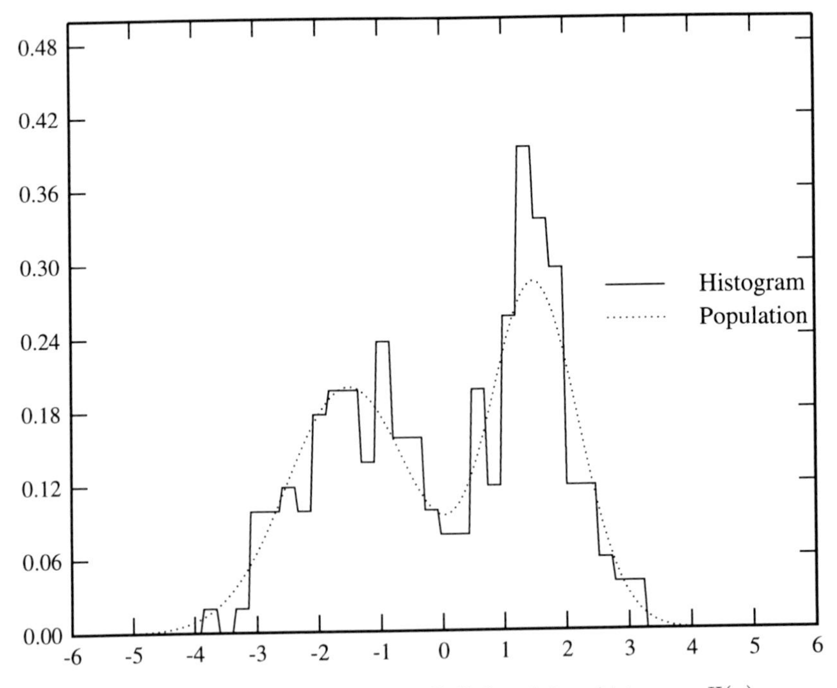

Figure 10.2-2: Population pdf $f(x)$ and data histogram $H(x)$.

Equation (10.2-15) represents a first variation only $1/(l-1)$ as large as the first variation of the non-smooth histogram for the same increment (or decrement) over $[a, b]$.

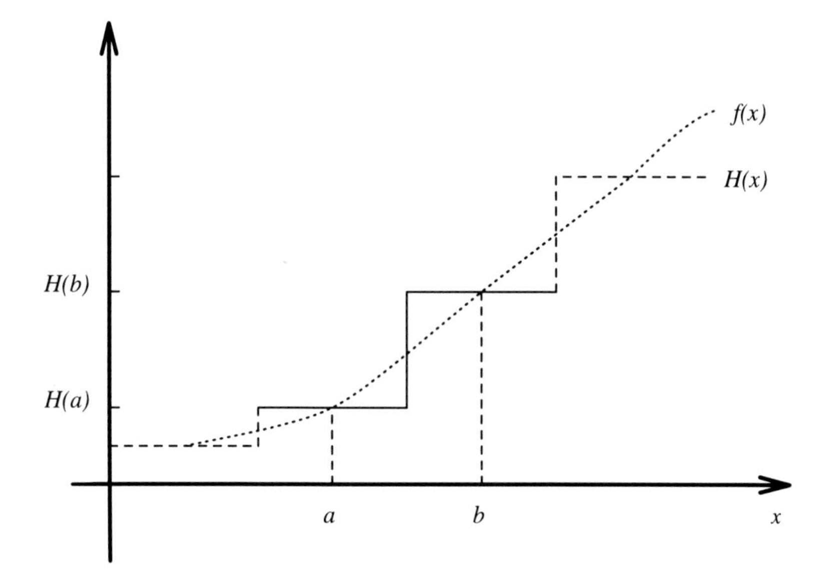

Figure 10.2-3: A small portion of a histogram $H(x)$ with a desired smooth curve $f(x)$ superimposed over the interval $[a, b]$.

In deriving T_{min} we have assumed that $f(x)$ and $H(x)$ have the same increment (or decrement) over the interval $[a, b]$. However, this does not necessarily hold in general. In Fig. 10.2-2 for example, in some intervals $H(x)$ has much greater variation than $f(x)$. In this case, the ratio between the two variations could be far less than $1/(l-1)$. Nevertheless, $1/(l-1)$ can still be used as a guideline. Therefore, to estimate E^2 we suggest the following procedure which is based on the above reasoning:

(i) compute the first variation of the histogram, call this E_H^2;

(ii) compute the first variation of the desired pdf from

$$E^2 \simeq \frac{E_H^2}{l}, \tag{10.2-16}$$

where l is the average number of grid points on which $\mathbf{f}$ is defined per histogram interval;

(iii) there being no reason why E^2_{even} and E^2_{odd}, in Eqs. (10.1-64)–(10.1-65) should be significantly different, choose

$$E^2_{even} = E^2_{odd} = \frac{1}{2}\left(\frac{E_H^2}{l}\right); \tag{10.2-17}$$

(iv) finally, allow for flexibility by replacing $\frac{1}{l}$ with a coefficient r whose initial value is $r \simeq \frac{1}{l}$. The value of r is varied until a satisfying result is achieved.

D. Choosing Bounds on the Second-Derivative Smoothing Constraint

Equation (10.1-78) defines the second-derivative smoothing constraint as

$$C_{SD}(\varepsilon) \triangleq \left\{ \mathbf{f} : \sum_i (2f_i - f_{i+1} - f_{i-1})^2 \leq \varepsilon^2 \right\}, \tag{10.2-18}$$

where the notation has been slightly changed, i.e., $\mathbf{y} \to \mathbf{f}$ to emphasize that it is a pdf we seek to smooth. How do we choose ε^2? We use a similar approach as in deriving E^2 for the first-derivative smoothness set in Eq. (10.1-62). In Fig. 10.2-4 a portion of a histogram $H(x)$ is shown superimposed with a small portion of the desired smooth reconstructed pdf $f(x)$.

Assume that $f(x)$ passes through the midpoint of each interval of the histogram $H(x)$, i.e, we impose the constraint $H(a) = f(a)$, $H(b) = f(b)$ and $H(c) = f(c)$. For simplicity we assume uniform intervals for $H(x)$. Assume further that in the interval $[a,\ c]$ the "smooth" pdf $f(x)$ is expressed by a vector, say, $(f_1, f_2, \cdots, f_{2l})^T$, where f_j are values of the pdf on the grid. Then the second-difference variation is given by

$$T = \sum_{i=2}^{2l-1} (2f_i - f_{i+1} - f_{i-1})^2. \tag{10.2-19}$$

What is the minimal value of T with $f(x)$ satisfying its constraint? To solve this problem, we use the following result [10-9]:

Proposition Let $f(t)$ be a function twice differentiable on the interval $[a,\ c]$ with $f(a) = H(a), f(b) = H(b)$ and $f(c) = H(c)$ for $a < b < c$, then the quantity

$$I \triangleq \int_a^c |f''(t)|^2 dt \tag{10.2-20}$$

is minimized by the quadratic curve $y(t)$ determined *uniquely* by the three points $(a,\ H(a))$, $(b,\ H(b))$, and $(c,\ H(c))$.

In the interests of brevity, the proof is omitted here. ∎

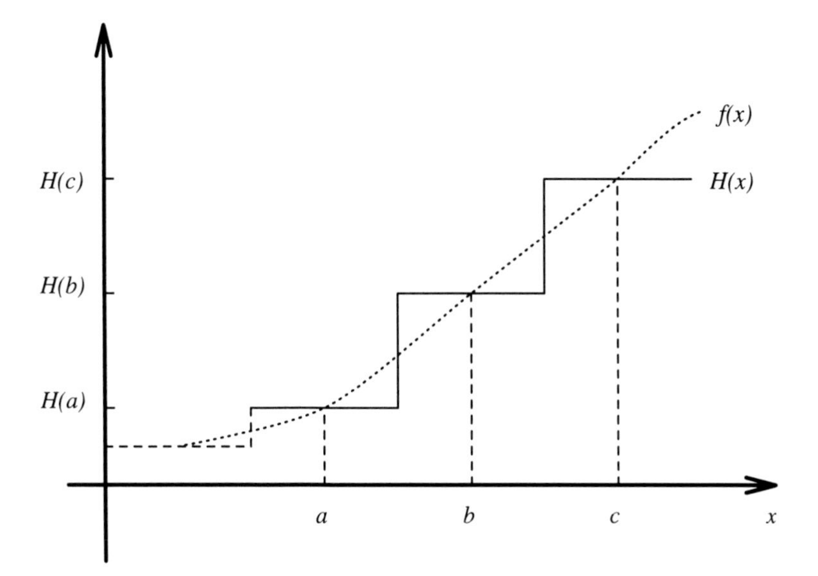

Figure 10.2-4: A small portion of a histogram $H(x)$ with a desired smooth curve $f(x)$ superimposed over the interval $[a, c]$.

The above proposition says that the smooth curve $f(x)$ is of the form

$$f(x) = Ax^2 + Bx + C. \tag{10.2-21}$$

Then

$$f''(x) = A \quad \text{and} \quad I = \int_a^c |f''(x)|^2 dx = A^2(c-a). \tag{10.2-22}$$

Furthermore, from the constraints we have

$$\begin{aligned} H(a) &= Aa^2 + Ba + C & (10.2\text{-}23) \\ H(b) &= Ab^2 + Bb + C & (10.2\text{-}24) \\ H(c) &= Ac^2 + Bc + C. & (10.2\text{-}25) \end{aligned}$$

Note that from $b - a = c - b = (c-a)/2$, it follows that

$$\begin{aligned} 2H(b) - H(a) - H(c) &= A(2b^2 - a^2 - c^2) + B(2b - a - c) & (10.2\text{-}26) \\ &= A\left[(b-a)(b+a) - (c-b)(b+c)\right] & (10.2\text{-}27) \\ &= A(b-a)(b+a-b-c) & (10.2\text{-}28) \\ &= -\frac{1}{2}A(c-a)^2. & (10.2\text{-}29) \end{aligned}$$

Thus,

$$A = -\frac{2[2H(b) - H(a) - H(c)]}{(c-a)^2}. \tag{10.2-30}$$

From Eq. (10.2-22),

$$I = \frac{4\left[2H(b) - H(a) - H(c)\right]^2}{(c-a)^3}. \tag{10.2-31}$$

On the other hand, by numerical differentiation and integration we have

$$\begin{aligned} I &= \int_a^c |f''(x)|^2 dx & (10.2\text{-}32) \\ &\simeq \sum_{i=2}^{2l-1} \left| \frac{2f_i - f_{i+1} - f_{i-1}}{(\Delta t)^2} \right|^2 \cdot \Delta t & (10.2\text{-}33) \\ &= \frac{1}{(\Delta t)^3} \sum_{i=2}^{2l-1} (2f_i - f_{i+1} - f_{i-1})^2 & (10.2\text{-}34) \\ &= \frac{T}{(\Delta t)^3}, & (10.2\text{-}35) \end{aligned}$$

where $\Delta t \triangleq (c-a)/2l$. Comparing Eq. (10.2-31) with Eq. (10.2-35), we have

$$\begin{aligned} T &= \frac{4\left[2H(b) - H(a) - H(c)\right]^2}{(c-a)^3} \cdot (\Delta t)^3 & (10.2\text{-}36) \\ &= \frac{\left[2H(b) - H(a) - H(c)\right]^2}{2l^3}. & (10.2\text{-}37) \end{aligned}$$

Thus over the interval $[a, c]$, the variation of the histogram is approximately reduced by the factor of $1/(2l^3)$ in a smooth pdf.

The above reasoning suggests the following procedure for estimating ε^2:

(i) compute the total histogram variation using

$$E_H^2 \triangleq \sum_i (2H(y_i) - H(y_{i+1}) - H(y_{i-1}))^2, \qquad (10.2\text{-}38)$$

where the y_i are the midpoint of each histogram interval;

(ii) estimate ε^2 as rE_H^2 where $r \simeq 1/(2l^3)$, where l is the average number of grid points per histogram interval. In general r will be a very small number;

(iii) finally, there being no reason why ε_1^2, ε_2^2 and ε_3^2 in Eqs. (10.1-83)–(10.1-85) should not satisfy $\varepsilon_1^2 \simeq \varepsilon_2^2 \simeq \varepsilon_3^2$, we set $\varepsilon_i^2, i = 1, 2, 3$, as

$$\varepsilon_i^2 = r\frac{\varepsilon^2}{3}. \qquad (10.2\text{-}39)$$

As in the case of Eq. (10.2-16), the bounds computed using the above reasoning serve primarily as guidelines. The actual value of r might be determined by varying from the initial value $r \simeq 1/(2l^3)$ until a satisfactory result is achieved.

10.3 Competing Algorithms

In Section 10.4 we compare the performance of the POCS-based pdf reconstruction algorithm with two other algorithms. We describe these algorithms below.

A. The Maximum Penalized Likelihood Method [10-4]

The criterion function to be maximized is

$$\Theta \triangleq \prod_{i=1}^{n} f(x_i) \exp\left(-\beta \int |f''|^2 \, dt\right), \qquad (10.3\text{-}1)$$

where n is the number of observations, and $f(t)$ is the unknown density. In the discrete approximation, $f(t)$ is estimated at a set of equally spaced grid points t_j for $j = 1, 2, \cdots, m$ with restrictions $f(t_1) = f(t_m) = 0$. The density at each point x_i is then estimated using linear interpolation. An examination of Eq. (10.3-1) shows that Θ is made up of a likelihood term and a penalty term. We note that a pdf characterized by rapid changes will *increase* the penalty term $\beta\|f''\|^2$ while *reducing* the likelihood function. This approach favors smooth pdfs. Indeed, by taking the logarithm of Θ, the criterion can be separated into a sum of a likelihood term and a penalty term. The actual maximization of Eq. (10.3-1) is discussed in detail in [10-4]. A program, DESPL, that reconstructs a pdf using the maximum penalized likelihood method is furnished in the Statistical Library of IMSL [10-10].

The parameter β ($\beta > 0$) in Eq. (10.3-1) must be adjusted to give the best results. If β is too small, the estimator for $f(t)$ will be too bumpy; if β is too large, the estimator will be oversmoothed. Typically β is chosen small and then increased until "excessive" bumpiness is removed. In our numerical comparisons we use visual inspection to adjust β in order to obtain best results. It is these best results we used in the comparison with POCS.

B. The Windowed Kernel Method

The windowed kernel method is essentially the method of Parzen windows [10-1] in which the unknown pdf $f(t)$ is reconstructed as

$$\hat{f}(t) \triangleq \frac{1}{nh}\sum_{i=1}^{n} K[\frac{t-x_i}{h}], \tag{10.3-2}$$

where $\hat{f}(t)$ is the estimated density of $f(t)$, $K(t)$ is the kernel or window function, x_i denotes the ith observation, n is the number of observations, and h is a fixed constant, called the window width. In general, the user supplies h as well as the window function. Some common window functions are given in [10-10] and include the uniform window, $K_u(t) \triangleq 0.5 \cdot \mathrm{rect}[t/2]$, the triangle window $K_t(t) \triangleq (1-|t|)\cdot \mathrm{rect}[t/2]$, and the Gaussian window and others. A good choice is the bi-weight window defined by

$$K_b(t) = \begin{cases} 15(1-t^2)/16 & |t|<1 \\ 0 & \text{elsewhere.} \end{cases} \tag{10.3-3}$$

The reason $K_b(t)$ is a good choice is that:

(i) it has finite support (unlike the Gaussian window) leading to much more efficient computations [10-10];

(ii) it has a continuous derivative for $|t| < 1$ thereby allowing the reconstruction of smooth pdfs.

An IMSL program called DESKN (p. 821 in [10-10]) will furnish pdf estimates by the window kernel method. As in the case of choosing β in DESPL, it is best to compute $\hat{f}(t)$ in DESKN for several values of h and see which h gives best results. The best visual results were used in the numerical comparison with POCS below.

10.4 Numerical Results

Example 10.4-1

Two hundred data samples were generated from the bi-modal pdf

$$f(x) = \frac{1}{2}\left[(2\pi)^{-1/2}e^{-\frac{1}{2}(x+3/2)^2} + (\pi)^{-1/2}e^{-(x-3/2)^2}\right]. \tag{10.4-1}$$

There are several ways to generate such data. One way is to generate 200 random samples from a standard normal $N(0,1)$ and then convert the samples to $N(-3/2, 1)$ or $N(+3/2, 1/2)$ with probability 1/2. The graph of the pdf and a 40-cell histogram of the data is shown in Fig. 10.2-2. By our visual criterion, the best penalized likelihood estimate (PLE) is obtained when DESPL is used with $\beta = 10$. The results are shown in Fig. 10.4-1. Note that the match between the PLE reconstruction and the original pdf is excellent.

The windowed kernel estimate when DESKN is used with a bi-weight window of width $h = 1$ also gave excellent results (Fig. 10.4-2). We note however, that the WKE result is slightly less accurate than the PLE for values of $x < 0$. For $x > 0$, the WKE result is superior. The latter also gives slightly better results in the tails of the pdf.

All the POCS results were obtained at a confidence level of 95%, i.e., $\alpha = 0.05$. For percentile points x_ξ, levels of 0.25, 0.5 and 0.75 were used. Probability constraint sets were designed to constrain probabilities at the 95% confidence levels in intervals of the form $[\,a_i,\ b_i\,]$. The length of each interval was adaptively fitted to contain 10 points. The use of 10 points per interval is a modest departure from the $\sqrt{N}$ rule (i.e., $\sqrt{200} \simeq 14$) but 200 is integer-dividable by 10 and not by 14. The POCS result with a first-derivative smoothness constraint, Eqs. (10.1-64) and (10.1-65), with a weight $r = 0.035$ is shown in Fig. 10.4-3. The POCS reconstruction matches the best reconstruction of the other schemes. However, as in the other schemes, there are significant undershoots and overshoots. Small variations in r (10 % or so) do not significantly affect the result. Some lumpiness begins to appear when r gets larger.

When second-derivative smoothing is used with weight $r = 4.0 \times 10^{-5}$ the overshoots and undershoots are significantly reduced and a superior reconstruction of the underlying pdf is obtained (Fig. 10.4-4). Significant changes in r (50 % or so) do not affect the result much.

Example 10.4-2

The exponential distribution is pervasive in natural phenomena. For example, the lifetime of a light bulb, the waiting time of a customer in a service station,

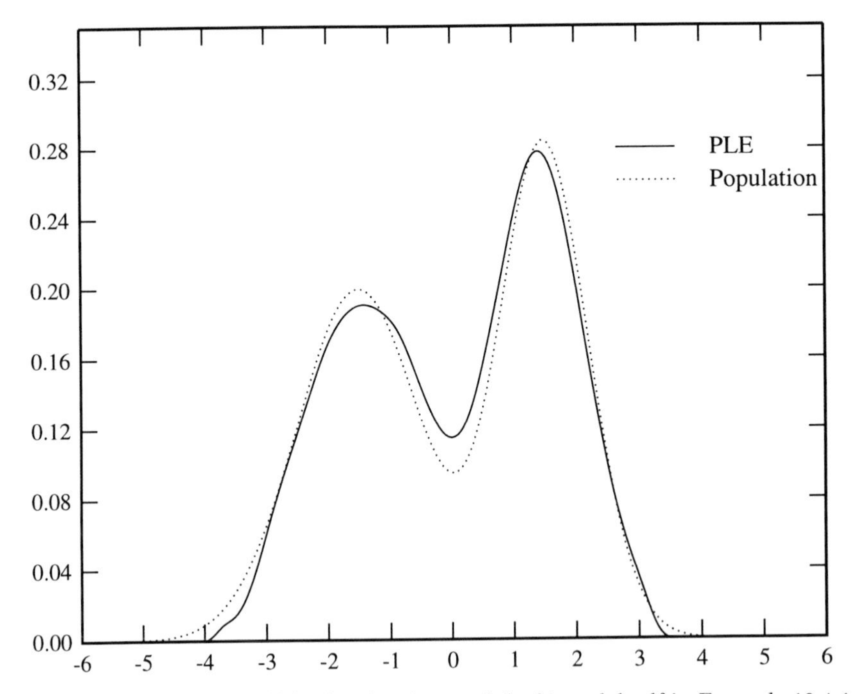

Figure 10.4-1: Penalized likelihood estimate of the bi-modal pdf in Example 10.4-1.

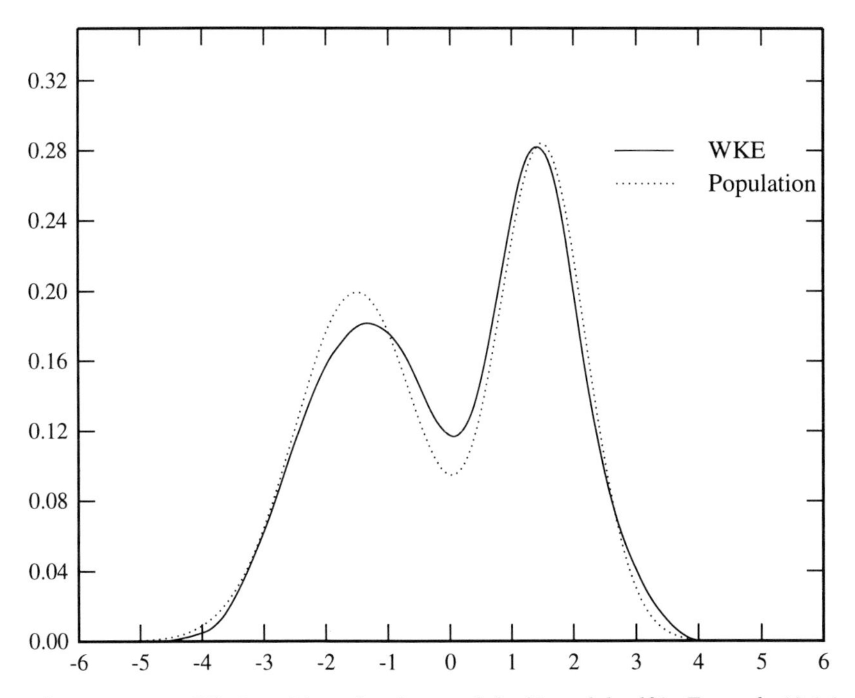

Figure 10.4-2: Windowed kernel estimate of the bi-modal pdf in Example 10.4-1.

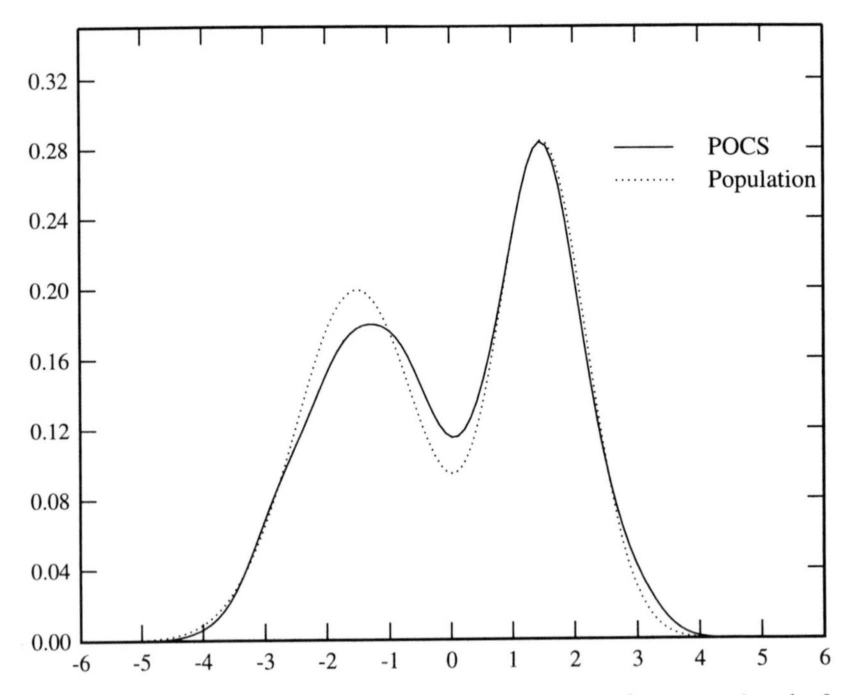

Figure 10.4-3: POCS estimate of the bi-modal pdf in Example 10.4-1 using the first-derivative smoothing constraint with $r = 0.035$.

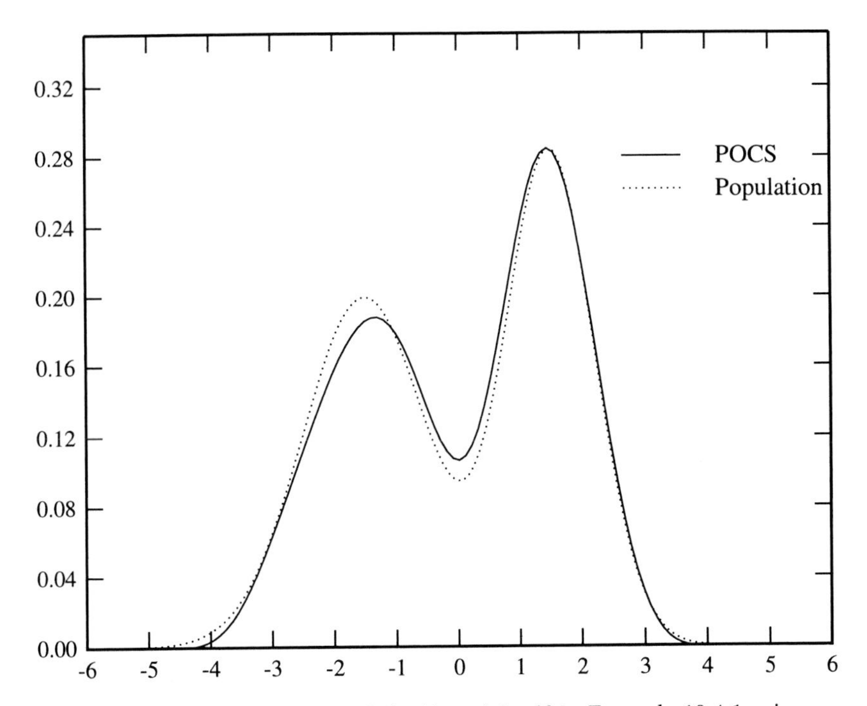

Figure 10.4-4: POCS estimate of the bi-modal pdf in Example 10.4-1 using second-derivative smoothing constraint with $r = 4.0 \times 10^{-5}$.

etc. are known to obey the exponential distribution. In these applications, the random variable of interest (lifetime, waiting time, etc.) cannot take on negative values. In other words, it is known *a priori* that the density function is zero for negative values. In this example we illustrate that the POCS algorithm can effectively incorporate this prior knowledge into the reconstruction process, leading to good estimate of a pdf.

In this example, one hundred samples were generated from the exponential pdf

$$f(x) = e^{-x}u(x). \tag{10.4-2}$$

The samples were generated using the IMSL routine RNEXP. A graph of the pdf and a 25-cell histogram are shown in Fig. 10.4-5. The best, or near best, PLE visual reconstruction with $\beta = 8$ using DESPL is shown in Fig. 10.4-6. Note the penalty-enforced smoothness at $x = 0$ leads to a serious mismatch between the true pdf and the PLE reconstruction.

The failure to reconstruct a pdf that is zero for $x < 0$ is apparent in the Parzen window WKE result as well (Fig. 10.4-7). While the WKE furnishes a significant improvement over the PLE result for $x > 0$, the smoothness inherent in the bi-weight window kernel causes a leakage of the reconstructed pdf for $x < 0$. Best visual results were obtained for $h = 0.85$.

The POCS algorithm eliminates the leakage of the reconstructed pdf by projecting onto the set C_4 in Eq. (10.1-53). Figure 10.4-8 shows the POCS reconstruction using a first-derivative smoothing constraint with weight $r = 0.04$. In general the reconstruction is superior to either the WKE or PLE methods, not only in enforcing $f(x) = 0$ for $x < 0$, but also in matching the shape of the true pdf. When the second-derivative smoothing constraint is applied with $r = 2.0 \times 10^{-5}$, the results (Fig. 10.4-9) are similar to those from first-derivative smoothing, i.e., no leakage and excellent shape matching. The shape of the reconstructed pdf is nearly coincident with the true pdf.

Summary

The method of projections onto convex sets (POCS), discussed in Chapter 6 and used several times in this book, was shown to be well-suited for estimating probability density functions from observations, i.e., data samples. The projection-based algorithm derived in this chapter uses experimental data as well as prior knowledge to reconstruct a pdf that is consistent with all the information fed into the algorithm. Existing estimating procedures cannot readily do this. We compared the POCS-based estimation procedure with two well-known and widely used methods: penalized likelihood and the method of windowed kernels. The numerical results showed that POCS furnishes results which were as good, if not better, as those of

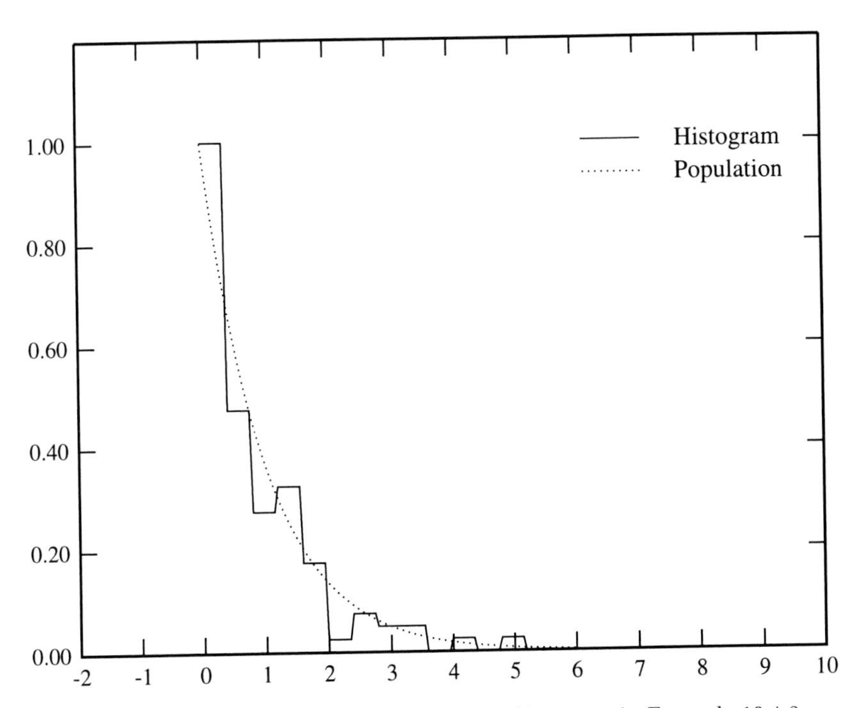

Figure 10.4-5: Population pdf and data histogram in Example 10.4-2.

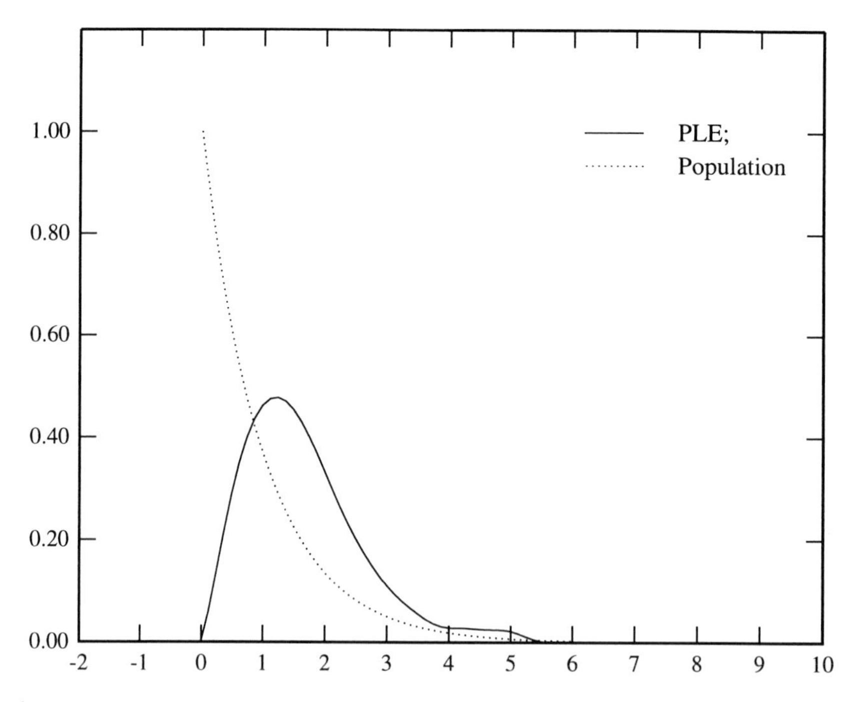

Figure 10.4-6: Penalized likelihood estimate of the exponential distribution in Example 10.4-2.

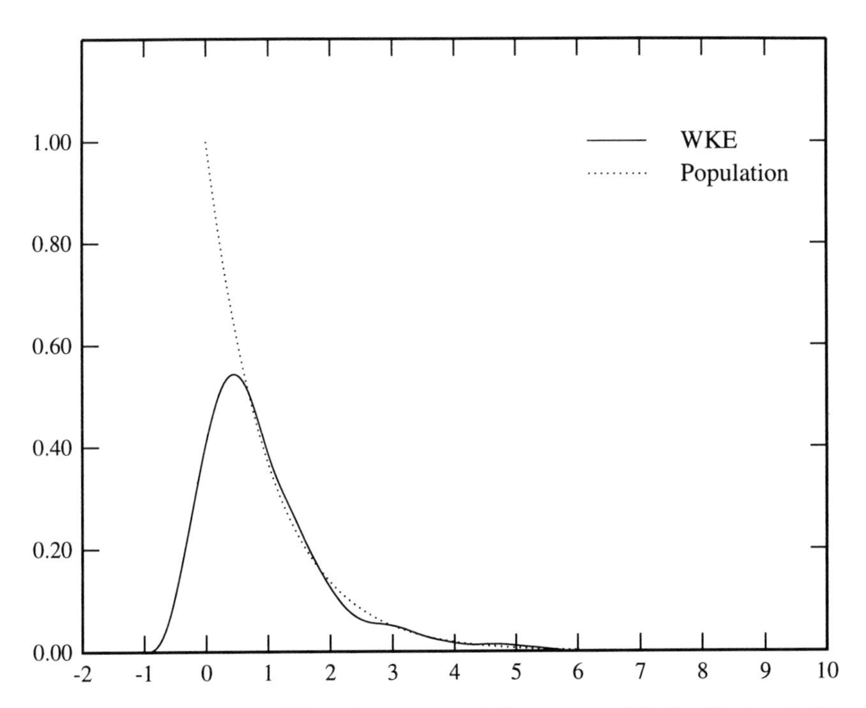

Figure 10.4-7: Windowed kernel estimate of the exponential distribution in Example 10.4-2.

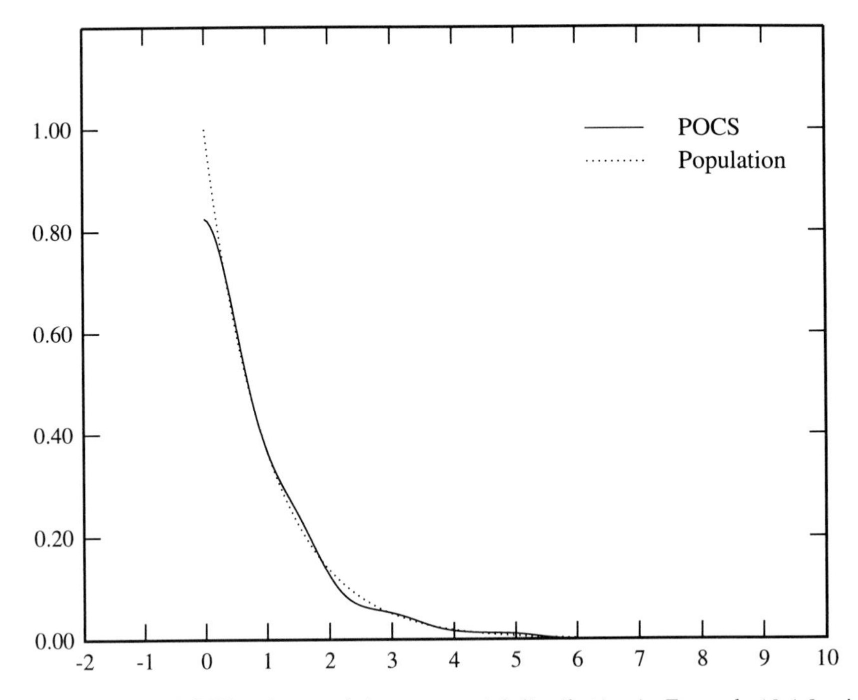

Figure 10.4-8: POCS estimate of the exponential distribution in Example 10.4-2 using first-derivative smoothing constraint with $r = 0.04$.

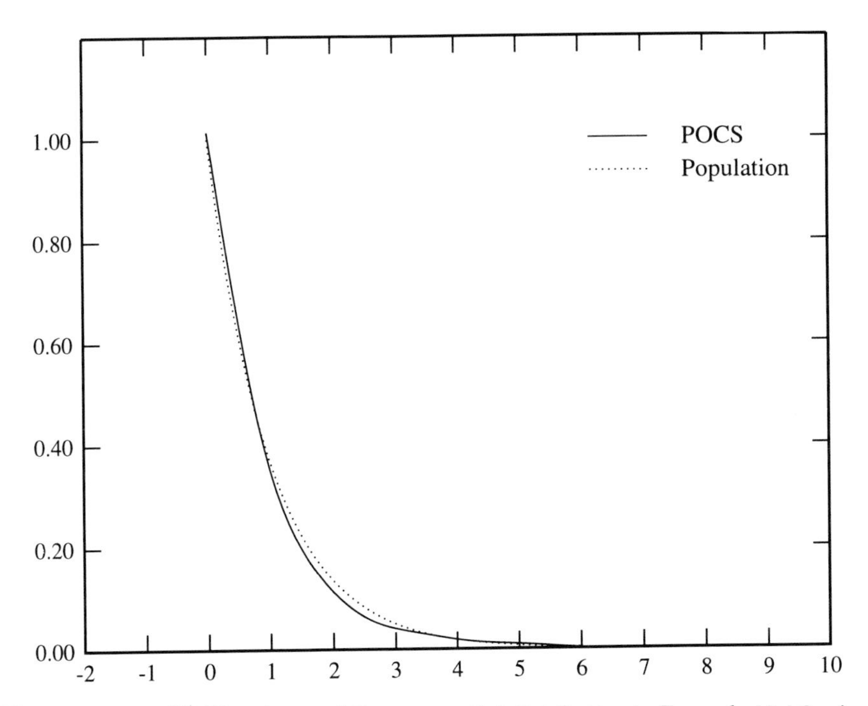

Figure 10.4-9: POCS estimate of the exponential distribution in Example 10.4-2 using the second-derivative smoothing constraint with $r = 2.0 \times 10^{-5}$.

the other methods. In the case of estimating the exponential pdf, POCS gave better results than either of the other methods and demonstrated once more that this remarkable technique is an important new tool in solving engineering problems.

Appendix 10.A: Review of a Few Results from Mathematical Statistics

In this appendix we briefly review some basic results on convergence of sequences of random variables that we shall use in the subsequent appendices. For a complete discussion, see [10-11] for example.

Convergence of a Sequence of Random Variables

Let $Z_1, Z_2, \cdots, Z_n$ be a sequence of jointly distributed random variables defined on the same sample space Ω. Let Z be another random variable defined on this same sample space. We define the following convergence properties.

Definition We say Z_n *converges to* Z *in probability* ($Z_n \xrightarrow{Pr} Z$) if for every $\epsilon > 0$

$$\lim_{n \to \infty} P[|Z_n - Z| > \epsilon] = 0. \tag{10.A-1}$$

∎

Definition Let $F_{Z_n}(\cdot)$ and $F_Z(\cdot)$ denote the cumulative distribution function of Z_n and Z, respectively. We say Z_n *converges to* Z *in distribution* ($Z_n \xrightarrow{d} Z$) if

$$\lim_{n \to \infty} F_{Z_n}(t) = F_Z(t) \tag{10.A-2}$$

at each point t where $F_Z(\cdot)$ is continuous. ∎

Theorem 10.A- 1 (Weak law of large numbers) *Let $X_1, X_2, \cdots$ be independent and identically distributed random variables with mean μ and variance σ^2. Then we have*

$$\overline{X} \triangleq \frac{X_1 + X_2 + \cdots + X_n}{n} \xrightarrow{Pr} \mu. \tag{10.A-3}$$

∎

Theorem 10.A- 2 (Central limit theorem) *Let $X_1, X_2, \cdots$ be independent and identically distributed random variables with mean μ and variance σ^2 (both finite). Then*

$$\frac{\sqrt{n}(\overline{X} - \mu)}{\sigma} \xrightarrow{d} N(0, 1) \tag{10.A-4}$$

where $N(0, 1)$ denotes the random variable with standard normal distribution. ∎

Theorem 10.A- 3 (Mixed convergence theorem) *If $X_n \xrightarrow{d} X$ and $Y_n \xrightarrow{Pr} c$, where c is a non-zero constant, and also $g(\cdot)$ is a function continuous at c with $|g(c)| < \infty$, then $X_n g(Y_n) \xrightarrow{d} g(c)X$.* ∎

Proposition 10.A- 1 Let $X_1, X_2, \cdots$ be independent and identically distributed random variables with mean μ, variance σ^2, and $E(X_1 - \mu)^4 < +\infty$. Then

$$S \triangleq \sqrt{\sum_{i=1}^{n} \frac{(X_i - \overline{X})^2}{n-1}} \xrightarrow{Pr} \sigma. \tag{10.A-5}$$

∎

Appendix 10.B: Determining the Bounds μ_L and μ_H for the Constraint Set C_1

Let $X_1, X_2, \cdots, X_n$ be independent identically distributed random variables with mean μ, variance σ^2, and $E(X_1 - \mu)^4 < +\infty$. Repeating Theorem 10.A-2 we have

$$\frac{\sqrt{n}(\overline{X} - \mu)}{\sigma} \xrightarrow{d} N(0, 1). \tag{10.B-1}$$

Furthermore, applying Propostion 10.A-1 we have

$$\frac{\sigma}{S} \xrightarrow{Pr} 1. \tag{10.B-2}$$

Finally, applying Theorem 10.A-3 to Eqs. (10.B-1) and (10.B-2) we have

$$\frac{\sqrt{n}(\overline{X} - \mu)}{S} \xrightarrow{d} N(0, 1). \tag{10.B-3}$$

Hence

$$P\left[-z_{\alpha/2} \leq \frac{\sqrt{n}(\overline{X} - \mu)}{S} \leq z_{\alpha/2}\right] \doteq 1 - \alpha. \tag{10.B-4}$$

Therefore an approximate $100(1-\alpha)\%$ confidence interval for μ follows:

$$\left[\overline{X} - z_{\alpha/2}\frac{S}{\sqrt{n}}, \quad \overline{X} + z_{\alpha/2}\frac{S}{\sqrt{n}}\right]. \tag{10.B-5}$$

Appendix 10.C: Determining the Bounds p^L and p^H for the Constraint Set C_2

Let $X_1, X_2, \cdots, X_n$ be independent identically distributed random variables with density function $f(x)$. Let $[a, b]$ be an interval in R. For each $X_i, i = 1, 2, \cdots, n$, define

$$Y_i = \begin{cases} 1 & \text{if } X_i \in [a, b] \\ 0 & \text{otherwise.} \end{cases} \tag{10.C-1}$$

Then $Y_1, Y_2, \cdots, Y_n$ are independent and identically distributed Bernoulli random variables with success probability

$$p = \int_a^b f(x)\,dx. \tag{10.C-2}$$

We have

$$E[Y_1] = p \quad \text{and} \quad \sigma^2 = p(1-p). \tag{10.C-3}$$

Define

$$\overline{Y} \triangleq \frac{Y_1 + Y_2 + \cdots + Y_n}{n}. \tag{10.C-4}$$

From Theorem 10.A-1 we have

$$\overline{Y} \xrightarrow{Pr} p. \tag{10.C-5}$$

Also by applying Theorem 10.A-2 we obtain

$$\frac{\sqrt{n}(\overline{Y} - p)}{\sqrt{p(1-p)}} \xrightarrow{d} N(0, 1). \tag{10.C-6}$$

Furthermore, applying Theorem 10.A-3 to Eqs. (10.C-5) and (10.C-6) we get

$$\frac{\sqrt{n}(\overline{Y} - p)}{\sqrt{\overline{Y}(1 - \overline{Y})}} \xrightarrow{d} N(0, 1) \tag{10.C-7}$$

Hence

$$P\left[-z_{\alpha/2} \le \frac{\sqrt{n}(\overline{Y} - p)}{\sqrt{\overline{Y}(1 - \overline{Y})}} \le z_{\alpha/2}\right] \doteq 1 - \alpha. \tag{10.C-8}$$

Therefore an approximate $100(1-\alpha)\%$ confidence interval for p follows:

$$\left[\overline{Y} - z_{\alpha/2}\frac{\sqrt{\overline{Y}(1 - \overline{Y})}}{\sqrt{n}}, \quad \overline{Y} + z_{\alpha/2}\frac{\sqrt{\overline{Y}(1 - \overline{Y})}}{\sqrt{n}}\right]. \tag{10.C-9}$$

The confidence level for the interval in Eq. (10.C-9) is approximately $(1 - \alpha)$ in that it is derived through the central limit theorem. For a moderately large sample size, say more than 100, it is reasonable to expect it to be accurate. In the literature, however, there is a way to derive an exact $100(1-\alpha)\%$ confidence interval for p in a Bernoulli trial. Some texts even give look-up tables for a number of sample sizes (see [10-11] for example). To give the reader a taste of how this is done, we derive a confidence interval for a special simple case in the following.

Example

Let's consider the case that $\sum_{i=0}^{n} Y_i = 0$, i.e., no observation occurs in the interval $[a, b]$. Clearly, the result given by Eq. (10.C-9) doesn't apply in this case. A confidence interval for p in such a case would be expected to be of the form $[0, p_0]$ where p_0 is to be determined by the confidence level.

Suppose that we require that

$$P\left[X_1 \notin [a, b], \cdots, X_n \notin [a, b]\right] \leq \alpha. \tag{10.C-10}$$

Since

$$P\left[X_1 \notin [a, b], \cdots, X_n \notin [a, b]\right] = (1-p)^n, \tag{10.C-11}$$

we require that

$$(1-p)^n \leq \alpha, \tag{10.C-12}$$

or

$$p \geq 1 - \alpha^{\frac{1}{n}}. \tag{10.C-13}$$

Therefore at confidence level $(1-\alpha)$, p_0 can be chosen as $1 - \alpha^{\frac{1}{n}}$.

Appendix 10.D: Determining the Bounds in Set C_3

Let $X_1, X_2, \cdots, X_n$ be independent identically distributed random variables with density function $f(x)$. For $0 < \xi < 1$, let x_ξ be such that

$$\xi = \int_{-\infty}^{x_\xi} f(x)\, dx. \tag{10.D-1}$$

For each $X_i, i = 1, 2, \cdots, n$, define

$$Y_i = \begin{cases} 1 & \text{if } X_i \leq x_\xi \\ 0 & \text{otherwise.} \end{cases} \tag{10.D-2}$$

Then $Y_1, Y_2, \cdots, Y_n$ are independent and identically distributed Bernoulli random variables with success probability $p = \xi$.

Applying Theorem 10.A-2 we obtain

$$\frac{\sqrt{n}(\frac{\sum_{i=1}^{n} Y_i}{n} - p)}{\sqrt{\xi(1-\xi)}} \xrightarrow{d} N(0,1). \qquad (10.D\text{-}3)$$

We have

$$P\left[-z_{\alpha/2} \le \frac{\sqrt{n}(\frac{\sum_{i=1}^{n} Y_i}{n} - p)}{\sqrt{\xi(1-\xi)}} \le z_{\alpha/2}\right] \doteq 1-\alpha, \qquad (10.D\text{-}4)$$

i.e.,

$$P\left[n\xi - z_{\alpha/2}\sqrt{n\xi(1-\xi)} \le \sum_{i=1}^{n} Y_i \le n\xi + z_{\alpha/2}\sqrt{n\xi(1-\xi)}\right] \doteq 1-\alpha. \qquad (10.D\text{-}5)$$

Observe that

i. the event $n\xi - z_{\alpha/2}\sqrt{n\xi(1-\xi)} \le \sum_{i=1}^{n} Y_i$ is true if and only if there are at least $\left\lceil n\xi - z_{\alpha/2}\sqrt{n\xi(1-\xi)} \right\rceil$ X_i's less than x_ξ;

ii. the event $\sum_{i=1}^{n} Y_i \le n\xi + z_{\alpha/2}\sqrt{n\xi(1-\xi)}$ is true if and only if there are at most $\left\lfloor n\xi + z_{\alpha/2}\sqrt{n\xi(1-\xi)} \right\rfloor$ X_i's greater than x_ξ.

Combining the above two statements we have

$$P\left[X^{\left\lceil n\xi - z_{\alpha/2}\sqrt{n\xi(1-\xi)} \right\rceil} \le x_\xi \le X^{\left\lfloor n\xi + z_{\alpha/2}\sqrt{n\xi(1-\xi)} \right\rfloor}\right] \doteq 1-\alpha. \qquad (10.D\text{-}6)$$

Therefore an approximate $100(1-\alpha)\%$ confidence interval for x_ξ is established immediately.

Appendix 10.E: Projecting onto sets C_{6e} and C_{6o}

Assume that $\mathbf{y}$ is a vector not in the set C_{6e}. Consider the Lagrange functional

$$J = \| \mathbf{y} - \mathbf{g} \|^2 + \lambda \left[\sum_{i \text{ even}} (y_{i+1} - y_i)^2 - E^2_{\text{even}} \right]. \qquad (10.E\text{-}1)$$

For i even,

$$\begin{aligned} \frac{\partial J}{\partial y_i} &= 2(y_i - g_i) - 2\lambda(y_{i+1} - y_i) = 0 \\ \frac{\partial J}{\partial y_{i+1}} &= 2(y_{i+1} - g_{i+1}) + 2\lambda(y_{i+1} - y_i) = 0. \end{aligned}$$

We obtain

$$\mathbf{g}^* \triangleq \begin{pmatrix} y_i \\ y_{i+1} \end{pmatrix} = \begin{pmatrix} 1-\alpha_e & \alpha_e \\ \alpha_e & 1-\alpha_e \end{pmatrix} \begin{pmatrix} g_i \\ g_{i+1} \end{pmatrix}, \qquad (10.\text{E-2})$$

where $\alpha_e = \frac{\lambda}{1+2\lambda}$. By enforcing the constraint of set C_{6e}, we have

$$\alpha_e = \frac{1}{2}\left[1 - \frac{E\text{even}}{\sqrt{\sum_{i \text{ even}}(y_{i+1}-y_i)^2}}\right]. \qquad (10.\text{E-3})$$

In a similar fashion, the projection onto set C_{6o} can be derived.

Appendix 10.F: Projecting onto sets $C_{7,1}$, $C_{7,2}$ and $C_{7,3}$

Assume that $\mathbf{y}$ is not in the set $C_{7,1}$. Consider the Lagrange functional

$$J = \| \mathbf{y} - \mathbf{g} \|^2 + \lambda \left[\sum_{j=1}^{\lfloor M/3 \rfloor} (y_{3j-2} - 2y_{3j-1} + y_{3j})^2 - \varepsilon_1^2 \right]. \qquad (10.\text{F-1})$$

For $j = 1, 2, \cdots, \lfloor M/3 \rfloor$, we have

$$\begin{aligned} \frac{\partial J}{\partial y_{3j-2}} &= 2(y_{3j-2} - g_{3j-2}) + 2\lambda(y_{3j-2} - 2y_{3j-1} + y_{3j}) = 0 \\ \frac{\partial J}{\partial y_{3j-1}} &= 2(y_{3j-1} - g_{3j-1}) - 4\lambda(y_{3j-2} - 2y_{3j-1} + y_{3j}) = 0 \\ \frac{\partial J}{\partial y_{3j}} &= 2(y_{3j} - g_{3j}) + 2\lambda(y_{3j-2} - 2y_{3j-1} + y_{3j}) = 0. \end{aligned}$$

We have

$$\mathbf{g}^* \triangleq \begin{pmatrix} y_{3j-2} \\ y_{3j-1} \\ y_{3j} \end{pmatrix} = \begin{pmatrix} 1-\theta & 2\theta & -\theta \\ 2\theta & 1-4\theta & 2\theta \\ -\theta & 2\theta & 1-\theta \end{pmatrix} \begin{pmatrix} g_{3j-2} \\ g_{3j-1} \\ g_{3j} \end{pmatrix}, \qquad (10.\text{F-2})$$

where $\theta = \frac{\lambda}{1+6\lambda}$. By enforcing the constraint of the set $C_{7,1}$, we have

$$\theta = \frac{1}{6}\left[1 - \frac{\varepsilon_1}{\sqrt{\sum_{j=1}^{\lfloor M/3 \rfloor}(y_{3j-2} - 2y_{3j-1} + y_{3j})^2}}\right]. \qquad (10.\text{F-3})$$

In a similar fashion the projections onto sets $C_{7,2}$ and $C_{7,3}$ can be derived.

References

[10-1] R.O. Duda and P.E. Hart, *Pattern Classification and Scene Analysis,* John Wiley, 1973.

[10-2] H. Stark, F.B. Tuteur and J.B. Anderson, *Modern Electrical Communications, 2nd edition*, Prentice-Hall,1988.

[10-3] N.S. Jayant and P. Noll, *Digital Coding of Waveforms*, Prentice-Hall, 1984.

[10-4] R.A. Tapia and J.R. Thompson, *Nonparametric Probability Density Estimation,* Johns Hopkins University Press, Baltimore, 1978.

[10-5] H. Stark and J.W. Woods, *Probability, Random Process, and Estimation Theory for Engineers*, Prentice-Hall, Englewood Cliffs, N.J. 1986.

[10-6] H. Stark and M.I. Sezan, "Incorporation of *a priori* moment information into signal recovery and synthetic problems," *Jour. of Math. Analysis and Applications,* vol. 122, pp. 172-177, Feb., 1987.

[10-7] Y. Yang, "Set-theoretic approach to visual communication problems," Ph.D. dissertation, Department of Electrical and Computer Engineering, Illinois Institute of Technology, 1994.

[10-8] H. Stark and E. Olsen, "Projection-based image restoration," *Jour. Opt. Soc. Am.,* A, vol. 9, No. 11, Nov. 1992.

[10-9] C.D. Boor, *A Practical Guide to Splines,* Springer-Verlag, 1978.

[10-10] *STAT/LIBRARY: FORTRAN Subroutines for Statistical Analysis,* IMSL, April 1987.

[10-11] E.J. Dudewicz and S.N. Mishra, *Modern Mathematical Statistics*, John Wiley & Sons, 1988.

Index